D1090461

Operation Epsilon: The Farm Hall Transcripts

Sketch by Erich Bagge

Operation Epsilon
The Farm Hall Transcripts

Introduction by Sir Charles Frank, OBE, FRS

UNIVERSITY OF CALIFORNIA PRESS
Berkeley · Los Angeles · Oxford

University of California Press 1993
Berkeley and Los Angeles, California

Published in arrangement with Institute of Physics Publishing, Bristol, England

Library of Congress Cataloging-in-Publication Data

Operation Epsilon: the Farm Hall transcripts/introduced by Sir
Charles Frank.
p. cm.
Includes index.
ISBN 0-520-08499-3 (alk. paper)
1. Scientists – Germany – Interviews. 2. Godmanchester (England) –
Intellectual life. I. Frank, Charles, Sir.
O141.0556 1993b
530'.092'2 – dc20 93-12789
CIP

Cover photographs are of the "guests" taken whilst detained at Farm Hall. Supplied with the kind permission of the U.S. National Archives.

Cover designed by Tom Kelly Designs, Bristol
Typeset in Great Britain by Macreth Media Services, Hemel Hempstead
Printed in Great Britain at the University Press, Cambridge

Publisher's foreword

From June to December 1945 ten leading German scientists were detained at Farm Hall, Godmanchester, near Cambridge, England. Their conversations were secretly recorded by British Intelligence. This book contains the complete transcripts of these conversations mostly translated into English. A summary of the background to how the detention came about is given by Sir Charles Frank in the Introduction to this book. The scientists detained were Erich Bagge, Kurt Diebner, Walther Gerlach, Otto Hahn, Paul Harteck, Werner Heisenberg, Horst Korsching, Max von Laue, Carl Friedrich von Weizsäcker and Karl Wirtz. Of the ten detained, Bagge, Korsching, von Weizsäcker and Wirtz survive. Our aim in publishing this book is to make an important and interesting historical document more easily accessible. For historical accuracy we have not corrected typing errors that appear in the Transcripts for the book printing. The Archival note gives further information on these and related matters.

We are fortunate that Sir Charles Frank agreed to read and comment on the Transcripts for us. He also provided a revised translation of Heisenberg's lecture of 14 August 1945 on the technical aspects of the Atomic Bomb, perhaps one of the key sections of the Transcripts. Sir Charles' personal knowledge and involvement with both the scientific and historical aspects of the Transcripts have been invaluable in the preparation of the book for publication. Professor Erich Bagge kindly allowed us to reproduce his sketch of Farm Hall as Frontispiece to the book. The photograph of the Farm Hall 'team' of listeners was kindly provided by Peter Ganz, a member of the team. The pictures of the 'guests' and house that appear in the Transcripts were supplied by The US National Archives via Professor Mark Walker. Mrs M Louisa de Echenique, the present owner of Farm Hall, kindly provided the photograph of the house as it is today.

Although it is not part of the Transcripts we felt readers would be interested in the biographical album compiled by the detainees and given to Captain Brodie for Christmas 1945 and therefore we have included an English Translation as an epilogue. We are grateful to Mrs P L C Brodie, widow of the late Captain Brodie, for permission to reproduce the album. A brief summary of what happened to the detainees after their return to Germany is given in the second part of the epilogue.

J A Revill
Publisher

Contents

Introduction

Ten German scientists (Bagge, Diebner, Gerlach, Hahn, Harteck, Heisenberg, Korsching, von Laue, von Weizsäcker and Wirtz), variously connected with Atomic Energy research, were detained, incommunicado, in Farm Hall, Godmanchester, fifteen miles from Cambridge, through July to December, 1945: a period covering the nuclear bombing of Hiroshima on August 6th, 1945. The 24 weekly Farm Hall reports, here reproduced for the first time in their entirety, contain the translations of their clandestinely recorded conversations. These are generally known as the Farm Hall Transcripts, though they only include one transcript in the strict sense, material transcribed from the recordings, in the original language, namely the Appendix to FH5.

How their detention came about has been described in Samuel A Goudsmit's book *ALSOS* and in R V Jones' introduction to its second edition (New York: Tomash/American Institute of Physics co-publication 1983). Throughout the work on the Manhattan Project directed by Major General Leslie R Groves, leading to the production of 'atomic' bombs in America, those engaged saw themselves as in a race with the Germans who had a head-start since nuclear fission had been discovered by Hahn in Germany at the end of 1938. The ALSOS mission (Greek ''ΑΛΣΟΣ: a sacred grove) under the scientific leadership of Goudsmit was charged with following closely behind the Western Allied invading forces in 1944 to locate and seize personnel, documents and material concerned with the German Atomic Bomb programme. Though the evidence collected by November 1944 was enough to convince Goudsmit that there was no German Atom Bomb in the making, there continued to be many, particularly in America, who would not believe it, and the mission continued with much the same target, at least for Intelligence purposes. These ten individuals were selected by Goudsmit from among those picked up, mostly at Hechingen, by an Anglo-American raiding party which had made its way through a gap in the crumbling German front, under the leadership of Colonel Boris T Pash, the principal military officer of ALSOS. Hechingen is on the Eastern edge of the Black Forest, and it was there that the greater part of the Kaiser Wilhelm Institut für Physik, in particular with its uncompleted nuclear reactor 'pile', had been relocated after being bombed out from Berlin.

Once they had been taken, what was to be done with them must have been the subject of much high-level transatlantic discussion. Any action, whether to release any of them, or to arrest more, or even to admit their existence, could focus attention on atom bombs at a time when two kinds were almost ready for test in America. Holding them in France was slightly embarrassing: they had been taken from a part of the front properly to be overrun by the French forces and France itself had interests in atomic energy. One American general even suggested that the simplest solution was to shoot them. It was R V Jones who pointed out that Farm Hall, Godmanchester, in England, which belonged to the

British Secret Service, was now free, and would be a suitable place to house them. He also suggested that microphones should be installed there before they arrived. This had become standard practice with senior prisoners-of-war. Experience had shown that their private conversations could be more revealing than interrogation. Any ethical doubts about this had faded during five years of war. These people, all civilians of course, were distinguishable from prisoners-of-war: but they were formally employed in an organization classified as 'important for war' (*kriegswichtig*) under the overall authority of Reichsmarschall Goering.

The best of reasons for selecting me to write an introduction for this material is that I am, I presume, the one surviving person in Britain who had lunch with the detainees while they were there. I knew most of them, from a period of eighteen months I spent at the Kaiser Wilhelm Institut (KWI) für Physik in Berlin (which had the subsidiary title Max Planck Institut: a name to be inherited, post-war, by all the various Kaiser Wilhelm Institutes), from Autumn 1936 to Summer 1938. Thus I had worked in the same laboratory as von Laue, von Weizsäcker and Wirtz. Karl Wirtz was a good friend of mine. We had a joint scientific publication, and had even undertaken to write a book together (never actually written) and had travelled together in South Germany, Czechoslovakia and Austria. Hahn's KWI für Anorganische Chemie was near the KWI für Physik in Berlin-Dahlem, and more than once I think we from the Physics Institute attended colloquia there. Heisenberg did not become Director of the KWI für Physik until after the war started, but on one occasion a party of us from Berlin had gone over for a two-day seminar series at his Institute in Leipzig, so I had at least eaten a meal in his house. Gerlach I had met at a conference (on Order-Disorder Phenomena, at Darmstadt) and he reminded me, when, presumably in the rôle of senior host, welcoming me to the Farm Hall lunch, he had there paid me the remarkable compliment of saying 'Was der Herr Frank sagt is gar nicht so dumm'. ('What Mr Frank says is not at all stupid' – this, though I did not know it at the time, is a famous German quotation, said by someone at a conference (about 1905?) with reference to Herr Einstein). I think I had only seen Harteck at a conference, without actually meeting him, and had not met Diebner, Korsching or Bagge before.

As FH16 for 29 October-4 November records, my visit on 2 November 1945 gave the listeners nothing much in the way of conversation which they found worth recording. In fact, after lunch Wirtz took me out into the garden, explaining that he would feel more sure that we could talk privately there: they had found some unexplained wires in the back of a cupboard in the house. I forget now how I responded to that. No doubt I had to exercise some economy with the truth because I don't like lying in any circumstances. I knew, of course, that there were microphones there: Commander Welsh had shown me the transcripts – in German if I remember rightly – of the conversations of 6 August 1945 (after the Hiroshima news – FH4) and 14 August 1945 (Heisenberg's seminar on the critical mass): equally of course I couldn't admit to knowing it.

Our conversation was in part about the fate of individuals. Wirtz's sister Gundi was well. Wefelmeier, I learnt, had died (of natural causes) during the war. Then we talked about our wartime activities. Anything I told him was true. Certainly I couldn't tell him about everything in which I had been involved. It might well be the same with him, I supposed. He gave me an account of their atomic energy work which was, unsurprisingly, totally consistent with the declaration signed by all the detainees on 8 August 1945 (FH4). In two respects it went further.

One was that the German assessment of the prospects for nuclear energy, early in the war, reached the conclusion that not only would it be impossible for Germany to make a nuclear bomb within the maximum putative duration of the war (because of the large critical mass for ^{235}U): it would also be impossible for anyone else. Hence, though it would be necessary to work on nuclear energy, to save the lives of good physicists from being squandered at the front, the bomb problem could be disregarded, and they would concentrate on the development of moderated slow neutron reactors for power generation, a project estimated to be within their reach, so as to lead the world into the use of civil nuclear power when the war ended.

The other was that his heavy-water-moderated natural uranium pile had been just sub-critical for two months before they were captured. The addition of two bricks of uranium oxide, he said, would have changed it from having a slow exponential decrease to having a slow exponential growth in neutron count after injection of a neutron pulse. (This, incidentally would have put him at the point reached by Fermi in Chicago rather more than two years before.) He had authority to take that uranium oxide from another research group, but that was about 200 miles away, and the railway was continuously bombed out throughout those two months: so he never reached that point.

The estimate that no-one would make a nuclear bomb within the duration of the war was actually correct (though with little margin) so far as Germany was concerned: Germany was out of the war by the date of the first American nuclear bomb test ('Trinity' on 16 July 1945).

Knowing from what Wirtz had told me that at least by November the detainees had suspicions that Farm Hall was 'bugged' (a thought which Heisenberg had pooh-poohed in July) I have been careful in reading through these transcripts to look for any evidence of a concerted deception or concealment plan, and found none. Indeed had there been one it must have been largely nullified by the change of circumstances with news of the Hiroshima bomb, which unquestionably astounded them. In their reticences, as I judge, they were much more concerned with the reactions of each other than with those of possible eavesdroppers.

I can hardly, however, avoid the thought that some of them display a surprising degree of naivety, ignorance, or reticence about the plutonium – $^{239}_{94}$Pu, the properties of which, though not directly known to them, had been well foretold by Bohr and Wheeler – which their peaceful uranium machine would generate if it ever operated: and most of them should have known that something said by Bohr, though it might be wrong must deserve to be taken seriously.

Passages from the official report of the conference† of 26 to 28 February 1942, held in the Kaiser-Wilhelm Institut für Physik, and involving nearly all of the physicists of the uranium project together with representatives of the Army High Command, the Reich Research Council and relevant Government Ministries, which are quoted *verbatim* (pp 30 and 39) in the book by Erich Bagge, Kurt Diebner and Kenneth Jay, *Von der Uranspaltung bis Calder Hall* (Hamburg: Rowohlt 1957) – a book also containing (pp 43–72) Bagge's diary from 22 April 1945 to 12 March 1946, covering the period of detention in Farm Hall – clearly state that operation of the '*Uranmaschine*' (then confidently expected to reach success) would generate element 94, easily separable and expected to be a material from which a nuclear bomb could be made (a recognition ascribed to von Weizsäcker 1940). There was even an estimate (p 39) that the required amount of element 94 for a bomb would lie between 10 and 100 kg. Of course, these plain statements were selected with hindsight in 1957 from a lengthy report, the whole of which could present a more confusing picture.

What is of central importance in the greater part of these conversations is the minimal amount of a fissile substance needed before it can possibly undergo a nuclear explosion – known as the *critical mass* (a term which has since entered the language metaphorically with other applications). At the outbreak of World War II, while it was common knowledge among scientists that the uranium isotope ^{235}U was capable of producing a devastating explosion, no-one knew within wide limits what was the critical mass for ^{235}U and many scientists would have made a guess in the order of tons.

A complication in the way of saying what it actually is is the need to distinguish several different quantities: the critical mass for ^{235}U alone; the considerably smaller critical mass for ^{235}U surrounded by a 'tamper' of material able to scatter neutrons back into the ^{235}U; and the amount required for an effective bomb (an amount only just larger than the critical mass would only make a trivial explosion). For actual knowledge about these we can hardly do better than refer to the Los Alamos Primer, a summary of five lectures given by Robert Serber in April 1943, thereafter handed to all new recruits to the Los Alamos 'Secret Limited' Tech Area, which has recently been published, with updating annotation by the original author (Robert Serber 1992: *The Los Alamos Primer* (Berkeley, CA: University of California Press) see pp 27–33). According to the original primer the critical mass for bare ^{235}U would be about

†The document here cited as a conference report must be the same as, or a document very closely related to, the February 1942 *Heereswaffenamt* report *Energiegewinnung aus Uran* cited by Mark Walker in his book *German National Socialism and the Quest for Nuclear Power 1939–1949* (pp 48,49 and elsewhere; Cambridge: Cambridge University Press 1989); Mark Walker tells us that the excerpts correspond.

60 kg, coming down to 15 kg when tamped with a thick shell of natural uranium. More doubtful estimates for ^{239}Pu were about 20 kg (bare) and 5 kg (with uranium tamper). These estimates were of course, at the time, stated as subject to test, but by reason of an extraordinary cancellation of errors in the earlier estimates for the nuclear parameters, the updating only changes the figures for ^{235}U (bare) from 60 kg to 56 kg and for ^{239}Pu from ~20 to 11 kg, leaving the figures for either with uranium tamper unchanged. Regarding the amount for an effective bomb, the first ^{235}U bomb, 'Little Boy', dropped over Hiroshima, incorporated about 50 kg of ^{235}U, of which about 2% underwent fission (Serber, pp xv and 60). The first nuclear bomb of all, that for the Trinity test, had an almost critical uranium-tamped sphere of plutonium which was compressed by a spherically converging shock-wave produced with ordinary explosive, which, so Serber suggests, doubled its density and thus reduced its critical mass by a factor of four (Serber, p 60). In this case about 20% of the charge underwent nuclear explosion. Because of its high radioactivity the use of plutonium required a more rapid bringing together of the parts of the critical mass, so necessitating this more sophisticated 'implosion' method of initiation, which incidentally increased the efficiency.

It is clear that Heisenberg's estimate of the critical mass for a nuclear explosion in ^{235}U was of crucial importance for determining German nuclear energy policy during the war, as was that of Frisch and Peierls in Britain and America: the former a gross overestimate, the latter a serious underestimate (chiefly from taking an erroneous value for the 'fission cross section'). Heisenberg probably gave various quantitative answers at various times. On 6 August 1945 (FH4) he was saying much more than 50 kg, and appears to have said 2 tons: he is prepared to discuss 50, 500 or 5000 kg. He then admits to Hahn on the same date that he never worked it out properly. Within days he examines the problem more seriously and presents his conclusions in the seminar he gave on 14 August 1945 (FH5 and its appendix), arriving at a critical mass lying between 20 kg (he actually says 16) and 210 kg. Prior to that his estimates were doubtless made by the 'back-of-the-envelope' calculation he outlines to Hahn on 6 August (FH4). This starts with the tacit assumption that one would wish to explode a mole of uranium, in round numbers 10^{24} atoms (about 390 grams: that would release enough energy to qualify as a 'super-bomb' but it would not explode by itself). To achieve explosion there must be enough room in fissile material for development of the branching chain-reaction. 10^{24} is about 2^{80}, so supposing that each fission yields two neutrons in place of one there must be 80 generations of successive fissions (and most of the energy is released in the last few generations). Neutrons go in random directions after a fission, so the final disintegrations will be distributed mainly over a sphere of radius $\sqrt{80}$ (near enough 9) times the mean free path between fissions. Taking that mean free path to be 6 cm this is a sphere of radius 54 cm and since the specific gravity of uranium is 19.6 its mass is about 6 tons (though he actually says about a ton).

What Heisenberg tells Hahn on 7 August about the time for development of the chain reaction (10^{-8} s) and the fraction of the charge undergoing fission (100%) is all of a piece with the same argument.

What Harteck tells Bagge during the night 6/7 August appears to be a hurriedly ill-remembered version of the same argument (he knows he has to take the square root of something, but takes the square root of the wrong quantity: with a little time to think he would have doubtless produced the Heisenberg argument correctly). He takes a larger multiplication factor of 2.8 instead of 2 and a smaller mean free path of 4 instead of 6 cm, to arrive at a critical mass of 200 kg.

It should have taken little time or extra thought for Heisenberg, or for several of his colleagues, to perceive a basic flaw in the argument of his 'back-of-the-envelope' calculation. In a smaller sphere there would be a significant loss of neutrons from the surface: but so long as the effective multiplication factor, after allowing for surface loss, remained greater than one the branching chain reaction would continue, only requiring a larger number of generations of multiplication for the fission rate to become very large. Whoever thought about it would see that Heisenberg's method of calculation should therefore in principle yield an upper bound for the critical mass: and that more work would be needed to estimate its factor of error. He could also see that for a mass nearer to the true critical mass the time for development of the nuclear explosion would be substantially larger, and in fact limited by the disintegration of the uranium mass, making the fraction of the charge brought to fission less than 100%.

In fact, a solution of the surface escape problem had already been published on 1 May 1939 by Francis Perrin (1939 *C. R. Acad. Sci., Paris* **208** 1394). Unaware at the time of the special rôle of ^{235}U he arrived at a radius of 130 cm for a critical mass of U_3O_8, hence a mass of 40 tons: which could be reduced to 12 tons by enclosing it in a blanket of neutron scattering but non-absorbing material. In a second paper, of 15 May 1939 (1939 *C. R. Acad. Sci., Paris* **208** 1573) he considered a self-stabilizing slow neutron reactor (not a bomb), moderated by a small water content and a very little cadmium, with the critical radius reduced by a half and therefore having one eighth of the critical mass. Siegfried Flügge (the theoretical physicist from Hahn's chemical laboratory) in his article of 9 June 1939 (1939 *Naturwissenschaften* **27** 402–10) argued merely that the diameter must be large compared with a mean free path (not really true, given a sufficiently large multiplication factor), and thus some metres, and proceeded to calculate for 1 cubic metre of U_3O_8 powder (4.2 tons).

Rudolf Peierls, in a paper (1939 *Proc. Camb. Phil. Soc.* **35** 610–15) submitted on 14 June 1939 and published in the September issue of the journal, which however did not appear until the December and presumably did not reach Germany, refined Perrin's treatment and extended it to cope with the case of relatively large multiplication factor, for which the critical size could be smaller than or comparable with the mean free path. He gave no numerical estimate, leaving his result in the form of algebraic expressions, and a graph, depending on nuclear parameters. Insertion of these parameters, within the range of uncertainty of knowledge at the time, would have shown that the critical mass (for ^{235}U) came down from the order of tons to the order of kilograms: as emerged in the Frisch-Peierls memorandum (M Gowing 1964 *Britain and Atomic Energy 1939–1945*, London: Macmillan, Appendix 1, pp 389–393: also, with a second part omitted by Gowing, in Serber, Appendix 1) which reached Sir Henry Tizard, via Oliphant, in March 1940. This paper, of course unpublished at the time, gave the numerical estimate for critical mass of ^{235}U of 600 g, suggesting the mass for a significant bomb (which would have to be larger than the critical mass) of 1 kilogram. This was too low, chiefly because the nuclear parameters were not yet well known. In the MAUD Committee Report (Gowing, Appendix 2), the British assessment of greatest importance for determining Anglo–American policy for atomic energy and the bomb, the estimates were raised by a factor of about 10.

Prior to that, informed opinion in Britain and America would have reached the same conclusion as that reached in Germany, that the required amount of ^{235}U was so great as to put the production of a nuclear bomb by anybody within the duration of the war beyond the range of possibility, and received that conclusion with welcome relief.

Heisenberg would undoubtedly have known Perrin's publications of just before the war, and how to apply their conclusions at least approximately to the case of ^{235}U. He or his colleagues presumably also had access to Fritz Houtermans' unpublished paper of August 1941 'On the question of the initiation of a nuclear chain reaction' (until 1970 in an archive of captured German documents at Oak Ridge: now in Department 1.1BK/Zentralbibliothek, Kernforschungszentrum Karlsruhe, Germany: Reference G94 (Ref. 14)). This excellent paper sets out *in extenso* the considerations leading up to Perrin's formula. It gives no numerical estimate, but concentrates on listing the nuclear parameters which need to be measured for an accurate calculation. It is less concerned with the fast neutron bomb problem than with the slow neutron moderated reactor, and especially the use of that as a generator of new radioactive nuclides rather than power. Two points of particular interest which it makes are that (for the latter application) one could economize in the effort for fuel enrichment, if needed, by working at reduced temperature (a matter discussed by Harteck (without mention of Houtermans) in FH5 section III) and that one of the products would be element 94 (plutonium) which would probably have similar fissile properties to ^{235}U, and be much easier to separate. An important point he did not spot, however, was the advantage of an assembly of fissile material and moderator heterogeneous on a scale comparable with the mean free path.

It is remarkable that Houtermans is not mentioned in these connections in the Farm Hall conversations. Houtermans, having worked in Russia, was perhaps excluded from Most Secret work. His communication was made from the private laboratory of Manfred von Ardenne, at Berlin-Lichterfelde Ost. Von Ardenne, whom Heisenberg and his colleagues would have regarded as an able inventor rather than a physicist, was then engaged on a contract with the German Post Office (!) to make an electromagnetic isotope separator, which evidently came to nothing.

Remarkable also is the fact that it took Hahn, the truly experienced chemist of the radioactive elements, a week (FH5, 13 August) to puzzle out that plutonium, which had been mentioned in the news, must be element 94. Mostly they speculated about element 93. Gerlach (an able, but non-nuclear, physicist) even suggested (FH4, on 6 August) that the Americans might have stabilized element 93 (against β-decay!) by storing it at low temperature.

There can be small doubt that the Germans would have made more progress with atomic energy had they been able to bring all their available talents to work together: not, however, that they could have made a nuclear bomb within the duration of their war, even if they had tried, and even if they had not squandered industrial energy on the V2 project. Even the Americans, with substantial industrial and scientific advantage, and important assistance from Britain and from ex-Germans in Britain did not achieve that (VE-Day, 8 May 1945, Trinity test, Alamogordo, 6 July 1945).

One has to wonder why Heisenberg, or someone else in Germany, did not give more serious consideration to the bomb problem before the Hiroshima demonstration. Heisenberg may well have thought that it was not worth spending effort on an accurate calculation before he had better knowledge of the nuclear parameters. All were probably so greatly relieved by the conclusion that there would be no bomb within the duration of the war, absolving them from very difficult decisions, that they were too willing to accept it. To some it would have been almost tantamount to *lèse-majesté* to query a pronouncement by Heisenberg on a matter of physics. And having, for whatever reason, renounced work on the bomb it would be only human to reinforce that decision with a sense of virtue.

Whether Heisenberg did right in putting serious consideration of the bomb problem aside is another question. Had he, in 1944 (FH4, 6 August) been less positive in assuring the German Foreign Office that there could be no reality behind an alleged American threat to drop a nuclear bomb on Dresden, it is just possible that the German surrender could have come a little sooner, to the benefit of everybody concerned.

The greatest fascination in these conversations lies in the insight they give into the minds of ten individuals, each weighing for himself a variety of competing loyalties, to humanity, to science, to his country and duty, to his group, to his family, to his *amour propre* and to his career: with knowledge of a truly terrible weapon to give importance to their thoughts. One must regret that this material was not released while Dürrenmatt was available to make it into a play. One may wonder whether in the hands of sufficiently skilful actors these conversations could be successfully enacted on the stage: and the answer must presumably be 'only to an audience of physicists'; with its main dramatic climax lying concealed in Heisenberg's seminar – and he not always the plainest of expositors – even von Laue found some difficulty in following him (FH5). The physicist reader, at least, may be able to find the sympathy to follow the drama in his own mind, recognizing that thoughts unspoken were no less important than what was said.

As a coda to the Farm Hall story and some relief to the growing tedium of detention, came the news on 16 November 1945 of the award to Hahn, the discoverer of nuclear fission, of a Nobel prize: which was duly celebrated that evening with a dinner (FH18 and its Appendix). Captain Brodie calls von Laue's speech on this occasion 'unfortunate'. In fact it was virtually the same speech as he would have made at short notice for this ceremonious occasion in any other circumstances, and naturally it mentioned Hahn's wife. This reminded them all of the families from which they were separated, and evoked some tears. This embarrassed Captain Brodie, whose efforts to maintain communications between them and their families had been less successful than could have been wished. (I am sorry that ADI (Science), Air Ministry, had not been asked to help in this: I think we could have done better.) However, sentimental sadness was presently submerged in the kind of light-hearted intellectual humour characteristic of German academics. It is a pity von Weizsäcker did not produce a *Schüttelreim* for the occasion (or if he did, it escaped the record). He used to be able to compose one at the drop of a hat.

F C Frank

Farm Hall as it is today.

Archival note

The Farm Hall Transcripts (comprising for the most part English translations of the conversations of ten German scientists detained from June to December 1945 in Farm Hall, Godmanchester, near Huntingdon) are Crown Copyright and now published by permission given to the Institute of Physics. They were classified 'Top Secret' and, though their existence was publicly known from publication of excerpts by General Groves in *Now it can be told: The story of the Manhattan project* (New York: Harper & Row 1962; 2nd edn, New York: DaCapo 1983) remained inaccessible until released into the public domain on 14 February 1992 in response to a letter to the Lord Chancellor signed by the presidents of the Royal Society and the British Academy and thirteen other scientists or historians. The letter and the Lord Chancellor's response are reproduced here with permission of the Lord Chancellor and signatories. Professor Nicholas Kurti, FRS drafted the letter in consultation with Professor Margaret Gowing, FBA, FRS and Sir Rudolf Peierls, FRS. Although the names of Gowing and Peierls appear on the letter they did not sign it. However, it is certain they supported the enterprise.

The Public Record Office (PRO) set of the Farm Hall Transcripts, for which the official reference is WO208/5019, appears to have been assembled from more sources than one. It bears Ministry of Defence Registry Stamps (MOD (PE) DSy (PE) Registry) dated 14 February and 5 May 1978, and what must be the final release date 6 February 1992. Its title page – TOP SECRET: Epsilon: 1 May to 30 December 1945. This document is of historical interest and must not be destroyed: TOP SECRET – denotes it as copy No 1 of copy No 2 of 3 copies. Nevertheless whereas the preamble and Farm Hall Report FH1 (reporting up to 18 July 1945) are designated copy No 2 those from FH4 to FH6 (of 5 September 1945) are copy No 4 being addressed to M W Perrin (called Lt. Col. in the preamble, but thereafter Mr) and Lt.-Cdr. E Welsh, these being the copies for Perrin; FH7 and 8, both of 15 September, are copy No 6 in a named distribution list of 7 (No 6 designating Welsh); and FH9 of 27 September is copy No 8, for Prof. Sir James Chadwick. All are thereafter designated copy No 3 and addressed to Perrin and Welsh: but FH10 and 11 are marked for Chadwick.

Many pages are smudgy, sometimes barely legible, evidently *n*th carbon copies. The last page of FH18, its 11 page appendix, and FH19 (8 pp), FH20 (2 pp), FH21 (2 pp) and FH22 (1 p) are entirely missing from the PRO set.

We are indebted to Arnold Kramish for providing us (via Professor Kurti) a copy of the American National Archives set of the Farm Hall Transcripts filed in Washington in Record Group 77, Records of the Manhattan Engineer District (MED), Entry 22, box 163.

This set has two cover sheets, marked respectively 153 pages and 126 pages (126 pages would bring one to two pages short of the end of FH14 dated 23 August, 1945, and 153 pages up to p 4 of FH17, which continues for another 17 pages to its completion on 15 September 1945). There are actually 256 pages in the whole set, which contains the 32 pages missing from the PRO set.

The cover sheets are stamped TOP SECRET (as are all pages throughout). That for 153 pages bears a heavily deleted RESTRICTED DATA stamp. Both bear a stamp dated 8.25.61 of the United States Atomic Energy Commission, Oak Ridge, Tennessee certifying that this document has been reviewed and it has been determined that it does not contain Restricted Data as defined in the Atomic Energy Act of 1954. Finally, they are stamped with 2.25.92, DECLASSIFIED: Authority UK Note 24 FEB 92. This declassification label appears on every page. Occasionally it covers up a few words in the last line, but these are always recoverable from the PRO copy.

All the reports in the Washington set are copy No 1, being the copies sent to General Groves. Some of them are not replicas of the PRO copies, being typed with a different typewriter. Arnold Kramish informs us that General Groves had some of them re-typed by his confidential secretary, Mrs Jean O'Leary, possibly for limited distribution to appropriate parties. General Groves was very security conscious, and would not have sensitive papers photocopied as this would involve sending them out of his office. Retyping of sensitive papers was common practice in his office. She has evidently done the re-typing with extreme faithfulness, line for line and word for word, even repeating the deletion of passages by over-typing xxxxxx... . The one deviation we have noticed is a replacement of the German place name spelling REIMS by the French RHEIMS.

It is the more legible (and complete) Washington set which has been used as the principal basis for the present printing. Occasionally the (?) against doubtful names has been deleted, or . . . filled in, with marginal initialling S.G. or S.A.G., no doubt Dr Samuel A Goudsmit. These amendments have been accepted. No such corrections occur after FH4 (11 August 1945). FH5 cannot have been read by Goudsmit or he would not have maintained later on that Heisenberg did not understand the nature of an atom bomb. Circle marks against proper names were doubtless indicators for someone making a card-index. These have been ignored, as have various side-linings and underscorings when not in the PRO set.

We are also indebted to Arnold Kramish for the information that Sir James Chadwick turned in his set of the 'Epsilon' papers to R E R Lees, Director of Security for the UKAEA on 27 January 1964. This too was a mixed set as far as copy numbers were concerned ranging from copy 2 of FH1 to copy 8 of FH9.

We are indebted to Professor Peter Ganz, a member of the listening/recording/translating team at Farm Hall until November 1945 for an account of how they worked. The team, seen in the photograph which Professor Ganz has provided, eight in all (two names forgotten) were, with the exception of Major Rittner and Captain Brodie (neither of whom listened or transcribed) native German speakers with several years experience of this kind of work and well able to identify voices. As every guest had a room of his own (this is not quite true: some bedrooms were shared by two), it was not difficult to

The listening team. From left to right: Brodie, Heilbron, Rittner, Ganz, unknown, unknown, Pulay, Lehmann.

distinguish the occupant and his visitors. After a little practice individual voices in a group could be picked out so that by the time the 'news' broke everyone was able to identify the guests' voices. During the day all rooms were covered.

Recording was made on shellac-covered metal discs. Conversations were recorded only when they were on the point of touching on 'interesting' topics. The rule was to record too much rather than too little. Only relevant material was transcribed, i.e. technical matters, political conversations, etc. Personal matters only if they helped to fill in biographical details. Inaudible words were indicated in the transcription by dots (?). As far as he remembers repetitions were left out only at the translation stage. Professor Ganz does not remember who did the translations. The summaries were prepared by Major Rittner and Captain Brodie (working, in part at least, from German originals), as were the final reports. He does not remember what fraction of the German transcript was translated but considers that a suggested estimate that the transcript contains 10% of the words spoken may well be roughly correct. They probably had six to eight recording machines. The discs, after use, were re-coated and re-used.

From this we may conclude that no original voice recordings survive. We have no positive information that the German transcripts were destroyed but it seems somewhat likely. All references to 'the Farm Hall tapes' (of which there are many, e.g. *The Times*, 24 February 1992; *The American Institute of Physics History News Letter* XXIV No 1, May 1992†) are of course wrong. There was

†An editorial error: Mark Walker had written 'recordings'.

no tape-recording in Britain at the time, the two standard methods being either by cutting on disc or magnetic recording on wire.

The Album presented to Captain Brodie by the detainees at Christmas 1945 is deposited at the Imperial War Museum, London. It is filed as Brodie Album: Special Misc R6. We are indebted to Mrs P L C Brodie, widow of the late Captain Brodie, for permission to reproduce a translation of the Album, included as an Epilogue to the Transcripts.

The Farm Hall Reports are here reproduced without any corrections of spelling errors or the like except where explicitly indicated.

F C Frank

The Royal Society
6 Carlton House Terrace, London, SW1Y 5AG
Telex 917876 Telephone 071-839 5561 Fax 071-839 7718

From the President
Sir Michael Atiyah. P.R.S.

20 December 1991

Dear Lord Mackay

We are writing, as historians and scientists, to ask you to approve the release to the Public Record Office of a 46 year-old document of great historical importance.

At the end of the war, the leading German scientists who had been concerned with plans for atomic weapons were captured and interned at Farm Hall, near Cambridge. They included W Heisenberg (now dead) and C F von Weizsacker (the brother of the German President). The house was "bugged", and their conversations were recorded. The transcripts include, in particular their reaction to the news of the dropping of the first atom bomb on Japan.

These "Farm Hall transcripts" have so far been withheld from public access, though their existence has been acknowledged. Their historical interest arises, in part, because there still is controversy about the role of the scientists in the German atomic-energy project. Some historians and some journalists claim that moral scruples prevented the German scientists from completing a nuclear weapon. A few extracts from the transcripts have appeared (in translation) in books, e.g. by General Leslie Groves and Professor S A Goudsmit, but they are out of context, and it is impossible to check their accuracy. It is desirable to release the transcripts while four of the members of the German group, as well as some who knew the German scientists personally, are still alive. They could make useful comments, but they are all of advanced age.

We note that, in correspondence with one of us, none of the four surviving Farm Hall scientists objected to the release of the transcripts. We also remember the words of the German President, Richard von Weizsacker, in a speech on 8 May 1985: "We need, and we have the strength to look truth straight in the eye without embellishment and without distortion...anyone who closes his eyes to the past is blind to the present."

We therefore urge you to reconsider your ban, under the Public Records Act, on the release of the German transcripts and their English translation, and if still in existence, of the original recordings.

Yours sincerely

Michael Atiyah

President of the Royal Society

Anthony Kenny

President of the British Academy

Lord Blake, FBA

Lord Bullock, FBA

Lord Dacre of Glanton, FBA

Sir Sam Edwards, FRS

Lord Flowers, FRS

Sir Charles Frank, FRS

Professor Margaret Gowing, FBA, FRS

Professor R.V. Jones, FRS

Lord Kearton, FRS

Professor N. Kurti, FRS

Sir Ronald Mason, FRS

Professor Robert O'Neill, FASSA

Sir Rudolf Peierls, FRS

Professor E.A. Roberts, FBA

Lord Zuckerman, FRS

The Rt Hon. Lord Mackay
Lord Chancellor's Department
House of Lords
London SW1A 0PW

FROM THE RIGHT HONOURABLE THE LORD MACKAY OF CLASHFERN

HOUSE OF LORDS,
LONDON SW1A 0PW

13 February 1992

Dear Michael,

You and a number of interested colleagues wrote to me in December last year asking the Government to consider again the question of the release of the Farm Hall transcripts. I am sorry that it has taken so long to let you have a substantive reply.

I am informed that the tapes themselves no longer exist' but the transcripts and their English translation have certainly survived. I know that there has been a deep continuing academic interest in these transcripts over the years and I am pleased to be able to tell you that, following one of the regular re-reviews of such intelligence-related material, arrangements have now been made for their release. This will be announced in a brief press statement by the Public Record Office this Friday, 14 February. You may be interested to note that the transcripts have been assigned to class WP 208, piece number 5019, and that they will be available from tomorrow for public inspection at the Public Record Office at Kew.

Once again, I am sorry you have had to wait so long for an answer, but I hope you will agree with me that this news is worth the wait.

I am copying this letter to the President of the British Academy.

Yours ever,

James.

Sir Michael Atiyah
President
The Royal Society
6 Carlton House terrace
LONDON
SW1Y 5AG

THE TRANSCRIPTS: OPERATION EPSILON

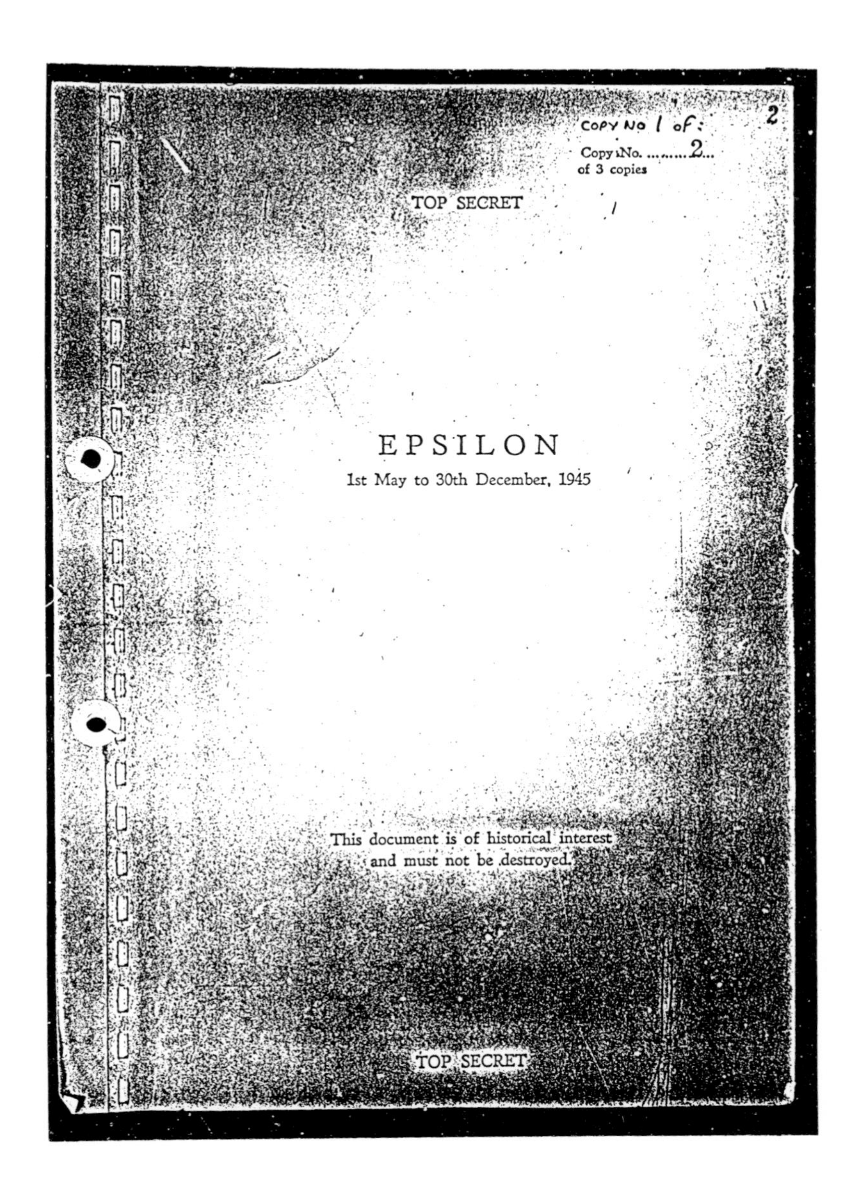
COPY NO 1 of:

Copy No. ....2...
of 3 copies

TOP SECRET

EPSILON
1st May to 30th December, 1945

This document is of historical interest
and must not be destroyed.

TOP SECRET

COPY NO. 1

To Lieut.-Col. M. W. Perrin
and Lieut.-Cdr. E. Welsh
From Major T.H. Rittner.

OPERATION "EPSILON"

I. PREAMBLE

1st May 1945

I received at H.Q. from Lieut. Cdr. WELSH instructions to proceed to RHEIMS (France), to report to G2 SHAEF and collect a party of German Scientists. A Chateau at SPA (Belgium) had been prepared for their detention. A number of distinguished British and American Scientists would be visiting them in the near future and my instructions were that these Germans were to be treated as guests. No one, repeat no one, was to contact them except on instructions from H.Q.

II. RHEIMS (2nd–7th May)

2nd May 1945

I proceeded by air to RHEIMS and reported to SHAEF where I was informed by Major KEITH, P.A. to A.C. of S. G2 that the Chateau at SPA was no longer available and that the party was to be held at RHEIMS at 75 Rue Gambetta until other arrangements could be made.

Arrangements had been made to draw American "A" Rations ready cooked and a staff of two British Orderlies and an American cook had been provided by SHAEF in addition to the necessary guards.

The same evening, the following arrived at 75, Rue Gambetta, escorted by Major FURMAN, U.S. Army:–

Professor	HAHN
"	VON LAUE
Doctor	VON WEIZSACKER
Doctor	WIRTZ
"	BAGGE
"	KORSCHING†.

Professor MATTAUCH, whom I had been told to expect was not among the party.

† The name of von Weizsäcker was systematically misspelt Weiszacker here and up to FH2 and intermittently thereafter. The interchange of 'z' and 's' has been corrected, but the Umlaut on 'a' not restored.

The 'c' in Korsching was almost always omitted, and has been inserted.

The professors were friendly and settled down well. They expressed appreciation of the good treatment they were receiving and a very pleasant atmosphere prevailed. At my request they gave me their personal parole not to leave the house or that portion of the garden which I allotted to them.

3rd May 1945

The following day I telephoned H.Q. (Lieut. Cdr. WELSH) and informed him of the situation regarding SPA. In the meantime I was informed by SHAEF that arrangements were being made to accomodate the party at a Chateau at VERSAILLES where the original policy regarding these detainees could be carried out. I met Major CALVERT and explained the situation to him.

5th May 1945

I was asked by SHAEF H.Q. Commandant to release the Allied Personnel who were required in connection with the impending VE Day negotiations and who, in any case had shown some reluctance to wait on Germans. I pointed out that it was essential for me to have staff for this purpose and it was suggested that I take on German P.W. It seemed to me that this would solve the staff problem and I accordingly agreed and acquired from the RHEIMS P.W. stockade, a German P.W. waiter and a cook, but I stipulated that these men would have to remain with me as long as the Professors were detained.

The professors were by this time beginning to get restive and they were particularly worried about their families. They asked permission to write letters and after referring the matter to H.Q. (Lieut. Cdr. WELSH) and obtaining sanction, the letters were written and, after being censored by me, were handed to General STRONG's secretary (Jun. Cmdr. FRAZER) at SHAEF for transmission.

7th May 1945

SHAEF informed me that arrangements had been made to accomodate the party at VERSAILLES and that H.Q. had agreed to the move. A Dakota was put at our disposal and the party took off at 1700 hours in the expectation that at last the long awaited contact with their British and American colleagues was about to take place.

III. VERSAILLES (7th–11th May)

On arrival at VERSAILLES, I reported to G2 SHAEF and found that the party was to be accomodated in a detention centre known as "Dustbin" at the Chateau du CHESNAY. This centre had been set up for the purpose of interrogating German NAZI Scientists and Industrialists. The conditions were most unsatisfactory from my point of view as complete segregation was impossible and there was great danger of undesirable contacts being made with the professors. In addition, only camp beds were provided and there was scarcely any other furniture. The food was the ordinary P.W. rations. It was obviously impossible to carry out my mission in these surroundings but I was able to pacify the professors who accepted the situation with as good a grace as possible and I promised to do my best to get them moved as soon as possible. The Camp Commandant did his best to make them comfortable.

8th May 1945

In spite of the general holiday atmosphere at SHAEF and in LONDON consequent upon the declaration of VE Day, I managed to contact H.Q. (Lieut. Cdr. WELSH) by telephone and explain the new situation. He told me to contact Major FURMAN, U.S. Army, in PARIS and try and make other arrangements through him. This I did and we decided to try and get the party back to the rue Gambetta at RHEIMS.

9th May 1945

Major FURMAN informed me that he was arranging for us to return to RHEIMS and that in the meantime the following Germans were to join the party:–

Professor HEISENBERG
Doctor DIEBNER.

The situation at the Chateau du CHESNAY was becoming more and more difficult as the professors were highly indignant at being treated as "war criminals" as they put it; Professor von LAUE was almost in tears. In spite of the fact that other German Scientists including Professor OSENBERG were in the house, and were being interrogated by British and American Officers, I was able to prevent the identity of my party being revealed. I had refused to submit a nominal roll or to allow any contact with them. Dr. ROBERTSON, Scientific Advisor to SHAEF did, however, see Professor von LAUE out for exercise and spoke to him, but I was able to persuade him to break off the conversation and he accepted the situation well.

10th May 1945

The professors were becoming more and more restive and they begged me to contact Professor JOLIOT in PARIS whom they assured me would help them. This request was of course refused and I told them they must have patience and that everything possible was being done for them.

In order in some way to alleviate conditions I took them in parties by car to VERSAILLES to see the gardens and Palace. On one occasion a guide asked for their identity cards in order to visit the Hall of Mirrors and we left the premises hurriedly having pleaded a previous engagement.

11th May 1945

It was now clear that the difficulties which had arisen were due mainly to an order issued by the Supreme Commander stating that no preferential treatment was to be given to any captured German nationals. This order was given after reports had appeared in the press describing the good treatment being meted out to Reichsmarschal GÖRING.

General STRONG refused to agree to my party returning to RHEIMS but arrangements were made by MIS, on the instructions of Brig. General CONRAD, to accomodate the party for a short time at a villa at LE VESINET, near ST. GERMAIN. The professors were overjoyed at the prospect of leaving what they called the concentration camp and after I had inspected the villa, I left VERSAILLES with the party in command cars on the evening of the 11th May after informing H.Q. by cable of the arrangements which I had made.

IV. LE VESINET (11th May–4th June)

The Villa Argentina, 89 Allee du Lac Inferieure at LE VESINET was a large house standing in its own grounds. MIS provided a guard and arranged for us to draw American "B" rations in a semi–cooked condition from their mess. They also supplied us with canteen goods. During our whole stay at LE VESINET, MIS gave me every possible assistance.

The professors were delighted with their new surroundings and the old atmosphere of cordiality quickly returned. There was some trouble with the plumbing and electric light in the villa which had been empty for some time and the professors all helped to remedy the defects. When outside help was necessary, such as on an occasion when the basement was flooded owing to a burst pipe, they were confined to their rooms whilst a very voluble and inquisitive French plumber dealt with the matter.

On the evening we arrived, Major FURMAN brought Professor HARTECK to join the party. As the party had now grown from six to nine, I asked the professors to renew their parole which they did.

12th May 1945

As we could only have the use of the villa for a very limited period, I cabled H.Q. through Colonel Robin BROOK urging that efforts be made to bring the party to England as it was obvious that it would not be possible to arrange accomodation on the continent suitable for carrying out my mission.

There was considerable speculation amongst the U.S. troops and the French civilian population regarding the identity of the party. I was accused of harbouring Marshal Petain. A number of inquisitive people including the owner of the villa who came post haste from PARIS when he heard from the concierge that his house was being occupied, were dealt with. I was able to spread the story that my party consisted of active Anti-Nazis who were being kept by us for their own protection.

The information, inadvertantly let out by Major FURMAN, that French Colonial troops were still in occupation at HECHINGEN and TAILFINGEN caused consternation amongst the professors who had been told that American troops would be taking over. Professor HEISENBERG asked permission to write a letter to his friend Dr. GOUDSMITH of the U.S. Army, who he believed was in PARIS, asking him to get news of the families. On receipt of a cable from Mr. PERRIN sanctioning this, I handed the letter to Dr. GOUNDSMITH who offered to do what he could. He subsequently gave me a letter from Professor HEISENBERG's wife which I handed to him for which he was duly grateful.

<u>16th May 1945</u>

Major GATTIKER was sent by Mr. PERRIN to get some information from Professor HARTECK and Dr. WIRTZ regarding the whereabouts of certain apparatus. I was present at the interview. All questions were answered.

<u>17th May 1945</u>

Brig. General CONRAD came to the villa and I showed him over. He did not speak to any of the professors. He expressed his satisfaction with the arrangements that had been made and agreed that the "B" scale of rations should continue although MIS wanted to reduce them to the "C" scale which was that authorised for P.W. On the following day, however, the rations were cut without warning and I protested to Colonel FORD the C.O. of MIS who eventually agreed to restore the original scale.

Major FURMAN sent Lieut. DIETESHEIM, U.S. army, to ask Professor HARTECK about certain apparatus at CELLE. I was present at the interview.

The professors spent their time in LE VESINET working in their rooms or sun-bathing in the garden. They developed a passion for physical exercise and even the more aged Professors von LAUE and HAHN could be seen running solemnly round and round the garden at six o'clock in the morning clad only in thin underpants. On Tuesdays and Fridays they assembled in the common room to hear a lecture by one of themselves. I was able to supply them with books, technical journals, and games.

During this time, Major FURMAN was endeavouring to find suitable permanent accomodation. He had obtained a letter from Brig. General BETTS, D.A.C. of S. G2 SHAEF asking all concerned to assist in finding accomodation for the party. He informed me that he had found a Chateau near LIEGE (Belgium) and I arranged to inspect it.

20th May 1945

I received a cable from V.C.S.S. informing me that WASHINGTON had been asked to agree to the professors being brought to England, and I replied stating that I proposed to inspect the Chateau in Belgium, and asking for instructions pending a decision from WASHINGTON. V.C.S.S. replied telling me to proceed as though I knew nothing about these negotiations.

25th May 1945

I flew to LIEGE with Major FURMAN to inspect the Chateau de FACQUEVAL near HUY. Lieut. Colonel WATKINS, U.S. Army, the Area Intelligence Officer took us to the Chateau. He pointed out that the administration would come under the local American Military Authorities and that it would have to be strictly in accordance with General EISENHOWER's order regarding treatment of enemy nationals. Apart from this, the security appeared unsatisfactory as there were Belgian civilians working on the estate and I was also informed that the political situation was very tense and that serious disturbances were expected. In addition, the owner of the Chateau, a very wealthy Belgian Lawyer, Mr. GOLDSCHMIT had just returned from five years as a P.W. in Germany. I spoke to him and he was naturally very distressed at having his home requisitioned. Of course he did not know the purpose for which it was going to be used but he told me he proposed to contest the validity of the requisition as the house had not been occupied by the Germans, and, according to an agreement between the Allies and the Belgian Government only such houses could be requisitioned. I checked this later and found it to be the case. He told me he was a friend of General ERSKINE, Head of the SHAEF mission to Belgium and that he intended to appeal to him. As a matter of fact, a request for de-requisition was received from General ERSKINE later, shortly before we left.

Lieut. Colonel WATKINS wanted the American guard troops accomodated in the best rooms in the house and suggested that the professors should be put in the attics. I refused to agree to this.

I reported to H.Q. (V.C.S.S.) by cable.

In the meantime MIS were pressing me to vacate the villa ARGENTINA which was urgently required to accomodate their own staff who were passing through LE VESINET for re-deployment. I saw Colonel FORD and informed him that it was impossible for us to move until other suitable accomodation had been found and assured him we were doing our best in this respect.

28th May 1945

I received a cable from Lieut. Cdr. WELSH informing me that he proposed to visit me and telephoned him arranging to meet him in PARIS on 30th May.

30th May 1945

Lieut. Cdr. WELSH arrived and I took him to the villa. I explained the whole situation to him and arranged to take him to see the Chateau de FACQUEVAL. He was very cordially received by the professors but they were disappointed that he was unable to give them any news of their families or any information regarding their future.

31st May 1945

Lieut. Cdr. WELSH and I flew to LIEGE and were taken to the Chateau de FACQUEVAL by Captain MUELLER, U.S. Army, Colonel WATKINS' deputy. After inspecting the house and grounds and talking to Mr. GOLDSCHMIT, we agreed that the place was unsuitable and we returned to LE VESINET.

1st June 1945

A conference took place at Major FURMAN's office in PARIS at which Major CALVERT was present. Lieut. Cdr. WELSH said that we considered the Chateau de FACQUEVAL unsuitable and that we were trying to get sanction to take the party to England where suitable accomodation was available. It was agreed that efforts should be made to remain at LE VESINET pending a decision on this point and Major FURMAN proposed sending a cable to General GROVES which was drafted. I understand that after I left the conference it was decided not to send this cable. Lieut. Cdr. WELSH then returned to LONDON.

I saw Colonel FORD at MIS who informed me that the Villa ARGENTINA was required immediately for a party of WACS and after some discussion he agreed to put other accomodation at our disposal for a few days.

3rd June 1945

Major FURMAN told me that the party were to move at once to Belgium on the orders of Brig. General CONRAD. I refused to move before Monday 4th June and telephoned H.Q. (Lieut. Cdr. WELSH) who confirmed that the move should take place.

It was not possible to arrange air transport and MIS provided two command cars with trailers and a saloon car for the journey.

4th June 1945

A movement order was obtained from Colonel FORD and we left LE VESINET at 1300 hours on 4th June and arrived at the Chateau de FACQUEVAL at 2345 hours.

During the whole of my stay in France I received valuable assistance from Colonel BROOK and his staff.

V. HUY (4th June–3rd July)

Although Major FURMAN had assured me that the arrangements he had made at the Chateau de FACQUEVAL were in accordance with the agreed policy, I found on arrival that this was not the case. Only P.W. rations were available and no provision had been made for a meal for the professors who had been travelling since midday. Fortunately I had brought American "K" rations which we had eaten by the roadside.

Captain DAVIS, U.S. Army, had been temporarily placed in command pending the arrival of Lieut. TOEPEL of the A.L.S.O.S. Mission who had been appointed to command the unit. Mr. OATES, an American C.I.C. man was also attached to the unit. These officials did everything in their power to help me.

It was pointed out to me that the American troops would object to any signs of fraternisation and that I would not be allowed to provide any extra food or comforts for the professors. As a matter of fact I did later provide additional food and drink with the connivance of the American Officer in charge, but without the knowledge of the American G.I,s.

5th June 1945

I received a cable from H.Q. (Lieut. Cdr. WELSH) stating that the whole future policy regarding the professors were being examined and that I was to use all my endeavours to keep them in a good frame of mind pending a decision.

In view of the attitude of the American troops it was impossible for me to live with the professors as I had done up to now. Ordinary "A" rations were drawn for the Officers and troops, whereas only P.W. rations were drawn for the professors, and two separate messes had to be set up. I reported the unsatisfactory position to Lieut. Cdr. WELSH at H.Q.

The professors had no alternative but to accept the position but they were getting to the end of their tether. They had been promised contact with British and American Scientists and had been assured that full provision was being made for their families. They could not understand why they were being treated in this way. I explained to them some of the difficulties that had arisen and was able to reassure them and keep them reasonably happy.

The routine at the Chateau was much the same as it had been at LE VESINET and, the weather being mainly fine, the professors spent most of their time in the garden. The guard troops had been provided with a piano and as they rarely used it, I persuaded them to give it to the professors. This instrument was in a very bad condition, a number of notes were missing, but it did not take them long to take the whole thing to pieces and repair it with improvised tools. I borrowed a local piano tuner's kit and they soon had it tuned. I also bought a wireless set which proved a very welcome addition to the amenities of the house.

Speculation as to the identity of the professors was as great in Belgium as it had been in France. The most popular guess was that the party consisted of von RIBBENTROP and his staff. There was considerable danger to security owing to the fact that the American troops, who were not trained in intelligence work, mixed freely with the village girls. They also made nightly trips to LIEGE. There were three Belgian civilians engaged as cooks etc. for the mess and these people could not be confined to the premises.

Lieut. Colonel WATKINS insisted upon the establishment being run as an American Military Station, he even wanted the Stars and Stripes flown from a flag staff in the grounds; this suggestion I vetoed. The Chateau was officially designated "Special Detention Centre, Area No. 5., Channel Base Section, ETOUSA" and this had to appear on all correspondence and requisitions. This drew attention to the nature of the establishment and there was the obvious danger of Swiss or Red Cross representatives claiming the right of entry.

<u>9th June 1945</u>

The professors were very worried when they read in the newspapers that the Russians were extending their zone of occupation in Germany. Dr. DIEBNER was frantic as it appeared that the town of STADTILM (Thuringia) where his wife and son were, was to come under Russian Occupation. He begged me with tears in his eyes to get his wife and son moved into the British or American zone. I pointed out that his previous activities hardly warranted our doing him a favour but said I would see what could be done. In the meantime Professor HEISENBERG had told me that Mrs. DIEBNER had worked with her husband and knew about all his work and that of the others and he thought it would be unfortunate if she fell into Russian hands. I consequently cabled this information to H.Q. with the request for the family to be moved. During the next few days Dr. DIEBNER showed signs of mental aberration and threatened first to attempt to escape and when he realised that this was impossible, he threatened to commit suicide. It was a great relief when I was able to inform him that his family had been moved to NEUSTADT Nr. COBURG. The receipt of this news moved him to such an extent that he asked to be taken to Church although he admitted that he had no religion and had not been inside a church for many years. I took him to the village church to mass the following Sunday where he caused a sensation by appearing dressed up as though for a Church parade.

<u>10th June 1945</u>

Lieut. TOEPEL arrived to take over command of the unit. This Officer was an A.L.S.O.S. man and had been present at HECHINGEN and TAILFINGEN when the professors were taken into custody. He knew those who came from there and they recognised him.

Lieut. TOEPEL handled a difficult situation very well indeed, cooperating in every way and turning a blind eye to my fraternisation whilst maintaining his position as O.C. of the troops.

14th June 1945

Professor GERLACH was brought from PARIS to join the party. The professors were delighted to see their old colleague.

15th June 1945

By this time the professors were again becoming very, very restive and they hinted to me that the time might soon come when they would take desperate measures to let the world know of their situation. They did not say what action they contemplated but said they would give me due warning. They showed a certain loyalty to me personally as they appreciated what I had been able to do for them. They assured me that they would not break their parole without withdrawing it. I had a long talk with Professor HEISENBERG who is the most sensible of them and he told me that their main worry was the lack of information about their families. He also said that they suspected that their potential value was being judged by the documents found at their institutes. He said that these did not give a true picture of the extent of their experiments which had advanced much further than would appear from these documents and maintained that they had advanced still further as a result of pooling of information since their detention. He begged for an opportunity of discussing the whole matter with British and American Scientists in order to acquaint them with their latest theories and work out a scheme for future co-operation. Professor HEISENBERG and Dr. HARTECK suggested that Professor BONHOEFFER of LEIPZIG who they believed was at FRIEDERICHSBRUNN in the OSTHARZ should be brought to join this party. They said he was an active Anti-Nazi who had worked with them and that it would be unwise to let him fall into Russian hands. The above information was passed to H.Q. by cable.

The professors again asked to be allowed to write to their families and I said I would try and arrange for the letters to be delivered. Letters were written and after censorship which necessitated a lot of rewriting, these were handed by me to Lieut. Cdr. WELSH in LONDON.

Lieut. Cdr. WELSH told me on the telephone that permission had been given for the professors to be brought to England and he asked me to come over as soon as possible to inspect Farm Hall.

16th June

As we required additional staff, I got two more P.W. from the stockade at NAMUR, a second cook and a man who was a barber by profession. This enabled us to be more or less independant of outside domestic help. The laundry was done in the house. A group photograph was taken by Mr. OATES. (The negative and all copies of this photograph are in my possession.)

17th June

In order to get an air passage to the U.K., I had to get myself temporarily attached to a British unit stationed at BRUSSELS and I accordingly got myself attached to 21 Army Group and got an Authority from them and proceeded to LONDON.

Lieut. Cdr. WELSH and I went to Farm Hall where arrangements had already been made to instal microphones. I had asked for such an instalation from the day I took charge of the professors. We arranged with Colonel KENDRICK to transfer the necessary staff of technicians from CSDIC (U.K.) to man the installation. We were fortunate also in obtaining the services of Captain BRODIE from CSDIC (U.K.) to act as Administrative Officer.

26th June

I returned to Belgium leaving Captain BRODIE to complete the arrangements at Farm Hall.

On arrival at BRUSSELS airport I was informed that Lieut. Colonel WATKINS was making an inspection of the Chateau de FACQUEVAL that afternoon. He was allowed to inspect only the American troops and their quarters and the professors were confined to the house during his visit.

The professors received the news of the impending move to England with mixed feelings. On the one hand they looked on it as a step forward in that they expected to meet their British colleagues, but on the other hand England seemed much further away from home than Belgium.

Certain difficulties arose regarding the journey of the professors to England as Lieut. Colonel WATKINS insisted upon orders directing him to release them from American custody. Eventually orders were obtained in PARIS directing him to release all personnel detained at the Special Detention Centre to me personally or to my representative, at LIEGE airport.

30th June 1945

I left for England leaving Lieut. TOEPEL in charge. All the professors gave me their word to carry out any instructions given by Lieut. TOEPEL and they were handed over to Mr. OATES, whom I had designated for the task, at LIEGE airport on the 3rd July and flown by special aircraft to TEMPSFORD.

Lieut. Colonel PAGE and his staff were extremely helful during the whole of my stay in Belgium.

VI. SUMMARY

The operation has been successful to date in that,

(1) The professors have been detained for over ten weeks without any unauthorised person becoming aware of their identity or place of detention, and,

(2) They have, with considerable difficulty, been kept in a good frame of mind.

The following are brief character sketches of the professors:–

VON LAUE:	A shy mild mannered man. He cannot understand the reason for his detention. He has been extremely friendly and is very well disposed to England and America.
HAHN:	A man of the world. He has been the most helpful of the professors and his sense of humour and common sense has saved the day on many occasions. He is definately friendly disposed to England and America.
HEISENBERG:	He has been very friendly and helpful and is, I believe, genuinely anxious to co-operate with British and American Scientists although he has spoken of going over to the Russians.
GERLACH:	Has a very cheerful disposition and is easy to handle. He appears to be genuinely co-operative.
HARTECK:	A charming personality and has never caused any trouble. His one wish is to get on with his work. As he is a bachelor, he is less worried than the others about conditions in Germany.
DIEBNER:	Outwardly very friendly but has an unpleasant personality and is not to be trusted. He is disliked by all the others except BAGGE.
VON WEIZSACKER:	A diplomat. He has always been very friendly and co-operative and I believe he is genuinely prepared to work with England and America but he is a good German.

<u>WIRTZ</u>: An egoist. Very friendly on the surface but cannot be trusted. I doubt whether he will co-operate unless it is made worth his while.

<u>BAGGE</u>: A serious and very hardworking young man. He is completely German and is unlikely to co-operate. His friendship with DIEBNER lays him open to suspicion.

<u>KORSCHING</u>: A complete enigma. He appears to be morose and surly. He very rarely opens his mouth. He has, however, become more human since his arrival in England.

<u>14th July 1945</u>

(Sgd) T.H. Rittner
<u>MAJOR</u>

COPY NO. 1

Mr. M. Perrin for General Groves
through Lt. Cdr. E. Welsh

Ref. FH1

To: Mr. M. Perrin and Lt. Comdr. Welsh.
From: Major T.H. Rittner.

OPERATION "EPSILON"
(3rd–18th July 45)

1. General

1. A report covering the operation on the continent from May 2nd until 3rd July 1945 has already been submitted.

2. The arrangemetns for bringing the party to England went according to plan and the following landed at TEMPSFORD on the afternoon of 3rd July and were taken to FARM HALL by car.

Professor VON LAUE.
Professor HAHN.
Professor HEISENBERG.
Professor GERLACH.
Doctor HARTECK.
Doctor VON WEIZSACKER.
Doctor WIRTZ.
Doctor DIEBNER.
Doctor BAGGE.
Doctor KORSCHING

together with four PW orderlies. A further PW orderly has since been added to the party.

3. All the Professors have renewed their parole to me in writing in respect of FARM HALL and grounds and I have warned them that any attempt by any one of them or by the orderlies to escape or to communicate with anyone will result in them all having their liberty considerably restricted.

4. Ordinary army rations are drawn for the professors and the officers and troops and these are prepared for all by the PW cooks.

5. Microphones have been installed in all the bedrooms and living rooms used by the professors. This installation has proved invaluable as it has enabled us to follow the trend of their thoughts.

In the following conversation, DIEBNER and HEISENBERG discussed the possibility of there being microphones in the house. The conversation took place on 6th July in the presence of a number of their colleagues:–

DIEBNER: I wonder whether there are microphones installed here?

HEISENBERG: Microphones installed? (laughing) Oh no, they're not as cute as all that. I don't think they know the real Gestapo methods; they're a bit old fashioned in that respect.

II. MORALE

1. The party has settled down well at FARM HALL but they are becoming more and more restive. The question of their families is causing them the greatest anxiety and I believe that if it were possible to make arrangements for an exchange of messages with their families, the effect on general morale would be immediate.

2. Most of the recorded conversations are of a general nature and show that they are pleased with the treatment they are receiving but completely mystified about their future.

3. Lt. Comdr. WELSH visited FARM HALL on 7th July. The atmosphere was somewhat tense as can be seen from the following conversations:

(a) Conversation between HEISENBERG, HARTECK, WIRTZ, KIEBNER and KORSCHING after the announcement of Lt. Comdr. WELSH's visit:

HEISENBERG: I can see the time is coming when we must have a very serious talk with the Commander. Things can't go on like this.

HARTECK: It won't do. We have no legal position since they have to keep us hidden.

HEISENBERG: Apparently they feel guilty about their own scientists, otherwise one can't understand it. I tell you what we'll do; one evening we'll make the Commander drunk and then he'll talk. We'll play bridge and then talk seriously from one o'clock onwards.

WIRTZ: I think you should speak to the Commander and tell him we are very dissatisfied and then we can make him drunk one evening.

HEISENBERG: Yes, that is the right sequence of events. First there will be an afternoon when we will go for him and break him down and then an evening when we will make it up.

HARTECK Yes, and tell him in no uncertain terms that we are being wronged.

HEISENBERG: Yes, of course.

DIEBNER: You appear to have a certain influence on him and I think that you could achieve something with him.

HEISENBERG: Well, I think I am more or less in his good books. I will point out to him that he has let STARK and LENNARD go on living happily in GERMANY whilst we poor wretches have to let our wives and children starve.

HARTECK: In the meantime the British and American soldiers are looting everything at home.

WIRTZ: He doesn't mind that.

HEISENBERG: Oh, yes, he does.

DIEBNER: With a bit of cunning, we may get something out of this. First of all they are keeping this whole business here secret and secondly the idea seems to be to be friendly to us.

HEISENBERG: I should say that the point is that they don't yet know what they want. That's the whole trouble. They don't want us to take part in any discussion regarding our future as they don't want us to have any say in the matter. They want to consider what to do and they have not yet agreed among themselves.

HARTECK: But they can't say to you: "You must stay here." They can merely ask: "Do you want to stay here under these conditions"? Or can they say: "You must stay here".

HEISENBERG: Of course, they can if they want to. Of course it is possible they will agree to ask us whether we want to stay in England or go to America but that we cannot stay in GERMANY.

DIEBNER: When the Commander comes he is sure to bring some letters or some good news with him. As soon as he comes he will try and pacify us with all sorts of excuses.

KORSCHING: Then he will talk for hours and afterwards think to himself. "Well that's all right, now I've calmed them down for a bit".

(b) Conversation between VON WEIZSACKER, HAHN, HEISENBERG, and WIRTZ after Lt. Comdr. WELSH had had a talk with them.

WEIZSACKER: I was very annoyed with the way the conversation began. That was when you started about the letters. When he said: "Yes, they have gone but there has been no reply yet." It's all very mysterious.

HAHN: No, He said they had not yet been sent. That's what he said.

WEIZSACKER: That came out bit by bit after we had really questioned him. And then the remarks about "misfortunes" etc. of course it's easy to bring up things like that and I can understand an Englishman being annoyed at our bringing that sort of thing up but that was not the proper way to reply to your questions. I felt I didn't want to listen to it all and I didn't want to give the impression that I disagreed when the man said things like that. That's why I left the room; also of course to make it easier for you.

HAHN: I would be very pleased Mr. HEISENBERG if you would have a talk with him. You needn't know the details of my conversation with him.

HEISENBRG: He started of his own accord at lunch. I got the impression that he is rather depressed about the whole situation here and the fact that he got a somewhat hostile reception. He noticed it all right.

WIRTZ: HAHN told him that we are living here like princes, but what use is that to us when we have no news of our families, have no idea what is to happen to us, and are out of touch with our work. Although we are well treated, we are nevertheless prisoners.

HEISENBERG: It certainly made an impression on him and he wanted to talk to me about it at lunch but I refused and said: "We will have a private discussion afterwards."

WEIZSACKER: I don't think we ought to spoil our chances with this man as he may be the one who can help us against others who are more hostile to us. He probably came here expecting us to be cheerful and to receive him as well as we did last time at LE VESINET. He felt at once that that was not the case and was naturally annoyed.

4. The general lines upon which the professor's minds are working can be seen from the following conversations.

(a) Conversation between DIEBNER, HEISENBERG, HARTECK, WIRTZ and KORSCHING on 6th July.

DIEBNER: Suppose you were to escape and get to CAMBRIDGE; you have a lot of friends there. That would cause a terrific sensation. The whole thing would become known. Surely you would do that if they detain you here for a year.

HEISENBERG: If nothing happens now, I will certainly go to the Major (RITTNER) in a comparatively short time and say to him: "I ask permission to break my parole." Then he will be in the awkward position of having to post an armed sentry outside my door. That will cause trouble higher up.

DIEBNER: That would at any rate result in some action being taken. What could they do to you? If you escaped and really tried to get to CAMBRIDGE, they could do nothing. They could get the police to bring you back but the damage would be done.

HARTECK: They seem to be afraid that one might do something hostile to England but they are hiding us from their own people and that is the amazing thing. If it had been the other way round we never hid a foreign scientist in Germany, the other scientists all knew about it.

DIEBNER: The awful thing with the English is that it takes ages before they make up their minds to do something.

WIRTZ: The Empire has been built up through centuries; they have plenty of time. They can't understand it when someone is in a hurry.

HEISENBERG: One can say that they do things better than others because they take their time.

DIEBNER: They have money and in consequence have time.

HARTECK: The longer one is here, the more anxious one is to get home. In addition, it annoys one to be left in doubt. One gets terribly bitter.

DIEBNER: That's it – terribly bitter.

HEISENBERG: It may be that the British Government are frightened of the communist professors, DIRAC and so on. They say: "If we tell DIRAC or BLACKETT where they are, they will report it immediately to their Russian friends, KAPITZA and Comrade STALIN will come and say: "What about the BERLIN University Professors? They belong in BERLIN."

DIEBNER: It's quite possible they just don't want to say anything.

KORSCHING: Then of course they will have to wait until everything has been settled by the "Big Three".

DIEBNER: I think the right thing in that case would be for the English to give us a hint in some way. They may not be able to say it openly because of Comrade STALIN.

HEISENBERG: It is possible that the "Big Three" will decide it at POTSDAM and that CHURCHILL will come back and say: "Off you go, the whole group is to return to BERLIN" and then we'll be in the soup.

(b) Conversation between WIRTZ and VON WEIZSACKER on 7th July:

WEIZSACKER: These people have "detained" us firstly because they think we are dangerous; that we have really done a lot with Uranium. Secondly, there were important people who spoke in our favour and they wanted to treat us well. These two facts were mixed up. Now they have got into this awful political muddle.

The decent thing for them to do now would be to say to us: "It is not possible to come to a decision about you so quickly. What shall we do? Would you like to remain with your families for the time being or – ?" They don't do that but prefer to keep us on ice. That's not nice of them. As a matter of fact I believe them when they say it has to do with the election and all sorts of political muddles. I don't believe it is due to malice that they do nothing with us but it is just that they cannot come to a decision about us.

WIRTZ: Yes, I could quite understand that, but they could say: "We will come to some arrangement now about your families." What's the idea of the whole thing?

WEIZSACKER: Yes but of course that is difficult – the French zone of occupation. But the damnable thing is that they won't let one have any say in what is to happen to one or one's family or give one any hopeful indication of what is going to happen.

(c) Conversation between WIRTZ, HAHN, and DIEBNER on 16th July after reading in the newspaper that LORD CHERWELL was attending the POTSDAM conference:

WIRTZ: That's the man who has had us detained.

HAHN: If CHERWELL knew we were detained here, something would happen. He doesn't know; he would certainly speak to one and discuss what he should or could do.

DIEBNER: Things like that will certainly be discussed. I imagine that they will decide at the "Big Three" conference which scientists are to go to RUSSIA.

HAHN: How should CHERWELL know anything? He doesn't know anything about us; that's the stupid part about it – But perhaps he does know.

III. ATTITUDE TOWARDS BRITISH AND AMERICANS

1. Some interesting sidelights on the attitude of some of the professors towards Britain and America appear from the following conversations between BAGGE and KORSCHING.

(a) 8th July.

KORSCHING: It makes me furious when people are so childishly anglophil. It was just the same in HECHINGEN.

BAGGE: How do you mean?

KORSCHING: They handed them the water on a platter, they did the same with the Uranium and all the instruments and all the secret files in duplicate and – I don't know – 20 gr. of radium. That's awful.

BAGGE: WIRTZ and BOPP buried 2 (10?) gr. of Radium which they will sell privately later.

KORSCHING: WEIZSACKER, although he is clever, imagines he can negotiate with them regarding the handing over of the water and on what conditions. They discussed it with the Commander. At first they wanted to say: "We will only tell you on condition that you let us go on working on it." They imagined they could get away with that. He need only threaten them with bread and water and they will give way.

(b) 11th July

BAGGE: If we want to continue working on our subject, we will certainly have to work together with the Anglo-Americans. No one has any money in Germany.

KORSCHING: If one is convinced that Germany will be occupied by the Russians for a long time and you work on the production of weapons for the English, the end result will be that you will make GERMANY into the (future) battlefield. The English, are, of course, really much too careful to think of fighting RUSSIA. Of course I would have no pangs of conscience in making neutron sources for the Americans. Of course we could not separate Uranium for them with the existing separation apparatus. I would be perfectly willing to carry on working with that as it is completely harmless (laughs).

From what I know of the Anglo-Americans, I don't relish the idea of their assimilating us as easily as all that. The result will be that all the good work we may do in our lives will, one could almost say, go to the credit of Anglo-American brains. You can't imagine WEIZSACKER and WIRTZ doing anything but remaining in GERMANY for the rest of their lives.

BAGGE: What do you mean? The first thing WIRTZ did was to ask "will we be given British nationality?".

KORSCHING: He had had all sorts of other discussions beforehand. Don't imagine it was his idea. I was once talking to WIRTZ and HEISENBERG and I said: "It would certainly be a clever move for anyone who is thinking of working in England to acquire British nationality as otherwise he would be shot if he fell into Russian hands". They both agreed that one would have to do that. If one is taken to England, one may have to stay there.

I would rather take Swedish nationality than stay in GERMANY and wait for the next war. On the other hand I would not make any effort to become British. If there is nothing more to be made out of GERMANY, one should at anyrate get away from RUSSIA. VON WEIZSACKER is more or less resigned to the idea of becoming Russian one day.

Suppose the English were to come and say: "You can carry on with your work, in fact you are to go on working on Uranium. We will take everything back to HECHINGEN but you must sign a paper." Then presumably one would have to sign in order to get away from here. But would you really do it?

BAGGE: I would say that even during the war I was able to carry on my scientific work freely and I would ask whether I could continue to do so.

KORSCHING: I would say the same, of course. If they said "No", I would sign all the same and do it in spite of that.

On the other hand, of course, they will not give us the heavy water any more. They may say: "Go back and work but not on the Uranium machine." They know we cannot get hold of two tons of Uranium secretly. And then of course they may say: "The Uranium machine people are to go back but the isotope-separators must carry on working at separating isotopes under American control."

BAGGE: Men like WIRTZ will want to do something too. WIRTZ may construct his curious machine again then.

KORSCHING: He will not be able to separate even 1 milligram of anything. Wirtz has the same problem as I have with my apparatus. It is a question of solubility. As long as he uses fluids which are not mercury, you get the solubility effect just the same with him as with me. The difference between his apparatus and mine is that his stages are single and mine are more compact. But of course he will try and play about with it even if it is no good. But I believe the English may be satisfied with the fact thay they have the apparatus. But I imagine we will have to sign one thing: that we must keep silence about all the apparatus they have taken away from us. I can't believe they will let us go; we could then publish the theory of both apparatus. We will certainly have to sign a declaration that we will not publish it. One must be very careful not to let ourselves in for anything.

2. DIEBNER and BAGGE somewhat surprisingly expressed a desire to acquire British nationality in the following conversation on 17th July:

DIEBNER: I would be glad BAGGE, if we could stay here.

BAGGE: It would be a wonderful thing if we could become English.

DIEBNER: And then have nothing more to do with the Party again. I would willingly take an oath never to have anything to do with the Party again.

IV TECHNICAL

1. The bi-weekly lectures are being continued. In fine weather these take place out of doors.

2. The following remarks were made by BAGGE in conversation with KORSCHING on 9th July:

BAGGE: I have now solved the wave equation. Now I have to calculate the correct distribution of the charge from the wave function and the quadrupole moment from the distribution of the charge. That is what I am doing just now. First of all it is known that the deuteron is near enough spherical, hence you can get the forces acting between proton and neutron from the intrinsic energy of the deuteron, i.e., you can find a force which gives the correct mass of the deuteron. This force, of course, corresponds to a certain relative direction of the spin of the particles. But in the deuteron a definite spin position is realized, the spins of the proton and the neutron are parallel, and for this relative psition of the spins you can calculate the force. A priori we know nothing about the spin position, but something can be calculated from scattering experiments. Namely, if the spins are anti-parallel, the force is (only) half as great. You can find a function, giving the force as function of the spins which has just this property. HEISENBERG has pointed this out. If you assume with HEISENBERG that the force depends on the spins in this way, the forces are twice as big in that position as compared with this position, and the spherical symmetry of the deuterons is preserved. Now we have the function of HEISENBERG's and the quantity of the forces, and we can take the scattering experiments correctly into account. In other words, with the help of HEISENBERG's functions, we can explain the experimental scattering results <u>and</u> the intrinsic energy without contradiction.

BETHE has shown that within the theory of YUKAWA you can make assumptions which will give HEISENBERG's function of the spins. To make the calculation invariant from the point of view of relativity, you have to introduce additional terms which also depend on the spin directions and which can explain the quadrupole moment of the deuteron. BETHE has assumed forces in such a way that 1) They agree with the results of the scattering experiments. 2) They give the correct mass of the deuteron. 3) They give the quadrupole moment of the deuteron correctly. That is all correct, but it is correct only, because he introduces a new term for every effect he wants to explain. The starting point was the mass of the deuteron which is obtained with a . . . Then come the scattering experiments which require HEISENBERG's term. To obtain the quadrupole moment correctly, you need the YUKAWA term which BETHE has used. Each term is introduced for a specific purpose. Of course, it is a possible theory. You put as much . . . into it, as you need to explain new (experimental) results.

HAHN: We don't know the properties of Uranium 94, but we know those of "93".

BAGGE: You wait until the "93" has completely disintegrated, and then you should really have pure "94".

HAHN: That is far too little, you can't do anything with that. You will get nothing of an element with a period of decay of 10000 years through the disintegration of an 2.3-day-element.

BAGGE: Why? You have the 93-element with a period of decay of 2.3 days, and now you wait for 20 days. Then there will be nothing left of the 93-element which will have completely transformed itself into the 94-element.

LAUE: That is too little.

HAHN: There are as many atoms as correspond to the "93". But you can prove the (existence of) "93" for the simple reason that in 2.3 days – that means actually (in) seconds it disintegrates by one five-thousandth.

BAGGE: Now KORSCHING does the following: he takes your trace of the 93-element which you have concentrated.

HAHN: Every ten years one "alpha" ray will be emitted. How can you demonstrate that?.

BAGGE: If so far you have been able to demonstrate 10000 years by alpha counting methods i.e., to confirm 10000 years as a lower limit, then you should be able to improve on this by approximately another 1000 by the use of a "Plattenmethode" as KORSCHING has stated; but the zero-effect (Null Effekt) will upset the measurements.

V. FINANCE

The professors told me some time ago that they all had German money with them which they would like to send back to their families. In consequence I asked them on 7 July to let me know how much each individual had. The following conversation took place between DIEBNER and GERLACH:

DIEBNER: I wanted to put down that I am carrying a certain sum.

GERLACH: I would just write; "I have so many thousand marks; it was money to pay –".

DIEBNER: Funds of the Reich Research Board (Reichsforschungsrat).

GERLACH: No not Reich Research Board but Research Society.

DIEBNER: Yes.

GERLACH: For the payment of the salaries of assistants and technical personnel. The money was at my home and I took it with me; I had no chance of banking it.

DIEBNER: I have just counted it. I should have had RM 95,000 with me but it is only RM 79,000 and something. I gave some of it to KREMER (?). I should have RM 35,000 of my own money and RM 60,000 belonging to the Research Board. But I have only got a total of RM 79,000. Perhaps I gave some to my wife.

Subsequently I was given the following list of money carried by each individual:

VON LAUE	RM	201
HAHN		785
HEISENBERG		1,809
HARTECK		10,400
GERLACH		400
VON WEIZSACKER		550
WIRTZ		726
DIEBNER		79,246
BAGGE		1,238
KORSCHING		1,034
	RM	96,389

VI. PERSONALITIES

1. The Professors

(a) VON LAUE Appears, from monitored conversations, to be disliked by his colleagues.

(b) HAHN Unpopular with the younger members of the party who consider him dictatorial.

(c) HEISENBERG Has been accused by the younger members of the party of trying to keep information on his experiments, to himself.

(d) VON WEIZSACKER Told WIRTZ that he had no objection to fraternising with pleasant Englishman but felt a certain reluctance in doing so "this year when so many of our women and children have been killed".

(e) DIEBNER Is very worried about his future and has told BAGGE that he intends to send in a formal request to be reinstated as a civil servant. He hopes we will forget that he was a member of the Nazi Party. He says he only stayed in the Party as, if Germany had won the war, only Party members would have been given good jobs.

2. Others

(a) BOTHE There has been a lot of speculation as to why Professor BOTHE has not joined the Party as expected. They imagine he has been clever enough to be able to stay in Germany and carry on with his work!

(b) EWALD (?) Stated by GERLACH to have possessed an exceedingly good mass-spectrograph able to produce an unusually large number of lines.

(c) MAUER (?) One of the professors in conversation with GERLACH said he was afraid of a physicist named MAUER who was an ardent Nazi but a poor research worker. MAURER worked with STRASSMAN (?) on the disintegration of molybdenium and Uranium.

(d) MEYER A physicist, head of the Developement Section of the torpedo experimental station. He is in his middle thirties and is a graduate of KARLSRUHE University. He is an ardent Nazi.

(e) STRASSMANN (?) Worked with MAURER on the disintegration of molybdenum and Uranium. (See above).

Farm Hall
19th July, 1945.

(Sgd) T.H. Rittner
Major

AMERICAN EMBASSY
OFFICE OF THE MILITARY ATTACHE
1, GROSVENOR SQUARE, W. 1.
LONDON, ENGLAND

11 August 1945

Subject: Transmittal of report F.H. 2 of Operation "Epsilon" - dated 1 August 1945.

To: Major Francis J. Smith,
Room 5119, New War Dept. Bldg., Washington, D. C.

Attached is report No. 2 of Operation "Epsilon". Report No. 1 has been furnished your office through British channels. Report No. 3 indicating the reaction of the guests to Valhalla Day will follow in the near future.

For the Military Attache:

HKCalvert
H. K. CALVERT,
Major, F.A.
Assistant to the Military Attache.

Inclosure - 1.
Cy No. 1 of above report.

COPY NO. 1

Captain Davis for
General Groves.

Ref. FH2

To: Mr. M. PERRIN and Lt. Cdr. WELSH
From: Major T.H. RITTNER

OPERATION "EPSILON"
(18–31 July 45)

I. General

There has been very little change in the position at FARM HALL since the last report. Outwardly the guests are serene and calm, but it is clear that their restiveness is increasing. Suggestions have been made that one of the guests should attempt to get a letter to CAMBRIDGE. Steps have been taken to prevent this.

II. Morale

The following conversations show the general trend of morale:

1. Conversation between HEISENBERG, VON WEIZSACKER, WIRTZ, HARTECH and DIEBNER on 18 July:

WEIZSACKER: I would say we must wait for the "Big Three". The whole thing is connected with that.

WIRTZ: This is the position. Why don't they want to send letters? Not because there is no post; that's all rot; of course they could send a letter. For some reason or other no one must know that Professor HEISENBERG etc. are here. That's the point. The moment anyone, even your wife is told "Professor HEISENBERG is well and happy", they will realise that he is still in captivity.

HEISENBERG: Everyone in HECHINGEN knows that I have been arrested, but the moment news gets through, they will know: "Ha! They are still alive".

WEIZSACKER: They know that in any case.

HEISENBERG: I could also imagine that they are afraid of the following: Assume that it became known that we are here; some clever journalist would turn up and, of course he would not be allowed in. He would have a look at the place from outside, see us playing all sorts of games in the garden, sun bathing etc. The next day there would be a terrific article in the newspaper just like it was with GOERING: "German Nazi Scientists enjoying life in England. For lunch they have –" He could write a wonderful article like that and that would of course be very awkward for everyone concerned. I could well understand that that is the reason they want to keep it secret here. Of course if our colleagues who know something about the business – GOUSMIT for instance – were clever, they would put another article in the newspaper, about anti-Nazis. It could start with Pastor NIEMOELLER and Bishop GAHLEN.

WIRTZ: A man like GOUDSMIT doesn't really want to help us; he has lost his parents.

HARTECK: Of course GOUDSMIT can't forget that we murdered his parents. That's true too and it doesn't make it easy for him.

DIEBNER: I would imagine that we will be given more freedom the moment the Russians say: "We agree, you will take over the scientists". They are negotiating with the Russians as to who shall be handed over to Russia and who shall not. Presumably that is being discussed in Berlin now.

WIRTZ: Surely the Major must have noticed that our morale has sunk.

HARTECK: He's noticed that all right.

WIRTZ: It's another question whether our attitude is directed against him personally.

HEISENBERG: No, he knows it is not against him personally.

WIRTZ: You can see it in the William Joyce case which has been postponed until 11 September. The English are like that.

HARTECK: Yes. They've got plenty of time.

WIRTZ: If I were ever to land with airborne troops in England I would have all the men arrested straight away and they would be separated from their wives for two years just to show them what it's like.

HEISENBERG: I think there is a 90% chance of our getting back to Germany.

HARTECK: Yes. I think that is most likely. At first I thought they would really be more interested in getting information out of us. But they don't do that.

HEISENBERG: Perhaps they won't do so.

HARTECK: Apparently not. They will wait until they can do it better themselves. Then we will have to swear on oath not to talk about the thing etc. and then perhaps they will pay each of us £500.

WIRTZ: Not on your life! We will have to pay for having been here.

2. Conversation between WIRTZ, HARTECK, HEISENBERG on 21 July:

WIRTZ: I think there is a very good chance we will get back to Germany. There is a 25% chance we will get back before 1 December. The chance of getting back between 1 December and the end of next year, I would put at 70%. I think there is a 40% chance that we will never get back at all. Of course the percentages don't add up to 100. I think there is a 15% chance that we will never see our wives again.

HEISENBERG: That's all much too pessimistic. I think there is a 35% chance that we will get back before 1 December. The chance of our getting back within a reasonable time after that date, I would put at 50%. The chance of our never getting back except perhaps in totally different circumstances after many, many years, I would put at 14%. There is a 1% chance that we will never see our wives again. I can see no reason to assume that they want to treat us badly, but I can see a reason to assume that they don't want to have us in Germany as they don't want us to pass on our knowledge to other people.

HARTECK: That is one point but on the other hand we may be shot; not by the English but by the people there. If one of us went to Hamburg University some mad student might come and shoot one.

HEISENBERG: I still feel very strongly that they are making an exception in our case in that they are treating us better than most others and therefore I should say we will see our wives again even if we don't return to Germany. That would only be prevented if something unforeseen occurred. Of course one never knows, something astounding may suddenly happen.

WIRTZ: That's what I think. I consider there is a 15% chance of that.

3. The thing which is worrying the guests more than anything else is the fact that they are unable to send news to or get news from their families. The following conversation between WIRTZ, KORSCHING and HEISENBERG took place shortly after I had discussed this question with HEISENBERG on 26 July.

WIRTZ: I can't understand that. My wife will tell every Frenchman that the English have taken me away.

KORSCHING: I don't believe that is the real reason.

HEISENBERG: Then what do you think is the real reason?

KORSCHING: They want to keep us as long as possible from contact with anyone.

WIRTZ: I don't quite understand that because, if that were really the case they ought to have taken our wives too. But in any case everyone in HECHINGEN knows we were taken away. I can't understand it.

HEISENBERG: The whole position with regard to Russia depends upon the outcome of this election. It is obvious that if Atlee becomes Prime Minister –

KORSCHING: We will be handed over to Russia. That's just it.

HEISENBERG: That would change the whole political situation.

WIRTZ: They have done wrong in detaining us and now they can't get out of it. It is unpleasant for them. I can see that one of us will have to get to Cambridge one day.

HEISENBERG: Yes in certain circumstances.

WIRTZ: We'll have to fix that, or send a letter to Cambridge. That should be possible.

KORSCHING: Of course, I will put it in the letter box.

HEISENBERG: That's all right but so far you have not been able to do it because you have given your parole.

KORSCHING: That's why I always said we should give it for a limited time.

WIRTZ: We will just say: "We take it back" and then one day –

HEISENBERG: The first thing they will do will be to post a sentry with a tommy gun.

KORSCHING: They can't do that so quickly; if we do it cleverly, it can be done at 10 o'clock in the evening. (laughter)

HEISENBERG: We could just throw it out of the window over the wall. You might do that in any case but let's wait a bit.

4. Speculation as to the reason for their detention is still a favourite topic of conversation as can be seen from the following talk between HEISENBERG, HARTECK and GERLACH on 26 July:

HEISENBERG: It looks as though the Americans fear nothing so much as the possibility of the French getting even an inkling of the Uranium business – very odd. The Americans know that JOLIOT is interested in the business and they are afraid that JOLIOT, who is a communist, will do something with the Russians. At any rate, if JOLIOT gets to know all about it, the Americans can't prevent the Russians from finding out all about it. If they were forbidding us to write letters merely in order to annoy us, there would be no reason for treating us so well here; and they have always treated our families well.

HARTECK: They are probably not really frightened of the French but only of the Russians.

GERLACH: Certainly.

HEISENBERG: The Russians are certainly two years behind us in the separation of Uranium but if they put people like FRENKEL and LANDAU etc. on to it they will most certainly succeed.

HARTECK: Is that the LANDAU from GOETTINGEN?

HEISENBERG: No, that is the man who was often in Copenhagen. He worked on –

GERLACH: Geomagnetism.

HEISENBERG: He worked with me at Leipzig. He's a very clever Russian Jew.

HARTECK: Doesn't Joffe have anything to do with it?

HEISENBERG: He deals with the political side. FRENKEL is a good man too.

GERLACH (?): The whole thing as far as we are concerned is really a political question. They're not interested in us as physicists.

GERLACH: LAUE has only heard about the Uranium machine since we have been in detention.

HARTECK: He knew absolutely nothing.

5. The following conversation between BAGGE and DIEBNER on 26 July shows their respective attitudes:

DIEBNER: Do you think GERLACH wants to stay here for five years?

BAGGE: We want to get the position clear.

DIEBNER: Do you thing VON WEIZSACKER wants to stay here for five years?

BAGGE: Oh yes, he wants to stay here. He likes it here. He says every day that he has never had such a good opportunity to think and work as he has here. You must see that the situation is getting worse. Up to now I always hoped that the thing would come to end in some sensible way but I have lost hope, that is the tragedy. When I see how slowly everything goes, how it is being kept more and more secret, the fact that even here in England they have to hide us from their own people, from their Lord Cherwell, from Churchill, and everybody, that's what I can't understand.

DIEBNER: They can't do that for ever. They must realise that something will happen if we don't acquiesce.

BAGGE: I'm frightened. I'm reaching the end of my tether. (half sobbing)

DIEBNER: About your family?

BAGGE: Yes of course that's one reason.

DIEBNER: If I have to stay here for a year and then go back to Germany, then I shall have the support of these people in some way.

BAGGE: And in the meantime my family will be dead. After all I feel responsible for my family. I saw it for myself. The first day the French arrived in HECHINGEN and raped the women one after the other and a few days later they took me away. The day I had to leave, three Moroccans were billeted in the house – that's been going on for three months and I'm supposed to look happy here. I shall go mad. I can't stand it much longer.

DIEBNER: You must stick it.

BAGGE: I shall refuse to go downstairs. I shall eat nothing. I shall go on hunger strike. (Note: BAGGE is much too fat and a course of bread and water would be good for his health)

DIEBNER: BAGGE, you mustn't think we're all complete fools. HEISENBERG is no fool. Do you think men who have wangled things to their own advantage all the time are going to let themselves be fooled.

BAGGE: You must also realise that if, during the war, we (put) people in concentration camps – I didn't do it, I knew nothing about it and I always condemned it when I heard about it – if Hitler ordered a few atrocities in concentration camps during the last few years, one can always say that these occurred under the stress of war but now we have peace and Germany has surrendered unconditionally and they can't do the same things to us now.

5. HAHN and DIEBNER had a long talk on 30 July part of which is reproduced elsewhere in this report. The following extract shows their attitude to the letter question and HAHN's philosophical acceptance of the situation:

HAHN: I read an article in the Picture Post about the Uranium bomb; it said that the newspapers had mentioned that such a bomb was being made in Germany. Now you can understand that we are being "detained" because we are such men. They will not let us go until they are absolutely certain that no harm can be done or that we will not fall into Russian hands or anything like that. To my mind it is a mistake to do anything. All my hopes and efforts are now directed towards getting into touch with my family. Of course I also think of my Institute as I am actually the only original member of the KAISER WILHELM GESELLSCHAFT left who was there when it was formed. Of course one is sad when one sees it all disappear but I can't do anything about it. One must be fatalist here. The longer one is "detained" here and knows nothing, the more one gets into a state where one racks one's brain to discover what is going to happen. I fight against it and make jokes. Also I don't take life too seriously in that I always look on the bright side of things.

DIEBNER: I would have been just the same in Germany. The day before I went away I said to my wife; "I suggest we commit suicide." I had reached that stage then.

HAHN: My wife was like that sometimes and that is why I am worried whether she will hold out without news. See what LAUE did against National Socialism and I think I worked against it too. We are both innocent but I am not allowed to write to my wife. I have told the Major: "If my American and English friends knew how I am being repaid for all my work since 1933, that I am not even allowed to write to my wife, they would be very surprised." We are being well treated here, our slightest wish is granted if it is possible, everything except writing letters.

DIEBNER: It is the future that worries me.

HAHN: The outlook for the future is dark for all of us. I have not got a long future to look forward to. Suppose you want to work later with GERLACH; do you think he will work on the Uranium machine? Men are not idealists and everyone will not agree not to work on such a dangerous thing. Every country will work on it in secret. Especially as they will assume that it can be used as a weapon of war.

We have no contacts abroad now. No foreigner can find out where we are and they will wonder. My Swedish friends with whom I used to correspond will wonder what has happened and will assume I am dead.

DIEBNER: I am becoming more and more pro-English. They do everything very decently. The Major takes great trouble.

HAHN: He takes great trouble and he would probably consider us ungrateful if we suddenly sabotaged everything. We can't do that.

DIEBNER: No, no, that's out of the question.

III. The Nazi Party

Some of the guests appear to be worried about their previous adherance to the Nazi Party and its effect upon their future. The following conversations show their fears.

1. Conversation between BAGGE and GERLACH on 30 July:

BAGGE: All the young assistants I knew had to join the Party; those from

Munich too, RENNER (?) and WELKE (?).

GERLACH: They didn't all do it.

BAGGE: Those who wanted to go to the University had to.

GERLACH: KAPPLER (?) and BUHL (?) who were with me didn't.

BAGGE: Do you know EULER who was one of HEISENBERG's assistants? He did not get a job at Leipzig because he wasn't a member of the Party. The fight lasted 18 months and HEISENBERG and HUND and Heaven knows who else couldn't manage it.

GERLACH: I managed it. BLUMENTHAL (?) was not in the Party. He had to go in 1937 because they said his wife was partly of Jewish extraction. He went into business. GRIMMEN (?) was not in the Party either.

BAGGE: MEYER (?)?

GERLACH: MEYER (?) was in the Party. He was at one time a big man in the SS but got fed up afterwards. We cured him. I don't know whether DUHN (?) was in or not.

BAGGE: I was not in the Party. In 1933 I was taken by the High School SA people and pushed into the SA just like all the other young assistants I know. For instance WIRTZ – I don't know about VON WEIZSACKER – and BOPP, they were all in the SA. It was compulsory and one could do nothing about it.

GERLACH: I didn't join the Teachers Union (Lehrerbund).

BAGGE: In our Institute all the assistants had to join the Lecturers Union (Dozentenbund).

GERLACH: RUECHERZ (?) didn't join. They tried to force us and we got letters and they made difficulties. We just threw everything into the wastepaper basket and didn't answer.

BAGGE: That is one way of doing it.

GERLACH: I maintain that it is not right to say that one <u>had</u> to do it. I never put anything in writing. Dr. BARTH was not a member of the Party. He had been an assistant in Russia for three years and was a proper assistant in the Institute. SCHUETZ (?) was a Party member without realising it.

BAGGE: That's what happened to me. In the autumn of 1936 my mother wrote to me to Leipzig asking whether I wanted to join the Party. Someone had asked. My mother thought it was a good thing and had sent my name in. A few months later I received my Party book which stated that I had been in the Party since 1 May, 1935. It had been back-dated 12 months. It also said that I had sworn an oath to the FUHRER in May 1935. Not one word of it was true.

GERLACH: I don't believe HILSCHI (?) was a Party member or MEISSNER (?) either, but I'm not sure. Only a few of the Munich men were members. They kept on complaining and making their silly speeches. I let them make them and occasionally I was really rude as, for instance when I said in the Faculty; "I don't care a damn what the Reichs Chancellery says".

DIEBNER: Taking the line of least resistance as so many did was of course not the right course.

GERLACH: I had a half Jew as assistant until the autumn of 1944; I kept on saying: "It's impossible to remove the man as so much depends on him". There was a girl who got into trouble later. We lost the assistant NEUMANN (?) who went into business later. None of the female personnel I had were Party members. I had no picture of Hitler in my Institute. They kept on coming and saying we should buy a picture of Hitler. I always said: "No, I already have one". I had a very small picture I had bought for 5 pfennig. The Nazis treated me badly. They reduced my salary and withdrew my allowances.

BAGGE: Didn't that happen to other people too?

GERLACH: No. Then they brought an action against me and I didn't go to the Institute any more. I said: "I won't go back until you withdraw the case". That was my trump card.

BAGGE: On what grounds did they reduce your salary? You had an agreement.

GERLACH: I just got a letter saying: "The agreement made between the Bavarian State and yourself is cancelled; from now on your salary will be as follows". And that was that.

(GERLACH leaves the room)

BAGGE: They could no nothing against him. He knew GOERING personally. His brother was in the SS and that's how he managed to stay on.+ GERLACH gets a certain personal amusement out of annoying people. It wasn't just his convictions.

+ Note. In a conversation with HAHN, GERLACH said that his brother was involved in certain big money deals with the SS. He found this out when a sum of money was once transferred to his account in Berlin instead of his brother's. He expressed his disapproval of his brother's association with the SS to HAHN.

DIEBNER: He has rows with everyone.

BAGGE: There's something behind it. Why do they keep on talking to us about the Party. HEISENBERG started it and now GERLACH has brought it up.

2. Part of a conversation between DIEBNER and HAHN on 30 July:

DIEBNER: I wanted to tell you how I came to join the Party and how I have suffered under the Nazis. In 1933 I became a Freemason in opposition to National Socialism. I <u>never</u> voted for Hitler. That became known in HALLE and the result was that I got into difficulties at the Institute. Then I went to the "Waffenamt" and was to have become a civil servant, but I did not. SCHUHMANN didn't forward my application. He said he couldn't do it because I was a Freemason. SCHUHMANN did his best for me and sent me to a man in Munich and after a year the thing went through and I became a civil servant, a "Regierungsrat".

HAHN: The fact of being a Party member does not necessarily tell against a man. The newspapers say that.

DIEBNER: Everyone knows my views. GERLACH knows them; I was never a National Socialist and never took any part in politics. WIRTZ knows my views. I told him: "I am a Party member. We'll see what happens. If the Nazis win, I shall still be a Party member and that will help us and if things go the other way, you will have to help me." That's what we arranged at that time. Now I feel rather isolated here.

HAHN: Do you feel that you are treated here differently to the others?

DIEBNER: That's just it. WIRTZ knows that HEISENBERG will help him no matter what happens. I am sure GERLACH would help me, he has always been very decent to me.

HAHN: The fact that you were in the Party hasn't really done you any harm.

DIEBNER: When I get back to Germany now everyone will say: "Party man. Party man!"

HAHN: None of us know what will happen to us. In my opinion it's no good worrying too much about the future as we have <u>no idea</u> what will happen to us. You got on quite well with JOLIOT didn't you?

DIEBNER: I have helped so many people. I persuaded SCHUHMANN to see that Professor PIETERKOWSKI (?) in Poland should be given facilities to go to Germany before the SS came. I often helped JOLIOT vis a vis the Gestapo.

HAHN: What happened to the Pole?

DIEBNER: I don't know. He didn't come. At Copenhagen SCHUHMANN wanted to remove the Cyclotron. I prevented Copenhagen from being touched. I have done <u>so much</u> against these people. For instance we prevented people being arrested in Norway.

HAHN: Then I don't understand why you are worried. We can only hope that we will be able to send letters home but I don't think we can expect anything else just yet.

3. In the following conversation on 18 July, HEISENBERG relates how he tried to help some of his colleagues and WIRTZ admits German atrocities:

HEISENBERG: During the war I had five calls for help in cases where people were murdered by our people. One was SOLOMAN (?), HOFFMAN's (?) son-in-law. I could do nothing in his case as he had already been killed when I got the letter. The second one was COUSYNS the Belgian cosmic ray man; he disappeared in a Gestapo Camp and I couldn't even find out through HIMMLER's staff whether he was alive or dead. I presume he is dead too. Then there was the mathematician CAMMAILLE; I tried to do something about him through SETHEL (?) but it was no good and he was shot. Then from among the Polish professors there was a logistician with a Jewish name – and then with the other Poles, the following happend; his name was SCHOUDER, a mathematician. He had written to me and I had put out feelers in order to see what could be done. I wrote to SCHOLZ (?) who had had something to do with Poland. Then SCHERRER wrote me the following ridiculous letter saying he had also had something to do with the case. He wrote: "Dear HEISENBERG, I have just heard that the mathematician SCHOUDER is in great danger. He is now living in the little Polish town and so-and-so under the flase name of so-and-so." That came in a letter which was of course opened at the frontier. It is unbelievable how anyone can write that from Switzerland. I heard nothing more about SCHOUDER and I have now been told that he was murdered.

WIRTZ: We have done things which are unique in the world. We went to Poland and not only murdered the Jews in Poland, but for instance, the SS drove up to a girls' school, fetched out the top class and shot them simply because the girls were High School girls and the intelligentia were to be wiped out. Just imagine if they arrived in HECHINGEN, drove up to the girls' school and shot all the girls! That's what we did.

IV. The Future

Speculation by the guests as to the future in general has been dealt with under the heading "Morale", but the following conversation between DIEBNER, KORSCHING and BAGGE on 21 July goes rather further:

BAGGE: For the sake of the money, I should like to work on the Uranium-engine; on the other hand, I should like to work on cosmic rays. I feel like DIEBNER about this.

KORSCHING: Would you both like to construct a Uranium-engine†?

DIEBNER: This is the chance to earn a living.

KORSCHING: Every layman can see that these ideas are exceedingly

† 'Engine' is not a German word. They will undoubtedly have said '*Uranmaschine*' and 'uranium-machine' would be a better translation. FCF

important. Hence there won't be any money in it. You only make money on ideas which have escaped the general public. If you invent something like artificial rubies for the watch making industry, you will make more money than with the Uranium-engine. Well, DIEBNER, we'll both go to the Argentine.

DIEBNER: I shall come with you.

KORSCHING: I know MERKADA (?) there. I could write to him. Of course the letter must not be opened on the way.

BAGGE: Who is he, a physicist?

KORSCHING: Yes, he has worked with SCHUELER. He came over to look around a bit. He came from the university of La Plata; not stupid, but of course he could not compete with SCHUELER. You can only build a Uranium-engine of your own in the Argentine.

DIEBNER: That is right, there are advantages in that.

BAGGE: I think, we should approach the Argentinian ambassador.

KORSCHING: The man ought to understand something about physics and that is always difficult as such people know nothing about it.

BAGGE: He knows nothing about it, but the Argentinian ambassador will know that there is something in it.

KORSCHING: But you have to consider, that the Argentian ambassador has to be careful that the British and Americans don't put one over on him somehow. They set one of their agents to work for instance, if you can talk to the Argentinian ambassador in Madrid or so, you might perhaps succeed, but I don't think you would here in England.

BAGGE: But if you disclose your identity and explain to him the whole situation?

KORSCHING: Yes, but then you will not be in a very strong position.

BAGGE: Then I get to La Plata and if I get the job as an assistant, let us say, that would not be bad at all. (Pause)

KORSCHING: Actually I find it somehow very typical perhaps, but quite possible that HEISENBERG really continues to work on the Uranium-engine, in the end several really productive ideas will have been contributed by all sorts of people but people will say in the end: "It has been HEISENBERG's work."

Is there any Uranium ore in the Argentine at all?

DIEBNER: I don't think so.

KORSCHING: It again makes it awkward if they have to import it – to have to import ten tons of Uranium!

BAGGE: You can't get that at all; only from the Russians perhaps and they will not part with it either.

KORSCHING: Still, I should like to get to HECHINGEN once more to collect the rest of my things. After all I still have all my books there and the telescope – though mind you I have hidden it from the French. Of course I did not hand that over. I have got all my glass prisms, lenses, etc. I lifted a floorboard, hid the stuff and nailed the board down again.

BAGGE: In the Institute?

KORSCHING: In my private lodgings.

(DIEBNER leaves the room)

If you work together with HEISENBERG on a Uranium-engine then you can write off your share. If you want to work on a Uranium-engine, then you would have to do it somewhere else. Of course it would be an idea to go to the Argentine with 2 people and say: "Here we are, we know how to do this and that; we have a good method for the separation of isotopes, we do not need to produce heavy water." Somehow in this fashion we have to do it. It would not come to anything if you collaborated with HEISENBERG on a Uranium-engine. They did not even bring along the small fry to this place; that is how outsiders judge the work. They get there and read all the secret reports before they take the people away from there.

BAGGE: How long before did they have the secret reports?

KORSCHING: Two or three days before. The principal question which GOUDSMIT put to me, was: "Is that your idea? Has that been published already is that anything new?" – that is all he wanted to know. And BOPP and FISCHER they just ignore one and say "Oh well, they just made some calculations for HEISENBERG." Apart from that for instance, the ordering of apparatus from the firms and all the other various things which we have done, WIRTZ just told him (GOUDSMIT): "I have done that." Do you think WIRTZ is going to be modest in front of Mr. GOUDSMIT? No, He says: "I have built this here, I conducted the negotiations with the firms, I had that built here and I have done the experimental work and as far as the countings are concerned – everybody knows only too well how easy it is to count particles – Messrs. FISCHER & BOPP did that." And that is how WIRTZ has excluded them. GOUDSMIT takes his word for it. BOPP was quite disgusted and astonished that suddenly he was dropped like that. And that is how it is all over the world. A scientist is asked "What have you thought out, where is your idea?" If you then make the strategic mistake of moving in the shadow of a man who is already world famous, then you are out of the limelight for the rest of your life and if you then raise your voice against that, then on top of it you will be called a trouble maker.

BAGGE: Did you notice how HEISENBERG wiped the floor with WEIZSACKER?

KORSCHING: And how! I rubbed my hands with joy. It is of course very degrading that he (WEIZSACKER) cannot even do a few simple calculations.

BAGGE: HEISENBERG can now of course make it up with him, if later he publishes the thing together with WEIZSACKER.

KORSCHING: As far as I know HEISENBERG, he will not do that.

BAGGE: I don't think he will either.

KORSCHING: He will publish it and mention WEIZSACKER etc. and in the end the whole effort of WEIZSACKER will have been in vain because it will be said "HEISENBERG is behind this."

BAGGE: For what remains in the end is the mathematical structure. The little bit of roundabout thinking which WEIZSACKER did will be forgotten.

KORSCHING: If WEIZSACKER does not now try hard to write down a few more formula then he is squashed altogether. I think it serves him right for WEIZSACKER has unlimited ambition. (Pause) Now the really positive point about the Chief (HEISENBERG) is the following: If you do some work of your own, which he acknowledges to be sound and worthwhile, then you have complete liberty to do it. In WEIZSACKER's institute you become a slave – "Now you do this, what you are doing is ridiculous, etc." WEIZSACKER would never let his people work in his institute as the Chief would.

BAGGE: That you can see from HOECKER.

KORSCHING: HOECKER is clever enough to wriggle out of it as a rule. But as we have said, if you want to work on the Uranium-engine, it is obviously completely useless to do it <u>with</u> the Chief.

BAGGE: If you want to build an aircraft today, then first you have to ignore your own interests, because the state is too much interested in it, to grant you liberty to work on it as you please. I would say, the aircraft is today comparable to the Uranium-engine. That is why, if one has purely scientific interests, one should slowly withdraw from it.

KORSCHING: On the other hand HEISENBERG will say, if we cannot build a cyclotron anyway – and it is obvious that we cannot build one in Germany with the American . . . – then we will have to hold back as a source of neutrons, at least a neutron-generator, for the production of artificial radioactive elements etc.

BAGGE: Why can't we build a cyclotron?

KORSCHING: Because we have no money. It takes too long – over there they have them readymade and if we do not now make some progress in nuclear physics, then Germany will slowly lose her place, where nuclear physics is concerned. A 2.50 meter cyclotron even if you could . . . start on it, would . . . . only be ready when the Americans would have completed all their work on the "2.50m". One can of course still build a small cyclotron, 1m, or 80m. It is obvious, that you can do a lot of things with the engine, enormous quantities, enormous concentrations of neutrons, in fact there are any amount of possibilities. I think the Chief has the right ideas slowly to wangle permission to run his own Uranium-engine for scientific purposes. He will probably obtain it, if the others do not in the meantime study the heavy water.

BAGGE: I am convinced, they (Anglo-Americans) have used these last 3 months mainly to imitate our experiments.

KORSCHING: Not even that. They used them to discuss with their experts their possibilities and to study the secret documents. They probably examined a few specimens of our Uranium-blocks. From these specimens they can see for instance, whether the engine has been running already. It could have been run; the blocks must have undergone some internal chemical change.

BAGGE: But they know already, that it did not run; that they were told.

KORSCHING: That is just it. They were told practically everything up to approximately the last series of measurements. It is the same to them whether it ever came to an increase in Neutrons of 5 or 50. The issue must be quite clear to them.

BAGGE: But they will certainly have the ambition to imitate our experiments as soon as possible and for that purpose they need the D_2O. Once they have worked with that – (int.)

KORSCHING: They'll obviously never again let go of it. If that is so, then a Uranium-engine can only run in Germany without the production of heavy water – which as HARTECK thinks is so frightfully easy, but connected with great expense, but can be run only with an efficient method for the separation of isotopes, which is technically workable with ordinary water.

BAGGE: Quite so. But with ordinary water you need 15 tons of Uranium even with an increase in concentration of 5%.

KORSCHING: No. Just consider, you can increase the concentration of Uranium from 0.7% to 1 or 2%. If they will not let us work on Uranium and we must sign the following statement: "I pledge myself, not to run a Uranium-engine for anybody anywhere in this world", then you must sign it.

BAGGE: I would only sign that under one condition: That they grant me enough money for other purposes, so that I have the possibility to carry on with my experiments.

KORSCHING: Of course we can say that. But then they will say: "Then we will contribute to your funds."

BAGGE: That would have to be a contribution of RM 100,000 per year.

KORSCHING: That we will never get, but perhaps we may get RM 30,000 a year. They do not want to destroy Germany but what England wants is to weaken her, otherwise they will never be able to achieve hegemony in Europe, if they immediately boost us up again.

BAGGE: They now seem to plan a "United States of Europe".

KORSCHING: Yes, if Russia would not constantly interfere. They know perfectly well, that once they have let us go back to Germany they'll only have 50% control over us. They can put somebody in my room, and I guarantee you, that without that fellow noticing it, I'll be able to make an experiment. I just know he goes to see his girl friend on Saturday, so I'll just work on Saturday night. It is possible that they themselves have already great quantities of heavy water and Uranium.

BAGGE: That I do not believe.

KORSCHING: But there are many military men in England, who say "Once we let those swine go back then they'll construct the Uranium-engine and in the end they'll blow it up." They might also say: "These people are so clever that our guard troops will be blown up with it, but not they themselves." There are also many people in England who say: "On no account must these people be treated generously; they must be made to work constantly under the threat of machine-guns." I do not believe that the Commander will achieve so much, that he will be able to say: "Here is your heavy water, here is your Uranium, now carry on with your work."

BAGGE: There is also the question, whether the Commander wants that.

KORSCHING: Quite, if the man says: "I assure you on my word of honour." What does it mean? He did not give it to us in writing. Also he has never said: "I shall take care that your position as scientists is safeguarded." He has not even done that, but all he has said was: "I assure you on my word of honour that I –" (int.)

BAGGE: You have heard that yourself?

KORSCHING: No, not the "word of honour", but the word "assure". Of course he will not have us beheaded, that is quite clear. After all he is more or less favourably disposed towards us. I am sure there are also people who say "Behead them!" There you have to be glad, that there is such a man as the Commander. If they put a piece of paper before you: "Here, please sign" there is nothing else left for you to do, but sign. You cannot write: "I pledge myself not work on the Uranium-engine in any state, except the Argentine." In that case you would find yourself in gaol for the next hundred years (laughs) I do not believe that they will send us away without our signatures or without any assurance. The Argentine would perhaps be quite nice as a sort of bold adventure; as I said before, if one were so far advanced with the separation of isotopes, that one knew for sure one can increase the concentration of Uranium by 1% with a certain small expenditure of energy, then it would have sense, but otherwise to do the same all over again would of course not make much sense.

I shall be glad when I have liberty of movement again and be able to walk in the street and buy a scientific book, when I can do anything at all, and can write letters to friends, who have survived the war.

BAGGE: I got into contact with the Uranium-engine only through the war, and I have always felt an outsider and for me it would mean to take a step which I do not want to take at all, because if it had been my endeavour to make a lot of money, I could have stayed at home with my parents. I would have probably kept clear of the war equally easily. If I had joined my father's business, I do not like to think how much money we could have earned.

KORSCHING: That you can also do with the Uranium-engine; if you really put a Uranium-engine before the Argentines, then you can say: "I am a scientist, I only want to build up a laboratory for myself; pay me 500.000 pesetas, but otherwise leave me in peace." Then you can of course work on cosmic rays at the University of La Plata as much as you like and on top of that you have the 500.000 pesetas. You would get them, if you got into the good books of the right professors and politicians.

BAGGE: But it could easily be, over there that there is an awful lot of intrigueing as well. Perhaps there are a lot of people like WIRTZ.

KORSCHING: Of course you will not get the amount of pesetas which you should get according to the value of the proposition, but even so, if you get only 3% it would be a fortune.

BAGGE: Actually you derive no benefits from your patent either.

KORSCHING: I did not tell them at all that it is a patent. I could have done so but then I would have lost everything. As it is now, if I find some third person in Sweden – if I say "This is the position, I have the patent, they do not know anything about it, take it out in Sweden; all I want for myself is 5% of what you can get from any firm." Then, when the Americans suddenly see that the patent has been taken out, they cannot do anything about it. Perhaps it will already be superflous in a year or so. That is why I have not given it to them. I admit it is still with the Patent agent in ZEHLENDORF. Fortunately there is nothing in the Institute; there I have, of course, hidden everything. In ZEHLENDORF it is in his private flat, thank God. So if the Russians have not pilfered everything there

– it may be that it has been burnt, then it is lost anyway – and if the British do not search every private house in ZEHLENDORF now, then they will not find it at all. About these 20 grams of radium, of which DIEBNER talked and which seems to belong to the German Radium Institute – I ask myself why do these people do that? If I had been there, I would have said: "Do you know that I have radium at all?" Then: "Do you know the exact amount?" Then I would have hidden at least 1 or 2 grams somewhere.

BAGGE: WIRTZ has hidden 2 grams. Only WIRTZ knows where these 2 grams are and then DIEBNER has some as well.

KORSCHING: But even so, it is too much that 20 grams still fell into their hands. One could have done it like that everywhere. I saw it myself, there they pinched some measuring apparatus. Those two apparatus which I took along, they could not pinch. On the other hand of course they must not notice it, because then they say: "All right, you starve in Germany, you will not get any money from us." But our two engines they need not have got of course. The childish thing is, we need only have put them on the lawn at the back and it would have been perfect. They did not even look into the . . . loft. I put umpteen things up there. They did not even notice the apparatus which was in that box in the Chemistry room – the box was two metres long.

V. Technical

The usual bi-weekly lectures have been given. These have been confined to general subjects.

FARM HALL
1 August, 1945.

(Sgd) T.H. Rittner
Major.

AMERICAN EMBASSY
OFFICE OF THE MILITARY ATTACHE
1. GROSVENOR SQUARE. W. 1.
LONDON. ENGLAND

1 September 1945

Subject: Farm Hall, "Epsilon" Report No. 2.- Re Goudsmit.

To: Major Francis J. Smith, Room 5004, New War Dept. Bldg., Washington, D. C.

Attention: Mr. Ryan.

1. Major Smith asked that this office get the original transcript of that part of Farm Hall Report No. 2 wherein Harteck mentioned Goudsmit's parents and made the statement, "Of course we murdered them". The German text of that statement is as follows:

WIRTZ: "Ein Mann wie GOUDSMIT will uns garnicht richtig helfen, der hat ja seine Eltern verloren.

HARTECK: Ja, also naturlich, ganz kann der GOUDSMIT nicht davon absehen, dass wir seine Eltern umgebracht haben. Das ist ja auch wahr, ich meine, dass ist nicht so ganz leicht für ihn."

2. The word "umgebracht" is probably best translated to mean "killed" rather than "murdered".

For the Military Attache:

H. K. Calvert

H. K. CALVERT,
Major, F.A.
Assistant to the Military Attache.

Note: See page 46 of FH2.

Copy No. 1.

Capt. Davis for General Groves.

Ref. FH3

To: Mr. M. PERRIN and Lt. Cdr. WELSH.
From: Major T.H. RITTNER.

OPERATION "EPSILON"
(1–6 August, 1945)

I. General

This report covers the period since my last report up to the evening of 6 August when the announcement of the use of the Atomic Bomb was made.

The effect of the announcement and the subsequent reation of the guests forms the subject of a separate report Ref. FH4.

II. Morale

In conversation with a British Officer regarding the position of communication with the families, HAHN completely broke down. BAGGE also came very near to tears when he described the fate worse than death which he pictured was that of his wife and children at the hands of the Moroccan troops.

General morale has however improved since I was able to tell the guests that permission had been granted for them to write letters to their families and that it was hoped to obtain answers. This permission was contained in a cable from Lt. Cdr. WELSH to Mr. PERRIN dated 1 August.

Letters were written and it was almost pathetic to see the efforts made by the guests to convey the information that they were in England. The look of discomfort on their faces when asked to delete certain sentences was obvious and subsequent monitored conversations showed that the sentences I had blue-pencilled were the ones which were intended to convey this information. The letters have all been rewritten, and I am trying to make arrangements through Captain DAVIS to have them delivered.

III. The Guests and the NAZIS

The Guests have been at great pains to clear themselves of any suggestion that they had any connection with the Nazis. GERLACH in particular has done his best to make this clear to his colleagues and one wonders whether this may not be due to a guilty conscience on his part. In this connection, GERLACH had a long conversation with me in the course of which I suggested that there must have been Gestapo agents working in their institutes. We also discussed the question of how much they had known of scientific work being carried out in other countries. This conversation had the desired effect and GERLACH proceeded to discuss these points with the other guests. The following conversations ensued:

1. Conversation between GERLACH, HAHN and HEISENBERG on 4 August.

GERLACH: The Major asked me what we had known about scientific work in enemy countries, especially on Uranium. I said "Absolutely nothing. All the information we got was absurd."

HEISENBERG: In that respect one should never mention any names even if one knew of a German who had anything to do with it.

GERLACH: For instance I never mentioned the name of that man ALBERS (?). The "Secret Service" people kept on asking me: "From whom did you get information?" and I always replied: "There was an official in SPEER's Ministry and in the Air Ministry who gave it out officially". I did not say it was ALBERS (?) who did it.

HEISENBERG: I had a special man who sent me amazing information from SWITZERLAND. That was some special office. Of course I have burnt all the correspondence and I have forgotten his name.

HAHN: Did you actually get any new information from him.

HEISENBERG: At that time I always knew exactly what was being discussed in the SCHERRER Institute regarding Uranium. Apparently he was often there when SCHERRER lectured and knew what they were talking about. It was nothing very exciting but, for instance he once reported that the Americans had just built a new heavy water plant and that sort of thing.

2. The following conversation took place between GERLACH and HEISENBERG on 5 August:

GERLACH: I have just been talking to DIEBNER about whether there may have been an SS man amongst our colleagues. There was that business with Dr. GRUENZIG (?); he was some sort of patent man who had once been a HITLER YOUTH leader.

HEISENBERG: I remember. He was at HECHINGEN.

GERLACH: Yes he was at HECHINGEN and we were always trying to get rid of him. ROSENBERG (?) sent him to us.

HEISENBERG: Oh yes. That's right. But that was right at the end, during the last six months perhaps.

GERLACH: Well the GRUENZIG (?) business was in the summer of 1944. At first he wanted me to send him to MUNICH. I mistrusted him and didn't let him see anything and we got rid of him with a lot of difficulty. I always said: "The man is too valuable to be used in an office job." Didn't GRUENZIG (?) once want to put someone in your Institute? We discussed it with you at the time and warned you. (Pause) Then DIEBNER told me he was always suspicious that someone from the British Secret Service had been with BOTHE, a certain Dr. GEHLEN (?). Did you ever know him?

HEISENBERG: Yes I knew GEHLEN (?). I must say I can understand your suspicion of GEHLEN (?). I knew GEHLEN (?) at LEIPZIG, he worked with DOEPEL, and I couldn't quite make him out. He was recommended to me by his cousin, Philosophy-Professor GEHLEN (?) who had been at LEIPZIG and was then moved to KOENIGSBERG. His wife was Swedish and I know he had contacts abroad.

GERLACH: He had been with the English Bank before.

HEISENBERG: What I didn't like about the man was the fact that he had had such a varied career. He was a man of about 36 or 37; he had worked in a bank in ITALY and then in an English bank; then he had had some technical job in Sweden and had had all sorts of other jobs. He had never really completed his studies. DOEPEL took him on as he made quite a good impression. I often discussed the matter with DOEPEL and we agreed that GEHLEN (?) should at any rate be told nothing about the Uranium business. Later he went to BOTHE. We could not make him out and I believe I spoke to BOTHE about it and told him I was not sure of him. On the other hand there was some business about his having been denounced to the Gestapo in LEIPZIG. He was supposed to have had contacts abroad and the matter was investigated and he was acquitted. I can't remember exactly what happened. I wouldn't mention the GRUENZIG (?) business to the Major as it might cost him (GURENZIG (?)) his life.

GERLACH: No I wouldn't do that. As I said, I didn't mention ALBERS (?). I didn't say anything about him in PARIS either.

HEISENBERG: I suspected two persons of belonging to the foreign "Secret Service". The first one is DELLENBACH and I am pretty certain about him and the second one is GEHLEN (?) but I am not sure about him.

GERLACH: I am quite certain about DELLENBACH. You know how he got his job?

HEISENBERG: I presume through his connections with BORMANN's cousin. I once discussed it with VOEGLER (?)

GERLACH: I also spoke to VOEGLER (?) about DELLENBACH and also spoke to SPEER's man GOERLER (?) about it. Then there was that unpleasant business about DELLENBACH and the Swiss telegram about the Uranium engine which our "Secret Service" intercepted. It all seemed to point to DELLENBACH. There were two other people mixed up in it, an engineer and a professor. He (DELLENBACH) wouldn't tell me his name but I heard afterwards that he had a very curious position and that he was continually at (HITLER's) Headquarters. His name was said to be SCHMITT. He was one of those mysterious people and HITLER made him a Professor during the war. Later that SD man POHL (?) came to me and told me the whole thing had been cleared up and that it had been an act of revenge on the part of one man who had pretended it was DELLENBACH in order to harm him. They didn't get the other man as he had got to SWITZERLAND in the meantime.

HEISENBERG: I'd like to talk to VOEGLER (?) about that as I raised so many objections to taking DELLENBACH into my Institute and VOEGLER (?) and I nearly had a row about it. Otherwise VOEGLER (?) is such a sensible man and he must have had his reasons.

IV. The POTSDAM Conference

The announcement of the Communique at the end of the POTSDAM Conference caused a certain amount of alarm and despondency among the guests. They were particularly upset about the clauses referring to the probable cession of eastern German provinces to POLAND. They noted the reference to German science but made very little comment. HEISENBERG's remark to the others was: "At any rate it would have been infinitely worse if we had won the war."

V. Professor BOTHE

1. HEISENBERG and GERLACH discussed BOTHE in the course of their conversation on Gestapo and other agents. The following is the text of their talk:

GERLACH: The BOTHE business is very odd.

HEISENBERG: I think the following may be the solution. BOTHE was a lot with JOLIOT. When BOTHE got ill and thereby escaped detainment, FLEISCHMANN managed to get into touch with JOLIOT and suggested that he should put in a word for him. JOLIOT would certainly have been delighted to do so partly because of the Uranium and partly for BOTHE's own sake. The others may then have been unable to get hold of BOTHE again. In the meantime, JOLIOT may have made them think that BOTHE was the only man who could really do anything in that line. It is obvious that BOTHE had contact with JOLIOT, he certainly treated him well during the war and that may be what has happened. But I wonder whether it will do BOTHE any good in the long run as I can't believe the Americans will allow JOLIOT much freedom.

2. BAGGE and DIEBNER also discussed BOTHE on the same day as follows:

DIEBNER: It doesn't look as though BOTHE will join us.

BAGGE: I think GEHLEN (?) is behind it. It looks as though GEHLEN (?) had the decency to keep BOTHE informed of what was going on so that BOTHE could make his plans as far as these people are concerned and act accordingly.

3. Another reference to BOTHE and GEHLEN (?) appear in Section III of this report.

VI. Technical

1. The usual bi-weekly lectures on general scientific subjects have taken place.

2. The following conversation between DIEBNER and BAGGE took place on 5 August.

DIEBNER: In the end we really had no more radium. There was an awful row as someone wanted some. I fetched another 3 grammes at the last moment.

BAGGE: Didn't firms like BRAUNSCHWEIGISCHE CHEMIEFABRIK have any more?

DIEBNER: I don't know. They may have had 1 gramme; all the rest had been requisitioned by the State. I got mine from the HARZ, I sent a car specially for it.

BAGGE: That was the Reichsstelle for radium?

DIEBNER: Yes. The Rechsstelle for Chemistry had the radium – 25 (?) lb. (sic).

BAGGE: It's a pity they didn't hide 10 grammes out of the 24 grammes.

DIEBNER: I wasn't there. If I had been there we wouldn't have handed it over.

The cars drove up and it disappeared. A pity, I had made up my mind not to hand it over.

FARM HALL,
GODMANCHESTER
8 August, 1945.

(Sgd) T.H. Rittner
Major.

Copy No. 1.

Capt. Davis for Gen. Groves.

Ref. FH4

To: Mr. M. PERRIN and Lt. Cdr. WELSH.
From: Major T.H. RITTNER.

OPERATION "EPSILON"
(6–7th August, 1945)

I. Preamble

1. This report covers the first reactions of the guests to the news that an atomic bomb had been perfected and used by the Allies.

2. The guests were completely staggered by the news. At first they refused to believe it and felt that it was bluff on our part, to induce the Japanese to surrender. After hearing the official announcement they realised that it was a fact. Their first reaction, which I believe was genuine, was an expression of horror that we should have used this invention for destruction.

3. The appendices to this report are:–

1. Declaration signed by all the guests setting our details of the work in which they were engaged in Germany.

2. Photographs of the guests with brief character sketched of each.

II. 6th August, 1945

1. Shortly before dinner on the 6th August I informed Professor HAHN that an announcement had been made by the B.B.C. that an atomic bomb had been dropped. HAHN was completely shattered by the news and said that he felt personally responsible for the deaths of hundreds of thousands of people, as it was his original discovery which had made the bomb possible. He told me that he had originally contemplated suicide when he realised the terrible potentialities of his discovery and he felt that now these had been realised and he was to blame. With the help of considerable alcoholic stimulant he was calmed down and we went down to dinner where he announced the news to the assembled guests.

2. As was to be expected, the announcement was greeted with incredulity. The following is a transcription of the conversation during dinner.

HAHN: They can only have done that if they have uranium isotope separation.

WIRTZ: They have it too.

HAHN: I remember SEGRE's, DUNNING's and my assistant GROSSE's work; they had separated a fraction of a milligramme before the war, in 1939.

LAUE: 235?

HAHN: Yes, 235.

HARTECK: That's not absolutely necessary. If they let a uranium engine run, they separate "93".

HAHN: For that they must have an engine which can make sufficient quantities of "93" to be weighed.

GERLACH: If they want to get that, they must use a whole ton.

HAHN: An extremely complicated business, for "93" they must have an engine which will run for a long time. If the Americans have a uranium bomb then you're all second-raters. Poor old HEISENBERG.

LAUE: The innocent!

HEISENBERG: Did they use the word uranium in connection with this atomic bomb?

ALL: No.

HEISENBERG: Then it's got nothing to do with atoms, but the equivalent of 20,000 tons of high explosive is terrific.

WEIZSACKER: It corresponds exactly to the factor 10^4.

GERLACH: Would it be possible that they have got an engine running fairly well, that they have had it long enough to separate "93".

HAHN: I don't believe it.

HEISENBERG: All I can suggest is that some dilettante in America who knows very little about it has bluffed them in saying "If you drop this it has the equivalent of 20,000 tons of high explosive" and in reality doesn't work at all.

HAHN: At any rate HEISENBERG you're just second-raters and you may as well pack up.

HEISENBERG: I quite agree.

HAHN: They are fifty years further advanced than we.

HEISENBERG: I don't believe a word of the whole thing. They must have spent the whole of their £500,000,000 in separating isotopes; and then it's possible.

WEIZSACKER: If it's easy and the Allies know its easy, then they know that we will soon find out how to do it if we go on working.

HAHN: I didn't think it would be possible for another twenty years.

WEIZSACKER: I don't think it has anything to do with uranium.

HAHN: It must have been a comparatively small atomic bomb – a hand one.

HEISENBERG: I am willing to believe that it is a high pressure bomb and I don't believe that it has anything to do with uranium but that it is a chemical thing where they have enormously increased the speed of the reaction and enormously increased the whole explosion.

GERLACH: They have got "93" and have been separating it for two years, somehow stabilised it at low temperature and separated "93" continuously.

HAHN: But you need the engine for that.

DIEBNER: We always thought we would need two years for one bomb.

HAHN: If they have really got it, they have been very clever in keeping it secret.

WIRTZ: I'm glad we didn't have it.

WEIZSACKER: That's another matter. How surprised BENZER (?) (MENZEL? SAG) would have been. They always looked upon it as a conjuring trick.

WIRTZ: DOEPEL, BENZER (?) and Company.

HAHN: DOEPEL was the first to discover the increase in neutrons.

HARTECK: Who is to blame.

(?) VOICE: HAHN is to blame.

WEIZSACKER: I think it's dreadful of the Americans to have done it. I think it is madness on their part.

HEISENBERG: One can't say that. One could equally well say "That's the quickest way of ending the war.

HAHN: That's what consoles me.

HEISENBERG: I still don't believe a word about the bomb but I may be wrong. I consider it perfectly possible that they have about ten tons of enriched uranium, but not that they can have ten tons of pure U. 235.

HAHN: I thought that one needed only very little 235.

HEISENBERG: If they only enrich it slightly, they can build an engine which will go but with that they can't make an explosive which will –

HAHN: But if they have, let us say, 30 kilogrammes of pure 235, couldn't they make a bomb with it?

HEISENBERG: But it still wouldn't go off, as the mean free path is still too big.

HAHN: But tell me why you used to tell me that one needed 50 kilogrammes of 235 in order to do anything. Now you say one needs two tons.

HEISENBERG: I wouldn't like to commit myself for the moment, but it is certainly a fact that the mean free paths are pretty big.

HARTECK: Do you want 4 or 5 centimetres, – then it would break up on the first or second collision.

HEISENBERG: But it needn't have the diameter of only 4 or 5 centimetres.

HAHN: I think it's absolutely impossible to produce one ton of uranium 235 by separating isotopes.

WEIZSACKER: What do you do with these centrifuges.

HARTECK: You can never get pure 235 with the centrifuge. But I don't believe that it can be done with the . . . centrifuge.

WIRTZ: No, certainly not.

HAHN: Yes, but they could do it too with the mass-spectrographs. EWALD has some patent.

DIEBNER: There is also a photo-chemical process.

HEISENBERG: There are so many possibilities, but there are none that we know, that's certain.

WIRTZ: None which we tried out.

HAHN: I was consoled when, I believe it was WEIZSACKER said that there was now this uranium – 23 – minutes – I found that in my institute too, this absorbing body which made the thing impossible consoled me because when they said at one time one could make bombs, I was shattered.

WEIZSACKER: I would say that, at the rate we were going, we would not have succeeded during this war.

HAHN: Yes.

WEIZSACKER: It is very cold comfort to think that one is personally in a position to do what other people would be able to do one day.

HAHN: Once I wanted to suggest that all uranium should be sunk to the bottom of the ocean. I always thought that one could only make a bomb of such a size that a whole province would be blown up.

HEISENBERG: If it has been done with uranium 235 then we should be able to work it out properly. It just depends upon whether it is done with 50, 500 or 5,000 kilogrammes and we don't know the order of magnitude. We can assume that they have some method of separating isotopes of which we have no idea.

WIRTZ: I would bet that it is a separation by diffusion with recycling.

HEISENBERG: Yes, but it is certain that no apparatus of that sort has ever separated isotopes before. KORSCHING might have been able to separate a few more isotopes with his apparatus.

WIRTZ: We only had one man working on it and they may have had ten thousand.

WEIZSACKER: Do you think it is impossible that they were able to get element "93" or "94" out of one or more running engines?

WIRTZ: I don't think that is very likely.

WEIZSACKER: I think the separation of isotopes is more likely because of the interest which they showed in it to us and the little interest they showed for the other things.

HAHN: Well, I think we'll bet on HEISENBERG's suggestion that it is bluff.

HEISENBERG: There is a great difference between discoveries and inventions. With discoveries one can always be sceptical and many surprises can take place. In the case of inventions, surprises can really only occur for people who have not had anything to do with it. It's a bit odd after we have been working on it for five years.

WEIZSACKER: Take the CLUSIUS' method of separation. Many people have worked on the separation of isotopes and one fine day CLUSIUS found out how to do it. It was just the question of the separation of isotopes which we neglected completely partly knowingly and partly unknowingly, apart from the centrifuges.

HEISENBERG: Yes, but only because there was no sensible method. The problem of separating "234" from "238" or "235" from "238" is such an extremely difficult business.

HARTECK: One would have had to have a complete staff and we had insufficient means. One would have had to produce hundreds of organic components of uranium, had them systematically examined by laboratory assistants and then had them chemically investigated. There was no one there to do it. But we were quite clear in our minds as to how it should be done. That would have meant employing a hundred people and that was impossible.

HAHN: From the many scientific things which my two American collaborators sent me up to 1940, I could see that the Americans were interested in the business.

WEIZSACKER: In 1940 VAN DER GRINTEN wrote to me saying that he was separating isotopes with General Electric.

HARTECK: Was VAN DER GRINTEN a good man?

WEIZSACKER: He wasn't really very good but the fact that he was being used showed that they were working on it.

HAHN: That wicked BOMKE was in my Institute.

HARTECK: I have never come across such a fantastic liar.

HAHN: That man came to me in 1938 when the non-aryan Fraulein MEITNER was still there – it wasn't easy to keep her in my Institute. I will never forget how BOMKE came to us and told me that he was being persecuted by the State because he was not a Nazi. We took him on and afterwards we found out that he was an old fighting member of the Party.

WEIZSACKER: Then we might speak of our "BOMKE-damaged" Institutes. (Laughter).

3. All the guests assembled to hear the official announcement at 9 o'clock. They were completely stunned when they realised that the news was genuine. They were left alone on the assumption that they would discuss the position and the following remarks were made.:–

HARTECK: They have managed it either with mass-spectrographs on a large scale or else they have been successful with a photo-chemical process.

WIRTZ: Well I would say photo-chemistry or diffusion. Ordinary diffusion. They irradiate it with a particular wave-length. – (all talking together).

HARTECK: Or using mass spectrographs in enormous quantities. It is perhaps possible for a mass-spectrograph to make one milligramme in one day – say of "235". They could make quite a cheap mass-spectrograph which, in very large quantities, might cost a hundred dollars. You could do it with a hundred thousand mass-spectrographs.

HEISENBERG: Yes, of course, if you do it like that; and they seem to have worked on that scale. 180,000 people were working on it.

HARTECK: Which is a hundred times more than we had.

BAGGE: GOUDSMIT led us up the garden path.

HEISENBERG: Yes, he did that very cleverly.

HAHN: CHADWICK and COCKROFT.

HARTECK: And SIMON too. He is the low temperature man.

KORSCHING: That shows at any rate that the Americans are capable of real cooperation on a tremendous scale. That would have been impossible in Germany. Each one said that the other was unimportant.

GERLACH: You really can't say that as far as the uranium group is concerned. You can't imagine any greater cooperation and trust than there was in that group. You can't say that any one of them said that the other was unimportant.

KORSCHING: Not officially of course.

GERLACH: (Shouting). Not unofficially either. Don't contradict me. There are far too many other people here who know.

HAHN: Of course we were unable to work on that scale.

HEISENBERG: One can say that the first time large funds were made available in Germany was in the spring of 1942 after that meeting with RUST when we convinced him that we had absolutely definite proof that it could be done.

BAGGE: It wasn't much earlier here either.

HARTECK: We really knew earlier that it could be done if we could get enough material. Take the heavy water. There were three methods, the most expensive of which cost 2 marks per gramme and the cheapest perhaps 50 pfennigs. And then they kept on arguing as to what to do because no one was prepared to spend 10 millions if it could be done for three millions.

HEISENBERG: On the other hand, the whole heavy water business which I did everything I could to further cannot produce an explosive.

HARTECK: Not until the engine is running.

HAHN: They seem to have made an explosive before making the engine and now they say: "in future we will build engines".

HARTECK: If it is a fact that an explosive can be produced either by means of the mass spectrograph – we would never have done it as we could never have employed 56,000 workmen. For instance, when we considered the CLUSIUS – LINDE business combined with our exchange cycle we would have needed to employ 50 workmen continuously in order to produce two tons a year. If we wanted to make ten tons we would have had to employ 250 men. We couldn't do that.

WEIZSACKER: How many people were working on V 1 and V 2?

DIEBNER: Thousands worked on that.

HEISENBERG: We wouldn't have had the moral courage to recommend to the Government in the spring of 1942 that they should employ 120,000 men just for building the thing up.

WEIZSACKER: I believe the reason we didn't do it was because all the physicists didn't want to do it, on principle. If we had all wanted Germany to win the war we would have succeeded.

HAHN: I don't believe that but I am thankful we didn't succeed.

HARTECK: Considering the figures involved I think it must have been mass-spectrographs. If they had had some other good method they wouldn't have needed to spend so much. One wouldn't have needed so many men.

WIRTZ: Assuming it was the CLUSIUS method they would never have been able to do anything with gas at high temperatures.

HARTECK: When one thinks how long it took for us to get the nickel separating tube I believe it took nine months.

KORSCHING: It was never done with spectrographs.

HEISENBERG: I must say I think your theory is right and that it is spectrographs.

WIRTZ: I am prepared to bet that it isn't.

HEISENBERG: What would one want 60,000 men for?

KORSCHING: You try and vaporise one ton of uranium.

HARTECK: You only need ten men for that. I was amazed at what I saw at I.G.

HEISENBERG: It is possible that the war will be over tomorrow.

HARTECK: The following day we will go home.

KORSCHING: We will never go home again.

HARTECK: If we had worked on an even larger scale we would have been killed by the "Secret Service". Let's be glad that we are still alive. Let us celebrate this evening in that spirit.

DIEBNER: Professor GERLACH would be an Obergruppenfuhrer and would be sitting in LUXEMBOURG as a war criminal.

KORSCHING: If one hasn't got the courage, it is better to give up straightaway.

GERLACH: Don't always make such aggressive remarks.

KORSCHING: The Americans could do it better than we could, that's clear.

(GERLACH leaves the room.)

HEISENBERG: The point is that the whole structure of the relationship between the scientist and the state in Germany was such that although we were not 100% anxious to do it, on the other hand we were so little trusted by the state that even if we had wanted to do it it would not have been easy to get it through.

DIEBNER: Because the official people were only interested in immediate results. They didn't want to work on a long-term policy as America did.

WEIZSACKER: Even if we had got everything that we wanted, it is by no means certain whether we would have got as far as the Americans and the English have now. It is not a question that we were very nearly as far as they were but it is a fact that we were all convinced that the thing could not be completed during this war.

HEISENBERG: Well that's not quite right. I would say that I was absolutely convinced of the possibility of our making an uranium engine but I never thought that we would make a bomb and at the bottom of my heart I was really glad that it was to be an engine and not a bomb. I must admit that.

WEIZSACKER: If you had wanted to make a bomb we would probably have concentrated more on the separation of isotopes and less on heavy water.

(HAHN leaves the room)

WEIZSACKER: If we had started this business soon enough we could have got somewhere. If they were able to complete it in the summer of 1945, we might have had the luck to complete it in the winter 1944/45.

WIRTZ: The result would have been that we would have obliterated LONDON but would still not have conquered the world, and then they would have dropped them on us.

WEIZSACKER: I don't think we ought to make excuses now because we did not succeed, but we must admit that we didn't want to succeed. If we had put the same energy into it as the Americans and had wanted it as they did, it is quite certain that we would not have succeeded as they would have smashed up the factories.

DIEBNER: Of course they were watching us all the time.

WEIZSACKER: One can say it might have been a much greater tragedy for the world if Germany had had the uranium bomb. Just imagine, if we had destroyed LONDON with uranium bombs it would not have ended the war, and when the war did end, it is still doubtful whether it would have been a good thing.

WIRTZ: We hadn't got enough uranium.

WEIZSACKER: We would have had to equip long distance aircraft with uranium engines to carry out airborne landings in the CONGO or NORTH WEST CANADA. We would have had to have held these areas by military force and produce the stuff from mines. That would have been impossible.

HARTECK: The uranium content in the stone in the radium mines near GASTEIN was said to be so great that the question of price does not come into it.

BAGGE: There must be enormous quantities of uranium in UPPER SILESIA. Mining experts have told me that.

DIEBNER: Those are quite small quantities.

HARTECK: If they have done it with mass-spectrographs, we cannot be blamed. We couldn't do that. But if they have done it through a trick, that would annoy me.

HEISENBERG: I think we ought to avoid squabbling amongst ourselves concerning a lost cause. In addition, we must not make things too difficult for HAHN.

HARTECK: We have probably considered a lot of things which the others cannot do and could use.

WEIZSACKER: It is a frightful position for HAHN. He really did do it.

HEISENBERG: Yes. (Pause) About a year ago, I heard from SEGNER (?) from the Foreign Office that the Americans had threatened to drop a uranium bomb on Dresden if we didn't surrender soon. At that time I was asked whether I thought it possible, and, with complete conviction, I replied: "No".

WIRTZ: I think it characteristic that the Germans made the discovery and didn't use it, whereas the Americans have used it. I must say I didn't think the Americans would dare to use it.

4. HAHN and LAUE discussed the situation together. HAHN described the news as a tremendous achievement without parallel in history and LAUE expressed the hope of speedy release from detention in the light of these new events.

5. When GERLACH left the room he went straight to his bedroom where he was heard to be sobbing. VON LAUE and HARTECK went up to see him and tried to comfort him. He appeared to consider himself in the position of a defeated General, the only alternative open to whom is to shoot himself. Fortunately he had no weapon and he was eventually sufficiently calmed by his colleagues. In the course of conversation with VON LAUE and HARTECK, he made the following remarks:–

GERLACH: When I took this thing over, I talked it over with HEISENBERG and HAHN, and I said to my wife: "The war is lost and the result will be that as soon as the enemy enter the country I shall be arrested and taken away". I only did it because, I said to myself, this is a German affair and we must see that German physics are preserved. I never for a moment thought of a bomb but I said to myself: "If HAHN has made this discovery, let us at least be the first to make use of it". When we get back to Germany we will have a dreadful time. We will be looked upon as the ones who have sabotaged everything. We won't remain alive long there. You can be certain that there are many people in Germany who say that it is our fault. Please leave me alone.

6. A little later, HAHN went up to comfort GERLACH when the following conversation ensued:–

HAHN: Are you upset because we did not make the uranium bomb? I thank God on my bended knees that we did not make an uranium bomb. Or are you depressed because the Americans could do it better than we could?

GERLACH: Yes.

HAHN: Surely you are not in favour of such an inhuman weapon as the uranium bomb?

GERLACH: No. We never worked on the bomb. I didn't believe that it would go so quickly. But I did think that we should do everything to make the sources of energy and exploit the possibilities for the future. When the first result, that the concentration was very increased with the cube method, appeared, I spoke to SPEER's right hand man, as SPEER was not available at the time, an Oberst GEIST (?) first, and later SAUCKEL at WEIMAR asked me: "What do you want to do with these things?", I replied: "In my opinion the politician who is in possession of such an engine can achieve anything he wants". About ten days or a fortnight before the final capitulation, GEIST (?) replied: "Unfortunately we have not got such a politician".

HAHN: I am thankful that we were not the first to drop the uranium bomb.

GERLACH: You cannot prevent its development. I was afraid to think of the bomb, but I did think of it as a thing of the future, and that the man who could threaten the use of the bomb would be able to achieve anything. That is exactly what I told GEIST, SAUCKEL and MURR. HEISENBERG was there at STUTTGART at the time.

(Enter HARTECK)

Tell me, HARTECK, isn't it a pity that the others have done it?

HAHN: I am delighted.

GERLACH: Yes, but what were we working for?

HAHN: To build an engine, to produce elements, to calculate the weight of atoms, to have a mass-spectrograph and radio-active elements to take the place of radium.

HARTECK: We could not have produced the bomb but we would have produced an engine and I am sorry about that. If you had come a year sooner,

GERLACH, we might have done it, if not with heavy water, then with low temperatures. But when you came it was already too late. The enemy's air superiority was too great and we could do nothing.

HAHN, GERLACH and HARTECK go on to discuss their position if they return to Germany and GERLACH considers that they will have to remain here another two years because they will be in danger. HAHN however feels that he could return to Germany without any danger to himself. GERLACH goes on to explain that the Nazi party seemed to think that they were working on a bomb and relates how the Party people in MUNICH were going round from house to house on the 27th or 28th April last telling everyone that the atomic bomb would be used the following day. GERLACH continues:

GERLACH: I fought for six months against ESAU and BEUTHE taking over all the heavy water and the uranium and having the engine made by the Reichs Anstalt. ESAU told me more than once: "The cube experiment is my experiment and I am going to see it through and I am going to take everything". And as I was stubborn and refused to give in. BEUTHE (?) sent that letter to HIMMLER through the S.D. regarding my political attitude. I know all about it and you have no idea the trouble I had with ESAU and what my position was in February and March of last year because of BEUTHE's (?) accusations. I wouldn't have given much for my chances of life at that time. That went on till September or October until ESAU eventually officially gave up his claim to the uranium and the heavy water.

HARTECK: Of course we didn't really do it properly. Theory was considered the most important thing and experiments were secondary, and then almost unintelligible formulae were written down. We did not carry out experiments with sufficient vigour. Suppose a man like HERTZ had made the experiments, he would have done it quite differently.

GERLACH: They did make experiments. They measured the emission of heat of uranium.

HARTECK: For instance if you measure the emission of heat and at the same time make the 23 – minute body.

GERLACH: What SCHÜTZE (?) was to have done later?

HARTECK: Why was that not done?

GERLACH: Perhaps it was.

HARTECK: You might perhaps have boiled the metal, so obtaining a large surface area which would behave towards neutrons as in STERN's experiments. Then you would see that in one case it was better by a few per cent and in another case worse. But such experiments were not made, or rather they wanted to persuade you against it.

HAHN: HERTZ did that.

GERLACH: Yes. He had all the material he could find.

HAHN: When was that – in 1944?

GERLACH: Yes, the end of 1944. But he had measured the emission of heat already two years before. I just went to HERTZ and said: "Look here, HERTZ, let's discuss the uranium business". He said: "I know nothing about it", so I told him all about it. Then he told me that SCHUTZE had made such heat experiments and then we discussed it and decided that that really was the best thing.

HAHN: So he (used) a small radium preparation and beryllium preparation . . .

GERLACH: 25 milligrammes and about a hundred grammes of uranium powder (?). He only used powder. When I heard about it, I said straightaway that that was the right method of examining small bodies.

HARTECK: We had 27 grammes of radium. If we had used – say – 5 grammes of radium as neutron sources we could easily have measured with the best shaped bodies.

GERLACH: We must not say in front of these two Englishmen that we ought to have done more about the thing. WIRTZ said that we ought to have worked more on the separation of isotopes. It's another matter to say that we did not have sufficient means but one cannot say in front of an Englishman that we didn't try hard enough. They were our enemies, although we sabotaged the war. There are some things that one knows and one can discuss together but that one cannot discuss in the presence of Englishmen.

HAHN: I must honestly say that I would have sabotaged the war if I had been in a position to do so.

7. HAHN and HEISENBERG discussed the matter alone together. HAHN explained to HEISENBERG that he was himself very upset about the whole thing. He said he could not really understand why GERLACH had taken it so badly. HEISENBERG said he could understand it because GERLACH was the only one of them who had really wanted a German victory, because although he realised the crimes of the Nazis and disapproved of them, he could not get away from the fact that he was working for GERMANY. HAHN replied that he too loved his country and that, strange as it might appear, it was for this reason that he had hoped for her defeat. HEISENBERG went on to say that he thought the possession of the uranium bomb would strengthen the position of the Americans vis-a-vis the Russians. They continued to discuss the same

theme as before that they had never wanted to work on a bomb and had been pleased when it was decided to concentrate everything on the engine. HEISENBERG stated that the people in Germany might say that they should have forced the authorities to put the necessary means at their disposal and to release 100,000 men in order to make the bomb and he feels himself that had they been in the same moral position as the Americans and had said to themselves that nothing mattered except that HITLER should win the war, they might have succeeded, whereas in fact they did not want him to win. HAHN admitted however that he had never thought that a German defeat would produce such terrible tragedy for his country. They then went on to discuss the feelings of the British and American scientists who had perfected the bomb and HEISENBERG said he felt it was a different matter in their case as they considered HITLER a criminal. They both hoped that the new discovery would in the long run be a benefit to mankind. HEISENBERG went on to speculate on the uses to which AMERICA would put the new discovery and wondered whether they would use it to obtain control of RUSSIA or wait until STALIN had copied it. They went on to wonder how many bombs existed. The following is the text of this part of the conversation:–

HAHN: They can't make a bomb like that once a week.

HEISENBERG: No. I rather think HARTECK was right and that they have just put up a hundred thousand mass-spectrographs or something like that. If each mass-spectrograph can make one milligramme a day, they they have got a hundred grammes a day.

HAHN: In 1939 they had only made a fraction of a milligramme. They had then identified the "235" through its radio-activity.

HEISENBERG: That would give them 30 kilos. a year.

HAHN: Do you think they would need as much as that?

HEISENBERG: I think so certainly, but quite honestly I have never worked it out as I never believed one could get pure "235". I always knew it could be done with "235" with fast neutrons. That's why "235" only can be used as an explosive. One can never make an explosive with slow neutrons, not even with the heavy water machine, as then the neutrons only go with thermal speed, with the result that the reaction is so slow that the thing explodes sooner, before the reaction is complete. It vaporises at 5,000° and then the reaction is already –

HAHN: How does the bomb explode?

HEISENBERG: In the case of the bomb it can only be done with the very fast neutrons. The fast neutrons in 235 immediately produce other neutrons so that the very fast neutrons which have a speed of – say – 1/30th that of light make the whole reaction. Then of course the reaction takes place much quicker so that in practice one can release these great energies. In ordinary uranium a fast neutron nearly always hits 238 and then gives no fission.

HAHN: I see, whereas the fast ones in the 235 do the same as the 238, but 130 times more.

HEISENBERG: Yes. If I get below 600,000 volts I can't do any more fission on the 238, but I can always split the 235 no matter what happens. If I have pure 235 each neutron will immediately beget two children and then there must be a chain reaction which goes very quickly. Then you can reckon as follows. One neutron always makes two others in pure 235. That is to say that in order to make 10^{24} neutrons I need 80 reactions one after the other. Therefore I need 80 collisions and the mean free path is about 6 centimetres. In order to make 80 collisions, I must have a lump of a radius of about 54 centimetres and that would be about a ton.

HAHN: Wouldn't that ton be stronger than 20,000 tons of explosive?

HEISENBERG: It would be about the same. It is conceivable that they could do it with less in the following manner. They would take only a quarter of the quantity but cover it with a reflector which would turn back the fast neutrons. For instance lead or carbon and in that way they could get the neutrons which go out, to come back again. It could be done in that way. It is possible for them to do it like that.

HAHN: How can they take it in an aircraft and make sure that it explodes at the right moment?

HEISENBERG: One way would be to make the bomb in two halves, each one of which would be too small to produce the explosion because of the mean free path. The two halves would be joined together at the moment of dropping when the reaction would start. They have probably done something like that.

> HEISENBERG went on to complain bitterly that GOUDSMIT had lied to them very cleverly and thinks that he might at least have told him that their experiments in AMERICA were further advanced. They agreed that the secret was kept very well. HAHN remarked on the fact that there had been no publication of work on uranium fission in British or American scientific journals since January, 1940, but he thought that there had been one published in RUSSIA on the spontaneous fission of uranium with dueterons. HEISENBERG repeated all his arguments saying that they had concentrated on the uranium engine, had never tried to make a bomb and had done nothing on the separation of isotopes because they had not been able to get the necessary means for this. He repeated his story of the alleged threat by America to drop a uranium bomb on DRESDEN and

said that he had been questioned by Geheimrat SEGNER (?) of the Foreign Office about this possibility. The conversation concluded as follows:

HEISENBERG: Perhaps they have done nothing more than produce 235 and make a bomb with it. Then there must be any number of scientific matters which it would be interesting to work on.

HAHN: Yes, but they must prevent the Russians from doing it.

HEISENBERG: I would like to know what STALIN is thinking this evening. Of course they have got good men like LANDAU, and these people can do it too. There is not much to it if you know the fission. The whole thing is the method of separating isotopes.

HAHN: No, in that respect the Americans and in fact all the Anglo-Saxons are vastly superior to them. I have a feeling that the Japanese war will end in the next few days and then we will probably be sent home fairly soon and everything will be much easier than it was before. Who knows that it may not be a blessing after all?

8. The guests decided among themselves that they must not outwardly show their concern. In consequence they insisted on playing cards as usual till after midnight. VON WEIZSACKER, WIRTZ, HARTECK, and BAGGE remained behind after the others had gone to bed. The following conversation took place:

BAGGE: We must take off our hats to these people for having the courage to risk so many millions.

HARTECK: We must have succeeded if the highest authorities had said "We are prepared to sacrifice everything".

WEIZSACKER: In our case even the scientists said it couldn't be done.

BAGGE: That's not true. You were there yourself at that conference in Berlin. I think it was on 8 September that everyone was asked – GEIGER, BOTHE and you HARTECK were there too – and everyone said that it must be done at once. Someone said "Of course it is an open question whether one ought to do a thing like that." Thereupon BOTHE got up and said "Gentlemen, it <u>must</u> be done." Then GEIGER got up and said "If there is the slightest chance that it is possible – it must be done." That was on 8 September '39.

WEIZSACKER: I don't know how you can say that. 50% of the people were against it.

HARTECK: All the scientists who understood nothing about it, all spoke against it, and of those who did understand it, one third spoke against it. As 90% of them didn't understand it, 90% spoke against it. We knew that it could be done in principal, but on the other hand we realised that it was a frightfully dangerous thing.

BAGGE: If the Germans had spent 10 milliard marks on it and it had not succeeded, all physicists would have had their heads cut off.

WIRTZ: The point is that in Germany very few people believed in it. And even those who were convinced it could be done did not all work on it.

HARTECK: For instance when we started that heavy water business the CLUSIUS method was apparantly too expensive, but I told ESAU that we should use various methods all at once; there was the one in NORWAY; and that we should have a CLUSIUS plant to produce 2–300 litres a year, that is a small one and then a hot–cold one. As far as I can see we could never have made a bomb, but we could certainly have got the engine to go.

WIRTZ: KORSCHING is really right when he said there wasn't very good co-operation in the uranium group as GERLACH said. GERLACH actually worked against us. He and DIEBNER worked against us the whole time. In the end they even tried to take the engine away from us. If a German Court were to investigate the whole question of why it did not succeed in Germany it would be a very very dangerous business. If we had started properly in 1939 and gone all out everything would have been alright.

HARTECK: Then we would have been killed by the British 'Secret Service'.

WIRTZ: I am glad that it wasn't like that otherwise we would all be dead.

(Pause)

BAGGE: It must be possible to work out at what temperature the thing explodes.

HARTECK: The multiplication factor with 235 is 2.8, and when one collides with the other how long is the path until it happens? 4 centimetres. Rx is the radius. Then you have to multiply that by the mean free path and divide it by the square root of the multiplication factor. That should be 3.2. R is about 14 centimetres, the weight is 200 kilogrammes, then it explodes.

9. GERLACH and HEISENBERG had a long discussion in GERLACH's room which went on half the night. In the course of this conversation they repeated most of the statements that had been made in the course of the general conversation downstairs and have been already reported. The following are extracts from the conversation:

GERLACH: I never thought of the bomb, all I wanted was that we should do everything possible to develope HAHN's discovery for our country.

HEISENBERG went on to stress the fact that they had concentrated on the developement of the engine and stated that although the Allies appeared to have concentrated on the bomb they could presumably also make the engine now. He attributed that they failed to perfect the engine to the attacks on the factories in NORWAY. He blamed HITLER for the fact that, as he puts it, "HAHN's invention has now been taken away from Germany." He went on:

HEISENBERG: I am still convinced that our objective was really the right one and that the fact that we concentrated on uranium may give us the chance of collaboration. I believe this uranium business will give the Anglo-Saxons such tremendous power that EUROPE will become a bloc under Anglo-Saxon domination. If that is the case it will be a very good thing. I wonder whether STALIN will be able to stand up to the others as he has done in the past.

GERLACH: It is not true that we neglected the separation of isotopes – on the contrary, we discussed the whole thing at TUEBINGEN in February, and there was a meeting at MUNICH. CLUSIUS, HARTECK and I said that this photo-chemical thing must be done. It took till the end of the year before the people who could do it were got together and the spectrograph obtained and special accommodation acquired, as the LITZ (?) Institute had been smashed up.

HEISENBERG: You shouldn't take remarks like the one KORSCHING made too seriously. He now thinks because the Americans have done it that <u>he</u> could have succeeded in separating isotopes if <u>he</u> had had more means at his disposal. That is of course sheer and utter nonsense. His experiment was interesting, that's why we carried it out, but I am convinced that the Americans have done it by completely other methods.

GERLACH: If Germany had had a weapon which would have won the war, then Germany would have been in the right and the others in the wrong, and whether conditions in Germany are better now than they would bave been after a HITLER victory –

HEISENBERG: I don't think so. On the other hand, the days of small countries are over. Suppose HITLER had succeeded in producing his EUROPE and there had been no uranium in EUROPE.

GERLACH: If we had really planned a uranium engine – in the summer of 1944 we would not have had a bomb – and that had been properly handled from a propaganda point of view –

HARTECK: That might have been a basis for negotiation. It would have been a basis for negotiation for any other German Government, but not for HITLER.

GERLACH: I went to my downfall with open eyes, but I thought I would try and save German physics and Germany physicists, and in that I succeeded.

HEISENBERG: Perhaps German physics will be able to collaborate as part of a great Western group.

> GERLACH then went on to repeat how ESAU had tried to get all the heavy water and uranium in order to have the experiments made at the REICHSANSTALT. HEISENBERG then continued.

HEISENBERG: Now that the whole thing has been made public, I assume that in a comparatively short time they will tell us what is to happen to us as I can't see the sense in keeping us detained as it is obvious that they are <u>much</u> further advanced than we were. There may be some details in which we could help them as they appear to have done very little in the heavy water line.

GERLACH: The only thing to do now would be to say: "We will get all the uranium people together CHADWICK, FERMI etc., and let them discuss it".

HEISENBERG: I wouldn't be surprised if in a comparatively short time we meet some of these people and perhaps something will come of it. It seems to me that the sensible thing for us to do is to try and work in collaboration with the Anglo-Saxons. We can do that now with a better conscience because one sees that they will probably dominate EUROPE. It is clear that people like CHADWICK and CHERWELL have considerable influence.

(Pause)

GERLACH: I would really like to know how they have done it.

HEISENBERG: It seems quite clear to me that it is the separation of isotopes. Although it is possible as HARTECK says that it is done with a hundred thousand mass-spectrographs.

GERLACH: I am not sure whether perhaps the BAGGE method –

HEISENBERG: That would never produce pure 235. The BAGGE method is not bad for enriching but the centrifuge is good for that too.

GERLACH: The BAGGE method enriches more.

HEISENBERG: Yes. It is a terrific lot to expect pure 235.

GERLACH: How pure must it be?

HEISENBERG: I should say 80% 235, and 20% 238 is alright, 50/50 would be alright, but there must not be much more 238 than 235.

> 10. WIRTZ and VON WEIZSACKER discussed the situation together in their room. VON WEIZSACKER expressed the opinion that none of them had really worked seriously on uranium with the exception of WIRTZ and HARTECK. He also accused GERLACH and DIEBNER of

sabotage. WIRTZ expressed horror that the Allies had used the new weapon. They went on to discuss the possibility of the Russians discovering the secret and came to the conclusion that they would not succeed under ten years. They went on as follows:

WIRTZ: It seems to me that the political situation for STALIN has changed completely now.

WEIZSACKER: I hope so. STALIN certainly has not got it yet. If the Americans and the British were good Imperialists they would attack STALIN with the thing tomorrow, but they won't do that, they will use it as a political weapon. Of course that is good, but the result will be a peace which will last until the Russians have it, and then there is bound to be war.

At this point HEISENBERG joined WIRTZ and VON WEIZSACKER. The following remarks were passed:

WIRTZ: These fellows have succeeded in separating isotopes. What is there left for us to do?

HEISENBERG: I feel convinced that something will happen to us in the next few days or weeks. I should imagine that we no longer appear to them as dangerous enemies.

WEIZSACKER: No, but the moment we are no longer dangerous we are also no longer interesting. It appears that they can get along perfectly well by themselves.

HEISENBERG: Perhaps they can learn something about heavy water from us. But it can't be much – they know everything.

WEIZSACKER: Our strength is now the fact that we are 'un-Nazi'.

HEISENBERG: Yes, and in addition, uranium was discovered by HAHN and not by the Americans.

WEIZSACKER: I admit that after this business I am more ready to go back to GERMANY, in spite of the Russian advance.

WIRTZ: My worst fears have been realised with regard to the complications which will now arise about us.

HEISENBERG: I believe that we are now far more bound up with the Anglo-Saxons than we were before as we have no possibility of switching over to the Russians even if we wanted to.

WIRTZ: They won't let us.

HEISENBERG: On the other hand we can do it with a good conscience because we can see that in the immediate future GERMANY will be under Anglo-Saxon influence.

WIRTZ: That is an opportunist attitude.

HEISENBERG: But at the moment it is very difficult to think otherwise because one does not know what is better.

WEIZSACKER: If I ask myself for which side I would prefer to work of course I would say for neither of them.

11. DIEBNER and BAGGE also discussed the situation alone together as follows:

BAGGE: What do you think will happen to us now?

DIEBNER: They won't let us go back to GERMANY. Otherwise the Russians will take us. It is quite obvious what they have done, they have just got some system other than ours. If a man like GERLACH had been there earlier, things would have been different.

BAGGE: GERLACH is not responsible, he took the thing over too late. On the other hand it is quite obvious that HEISENBERG was not the right man for it. The tragedy is that KORSCHING is right in the remarks he made to GERLACH. I think it is absurd for WEIZSACKER to say he did not want the thing to succeed. That may be so in his case, but not for all of us. WEIZSACKER was not the right man to have done it. HEISENBERG could not convince anyone that the whole thing depended on the separation of isotopes. The whole separation of isotopes was looked upon as a secondary thing. When I think of my own apparatus – it was done against HEISENBERG's wishes.

DIEBNER: Now the others are going to try and make up to the Major and sell themselves. Of course they can do what they like with us now, they don't need us at all.

BAGGE: I won't do it. I will work on cosmic rays. Do you remember how VON WEIZSACKER said in BELGIUM "When they come to us we will just say that the only man in the world who can do it is HEISENBERG." WEIZSACKER is very upset about the whole thing.

(Pause)

BAGGE: You can't blame SPEER as none of the scientists here forced the thing through. It was impossible as we had no one in GERMANY who had actually separated uranium. There were no mass-spectrographs in GERMANY.

DIEBNER: They all failed. WALCHER (?) and HERTZOG (?) wanted to build one, but they didn't succeed.

12. Although the guests retired to bed about 1.30, most of them appear to have spent a somewhat disturbed night judging by the deep sighs and occasional shouts which were heard during the night. There was also a considerable amount of coming and going along the corridors.

III. 7 August

1. On the morning of 7 August the guests read the newspapers with great avidity. Most of the morning was taken up reading these.

2. In a conversation with DIEBNER, BAGGE said he was convinced it had been done with mass-spectrographs.

3. HAHN, HEISENBERG and HARTECK discussed the matter in the following conversation:

HAHN: What can one imagine happens when an atomic bomb explodes? Is the fission of uranium 1‰, 1%, 10% or 100%?

HEISENBERG: If it is 235, then for all practical purposes it is the whole lot, as then the reaction goes much quicker than the vaporisation as for all practical purposes it goes with the speed of light. In order to produce fission in $1o^{25}$ atoms you need 80 steps in the chain so that the whole reaction is complete in 10^{-8} seconds. Then each neutron that flies out of one atom makes two more neutrons when it hits another uranium 235. Now I need 10^{25} neutrons and that is 2^{80}. I need 80 steps in the chain and then I have made 2^{80} neutrons. One step in the chain takes the same time as one neutron to go 5 centimetres, that is 10^{-9} seconds, so that I need about 10^{-8} seconds, so that the whole reaction is complete in 10^{-8} seconds. The whole thing probably explodes in that time.

(Pause)

HEISENBERG: They seem to have made the first test only on 16 July.

HAHN: But they must have had more material then. They could not make a 100 kilogrammes of new uranium 235 every fortnight.

HEISENBERG: They seem to have had two bombs, one for the test and the other for –

HARTECK: But in any case the next one will be ready in a few months. STALIN's hopes of victory will have been somewhat dashed.

HAHN: That's what pleases one about the whole thing. If Niels BOHR helped, then I must say he has gone down in my estimation.

4. GERLACH and VON LAUE discussed the position of Niels BOHR and the part he had played. GERLACH said he was very upset about this as he had personally vouched for BOHR to the German Government.

VON LAUE said that one could not believe everything that appeared in the newspapers.

5. In a conversation with VON LAUE, VON WEIZSACKER said it will not be long before the names of the German scientists appear in the newspapers and that it would be a long time before they would be able to clear themselves in the eyes of their own countrymen. He went on to quote from the newspaper that we were unable to control the energy, from which he assumed that we were not yet in possession of a uranium engine, so that their work would still be of considerable value. He ended by saying:

WEIZSACKER: History will record that the Americans and the English made a bomb, and that at the same time the Germans, under the HITLER regime, produced a workable engine. In other words, the peaceful developement of the uranium engine was made in GERMANY under the HITLER regime, whereas the Americans and the English developed this ghastly weapon of war.

6. GERLACH continued to complain to DIEBNER about the attitude of KORSCHING the evening before. They went on to discuss the methods by which information concerning their work may have leaked out. They reminded themselves that HEISENBERG and VON WEIZSACKER once spent four weeks in SWITZERLAND and had discussions with SCHERRER. GERLACH and DIEBNER went on to discuss the political aspects of the possession of the atomic bomb, and expressed satisfaction that the Russians appear not to have the secret.

7. In a conversation between WIRTZ, VON WEIZSACKER and HEISENBERG, HEISENBERG repeated that in July 1944 a senior SS official had come to him and asked him whether he seriously believed that the Americans could produce an atomic bomb. He said he had told him that in his opinion it was absolutely possible as the Americans could work much better and quicker than they could. VON WEIZSACKER again expressed horror at the use of the weapon and HEISENBERG replied that had they produced and dropped such a bomb they would certainly have been executed as War Criminals having made the "most devlish thing imaginable".

8. At 6 o'clock the guests all heard Sir John Anderson speak on the wireless. The subsequent conversation was merely a repetition of previous ones, and was chiefly concerned with somewhat caustic comments on the usage to which the new discovery had been put. HEISENBERG's final comment was:

HEISENBERG: If the Americans had not got so far with the engine as we did –

that's what it looks like – then we are in luck. There is a possibility of making money.

9. Later, GERLACH and HEISENBERG had a long discussion in which they discussed the future. GERLACH said he hoped they would be able to discuss the whole question with people like CHADWICK. HEISENBERG said he felt sure something of the sort would be done, but he felt they should wait and see what happened. They went on to discuss references in the newspapers to the alleged work which had been done in GERMANY on the bomb, and said they hoped it would be possible to prevent the newspapers from continuing to make such statements. They ended their conversation by expressing surprise that they had known nothing about the preparations that had been made in AMERICA. HEISENBERG said that someone from the German Foreign Office had told him that 70% of their Gestapo agents in SPAIN had just stayed there to work for the other side.

10. HEISENBERG, VON WEIZSACKER, WIRTZ and HARTECK also discussed the future and came to the conclusion that they would probably be sent back to HECHINGEN and allowed to continue their work. They realised however, that we might be afraid of their telling the Russians too much. In this connection they mentioned that BOPP, JENSEN and FLUEGGE could also tell them a lot if they wanted to. They came to the conclusion that GROTH was probably in ENGLAND.

IV. The Memorandum Signed by the Guests

All the guests have been extremely worried about the press reports of the alleged work carried out in GERMANY on the atomic bomb. As they were so insistent that no such work had been carried out, I suggested to them that they should prepare a memorandum setting out details of the work on which they were engaged, and that they should sign it. There was considerable discussion on the wording of this memorandum, in the course of which DIEBNER remarked that he had destroyed all his papers, but that there was great danger in the fact that SCHUMANN had made notes on everything. GERLACH wondered whether VOEGLER had also made notes. From the conversation it did however appear that they had really not worked on a bomb themselves, but they did state that the German Post Office had also worked on uranium, and had built a cyclotron at MIERSDORF. GERLACH stated that the SCHWAB Group also had some uranium, and he remembered that the SS had come to him once and tried to obtain large quantities of heavy water. HARTECK also mentioned an SS Colonel whose name he could not remember, who had previously been with MERK (?), who had shown considerable interest in the subject. WIRTZ remarked that they should remember that there was a patent for the production of such a bomb at the KAISER WILHELM Institute for Physics. This patent was taken out in 1941. Eventually, a memorandum was drawn up and a photostat copy of it is attached to this report. WIRTZ, VON WEIZSACKER, DIEBNER, BAGGE and KORSCHING at first did not want to sign it, but were eventually persuaded to do so by HEISENBERG.

FARM HALL,
GODMANCHESTER.
11 August, 1945.

(Sgd) T.H. Rittner
Major.

Copy No. 1.

FH4

APPENDIX 1.

PHOTOGRAPHS OF FARM HALL AND THE GUESTS DETAINED THERE

FARM HALL

PROFESSOR OTTO HAHN

The most friendly of the detained professors. Has a very keen sense of humour and is full of common sense. He is definitely friendly disposed to England and America. He has been very shattered by the announcement of the use of the atomic bomb as he feels responsible for the lives of so many people in view of his original discovery. He has taken the fact that Professor MEITNER has been credited by the press with the original discovery very well although he points out that she was in fact one of his assistants and had already left Berlin at the time of his discovery.

PROFESSOR
MAX von LAUE

A shy, mild-mannered man. He cannot understand the reason for his detention as he professes to have had nothing whatever to do with uranium or the experiments carried out at the Kaiser Wilhelm Institute. He is rather enjoying the discomfort of the others as he feels he is in no way involved. He is extremely friendly and is very well disposed to England and America.

PROFESSOR WALTHER GERLACH

Has always been very cheerful and friendly, but from his monitored conversations is open to suspicion because of his connections with the Gestapo. As the man appointed by the German Government to organise the research work on uranium, he considers himself in the position of a defeated general and appeared to be contemplating suicide when the announcement was made.

PROFESSOR W. HEISENBERG

Has been very friendly and helpful ever since his detention. He has taken the announcement of the atomic bomb very well indeed and seems to be genuinely anxious to cooperate with British and American scientists.

PROFESSOR P. HARTECK

A very charming personality. Appears to be interested only in his research work. He has taken the announcement of the atomic bomb very philosophically and has put forward a number of theories as to how it has been done.

PROFESSOR
C.F. von WEIZSACKER

Outwardly very friendly and appears to be genuinely cooperative. He has stated, both directly and in monitored conversations, that he was sincerely opposed to the Nazi regime and anxious not to work on an atomic bomb. Being the son of a diplomat he is something of one himself. It is difficult to say whether he is genuinely prepared to work with England and America.

A clever egoist. Very friendly on the surface, but cannot be trusted. He will cooperate only if it is made worth his while.

<u>DOCTOR K. WIRTZ</u>

A serious and very hard-working young man. He is completely German and is unlikely to cooperate.

<u>DOCTOR E. BAGGE</u>

DOCTOR H. KORSCHING

A complete enigma. On the announcement of the use of the atomic bomb he passed remarks upon the lack of courage among his colleagues which nearly drove GERLACH to suicide.

DOCTOR K. DIEBNER

Outwardly friendly but has an unpleasant personality and cannot be trusted.

Copy No. 1

FH4

APPENDIX 2.

8 August 1945

Da die Presseberichte der letzten Tage über die angeblichen Arbeiten an der Atombombe in Deutschland zum Teil unrichtige Angaben enthalten, möchten wir die Entwicklung der Arbeiten zum Uranproblem im Folgenden kurz beschreiben.

1) Die Atomkernspaltung beim Uran ist im Dezember 1938 von Hahn und Strassmann am Kaiser-Wilhelm-Institut für Chemie in Berlin entdeckt worden. Sie war die Frucht rein wissenschaftlicher Untersuchungen, die mit praktischen Zielen nichts zu tun hatten. Erst nach ihrer Veröffentlichung wurde ungefähr gleichzeitig in verschiedenen Ländern entdeckt, daß sie eine Kettenreaktion der Atomkerne und damit zum ersten Mal eine technische Ausnutzung der Kern-Energieen ermöglichen könnte.

2) Beim Beginn des Krieges wurde in Deutschland eine Gruppe von Forschern zusammengerufen, deren Aufgabe es war, die praktische Ausnutzbarkeit dieser Energieen zu untersuchen. Die wissenschaftlichen Vorarbeiten hatten gegen Ende 1941 zu dem Ergebnis geführt, daß es möglich sein werde, die Kern-Energieen zur Wärme-Erzeugung und damit zum Betrieb von Maschinen zu benutzen. Dagegen schienen die Voraussetzungen für die Herstellung einer Bombe im Rahmen der technischen Möglichkeiten, die Deutschland zur Verfügung standen, damals nicht gegeben zu sein. Die weiteren Arbeiten konzentrierten sich daher auf das Problem der Maschine, für die außer Uran "schweres" Wasser notwendig ist.

3) Für diesen Zweck wurden die Anlagen der Norsk Hydro in Rjukan zur Produktion von größeren Mengen von schwerem Wasser ausgebaut. Die Angriffe auf diese Anlagen zuerst durch ein Sprengkommando, dann durch die R.A.F., haben diese Produktion gegen Ende 1943 zum Erliegen gebracht.

4) Gleichzeitig wurden in Freiburg, später in Celle, Versuch angestellt, durch Anreicherung des seltenen Isotops 235 die Benutzung des schweren Wassers zu umgehen.

5) Mit den vorhandenen Mengen des schweren Wassers wurden zuerst in Berlin, später in Haigerloch (Württemberg), die Versuche über die Energie-Gewinnung fortgeführt. Gegen Ende des Krieges waren diese Arbeiten so weit gediehen, daß die Aufstellung einer Energie liefernden Apparatur wohl nur noch kurze Zeit in Anspruch genommen hätte.

Anmerkungen

Zu 1) Die Hahn'sche Entdeckung ist kurz nach ihrer Veröffentlichung in vielen Laboratorien, insbesondere in den Vereinigten Staaten, nachgeprüft worden. Auf die großen Energieen, die bei der Uranspaltung frei werden, wurde von verschiedenen Forschern, zuerst wohl von Meitner und Frisch, hingewiesen. Dagegen hatte Professor Meitner bereits ein halbes Jahr vor der Entdeckung Berlin verlassen und war an der Entdeckung selbst nicht beteilgt.

Zu 2) Die rein chemischen Arbeiten des Kaiser-Wilhelm-Institutes für Chemie über die Folgeprodukte der Uranspaltung sind im Kriege ungestört fortgeführt und veröffentlicht worden. Die unter 2) genannten wissenschaftlichen Vorarbeiten über die Energie-Gewinnung umfaßten Untersuchungen von folgender Art: Theoretische Abschätzung über den Ablauf der Reactionen in Gemischen aus Uran und schwerem Wasser. Messungen des Absorptionsvermögens von schwerem Wasser für Neutronen. Untersuchungen über die bei der Spaltung frei werdenden Neutronen. Untersuchungen über die Neutronenvermehrung in kleinen Anordnungen aus Uran and schwerem Wasser. - Zur Frage der Atombombe sei noch festgestellt, daß den Unterzeichneten keine ernst zu nehmenden Untersuchungen etwa anderer Gruppen in Deutschland über das Uranproblem bekannt geworden sind.

Zu 3) Die Schwer-Wasser-Produktion in Rjukan wurde zunächst durch den konsequenten Ausbau der schon vorhandenen Anlage und dann durch den Einbau von in Deutschland entwickelten katalytischen Austausch-Öfen auf etwa 220 Liter pro Monat gesteigert. Die Stickstoffproduktion des Werkes wurde hierdurch nur unwesentlich vermindert. Mit Uran und Radium ist in Rjukan nie gearbeitet worden.

Zu 4) Zur Isotopentrennung wurden verschiedene Verfahren angewandt. Das Clusius'sche Trennrohr erwies sich als ungeeignet. Die Ultra-Zentrifuge ergab eine geringe Anreicherung des Isotops 235. Die anderen Verfahren hatten bis zum Ende des Krieges noch keine sicheren positiven Ergebnisse geliefert. Eine Isotopentrennung in großem Maßstab ist nicht in Angriff genommen worden.

Zu 5) Ferner wurde eine Energie lieferende Apparatur vorbereitet, die zwar ohne schweres Wasser, jedoch bei sehr tiefer Temperatur künstlich-radioaktive Substanzen in großer Menge erzeugen sollte.

Zu 3) - 5) Im Ganzen sind von den deutschen Behörden (zuerst Heereswaffenamt, später Reichs-Forschungsrat) für das Uranvorhaben Mittel bereit gestellt worden, die gegenüber den von den Allieerten eingesetzten Mitteln verschwindend gering sind. Die Anzahl der Menschen, die an der Entwicklung beteiligt waren, (Wissenschaftler und Hilfskräfte in Instituten und Industrie), hat wohl in keiner Phase einige Hundert überschritten.

(Otto Hahn)

(M.v.Laue)*

(Walther Gerlach)

(W.Heisenberg)

(P.Harteck)

(C.F.v.Weizsäcker)

(K.Wirtz)

(E.Bagge)

(H.Korsching)

(K.Diebner)

*Meine Unterschrift bedeutet, daß ich mich für die Richtigkeit der obigen Darstellung mit - verbürge, nicht aber, daß ich an den darin erwähnten Arbeiten irgend welchen Anteil gehabt habe.

M.v.Laue.

TRANSLATION

8 August 1945

As the press reports during the last few days containly partly incorrect statements regarding the alleged work carried out in Germany on the atomic bomb, we would like to set out briefly the development of the work on the uranium problem.

1. The fission of the atomic nucleus in uranium was discovered by Hahn and Strassman in the Kaiser Wilhelm Institute for Chemistry in Berlin in December 1938. It was the result of pure scientific research which had nothing to do with practical uses. It was only after publication that it was discovered almost simultaneously in various countries that it made possible a chain reaction of the atomic nuclei and therefore for the first time a technical exploitation of nuclear energies.

2. At the beginning of the war a group of research workers was formed with instructions to investigate the practical application of these energies. Towards the end of 1941 the preliminary scientific work had shown that it would be possible to use the nuclear energies for the production of heat and thereby to drive machinery. On the other hand it did not appear fassible at the time to produce a bomb with the technical possibilities available in Germany. Therefore the subsequent work was concentrated on the problem of the engine for which, apart from uranium, heavy water is necessary.

3. For this purpose the plant of the Norsk Hydro at Rjukan was enlarged for the production of larger quantities of heavy water. The attacks on this plant, first by the Commando raid, and later by aircraft, stopped this production towards the end of 1943.

4. At the same time, at Freiburg and later at Celle, experiments were made to try and obviate the use of heavy water by the concentration of the rare isotope U 235.

5. With the existing supplies of heavy water the experiments for the production of energy were continued first in Berlin and later at Haigerloch (Wurtemburg). Towards the end of the war this work had progressed so far that the building of a power producing apparatus would presumably only have taken a short time.

REMARKS (referring to previous paragraphs)

Para. 1. The Hahn discovery was checked in many laboratories, particularly in the United States, shortly after publication. Various research workers, Meitner and Frisch were probably the first, pointed out the enormous energies which were released by the fission of uranium. On the other hand, Meitner had left Berlin six months before the discovery and was not concerned herself in the discovery.

Para. 2. The pure chemical researches of the Kaiser Wilhelm Institute for Chemistry on the elements produced by uranium fission continued without hindrance throughout the war and were published. The preliminary scientific work on the production of energy mentioned in paragraph 2 was on the following lines:-

Theoretical calculations concerning the reactions in mixtures or uranium and heavy water. Measuring the capacity of heavy water to absorb neutrons. Investigation of the neutrons set free by the fission. Investigation of the increase of neutrons in small quantities of uranium and heavy water. With regard to the atomic bomb the undersigned did not know of any other serious research work on uranium being carried out in Germany.

Para. 3. The heavy water production at Rjukan was brought up to 220 litres per month, first by enlarging the existing plant and then by the addition of catalytic exchange-furnaces which had been developed in Germany. The nitrogen production of the works was only slightly reduced by this. No work on uranium or radium was done at Rjukan.

Para. 4. Various methods were used for separating isotopes. The Clusius separating tubes proved unsuitable. The ultra-centrifuge gave a slight concentration of isotope 235. The other methods had produced no certain positive result up to the end of the war. No separation of isotopes on a large scale was attempted.

Para. 5. Further a power producing apparatus was prepared which was to produce radio-active substances in large quantities artificially without the use of heavy water but at very low temperatures.

Paras. 3 and 5. On the whole the funds made available by the German authorities (at first the Ordnance Department and later the Reichs Research Board) for uranium were extremely small compared to those employed by the Allies. The number of people engaged in the development (scientists and others, at institutes and in industry) hardly ever exceeded a few hundred.

Signed -	Otto Hahn	M.v. Laue
	Walther Gerlach	W. Heisenberg
	P. Harteck	C.F.c. Weizsacker
	K. Wirtz	E. Bagge
	H. Korsching	K. Diebner

(My signature signifies that I share responsibility for the accuracy of the above statement, but not that I took any part whatever in the above mentioned work. Signed - M.v. Laue)

AMERICAN EMBASSY
OFFICE OF THE MILITARY ATTACHE
1, GROSVENOR SQUARE, W. 1,
LONDON, ENGLAND

1 September 1945

Subject: Transmittal of Operation Epsilon.†

To : Major F. J. Smith, Room 5004, New War Dept. Bldg., Washington, D. C.

Transmitted herewith for your information is a copy of Operation Epsilon, Ref: F.H. 5., 8-22 August 1945.

For the Military Attache:

H K Calvert

H. K. CALVERT,
Major, F.A.,
Assistant to the Military Attache.

Incl:
As listed above.

† Letters of transmittal have not been included henceforth unless they provide additional information or interest. American circulation lists have also not been included.

Copy No. 1.

Capt. Davis
For General Groves.

Ref: F.H. 5.

To: Mr. M. PERRIN and Lt. Cdr. WELSH.
From: Major T.H. RITTNER.

OPERATION EPSILON
(8th - 22nd August, 1945).

I. GENERAL.

1. The guests have recovered from the initial shock they received when the news of the atomic bomb was announced. They are still speculating on the method used to make the bomb and their conversations on this subject, including a lecture by HEISENBERG, appear later in this report. The translation of the technical matter has been very kindly undertaken by a member of the Staff of D.S.I.R. The original German text of HEISENBERG's lecture has been reproduced as an appendix to this report.

2. There is a general air of expectancy as the guests now feel there is no further need for their detention and they assume that they will shortly be told what plans have been made for their future and that they will soon be reunited with their families. They are eagerly awaiting replies to their letters which have now been despatched.

3. The declaration of the surrender of JAPAN was greeted with relief rather than enthusiasm. The guests listened with great interest to the King's broadcast on "VJ" Day and all stood rigidly to attention during the playing of the National Anthem.

4. Sir CHARLES DARWIN paid a visit to FARM HALL on 18th August. This was the first time the guests had had contact with a scientist since their detention and they were delighted to have the opportunity of meeting him. Conversations during the visit and subsequent reactions are dealt with elsewhere in this report.

II. The Future.

1. A number of the guests have discussed their attitude towards co-operation with the Allies. The following conversation took place between HEISENBERG, VON WEIZSACKER and GERLACH on 10th August.

GERLACH: If you were faced with the opportunity of co-operation in order to make the bomb useful for mankind, would you do it?

HEISENBERG: It is unlikely to arise in that form, as it can't be done. "Useful for mankind" means only that the Russians shouldn't get it, but that can't be prevented as the Russians are certain to have the atom bomb in five year's time, possibly in a year. From what CHADWICK said in that interview I would say that the Allies will try and form a Control Commission with the Russians, and they will control the manufacture of uranium 235, and the uses to which it is put. They will try and come to a peaceful understanding with the Russians in some way. I would have no objection if one could be included in such an organisation so that we could share, in some way, control of it for GERMANY. I imagine that there will be some sort of organisation embracing all the nuclear physicists in the world.

WEIZSACKER: What are your feelings about that?

GERLACH: I would join.

2. HAHN and VON WEIZSACKER discussed their future in the following conversation on 11th August.

HAHN: I have been thinking and it is quite possible that they will not send us back to GERMANY unless we undertake to do quite different work, and they will say: "Stay here, if you like we will let your wives come over, but you must be a long way from RUSSIA."

WEIZSACKER: If I was faced with the alternative of working on uranium in ENGLAND or AMERICA, or not working on uranium in GERMANY, I would very quickly choose, not working on uranium in GERMANY.

HAHN: So I should think.

WEIZSACKER: In fact I think I should say "Even if you keep me here I would prefer not to work on uranium for the time being. I would like to wait a bit and see whether I can overcome the antipathy I have to the bomb."

HAHN: I have really come to the conclusion, sad as it is, that as far as I am concerned, I shall probably do nothing more. I might have been able to do something if they had let me carry on with my harmless chemistry; assuming I had been sent 10 grammes of this stuff which had in some way given off something without exploding, then one could have that stuff and use it as an indicator for chemistry or even biology; I would enjoy that and then perhaps I could help a bit, but I don't see anything for me.

WEIZSACKER: I would like it best if they just said to me "You needn't sign any paper, you can just go back to GERMANY."

HAHN: They won't do that. If they let you go back to GERMANY, you will have to give an undertaking on oath not to work on uranium.

WEIZSACKER: One could do that. If I was in GERMANY, I don't feel that I would want to carry on any opposition to the Anglo-Saxons there. There will be plenty of that in any case. On the contrary I would feel more inclined to persuade people to come to a sensible understanding with them. If I can help in this respect as far as the Germans are concerned, I feel I should be in a much stronger position if I could say that I was under no obligation to them. If I have to say I am working for their Armed Forces I would have no moral authority, and I consider that more important in GERMANY than for me to continue to work on uranium.

HAHN: If I could choose, although things look bad in GERMANY, I should prefer to go to GERMANY than stay here, provided I had some means of livelihood.

WEIZSACKER: Of course.

> 3. HEISENBERG AND VON WEIZSACKER also discussed international scientific co-operation in the future. The first part of the conversation could not be recorded, but it appeared that they would consent to co-operation only if they were treated as equal partners personally. They seem to consider international physics as being almost synonymous with work under the leadership of NILS BOHR. The conversation continued:

HEISENBERG: I think each of us must be very careful to see that he gets into a proper position.

WEIZSACKER: That was not possible under the Nazis. The right position would really have been in a concentration camp, and there are people who chose that. Of course it is a question how one can get into the right position with the present regime, but one can try.

HEISENBERG: It will certainly be easier than it was before, there is no doubt about that.

WEIZSACKER: I have a lot in common with the Anglo-Saxons, but not with their Governments.

HEISENBERG: I believe that BOHR and his friends, particularly the politicians, might have certain doubts about you - partly because of your father, and possibly even more because of the STRASSBURG business.

(Pause)

There is also the difference between you and me that a man of my age must be in an organisation of that kind whether he wants to or not.

WEIZSACKER: If you are in GERMANY, a great deal of the responsibility for the continuation of physics in GERMANY will be yours.

(Pause)

WEIZSACKER: WIRTZ is no longer keen on working on uranium.

HEISENBERG: I think it more likely that he is no longer interested because he sees that the others were able to do it better than we were.

WEIZSACKER: I think it is more than that. I think that he and I think much the same way about it.

4. VON WEIZSACKER also discussed the future with WIRTZ and repeated his previously expressed views. The conversation then continued:

WIRTZ: So you don't want to work on uranium in GERMANY under Allied control.

WEIZSACKER: I wouldn't like to do that at the moment. I would not like to be less free than other Germans, but I would not like to work for the Allied Governments on something which is so closely connected with the means by which they finished the war with JAPAN. What I would like to do would be to lecture on physics at some GERMAN University and to study cosmology and philosophy. What I would not like to do would be for us all, including HEISENBERG, to go back to GERMANY, and for none of us to remain in touch with this business. I have a feeling that HEISENBERG has quite different ideas to me, and that he would very much like to continue working on the uranium engine. Personally, I would like very much if HEISENBERG were to work on the machine in GERMANY and keep personal contact with the English and Americans, and that I should have contact with HEISENBERG, but not work on the engine myself. I believe that there will probably be a long period of peace, during which time the Anglo-Saxons will be the strongest people in the world. The Russians will not be overrun by them, but they will probably have to curtail their expansionist policy in EUROPE because of the bomb. The only alternative will be a war very soon between the Russians and the Anglo-Saxons, in which I will be just as unwilling to help the Russians as the Anglo-Saxons.

5. DIEBNER and BAGGE had the following character revealing conversation on 12th August:

DIEBNER: If I knew anything I would not be here.

BAGGE: If you run away, or if you escape you will be shot. You must realise these people are not interested in Mr. DIEBNER's life when you consider the danger that might arise if Mr. DIEBNER got to Russia.

DIEBNER: There is no danger because I don't know anything. For instance, I don't know why it is easier to produce fission in element 94. I didn't know all that. It is not in any book.

BAGGE: People like DOEPEL are sure to be alreay in RUSSIA and working there. If people like DOEPEL and BOHSE(?) realise that by this mans they cannot only save their lives, but improve their position, I am sure they will work over there. I will tell you something, DIEBNER, during the war you (helped) BOHSE and now if they are making the atomic bomb for the Russians and have got themselves a good -

DIEBNER: They won't forget me.

BAGGE: That's it.

DIEBNER: I have got good connections there.

BAGGE: After all you did the best you could for them.

6. The views of the guests on co-operation can be summarised as follows:

(a) BAGGE: In previous reports BAGGE has shown himself to be thoroughly German and he has expressed a hearty dislike of the Allies. He has shown slightly pro-Russian sentiments. He has suggested to DIEBNER that some of his (DIEBNER's) friends, who may be working for the Russians, might help him.

(b) DIEBNER: Has suggested in previous reports that he would like to stay in England but he has not expressed himself on the subject of co-operation with the Allies. He welcomed KORSCHING's suggestion of working in the ARGENTINE on uranium. In a conversation with BAGGE DIEBNER agreed that some of his friends, who may be working for the Russians, might help him.

(c) GERLACH: Told HEISENBERG that he would be prepared to join an International body of physicists.

(d) HAHN: Has not expressed himself in the past on the subject of co-operation with the Allies. In conversation with WEIZSACKER he said that he could see no future for himself and that he wanted to go back to Germany at all costs.

(e) HARTECK: Has not expressed himself on the subject of co-operation.

(f) HEISENBERG: In previous reports HEISENBERG has expressed the hope of being able to continue his work as part of an International group of physicists.

He repeated this hope in conversation with GERLACH and said he would like to be responsible, as part of an International body, for the work in Germany.

(g) KORSCHING: Has often expressed annoyance at anglophil behaviour of some of his colleagues. He does not appear to be willing to work for the Allies and has suggested going to the Argentine.

(h) VON LAUE: Has not expressed himself on the subject of co-operation.

(i)WIRTZ: In a previous report WIRTZ said that he considered it useless to continue work on nuclear physics in Germany.

(j) VON WEIZSACKER: In previous reports VON WEIZSACKER has shown that he is somewhat shocked at the idea of co-operation with the Allies so soon after the war. He has advocated State Control of Science.

In conversation with WIRTZ he said that he wants to return to Germany and is prepared to sign an undertaking not to work on uranium or to tell anyone about such work. He does not want to work for the Allies either in Germany or elsewhere but he says he has no intention of working for the Russians. He would like to lecture on physics at a German University and work on cosmology and philosophy.

III. TECHNICAL

1. There has been a lot of discussion of the technical aspects of the atomic bomb. Eventually HEISENBERG was asked to give a lecture on the subject which took place on 18th August†.

2. The following conversation took place between DIEBNER, HARTECK and BAGGE.

DIEBNER: We had to fetch the heavy water from Norway.

HARTECK: Just think of the hot and cold tube method, that would have cost RM. 15 million for which I.G. guaranteed 5 tons a year, though perhaps we would have got six tons. Then they said they couldn't build it and in their case it was only a few thousandths of their annual turnover. That was in fact the first method ready for industrial development though perhaps not the cheapest. I.G. did build the small apparatus for about 50 litres a year and that cost R.M. 300,000. Of that R.M. 100,000 was for measuring instruments and 200,000 for the apparatus. For an apparatus a hundred times as big which was needed industrially, these R.M. 200,000 would have been increased as the two-thirds power, that is they would have been R.M. 2½ million and then we should have

† An error: it was 14 August. FCF

got the five tons. Suddenly they said that was a mistake and that for a hundred times the size the cost would be a hundred times. They always said "We'll build it after the war". When I first talked with Herr BASCHE I wanted to bring DRAMM into it and it appeared that DRAMM could not take part because he was a quarter Jewish. He left RUHRCHEMIE because he had a row - the whole senior Staff was changed - and he was one who had followed the whole construction through. Just think if we had had him! He would have built the thing for us, with him we could have built it.

DIEBNER: I.G. had ESAU on their conscience.

HARTECK: In practical matters I was the stupid one I worked for almost six years on this thing alone and the people who've done other things turn out to be the lucky ones.

It seems clear to me that if we had had a machine for 10 or 12 tons, mixing with paraffin and cooling by liquid air would have been good in every case. If one has one part of uranium and one part of hydrogen and mixes them homogeneously and cools it to liquid air temperature then it is reduced by half and the absorption becomes very strong and might fall to 20%. Since there is so much uranium, the 2n process is very probable and on the other hand the absorption by water, if it is homogeneous, is almost zero, only 2 or 3%. Heavy water absorbs nothing at all, perhaps 1%, but paraffin absorbs 2%. In our machine absorption at the resonance position was about 10 to 12%, at lower energies it was 15 or 16% with heavy water. The 2 n multiplication was perhaps 12 to 14%, I can no longer remember. In the machine it was stronger. Taken altogether there were certainly regions where the multiplication factor was bigger than in the uranium machine. It seems to be beyond dispute that it works.

BAGGE: That would man that at these temperatures the apparatus would run.

HARTECK: Then we could get it running-quite simply, for we need only the metal and this big cask, you just put liquid air in on top and pump it out- then it starts.

BAGGE: Yes then it only makes neutrons, heats up and stops again.

HARTECK: Yes quite right. You let it run for a period for you can cool with a certain number of kilowatts, and when it has run a certain time, the others have been enriched and then you get them out. That is one special way in which you could do the thing. They wouldn't all run all the time but you need them all to start up, with a colossal expenditure of cold. The other way is the photochemical- that you just go on and on trying at it with a big staff. As soon as one has the photochemical process it is much cheaper than the mass spectrograph and in any case naturally for other purposes you would try out the isotope lock and the centrifuge and such methods, for one should not rely on one method. If you consider this whole programme of work with the heavy water and low temperatures and the photochemistry and all that it involves. I'm not certain whether these fellows can do it all.

3. The following conversation took place between GERLACH and HARTECK on August 8th, 1945.

HARTECK: We want chiefly to have a machine just as a means of studying it.

GERLACH: The technical difficulty in the present water machine is the preparation of the uranium and then the whole subsequent process like casting it to shape. Therefore, it would be better if it were something with a powder.

HARTECK: I first proposed the scheme 2½ years ago because I thought one would have to make other experiments on account of stabilisation and it has been proved step by step and that was the only sensible thing to do.

4. The following conversation took place between HAHN, GERLACH and BAGGE on August 11th, 1945.

HAHN: In GERMANY they could prepare a few grammes of ionium per annum.

GERLACH: No, we prepared 60 kg. of pure ionium.

HAHN: I reckon that that is impossible. Is that supposed to have been done in GERMANY?

GERLACH: No, in BELGIUM. I don't know where it was done but I do know that it played a big part in some espionage business. STETTER interested himself in it and there was some S.S. business, some agent - one doesn't know who he was - was sent to me, I was to work with him, an Indian. They said he was a Japanese spy. He was determined to find out where we got the ionoim from. That was 1944/5.

GERLACH: Did STETTER work with the S.S. on ionium?

HAHN: No.

The fission of ionium was experimentally proved in the Radium Institute at VIENNA, but for that one does not need to build a machine - they get fifty times as much ionium from pitchblende weight for weight as radium. If therefore, you prepare one gramme of radium you also prepare fifty grammes of ionium. The ionium must, therefore, correspond to a production of 1 kg. of radium if they have carried it our quantitatively.

GERLACH: Can one make a bomb with it?

HAHN: The neutron energy in ionium has only been approximately calculated. One had no pure ionium but one took this mixture of ionium and thorium in

VIENNA which contains about 50% ionium and in that they believed that they had demonstrated fission at a lower voltage than with thorium alone. In any case there is a report from the year 1941 from the VIENNA Radium Institute; who wrote it I do not know, probably STETTER.

GERLACH: Can you make a bomb out of 2 kg. ionium?

HAHN: I don't know.

BAGGE: Can one prepare ionium in large quantities?

HAHN: Yes, but not at all with thorium- 80% thorium would not be too bad.

BAGGE: You get so much more ionium than radium?

HAHN: Ionium has a half-life of 80,000 years and radium only 1,500 and since they are produced in the same series, they are proportional to their half-lives, but in all uranium pitchblende there is always some thorium and they prepare this thorium with it. Thorium has an exceedingly long life of 10^{10} years. Thorium undergoes fission, we did that by January 1939. Ionium is also fissile but it's still a question with what sort of neutrons.

GERLACH: That depends chiefly on the cross-section.

HAHN: It has the atomic number 90 and 232.

BAGGE: Then you see that it'll undergo fission with difficulty. All the heavy elements with even numbers undergo fission with difficulty.

GERLACH: If however it had a very short mean free path and a very big cross-section?

BAGGE: It wouldn't have those. If it's of even number then it will not be split by thermal neutrons but with neutrons of higher energy and then it cannot have anything like so large a cross-section; for the cross-section must be of the order of the geometrical cross-section of the nucleus. That is about 2.5 times 10^{-24}. And 2.5 x 10^{-24} gives I think a mean free path of 6 cms.

HAHN: I don't know how they did it because unlike us they had no pure radio-thorium.

5. After reading the newspapers on 8 August, the following comments were made:

BAGGE: It says in the newspapers that 236 will be used which results from the bombardment of 235. This they call "Pluto". This might be 93 but this would mean that they had a running stabilised machine. On the other hand they say that they have not got one. Therefore it is still difficult to believe.

(Pause)

HEISENBERG: But if they are working with heavy water the material will be lost to them.......Here they have however quite slow reaction times, since heavy water implies thermal neutrons and thermal neutrons indicate reaction times of the order of 10^{-3} secs. or so.

WIRTZ: They only need that to produce additional neutrons.

HEISENBERG: I see, simply for the increase of energy production. That I agree. But I do not see how the reaction can take place in 8 lbs. of something, since the mean free paths are fairly long. They have always got free paths of 4 cms. In 8 lbs. they will surely get no chain reaction whatsoever. But still it may not necessarily be true what is written in the newspaper.

I still do not understand what they have done. If they have this element 94, then it could be that this 94 has quite a short mean free path. We have done little research in the field of completed fast neutron reactions because we could not see how we could do it because we did not have this element, and we saw no prospect of being able to obtain it. How they have obtained this element is still a mystery.

WIRTZ: Yes, but I believe that they have got it and I feel sure that the bomb is not big.

HEISENBERG: That is to say it might be of the order of 400 kilos.

6. The following day the newspapers mentioned that the atomic bomb weighed 200 kilos and the following conversation took place between HARTECK and HEISENBERG.

HARTECK: Do you believe that it is true that this is the weight of the bomb or that they wish to bluff the Russians?

HEISENBERG: This has worried me considerably, and therefore this evening I have done a few calculations and have seen that it is more probable than we had thought on account of the substantial multiplication factors which one can have with fast neutrons. We have always calculated with a multiplication factor of 1.1 because we had found this in practise with uranium. If they have a multiplication factor of 3 or 5 then naturally it is a different matter. We said we need about 80 links in the chain reaction; now the mean free path is 4 cms. therefore we must have 80 long divisions (so was the rough estimate) and this would then come to about a ton. This calculation is right if the multiplication factor is 1.1 because even then we use really every neutron which "escapes" for multiplication. If on the other hand the multiplication factor is 3, things are quite different. Then I can say, if the whole thing is only as big as the mean free path, then one neutron which walks around therein once meets another and makes three neutrons. From these three neutrons one will already come back, the other two can go off; the one that comes back will for certain make another three. In practice therefore I need only the mean free path for the thing to work.

HARTECK: What are these 100-200 kilos which are around it?

HEISENBERG: This will be the reflector. For instance they might have lead as a reflector and of course part of the weight could be apparatus; you must remember that it is a dangerous business. They must arrange it so that it is at first taken apart into two pieces between which a reflector, of shall we say lead or some other material, is placed, and then at the right moment this lead will be pulled out and the thing clapped together.

(Pause)

Well how have they actually done it? I find it is a disgrace if we, the Professors who have worked on it, cannot at least work out how they did it.

HARTECK: I believe it would be technically possible to produce 2 kilogrammes of proto-actinium though of course an enormous amount of materials would have to be used. For ten such bombs this would mean 50 kilos of radium in three years which is unbelievable.

HEISENBERG; Nevertheless, they have quite obviously worked on a scale of quite fantastic proportions.

HARTECK: I believe rather in a machine than in the production of so many kilogrammes...... because it has never yet been found in the world. One finds always a......... Yes, but on these grounds..... they could not produce this. They would have to have a billion tons to get out sufficient material for man to weigh. RUTHERFORD in his time made it artifically and on that basis they have made their calculations. There are perhaps 10^{-15} grammes in the world. Or they must have had unbelievable luck and somewhere they found this element, in the same way that suddenly caesium in large quantities was found in SWEDEN.

HEISENBERG: In the newspapers is the statement that the essential element was discovered in 1941.

HARTECK: This could be the decay product of 23 min. half-life uranium from which they have made 93.

HEISENBERG: There is something else. If they have made it with a machine then there is the fantastically difficult problem that they have had to carry out chemical processes with this terrifically radio-active material.

HARTECK: Then they must have to start by letting it cool off for 6 months.

HEISENBERG: Right. They built a machine from which they must start by removing a large amount of energy so that the chemical change takes place;

therefore they must have let the machine run at a constant temperature so that they could take out the total energy which is in order of magnitude the equivalent of a power station working for ten years.

HARTECK: The machine need only work slowly.

HEISENBERG: It does work slowly.

HARTECK: That is again the difficulty. I believe that if 1% conversion has taken place then the machine stops.

HEISENBERG: Then if 1% conversion has taken place then they must dismantle the machine.

HARTECK: To refurbish?

HEISENBERG: Yes and thereby they will of course be able to take out 94, nevertheless one could say that if they have done it with a machine of one ton they will be able to take out not an inconsiderable amount of material. For instance, if 1% conversion has taken place then they can take out 1% - that is to say 10 kilos of 94 out of the machine.

HARTECK: It is of course not only 94 which is formed.

HEISENBERG: I do not believe that the Americans could have done it. They would have had to have had, shall we say, a machine running at least not later than 1942 and they would have had to have had this machine running for at least a year and then they would have had to have done all this chemistry.

HARTECK: Highly improbable.

HEISENBERG: I believe it almost more likely that they have done something quite original such as getting out proto-actinium in quantity from colossal quantities of maerial. It says in the newspaper that enormous quantities of material went in and practically nothing came out. This appears to me to fit in with such a chemical process or isotope separation, but one does not get an impression from the newspapers that 235 has been used for it. Do you know when the work was carried out by the Americans which showed that proto-actinium could be disintegrated at 150,000 volts?

HARTECK: That was about 1940-1941. It was certainly before 1942.

HEISENBERG: Perhaps the facts are that thereby they discovered the element "Pluto". "Pluto" is a code name. Proto-actinium also starts with "P". Perhaps they simply say to themselves "if we now extract the proto-actinium, then we can do the thing".

HARTECK: Then as by-product they will have about 50 kilos of radium. They will be able to warm their feet with it!

HEISENBERG: Is proto-actinium really so active?

HARTECK: No, not so active as the equivalent radium. Proto-actinium has a much weaker activity. It is in the ratio of 30,000:1,500, i.e., 20 times weaker. One has the same quantity of proto-actinium in pitchblende, on the basis of equivalent weight, although it is 20 times weaker because only 4 or 5% comes out of the neighbouring chain of uranium 235. Assuming, that proto-actinium had a parent substance, such as the parent substance of radium is present mixed in the ratio of 1:1, then on the basis of equivalent weight there would be 20 times as much proto-actinium in pitchblende as there is radium, because it has a half-life 20 times longer than radium. From uranium both are produced simultaneously. One is 235 which makes proto-actinium which builds up to 4.6% in the course of uranium decay, so that in fact 50 kilos of proto-actinium give also 50 kilos by weight of radium, and these 50 kilos by weight of radium have the same activity as 50 kilos of uranium, whilst the 50 kilos of proto-actinium only have the same activity as 2.5 kilos of radium.

HEISENBERG: Well I see now that it is possible with fast neutrons. Just now we have discussed proto-actinium. Perhaps the others have used proto-actinium; this is almost easier to imagine than all other methods.

HARTECK: If GROSSE helped them who has produced pure proto-actinium and is a very good chemist who is very interested in proto-actinium and has something, he has placed it at their disposal then they have tried to split it to establish whether proto-actinium undergoes fission.

HEISENBERG: We knew that proto-actinium undergoes fission below about 50,000 or 100,000 volts much lower than in the case of thorium, and this means that if one has pure proto-actinium in considerable quantity, then the whole thing would blow up. At one time we spoke about this I know. I once came to you and asked how much proto-actinium there was.

HARTECK: Then I said there was perhaps half a gramme.

HEISENBERG: Then I said "Well with that amount it certainly won't work". Perhaps they are making this stuff on a kilogramme scale, and if they then make a good reflector then they can work well with 1 or 2 kilogrammes.

HARTECK: I would estimate that during the last one to two years 1-1½ kilos of radium have been produced in the world. Naturally this could be increased by a factor of about 10, and one would obtain then in a year about a kilo, but they must have more.

7. Later the same day the guests discussed the new bomb in the following conversation:

WEIZSACKER: I find, having regard to the Russians, they are telling a little too much about the bomb.

GERLACH: But each time they make exactly different statements. I am of the opinion that there is as little truth in one as in another.

HEISENBERG: It is till not clear which method has been adopted. Either they have taken pure proto-actinium or they have really separated the isotopes, or they have made element 94 with machines.

HARTECK: If it had been that we had been in RUSSIA and had the necessary material and had to start from nothing, then we would undertake to make a machine work in nine months.

HAHN: No, no, no.

HEISENBERG: If one had the necessary material. The machine is now no more of a problem.

WIRTZ: If you had no heavy water, or if you had too little, how would you do it?

HARTECK: Production of heavy water would begin within six months in a properly designed chemical plant.

HEISENBERG: This is the question, is the fission cross-section for fast neutrons in 238 and 235 substantially different?

HARTECK: What indeed is the biggest fission cross-section that one can take?

HAHN: Doesn't this depend on the exploitation of 238, also for instance on the Japanese elements for which it is much smaller.

HEISENBERG: For 235 it is smaller.

HAHN: Yes, 235 is 0.7% of 238. 238 is obviously split by fast neutrons into exact halves - <u>also</u> in exact halves not <u>only</u>. But this exploitation into exact halves is considerably smaller than for the normal fission products with fast neutrons. In this instance the Japanese for once published something original, it appeared as if only these Japanese elements, as we have called them, came out. This is however not the case. They are very much less than the others.

HARTECK: I had only thought what is believed to be the highest.

HEISENBERG: What could one have as the highest?

HAHN: About 2½ cms.

HEISENBERG: No, what do you take as the cross-section?

HARTECK: Tell me, what is the geometrical diameter, how big is it?

HEISENBERG: The radius is 0.8×10^{-12}.

(DIEBNER produces a book)

I want to have the fission cross-section for fast neutrons.

DIEBNER: 0.5 x 10^{-24}.

HAHN: And what is the cross-section for slow ones?

DIEBNER: For medium fast ones it is 0.1 and for slow ones 1,000 or thereabouts.

HEISENBERG: Then the mean free path is 8 cms. If it is 0.5 then it is of course bigger by a factor of 4½.

HAHN: I do not understand that. How can then 0.5 x 10^{-4} - how big must it then be for slow ones? Obviously much bigger.

HEISENBERG: With pure 235, a couple of hundred.

HAHN: I see, times 10^{-24}?

HEISENBERG: Yes. In a mixture of 238 plus 235 the fission cross-section ratio is approximately 3.3. If I recalculate for 235 then I must multiply by 100 or 140 and then I get the value of 400. But for fast neutrons it is quite small.

(Pause)

If, for example, you put round it ordinary uranium as a reflector then things are quite different. You can of course have luck if you make element 94.

With uranium the mean free path is almost 40 cms.

KORSCHING: What would happen if there were a mixture of 94 and - Is it simply proportional to the mixture, or would it be more favourable?

HEISENBERG: I would say proportional.

HARTECK: It is usually always worse because the mean free path becomes longer and the weight goes in the Kubus.

HEISENBERG: The best arrangement is to place the pure element, which is quite pure, in the middle and to place round it the best possible reflector.

(Pause)

I would say in the most favourable circumstances one would obtain a geometrical cross-section of about 8 cms. and if one then, by some process, obtained out of it 5 neutrons, one need not make the radius any bigger than one mean free path. If it has moved one mean free path, then of the 5 neutrons at least one or two will come back and the process will be repeated. Then one can easily make it go with one mean free path.

HARTECK: Is there actually also a metal which will reflect back principally fast neutrons? Actually there is none, there is always a forward persistence.

HEISENBERG: Yes, but on the contrary, the heavy ones have the tendency to scatter under little corners.† In the case of light ones it is fairly spherically symmetric, but in the case of the heavy ones there is an elastic scattering, which goes in a strong forward direction, but they do not then move.

(Pause)

I would rather think if we are dealing with such small quantities that it must be that they will obtain so many neutrons.

HARTECK: This would again suggest the already working machine and "Pluto" - for the heavy - .....

HEISENBERG: For the 94, yes. I do not know, perhaps just as many come out in the case of proto-actinium.

GERLACH: What is the case then with ionium?

HEISENBERG: This is the parent substance of radium. This has no chances as a result of fission. Fission would most likely occur with fast neutrons. With fast neutrons I might believe it possible.

(Pause)

Let us assume that they have done it with proto-actinium, which to me at the moment appears to be the most likely. Proto-actinium can be separated chemically if one is willing to undertake the necessary effort. These stories of radiation screens fit in although proto-actinium is about 20 times weaker in activity than radium. If they are making kilos of it it will be terrible. I have the impression that for this expenditure (of £500,000,000) they could have produced this quantity of proto-actinium.

BAGGE: If they had made 20 kilos of proto-actinium how many tons of uranium would they have had to work with?

HEISENBERG: Approximately the same as with radium.

BAGGE: That is to say from 7 tons, 1 gramme.

HEISENBERG: They would have had to work 140,000 tons of material.

BAGGE: Have they got this?

HEISENBERG: But of course.

DIEBNER: There are still 10,000 tons lying in BELGIUM.

HEISENBERG: Was there ever so much proto-actinium in GERMANY?

---

† This should obviously read 'scatter through small angles'. FCF

HARTECK: No, no, out of the question.

HEISENBERG: At one time I spoke to HAHN. We dismissed the proto-actinium problem as being unpracticable.

HARTECK: What is the decay constant of proto-actinium?

HEISENBERG: The alpha decay constant is 3.2×10^4 years.

HARTECK: And how many neutrons...... proto-actinium.

HEISENBERG: It is the same thing as in the case of 235. Approximately the same will come out as in the case of 235. But the relationship of neutrons to protons is somewhat higher in the case of proto-actinium. In the case of uranium 235 two neutrons come out per fission. In the case of proto-actinium it is perhaps 2-5.

BAGGE: On the other hand, it does not fit in that it was only in 1941 that the element "Pluto" was discovered.

HEISENBERG: Yes, but in the first place "Pluto" may be a code name. At the end of 1940 or the beginning of 1941, the American work appeared to establish that proto-actinium was fissionable below about 50,000 volts. During the discovery they might also have thought that they had discovered that it was spontaneously fissionable.

BAGGE: Already in the work of BOHR and WHEELER† it was stated that it was fissionable.

HEISENBERG: No, the actual facts were that in the case of proto-actinium, I believe, BOHR and ..... had made a miscalculation. And then when the American work appeared, a memorandum came from BOHR in which he announced that there was a mathematical mistake in his formulae and if the formula was correctly set up, something came out of proto-actinium.

WIRTZ: Well, I am prepared to take on a bet that they have separated isotopes.

HEISENBERG: Yes, I will of course not deny it, but there are now three quite clear ways in which they can have done it and only three: isotope separation, proto-actinium and a machine with D_2O and element 94.

WIRTZ: My method is, a machine with light water and the separation of element 94. I do not believe that they have so much heavy water at their disposal that they could really have many working uranium machines.

BAGGE: For what purpose have they hitherto used the heavy water?

---

† Wheeler's name appears as dots in the transcript, but the reference is undoubtedly to the paper by Bohr N and Wheeler J A, July 1939: On the mechanism of nuclear fission *Physical Review* **56** 426–50. FCF

HEISENBERG: Yes, this would fit in with WIRTZ's hypotheses. If one has heavy water for only one machine, then from one machine, one can take out quite a lot of 94 in the course of time.

WIRTZ: Yes, and then I believe they also put a little heavy water in the bomb in order to raise the factor 2.

HEISENBERG: That may be so.

(Pause)

One has the impression that the English must have done something else other than the mass spectrograph.

8. On 13 August, HAHN and WIRTZ made the following remarks:

HAHN: Element 93 decays in 2-3 days into 94. They have of course 94. This is obviously plutonium. They have 93 as a nice stable conversion product resulting from the bombardment with fast neutrons of uranium 237.

WIRTZ: Today it states in the newspaper that they could not work with the rare element plutonium because they reckoned it would take 70,000 years to obtain sufficient before they could make a bomb. This means that until now they have not got a running machine.

HAHN: 93 can be quantitatively separated from 92. STRASSMANN and I have worked out the quantitative separation.

9. HEISENBERG gave the following lecture on 14 August in which he described his ideas of how the atomic bomb was made. As this is highly technical the full German text is reproduced in the appendix.†

HEISENBERG: I should like to consider the U235 bomb following the methods we have always used for our uranium machine. It then turns out in fact that we

---

What follows, as far as p 140, being the work of another translator from that (or those) of the remainder of the transcripts we have felt at liberty to make some revisions. There were some explicit errors of translation, particularly rendering of *viel* as little instead of much (just once) and *hinreichend* as enormously instead of sufficiently. On p 126 the name DOERRIG (with a question mark) from the German text was replaced by DIEBNER.

Some equations are now displayed, on the clearly justified presumption that Heisenberg wrote them out, elucidating his symbols in words as he did so.

The German text in the Appendix has been left unaltered, save for correction of those obvious copying errors rendering Greek ν as v: but it is to be noted that what was typed as 1 (one) is sometimes 1 (one), sometimes l (el). This gives us a sample of the German text produced by the listeners, after it has passed through the hands of a typist, for material not fully understood by either. FCF

can understand all the details of this bomb very well. I will begin then by recapitulating once more the chief data on U235. I may repeat briefly what happens in this bomb. If a neutron is present in U235, it travels on and soon meets a 235 nucleus. Then two things can happen. Either it is scattered elastically or it causes a fission. If it is scattered, the scattering may be either elastic or inelastic, and so the neutron goes on with the same or a lower velocity. This does not prejudice the power of the neutron to cause fission later. There is therefore no probability of neutrons being lost anywhere. The process naturally goes on; the neutron at some time meets a 235 and splits it. In the fission, some neutrons are produced. These neutrons behave like the first, and so the chain reaction goes on. If one had an indefinitely large amount of U235, the chain reaction could go on indefinitely, for 2 or 3 neutrons would always result from each one by fission. These two or three would repeat the process, and so it would go on. Thus the total number of neutrons would increase exponentially. The multiplication of neutrons is however in competition with the process by which neutrons escape from the mass. If in fact one has a finite mass of uranium, those neutrons at the surface which are moving outwards escape, and have no chance of taking part in fissions. So the question arises whether this loss of neutrons by escape from the mass, is greater or less than the gain of neutrons arising from the production inside. To calculate this, or convince oneself, it is necessary to have the cross sections and mean free paths. I have a handy formula for this, and here it is: the mean free path for any process in uranium 235 is 22 cm./cross-section, the cross-section being measured in 10^{-24} (cm^2). This means that if for example the cross-section for fission is 1 x 10^{-24}, then the mean free path (M.F.P.) for fission is 22 cm. The scattering cross-section is empirically about 6 x 10^{-24}, and so the M.F.P. is 3.7 cm. The most difficult part is the cross-section for fission. This naturally depends markedly on the velocity. For thermal neutrons, as far as we know, it is in the neighbourhood of 300 to 400 x 10^{-24}, i.e., very big. For fast neutrons I cannot now recall the figure exactly.

There is a figure of 0.5 x 10^{-24} in DOERRIG's(?) Tables, but perhaps it refers to 238 and not 235. I think though that we can only know this figure really accurately for 238, because with fast neutrons the 238 and 235 are both subject to fission, and so when you measure the fission it is that of 238 in most cases, and I don't quite see how the two fissions can be properly differentiated.

HAHN: Because there was only 0.7% of it?

WEIZSACKER: It will never be possible to measure that. I do not see how it could be done.

HEISENBERG: Perhaps you could do it like this: you would take neutrons of velocity slightly below the critical limit for 238, but not thermal; let us say, 10^5 volt neutrons. With these you would measure the cross-section and then calculate with the ratio.

WEIZSACKER: Yes, but then you don't know exactly how the cross-section is going to depend on energy.

HEISENBERG: One could perhaps say that at these very high energies the cross-section would be comparable with the geometrical cross-section of the nucleus. It would probably not depend so greatly on energy. BOHR's theory even suggests that it would be constant over the whole energy range.

HARTECK: If they have a good enough spectrograph in the U.S.A. to effect a complete separation, then you need do no more than to have a thin layer and to multiply correspondingly. For if a nucleus breaks up, it gives off much more than one alpha particle, which you find with counters, and thus get the ratio of thermals to fast neutrons. In this way you would solve it experimentally. They have certainly determined it exactly.

HEISENBERG: Yes. I have indeed made an estimate of 0.5 to 2.5 for the cross-section, since I argued that it was 0.5 for 238, which must be a lower limit as 235 is more fissile and 2.5 is the nuclear cross-section, the true collision cross-section, and it cannot be much greater. Admittedly, it might be a little more than πr^2, but it must lie within these limits. Therefore, if you take the limits between 0.5 and 2.5, the resulting M.F.P. for fission is from 9 to 44 cm. For scattering, again, we know that for all the heavy elements the cross-section is about 6 at high energies. In uranium we know it pretty exactly, for it is, I think, equal to 6.2×10^{-24} in 238. In lead it is slightly less, but they are all in this region. So we can say that for scattering the cross-section is pretty certainly in the neighbourhood of 6×10^{-24} and so the M.F.P. is about 3.7 cm. Then there is an important quantity we need, the multiplication factor, i.e., the number of neutrons produced from a collision which results in fission of 235. Since we know the multiplication factor for thermal neutrons very closely from our BERLIN experiments, which however, we calculate not per fission, but per thermal neutron absorbed. Now for thermals, the cross-section for fission is about 3 and that for absorption about 6.2, i.e., we get actually only half the true multiplication factor if we take our figures.

Our figure is 1.18, so we can say the multiplication factor is really 1.18 x 2 or, roughly speaking, between 2 and 2.5.

LAUE: I didn't understand that.

HEISENBERG: We have always experimented with ordinary uranium, in which 235 is rather less than 1% and the rest 238. In this ordinary uranium we have the following relations for thermal neutrons: we have a cross-section of 6×10^{-24} for the capture of a thermal neutron, which may be simply captured to form 239, which decays with a period of 23 mins., or may cause fission, which must be fission of 235 - (int.)

HAHN: Are you still thinking only of thermal neutrons?

HEISENBERG: For the moment I am speaking only of thermals. For thermals, the fission cross-section of 235 is naturally frightfully high — it is actually in the neighbourhood of several hundred — but because it only occurs with a frequency of 1%, it only plays a small role on the whole. Thus the effective cross-section calculated on 238 is only about 3. This figure of 3 was I think, determined by DROSTE in your lab.

HAHN: Yes.

HEISENBERG: I think DROSTE got a result of 3.1 or thereabouts. If two neutrons are absorbed, you have a fission in only one case, because the total cross-section is 6 and the fission cross-section 3. Hence if the multiplication factor for a fission is x, then the effective multiplication is simply *either* ½*x*, *or* $x/2$, because every second neutron is killed by being captured.

HARTECK: If one has the resonance... (passage a bit garbled — see German text) Would you think that in this case the BREIT-WIGNER formula, which gives $1/v$ exactly for low temperatures, held, or could there be a small difference at low temperature between the extrapolation from the BREIT-WIGNER formula and direct (Fission?)? You see, is there a relative deviation of 10% between them, or is the $1/v$ law exact at the low temperatures?

HEISENBERG: I should think that it would go very exactly as $1/v$.

WEIZSACKER: This formula simply means that the probability of the process per unit time is constant.

HARTECK: That is quite clear, but there is still the exact influence of the resonance level.

HEISENBERG: Yes, but it is so far away.

WEIZSACKER: All the same, I should say the smaller the v to which you go, the better ... (interrupted)

HEISENBERG: But the whole resonance affair plays no part at all in the bomb. I wanted to make this brief digression into the thermal region only to say that on the basis of our experiments we should conclude that the multiplication factor in fission of 235 is about 2 to 2.5, or more exactly we should say 2 x 1.18. That means that when a 235 is split, there are 2 to 2.5 neutrons produced. It seems to me that the Americans or the English are of the opinion that the figure is a bit higher, because in the paper there was a picture of the uranium process, and for each fission they showed three neutrons produced. So I think it is quite conceivable that in actual fact the number is about 3, and that the cross-section which DROSTE found is slightly high, the true one being a little less. The Americans had actually determined the fission cross-section as 2 sometime earlier, and then the number would come out as exactly 3.

DIEBNER: The values — the calculated values — were a little higher. In our last experiment, they came out somewhat higher.

HEISENBERG: Yes, that was a question of 1%, which doesn't count for much. I have therefore after that taken a multiplication factor of, I think, 2.5. So now we have collected the most important data. Now we must ask what happens in the uranium when a fast neutron sets off on its journey? We can say this: if we had an indefinitely large amount of uranium 235, and neutrons in it, it is clear that the number of neutrons would grow exponentially and the chain reaction would proceed indefinitely. Next we must consider how the exponential growth takes place. So I wrote down the equation, n proportional to $e^{\nu t}$, where ν is a characteristic reciprocal time for this exponential growth. This characteristic time, which we shall need later as the reaction time, can easily be estimated, if we note that the time before a neutron makes one fission is plainly the ratio of the M.F.P. for fission to the velocity. Thus, a neutron travels about 9 cm. to the next fission, and I have to calculate how long it travels before covering 9 cm. This time is λ_{fission}/velocity. In this time, the multiplication of neutrons is exactly the factor x, i.e., instead of 1 neutron I have now x neutrons. So for the determination of the quantity ν I get the equation

$$e^{\nu\lambda_{\text{fission}}/v} = x,$$

or in other words the characteristic reciprocal time

$$\nu = \frac{v}{\lambda_{\text{fission}}} \log x.$$

We can take the velocity as approximately 1/20 that of light, for they are neutrons of about 1,000,000 volts. We know then how a number of fast neutrons would increase in an indefinitely large block of uranium. But when the distribution of neutrons is not uniform, and if we ask, not how the total number changes, but how the number in a particular small volume changes, then naturally neutrons can also leave this small volume by diffusion. So I must write down a diffusion equation for the neutrons, and I must know their diffusion constant. According to the kinetic theory of gases this is always equal to the M.F.P. for scattering, multiplied by the velocity and the factor 1/3. The M.F.P. for scattering is known, it is 3.7 cm. The velocity is also known, it is 1/20 of the velocity of light. From these I get the diffusion constant. Finally I have for the rate of change of density at any point:

$$\dot{\rho} = D\Delta\rho + \nu\rho$$

where D is the diffusion constant and Δ the Laplace operator. One sees that if the term $\nu\rho$ wasn't there, so no neutron-multiplication, this would simply be the known equation of heat conduction. This term $\nu\rho$ signifies that the neutrons also multiply. One also sees dimensionally - that is important - that the ratio D/ν is evidently the square of a length, since if I divide through by ν I have D/ν times the Laplace operator applied to ρ and this must have the same dimensions as ρ. Since FERMI's work this has been known as the square of the diffusion length, l^2. This l comes out numerically as 6.2 cm divided by the square root of the fission cross-section, the latter in units of 10^{-24}. If the fission cross-section is precisely 1 x 10^{-24}, then this diffusion length is 6.2 cm. Perhaps at this point we can consider for a moment the comparison with our machines. In our machines the diffusion length was always much greater. I no longer have the figures exactly in my head, but I believe I can remember that in our best machine, the one we last had, it was 35 cm. For that reason the machines had to be so very large. That is for two reasons. Firstly, the multiplication factor is much smaller, here it is 2.5, there it is only 1.18, because the 238 takes so much away. Secondly - and this is still more important - a neutron travels a long way from where it is produced before it becomes thermal and is absorbed. Thus in our machines it was slowed down by D_2O and then captured as a thermal neutron - it travelled about 20 cm before it was captured. This distance naturally adds to l, it must be compounded with l. So here the diffusion length is only about 6.2 cm, there it was in the neighbourhood of 35 cm.

We now have the diffusion equation according to which the neutrons spread out. I now wish to solve this equation for a uranium sphere of given radius R, and I will assume this uranium sphere is further surrounded by some other substance. What sort of substance we can consider later. Naturally we can also take it as vacuum, but in any case we assume a sphere so as to solve the equation easily. There will certainly be solutions in which the neutron density everywhere increases or decreases exponentially. Namely we shall obviously have the following: if the uranium sphere is small more neutrons travel outwards than are produced in the interior and the neutron density diminishes exponentially. If on the other hand I make the sphere sufficiently large the outward neutron flux does not win against the neutron multiplication in the interior, and then the neutron density increases exponentially. So I arbitrarily put for the neutron density

$$\rho = \rho_0 e^{\mu t}$$

and seek to determine this characteristic factor μ in the exponent. If μ is positive the machine works and the bomb explodes. If μ is negative any neutrons which are in the mass travel away and are of no further interest. So putting $\rho = \rho_0 e^{\mu t}$ one gets the equation:

$$D\Delta\rho + \rho(\nu - \mu) = 0.$$

One has to solve this equation. For this one usefully introduces another characteristic length:

$$D/(\nu - \mu) = l_1^2.$$

The solution of this equation is known: it is simply the equation of known type $\Delta\phi + k^2\phi = 0$. The spherically symmetric solution is (sin x)x (x is thus the variable) or $1/l_1$.† The solution in the interior of this sphere is thus $(1/r)$ sin (r/l_1). So the neutron density in the interior must distribute itself according to this formula $(1/r)$ sin (r/l_1). This has to be the case if I do not assume there is some neutron source within it. If I had an internal source of neutrons it could naturally follow that I should need a solution singular at $r = 0$, for example, because there was a source there. But I will assume that there isn't any source so that the neutrons are distributed smoothly over the sphere and that can only be according to the formula $(1/r)$ sin (r/l_1).

LAUE: Is that a source independent of the uranium?

HEISENBERG: Yes. So inside the sphere we certainly have this dependence $(1/r)$ sin (r/l_1). What is it outside? There one can say the following: we have to do with fast neutrons which in the surrounding substance are scattered and slowed down. The slowing down does no harm because a slow neutron can also cause fission, so I can say that the neutron number does not alter outside. Thus outside I have no absorption, neither positive nor negative, and there is pure diffusion of fast neutrons. I can thus say that outside I have a medium in which neutrons only diffuse but are neither absorbed nor multiplied. In this external medium a diffusion equation like the one above holds, but with the simplification that $\nu = 0$, so that there is no multiplication. The solution of this simplest diffusion equation is naturally that ρ diminishes as $1/r$. Thus in the outer space ρ, the density, must diminish as $1/r$, because we are dealing with ordinary diffusion without accompanying absorption. I have there simplified it a bit by neglecting μ, which is small compared with ν.

HAHN: They come back from outside, I think.

HEISENBERG: Yes. A notable fraction of the neutrons flow out and a part will always be scattered back. So I must set down the boundary conditions. These are that at the separation surface between uranium sphere and mantle the density and the flux must both be continuous. I have already given the internal density as $(1/r)$ sin (r/l_1), and externally it is α/r. I must thus choose α so that at $r = R$ both expressions are equal.

LAUE: Why must the density be continuous?

HEISENBERG: There can be no jump in neutron density. The neutrons move very rapidly back and forth across the boundary. Hence I cannot maintain a density jump. Then beneath these two density terms I have written the flux. One obtains the flux easily by differentiating the density with respect to r: the flux is given by the density gradient multiplied by the diffusion constant. On the left we have the differential with respect to r, multiplied by the diffusion

† *sic*, from German text, but evidently a little garbled..

constant D_i, where I have used the suffix i to refer to the interior, and on the right we have the differential with respect to r of α/r, that is $-\alpha/r^2$, multiplied by the external diffusion constant. Then I write the condition that both should be equal, and you can believe me that then α drops out, it can be eliminated, and there remains for the ratio of the diffusion coefficients

$$D_a/D_i = 1 - (R/l_1) \cot (R/l_1)$$

BAGGE: How is it if there is vacuum or air outside?

HEISENBERG: You can still use this formula, but D_a must be taken as infinite. You must say that each neutron which comes out is immediately carried away. That leads to the neutron density at the surface being zero, because outside they are carried away with arbitrarily great speed. Now we see that to get a clear result we must know the diffusion constant outside as well as inside.

That means one must know something about the material one puts around the uranium sphere. We also see that it is expedient to surround it with something which scatters as many neutrons as possible back, that is, which has as high a scattering cross-section as possible. But here we are concerned with the scattering of fast neutrons, for which there is no great variation in the scattering cross-section. One can say that if I put, say, lead, or U238 around it the scattering cross-section is always of the order of 6×10^{-24}. If I put carbon there, for a light one, then I have about 4×10^{-24}. In compensation for that the atoms in carbon are rather more densely packed than in uranium. Do what you will, you always get M.F.P.s in the region of 4 cm. Thus I think there would be little to gain by looking for some particularly sophisticated element, but it probably always lies in the neighbourhood of 4 cm.

WEIZSACKER: The wording of the report was the uranium was mixed with graphite to slow the neutrons down.

HEISENBERG: I can think that they mixed it with graphite to increase the scattering cross-section, reducing D_i and so making the whole machine a little smaller. But you don't gain much that way because for fast neutrons carbon scatters just as well as 235 - (int.)

HARTECK: The volume goes up cubically, which is always bad.

HEISENBERG: You must always take one 235 out to plug in one carbon atom.

WEIZSACKER: Overall I get more material of a substance with the same M.F.P. for scattering.

HARTECK: But there is not only the M.F.P. for scattering to consider but also the M.F.P. for capture. If, let us say, I put in one half I have only half as much uranium per unit volume: so I must cover twice the distance before making a fission, and so I get four times the quantity of uranium. I gain a little bit by more scattering. It always gets worse.

WEIZSACKER: But in that case the neutrons are notably slower and have a considerably greater cross-section.

HEISENBERG: That has practically no effect. The $1/v$ law begins much lower. Here on the contrary the effective cross-section is independent of the velocity, simply the cross-section of the nucleus. But round the outside, as we said, one must put something, and if possible something with as small a M.F.P. for scattering as possible, and I have as a trial assumed that this external scattering M.F.P. was the same as the internal: carbon on the outside already does that approximately. Because of the uncertainty which enters here I have omitted allowance for this μ term for the outside since it is quite arbitrary whether I take 4 cm or 6 cm there.

GERLACH: Do you think they put carbon in to avoid melting? If you take uranium carbide powder and reduce it that is already a very high density.

HEISENBERG: It could be something like that, some such technical trick seems to me likely. Perhaps we can consider that a little later. So, if the external and internal diffusion constants are equal, $D_a/D_i = 1$ and the cotangent must be zero. It is zero when its argument is $\pi/2$, so the radius is $(\pi/2)l_1$. We can see that it would be less favourable to put no scattering material outside, for with empty space around the uranium sphere then D_a would be infinite and the cotangent has to be minus infinity, which is when its argument is π, so that the radius must be twice as large. Thus we can say quite roughly that if there is nothing around the uranium sphere the critical radius at which it goes off is twice as large as if we put carbon or lead or uranium 238 around it. So I have now assumed that the ratio of the scattering cross-sections outside and inside is 1 so that the sphere radius R is $(\pi/2)l_1$. Now I substitute from the equation for l_1, and get an equation for μ. It is

$$\mu = \nu - \frac{D\pi^2}{4R^2}, \text{ or } \mu = \nu\,(1 - R_K^2 R^{-2}),$$

where R_K is a critical length, $R_K = (\pi/2)l$, which is approximately equal to 9.7 cm^2/(fission cross-section)$^{1/2}$.† From this one sees that μ is in fact negative if R is very small. If I take a very small sphere then $R_K{}^2\,R^{-2}$ is greater than 1, and the neutron density decreases exponentially. If on the other hand R is greater than R_K then μ is positive, and the neutrons increase exponentially, that is the whole thing explodes. We see therefore that R_K is the critical size which the uranium sphere must have as a minimum if the affair is to explode.

The explosion then follows exponentially with this time factor μ. Now one sees that this critical radius is rather sensitive to the value of the fission cross-section: If I assume the smallest value, 0.5, I get a critical radius of 13.7 cm and if I assume the greatest, 2.5, I get 6.2 cm.

---

† The text was obviously corrupt here, reading "μ_K is approximately 9.7 cm/(fission cross-section)$^{-1/2}$". FCF

HARTECK: Is that with a back-scattering mantle?

HEISENBERG: With a back-scattering mantle for which the external M.F.P. is the same as for the interior. Now this 6 cm is really notably greater than the others say. The others state that the whole explosive mass weighed 4 kg, and this sphere of 6.2 cm weighs about 16 kg. Now there are two possibilities: one is that the fission cross-section is really still higher, and it is not quite unthinkable that it is practically as large as the scattering cross-section, so about 6×10^{-24}, and that doesn't quite get there. So I don't know what else can be altered in the data. They can also have a better back-scattering mantle, but that can't gain them much.

WEIZSACKER: You said before that the $1/v$ term begins much lower down, it begins as soon as the wave-length of the neutrons is larger than the geometrical cross-section.

HEISENBERG: Yes.

WEIZSACKER: Perhaps that is why they add graphite, to have slower neutrons and a larger fission cross-section.

HEISENBERG: But firstly carbon slows down very badly - (int.)

WIRTZ: But if you had some heavy hydrogen, say as many deuterium atoms in the form of heavy paraffin, as you have uranium atoms, pressed together.

HEISENBERG: In any case it would be still better light than heavy.

WEIZSACKER: At what energy for the neutrons is the wave-length equal to the nuclear cross-section?

HEISENBERG: At 100,000 volts it is 10^{-11} cm.

WEIZSACKER: For the nucleus you might take about 10^{-12} cm; that means that at a few million volts the nuclear cross-sections and the wave-lengths of the neutrons are comparable. If only they succeeded in bringing the neutron energy down to, say, half a million volts the effective cross-section would be a good deal greater.

HEISENBERG: I don't believe that at all, because we are not in the resonance region here. Naturally at the resonance level the cross-section can be simply the square of the wave-length.

WEIZSACKER: For thermal neutrons this cross-section is 50,000 or so.

HEISENBERG: We know that for thermals the fission cross-section is 400×10^{-24}; if you now go to 100,000 volts then according to the $1/v$ law, if it holds, the effective cross-section has already become 0.4. So one can say that according to the $1/v$ law something smaller comes out than is certainly there.

WEIZSACKER: Now comes the further question, how strongly are the neutrons slowed down, for is it indeed true that practically every collision is inelastic?

HEISENBERG: I would say that after three or four collisions it has perhaps already gone down to a tenth, perhaps even less.

WEIZSACKER: If one simply uses WEISSKOPF's formula one will always say that a neutron which enters the nucleus with a particular energy E comes out again with an energy of the order of magnitude perhaps $E/4$.

HEISENBERG: On the other hand, if you have a scattering cross-section of 6 then at least 50% of this scattering cross-section is pure elastic scattering, it goes by outside and is deflected.

HAHN: If one has a uranium mantle then there is some disintegration by fast neutrons, which does not give much, but still some multiplication.

HEISENBERG: So I would like to propose the following interpretation, that what they have in the middle is a mixture of 238 and 235: such as 50% 235 and 50% 238, since they are not likely to have produced really pure 235: from this mixture, which is surely not much worse than pure 235, they would have a sphere of radius say 8 cm, but in this sphere of radius 8 cm there is correspondingly less uranium 235 and they have stated that as the weight.

HAHN: LISE MEITNER says here something about heavy water.

HEISENBERG: She certainly doesn't know anything. Heavy water makes no sense at all, since only processes with fast neutrons are involved and for fast neutrons light water is just as good as heavy.

HAHN: If a fast neutron is shot into uranium metal it multiplies itself by 1.4 or 1.35 before it is slowed down.

HEISENBERG: That naturally depends on the size, and again is a kind of diffusion process, but as the most favourable, for an infinitely large sphere of good 238, it is increased by 40%. They could probably use a mixture in which there was only, say, 10% 235 and 90% 238. The 238 has anyway the advantage that if it does undergo fission many neutrons are produced, about 5. So, as we have just established it can have to do both with the fission cross-section and the use of such a mixture. Now one must enquire, if one has such a sphere, how does the explosion proceed? For this I have assumed a sphere of radius only 5% greater than the critical radius, since they must make sure that the halves of the sphere are certainly below the critical limit. Obviously they put the sphere together from two hemispheres.

KORSCHING: Couldn't they simply be cylinders?

HEISENBERG: They could be cylinders but I would say spheres really always give the simplest case since one finally wants the neutrons as far as possible in one space.

KORSCHING: Hemispheres are difficult to put together.

HEISENBERG: There is anyway an iron cylinder behind. A possible construction would be to have the hemisphere attached to an iron cylinder, able to move in a gun-barrel, to be fired in this gun against the other hemisphere held in the same way. If one assumes that the sphere after putting together has a radius 5% greater than the critical radius then the hemisphere, if I consider it as a sphere - actually it has a different shape - has a radius of about 85% of the critical radius. In other words for the hemispheres one must go rather close to the critical conditions, and they must have measured these very accurately so that nothing should happen to them. All the same they naturally have a safety factor in that the hemisphere is surrounded by air on the one side when it stands there empty, so that the back-scattering is there much poorer.

WEIZSACKER: I find this geometrical form unfavourable. Really one should cut out of this hemisphere a smaller shape with rounded edges, which is really effective, since the thing with sharp edges doesn't help at all.

HEISENBERG: Agreed. So a possible construction would be to have this hemisphere in an iron cylinder. Let us say at the bottom there is an iron cylinder with hemisphere upwards, then a gun-barrel, and at the top an iron cylinder with hemisphere downwards, the two flat faces of the hemispheres facing each other. Now above I have a small charge of explosive so that at a given time this upper cylinder is shot down on to the lower and the whole made into a complete sphere. I will now say why that must be done so fast. The process goes according to the equation $\rho = e^{\mu t}$ where μ is the reciprocal time which was calculated in the formula above. When one assumes that the radius is only 5% greater than the critical radius what comes out is $\mu = 0.5 \times (10^7 \text{ sec})^{-1}$. So in 10^{-7} sec. the intensity increases e-fold. That means the whole reaction is completed in about 10^{-5} seconds. Now the question is, in the first place, why must one put the hemispheres together so fast? Couldn't one bring them together slowly and say "All right, if I bring them together slowly, at some time it goes off, when I have them close enough together, and that is all right." That appears wrong to me for the following reason. If I put it together slowly the process begins when I am much less than this 5% but only, say 1/10 per thousand above the critical radius. Then the time factor with which it goes is much slower, thus not $(10^7 \text{ sec.})^{-1}$ but only 10^{-4} or so. Then the whole explosion needs not 10^{-7} but 10^{-4} seconds. In this time the affair would heat up so much as to come to 5,000° or 10,000° or 20,000°. As soon as I have these high temperatures the whole affair bursts apart. This bursting apart follows in times comparable with 10^{-4} seconds. That means that the reaction does not go to completion at all, but practically only till about 10,000° is reached and then it comes to an end. So I must choose a sufficiently short time that it does not fly apart before. I must estimate that, and what comes out is that the time should not be longer than 10^{-6} seconds, perhaps even $(1/2) \times 10^{-6}$, but that is about the limit. One can reach that if the two hemispheres are shot together, but not if they are merely brought together slowly. I can say that there is a critical separation, say 1 cm. When they are more than 1 cm apart nothing at all happens, it doesn't start. When I have them closer than a centimetre apart it goes off explosively. This last centimetre must be traversed in 10^{-6} seconds.

HARTECK: That won't do, it would be 10,000 metres per second.

HEISENBERG: Well 10^{5} (means 10^{-5} seconds?) is perhaps still possible. Then they won't use it up completely.

HARTECK: "As fast as possible" is not enough. It is important.

HEISENBERG: It is very important. There is still the possibility of arguing differently. One could say as follows: if one brings no neutron source into the neighbourhood and so to speak leaves it to the cosmic rays then I could say: "I bring them together relatively slowly then at first there will be no cosmic ray neutron there." But that is playing with chance. It could happen that there was a neutron there precisely when the critical radius was passed and then the efficiency of the bomb would be much worse.

WEIZSACKER: How often do cosmic ray neutrons occur in such a small region?

HEISENBERG: That is now the initiation question. For that I have estimated as follows: firstly, initiation could arise from spontaneous fission. One sees however that that won't suffice. The lifetime for spontaneous fission is about 10^{14} years if I remember rightly.

HARTECK: 10^{14} to 10^{16}.

HAHN: How does one know that 235 undergoes spontaneous fission?

HEISENBERG: Well, there is probably 238 there as well. The two are about comparable. That would be about 10^{22} seconds and if I want a neutron in 10^{-6} seconds I must have 10^{38} atoms: which of course I haven't.

WEIZSACKER: One could of course say conversely that if there is definitely no other initiation mechanism one puts it together slowly and it goes off.

HEISENBERG: Yes, I don't know. Perhaps they made some such trick. Then of course they could have bad luck: they must run the risk of the bomb giving them very low efficiency.

WEIZSACKER: If they then bring them together relatively fast the probability that it goes wrong is very small.

HEISENBERG: Only, if you shoot it together fast, you must take care that it doesn't bounce apart again.

HAHN: What would happen if they introduced 1/1000 mg radium?

HEISENBERG: That would certainly be enough.

GERLACH: If one half is 85% of the critical size, how long does a thermal neutron live in it?

HEISENBERG: You mean if one considers the thermal neutrons from the cosmic rays?

GERLACH: Yes, since they make new neutrons in the hemisphere.

HEISENBERG: That doesn't live long.

HARTECK: A thermal has a velocity of 10^5 cm. And if we have pure 235 the M.F.P. is only a few hundredths of a centimetre. So the order of magnitude will be 10^{-7} seconds.

HEISENBERG: Yes, agreed, that is much too small.

GERLACH: But again there is another neutron there.

HEISENBERG: That runs away, it is too fast.

GERLACH: A formula was given for such decay times.

WEIZSACKER: Here it is, worked out. Yes, in our big machines the decay times were much greater.

(Pause)

HEISENBERG: It can be that they do it so: they bring the two parts together as fast as possible and say: "We hope that we don't have the bad luck that precisely as the critical size is passed the first neutron is already there."

KORSCHING: If the decay times are so extremely sensitive one must be able, by screwing the things together quite slowly with a micrometre screw, to make it burn up as slowly as one likes.

HEISENBERG: Fluctuations then come in. Of course, in principle it would be possible.

I have also made an energetic consideration. Let us assume that 4 kg 235 is really "burnt up". That releases 3 x 10^{21} ergs. That is quite trivial: 3 x 10^{-4} ergs are released per fission, which I must multiply by certain factors. Now I can ask how high does the temperature become in this block of uranium? Here one can calculate simply that practically all the energy is converted into radiation, since the specific heat of the metal is independent of T, its energy goes with T.

LAUE: You are dealing with uranium gas.

HEISENBERG: Yes. On the other hand the radiation which simultaneously arises in this space has an energy content which according to BOLTZMANN's law goes with T^4. So at a high enough temperature the radiation predominates over everything else. So practically all the energy is converted into radiation and I can calculate the temperature by saying that it will increase exponentially, like the neutron density, according to an $e^{\mu t}$ law. It will rise to a maximum, namely when all the energy is used up. All the energy must by this time have flowed out through the surface, since the radiation goes with the velocity of light, so spends practically no time within the uranium. Now according to the STEFAN-BOLTZMANN law the energy of σT^4 per second per cm^2 passes through the surface, σ being the constant 5.8×10^{-5} erg cm^{-2} deg^{-4}. Thus the total energy must equal the time integral over the surface of the uranium of T^4, which with exponential increase is

$$\frac{F\sigma T_{max}}{4\mu}$$

and if I put in the data we assumed above one comes to a maximal temperature, T_{max}, of 3×10^7 degrees. That is somewhat higher than the newspapers give: I have read the figure 3.6×10^6 degrees in the newspapers. It seems as if all the energy is not converted. Agreement to the last power of 10 may not matter very much since we do not know the exact conditions of the bomb, and again it is a question whether the figure in the newspapers is right. If one really has the sphere together before the thing goes off it hardly expands at all in 10^{-7} seconds. I have calculated how much the sphere expands and one comes to an expansion of the order of a few millimetres at 10^{-7} seconds.

LAUE: Above all it is the radiation pressure which comes into question for the scattering mantle.

HEISENBERG: The radiation pressure at maximal temperature is about 10^9 atmospheres. In itself that is enormous, but it is not enough in this extremely short time to expand the affair noticeably.

HARTECK: So how then was the temperature calculated?

HEISENBERG: I have simply calculated thus: at the end of the reaction practically all of the energy released has gone away as radiation. That is of course not quite right, a fraction of the energy has been used to accelerate the particles, to move the molecules: but practically at these high temperatures radiation plays the main part. Most of the energy goes in radiation and only relatively little in particle motion. This radiation energy comes out with the velocity of light. Therefore I said that at the end of the reaction practically all the energy released has flowed out through the surface. The radiation at 3×10^7 degrees, being PLANCKian radiation, is X-rays, with a maximum at about 1 Ångstrom wave-length, more exactly 0.86 Ångstrom. Most of the radiation comes out as X-rays and is absorbed in the surrounding atmosphere. The first thing that happens will be that the first 10 metres of air are brought to white heat. The first thing one will see will be a glowing ball of about 20 metres diameter, glowing white because of the absorbed X-rays. This radiation sends out further radiation again and so the affair gradually spreads out.

WEIZSACKER: Whereby though then to be sure only a small part of the radiation is of that energy, as soon as the thing has come to reasonable temperatures the greater part of the pressure is gas pressure, and the rest goes on as a pressure wave.

HEISENBERG: At the beginning the affair has wonderfully spherical symmetry, and then gradually convection phenomena enter to complicate it.

WEIZSACKER: I would say there are three phases: a radiation phase, then a phase in which a pressure wave is first generated, then it expands with this pressure wave at the same time, and then convection will set in and carry the whole thing upwards.

LAUE: How much energy goes out at first as visible radiation?

HEISENBERG: I compare this inner uranium sphere with a body, say, at 1000°. Then it would radiate 10^4 times more in the visible than this glowing body. The surface of this uranium sphere thus radiates about 2000 times more than the surface of the sun.

WEIZSACKER: That means that if one sees it from a distance equal to that of the sun it radiates 1000 times more than the sun.

HEISENBERG: It would be interesting whether objects could be knocked down by the pressure of the visible radiation. That is quite thinkable.

(Pause)

Discussing the lecture

HEISENBERG: Every neutron in our machine has a chance of 1/20 to be captured by resonance, so for every fission practically 1/20 of 94 is made.

HAHN: I haven't understood that.

HEISENBERG: In our machine it was always so arranged that each neutron would have a chance of 1/20 of being captured by resonance before it became thermal.

HAHN: Oh, I see.

IV. VISIT OF SIR CHARLES DARWIN

1. Sir Charles DARWIN who was accompanied by Lt. Cdr. WELSH arrived at FARM HALL in time for tea on 18 August. After meeting the guests at tea he had a conversation alone with HAHN and they were later joined by HEISENBERG. Both expressed appreciation of the treatment they were receiving here and stated that their chief worry was the absence of news of their families. HEISENBERG gave Sir Charles some details of the work carried out at his Institute. The following are extracts from the conversation, which took place in English:

HEISENBERG: We have tried to make a machine which can be made out of ordinary uranium.

DARWIN: With a little bit of enrichment?

HEISENBERG: No, not at all.

DARWIN: With heavy water?

HEISENBERG: Yes. That worked out very nicely and so we were interested in it.

(Pause)

After our last experiments, if we had had 500 litres more heavy water, I don't doubt that we had got the machine going.

DARWIN: What heat could you have got?

HEISENBERG: Such an apparatus stabilizes itself at a certain temperature. One can simply fix the temperature by taking a certain amount of heavy water to it. If you have got enough uranium, more heavy water will give a rise in temperature.

As soon as we had the machine going, we could have made almost any intensity of radio-active isotopes. Because, just by taking enough energy out, you can raise the intensity as high as you want.

DARWIN: Had you got enough heavy water to do this?

HEISENBERG: We had actually in GERMANY two tons of heavy water. That would be just sufficient to do it. Actually we had only one-and-a-half tons, and the remaining 500 litres of heavy water had some impurities of light water in it. And so we could not use them and intended to make them pure again. By then the war was over.

DARWIN: Did you contemplate anything about developing an engine, a real engine?

HEISENBERG: There had been plans to use this heat and to make an engine.

DARWIN: Had you any ideas about that?

HEISENBERG: One plan was to enrich the isotope 235 a little bit, so that that thing would run with ordinary water, then you can simply pump the ordinary water through.

DARWIN: You were thinking of getting up (to?) about 400 degrees, that sort of thing?

HEISENBERG: Yes. But the main trouble is the corrosion of the uranium, the chemical change. We tried different schemes to perfect it.

DARWIN: You <u>tried</u> some, or were you only talking about it?

HEISENBERG: We have done some work on it, but it was not very successful.

HAHN: The prevention of corrosion was not finished yet. HARTECK knows about that and about the heavy water.

HEISENBERG: The other scheme would be to work with heavy water and ordinary uranium and then, of course, you can only go to low temperatures. Then that water should be pumped through and should give the heat off to ordinary water, but we have not got very far. We in our institute were only interested to make such a machine which would give us high neutron intensities with which we intended to play.

DARWIN: Have you thought much about '94'?

HAHN: You can produce it only by having a machine like this. There must be a fairly stable '93', too.

There is produced a '93' from the '237' by fast neutrons, which decays in seven days into this inactive '93'. You are probably more interested in the thing of the <u>slow</u> neutron. With ordinary uranium with slow neutrons you get an active '93' and an inactive '94'. But there is produced by fast neutrons a '93' by means of the decomposition of uranium of the atomic weight 237, not 239. If a neutron hits a (239?), two neutrons are sent off.

DARWIN: There is a '239' which gives off two neutrons and forces(?) the two......

HAHN: You have '238'. When this '238' is hit by neutrons, two neutrons are sent away. Therefore you get uranium 237. And this uranium 237 decomposes in seven days to - (int.)

DARWIN: Is that on a very narrow absorption-band? Or something of that kind, so that you cannot get much of it?

HAHN: We don't know anything about it, because it is not active (enough).

HEISENBERG: I should not expect any narrow band. I should say, that any high-speed neutron would do it, but it must be a very high-speed neutron, because it must also have the energy to knock another neutron out.

DARWIN: I see your point. It is not a resonance-level?

HAHN: No. In my laboratory, where I had no strong substances at all, I could make this '237', decaying in seven days into an inactive element.

DARWIN: That was 'beta-active'?

HAHN: '237' gives (off) 'beta-rays'.

DARWIN: Having done that, it is stable?

HAHN: Fairly stable. It may be even 10,000 years. Nobody knows – <u>we</u> don't know it, anyhow.

DARWIN: You were doing it with radium and/or beryllium?

HAHN: Yes. In the HEISENBERG laboratory in BERLIN, they had a high-tension plant, and there we got some stronger preparations.

DARWIN: What sort of high-tension was that?

HEISENBERG: We had about 10^6 volt.

HAHN: In BERLIN, we could have these preparations. In TAILFINGEN, we had not got them anymore, we tried to get them from the REICHPOST, they had something too. And we hoped to get something from the cyclotrone of BOTHE.

2. After dinner there was a general discussion. Sir Charles raised the question what they, the scientists, were going to do about things in the light of the atomic bomb. GERLACH replied that he did not think, there would 'free science' in the world from now on. GERLACH, HAHN, HEISENBERG and HARTECK strongly expressed the opinion that RUSSIA would never play the game. While scientists in other countries would publish their work, they did not think that the Russian Government would allow Russian scientific work to be made public. Von WEIZSACKER put forward the opinion that either every physicist in every country refuses to hand over the secret of atomic force to any government – which all agreed was impossible - or scientists must lead the governments themselves. HARTECK wondered whether RUSSIA would agree to international control of the atomic bomb, whereupon HAHN and GERLACH suggested that it would be a good thing perhaps, if every one of the big countries had 500 such bombs. Then everybody would be afraid of everybody else. HEISENBERG thought that, if there are two powers of equal strength, there is bound to be a war, because there are always problems when each power will think that the right is on their side and there would be no other way of settling it except by war. HAHN again emphasized that they were all very much afraid of RUSSIA and that they felt profound distrust of STALIN. If conditions in GERMANY should get very bad, even British and American controlled GERMANY would be driven into the arms of STALIN. They themselves would rather see GERMANY lean towards the West and they hoped that the Western Allies would help GERMANY to such an extent with food etc., that such bad conditions would not arise. HEISENBERG in the end stated that at present all scientists were too dependent on their governments and he thought, scientists must try and get some political influence.

3. Later in the evening the following conversation took place in the presence of Lt. Cdr. WELSH who had shown them a copy of the White

Paper dealing with the atomic bomb:

WEIZSACKER: They have decided to do it by gaseous diffusion.

HARTECK: HERTZ tried that and even he didn't scceed.

I persuaded ALBERS who was at CELLE to make a compound which would have a vapour tension the same as mercury, at room temperature - such a substance is suitable for (testing) differences in speed of vaporisation. If you make ten thousand kilogrammes of such a substance the whole thing becomes automatic; it is vaporised and condenses. You can make 10,000 one after another and then it must work.

GERLACH: We had discussed a sort of continuous belt, it would have to be done in a high vacuum. That is the difficulty. These differences in the speeds of vaporisation -

HARTECK: You can work out how it would go and you will see that you get quite fantastic results (?). It must work with the speed of vaporisation. We know that from HEVESY's experiments that is simply the kinetic gas theory.

GERLACH: But HEVESY only got fractions of thousandths.

HARTECK: Those were fraction of thousands in thickness. If the thing was to be any use, one would have to build thousands of them one behind the other.

GERLACH: I have never worked it out, but I know that the difficulty is that you can only get such very small quantities. I would do it, as I told you recently, by this variation in the MAXWELLIAN law. I think one ought to try the radiometer method. You have a radiometer which is bombarded from one side which produces a current round the radiometer from the warm to the cold side, and you can put them one behind the other so that you would have a whole lot of warm-cold surfaces warm-cold, warm-cold.

HARTECK: If I were allowed to co-operate on this problem with large means at my disposal I would go out for photo-chemistry. I have no doubt that it could be done.

4. HEISENBERG and KORSCHING discussed their position in the light of Sir Charles DARWIN's visit in the following conversation:

HEISENBERG: I don't see how war between RUSSIA and AMERICA can be prevented as RUSSIA appears determined to cut herself off from the rest of the world because they have a different ideology. The Anglo-Americans want the balance of power in EURASIA. The only balance of power they can achieve now is the whole of EUROPE against RUSSIA. The only choice for us is either to join this Western European bloc or join in with RUSSIA. My own feelings are that the Western European bloc is better, but I can understand someone saying that we ought to join up with RUSSIA. This is a standpoint which could be discussed.

KORSCHING: What depresses me is the thought that all the work we may do in GERMANY will, so to speak, fall into the laps of other people.

HEISENBERG: That attitude would not be correct if there were to be a united EUROPE, as we would then be part of it. I think that if a United States of EUROPE were to be formed it would be in our interests to fight for it as all our old traditions would remain in such a united EUROPE whereas if we were to start now as part of the Russian Empire, everything that had ever been in GERMANY would disappear. I admit that something would then start which augers well for centuries to come, but after all we have been brought up in the other tradition.

KORSCHING: It annoys me that GERMANY is not clever enough to work for ENGLAND without losing her soul so to speak. It is not so much a question of "Shall we let the Russians on to the RHINE", as no one can honestly want the Russians on the RHINE, but "Will this pan-EUROPA eventually be predominantly German or English", that is the question.

HEISENBERG: There is no doubt about it, it will be predominantly English. We can't do anything about that. When pan-EUROPA has existed for 300 years, then I should say GERMANY will play an important part, but the spirit of the whole thing will be English.

5. The following morning HEISENBERG and BAGGE had the following conversation:

HEISENBERG: My general impression is that apparantly a large proportion of the people who are handling this business are of the opinion that we should go back to GERMANY and even continue to work on this uranium business as that has really nothing directly to do with the bomb. I am not really very happy about the result of this visit. They don't want to tell us anything definite as long as they have not come to any definite decision, and they can't come to any definite decision, because it is all so completely new, and they don't know what to do. Apparantly, the tendency among these people like DARWIN is to consider very carefully a long term plan, to consider the state of EUROPE in four years time and how gradually to return to peace conditions. They are not so much concerned with the next three or four years the main thing being gradually to arrive at a stabilised EUROPE in a few years time. Of course that is right, that is the important thing. For that reason it doesn't matter to them if we are detained for three months or six months. They say to themselves "A few weeks one way or the other won't matter as it is a question of the whole problem".

They intend to let us play some part in it in the long run. They seem to be thinking along the lines of a United States of EUROPE and say to themselves "The German scientists are among the most important people as they have a certain amount of influence in GERMANY and because, in addition, they are sensible people with whom one can discuss things". But they are not in any hurry.

BAGGE: There can't be any hurry about it until they have made up their minds whether they want to bring the Russians in on the atomic bomb or not.

HEISENBERG: True.

BAGGE: Of course at the moment I am only interested in my own fate.

HEISENBERG: Yes, of course. That was the great difficulty in the discussion yesterday, the fact that our first interest is our future, whereas DARWIN and the others are interested firstly in the fate of the world. That is understandable as for them there is no difficulty about the next three or four years, but their big problem is what will happen after that, as they are naturally frightened that one fine day they will be attacked just as we were attacked (sic). Our most important problem is the question of our families.

FARM HALL,
GODMANCHESTER

(Sgd) T.H. Rittner
Major

23 August, 1945.

Copy No. 1

F.H.5

APPENDIX

GERMAN TEXT OF LECTURE
GIVEN BY PROFESSOR HEISENBERG
on AUGUST 14th, 1945

HEISENBERG: Ich moechte die U-235 Bombe nach den Methoden behandeln und besprechen die wir bisher immer bei unserer Uran-Maschine angewendet haben. Da stellt sich dann in der Tat heraus, dass man alle Einzelheiten dieser Bombe wirklich sehr gut verstehen kann. Ich will damit anfangen, noch einmal die wichtigsten Daten vom U-235 zu rekapitulieren. Ich will vielleicht ganz kurz noch mal wiederholen, was passiert in dieser Bombe. Wenn im Uran 235 ein Meutron vorhanden ist, dann laeuft dieses Neutron ein Stueck weit und trifft dann auf einen 235-Kern. Dann kann es zwei Sachen machen. Entweder es wird elastisch gestreut, oder es spaltet. Wenn es gestreut wird, dann kann die Streuung elastisch oder unelastisch sein; dann fliegt das Neutron also mit derselben oder einer niedrigeren Geschwindigkeit weg. Das tut aber der Faehigkeit dieses Neutrons weiterzuspalten keinen Eintrag. Es ist also keine Wahrscheinlichkeit, dass Neutronen irgendwie verloren gehen. Der Vorgang geht dann natuerlich so: es trifft also nach einiger Zeit auf ein 235 und spaltet es. Bei der Spaltung kommen wieder einige Neutronen 'raus. Diese Neutronen machen dasselbe, und so laeuft die Kettenreaktion. Wenn man beliebig viel Uran 235 haette, dann wuerde die Kettenreaktion auf jeden Fall beliebig lange weiterlaufen, denn es wuerden immer aus einem Neutron 2 bis 3 bei der Spaltung entstehen. Diese 2 bis 3 wuerden wieder den Prozess machen, und so wuerde es weitergehen. Also wuerde die Gesamtmenge von den Neutronen exponentiel anwachsen. Mit diesem Prozess der Vermehrung der Neutronen steht aber in Konkurrenz der Prozess, wo die Neutronen nach aussen aus dem Stueck weglaufen. Wenn man naemlich ein endliches Uran-Stueck hat, dann werden die Neutronen, die an der Oberflaeche sind und nach aussen gerichtet sind, eben auslaufen ohne, dass sie weiter an der Spaltung teilnehmem koennen. Also es fragt sich, ob der Neutronenverlust dadurch, dass die Neutronen aussen aus dem Stuech 'rauslaufen, groesser oder Kleiner ist als der Neutronengewinn, der durch die Vermehrung im Inneren kommt. Wenn man das nun rechnen oder sich ueberlegen will, dann muss man zunaechst mal die Wirkungsquerschnitte haben und die freien Weglaengen haben. Da habe ich zunaechst einmal eine Faustformel hingeschrieben. Die heisst so: die freie Weglaenge fuer irgendeinen Prozess im Uran 235 ist gleich 22 cm/Wirkungsquerschnitt, und den Wirkungsquerschnitt diesmal gemessen in 10^{-24}. Das heisst also, wenn der Wirkungsquerschnitt, z.B. fuer Spaltung 1×10^{-24} ist, dann ist die freie Weglaenge fuer Spaltung 22 cm. Der Wirkungsquerschmitt fuer Streuung ist empirisch etwa 6×10^{-24}, daher ist die freie Weglaenge 3,7 cm. Nun ist das

Schwierigste dabei der Wirkungsquerschnitt fuer Spaltung. Der Wirkungsquerschnitt fuer Spaltung haengt erstens natuerlich sehr stark ab von der Geschwindigkeit. Bei thermischen Neutronen ist er, soviel wir wissen, in der Gegend, von etwa 300 bis 400 x 10^{-24}, also sehr gross. Bei schnellem Neutronen ist mir die Zahl nicht mehr genau in Erinnerung. Eine Zahl, die in den DOERRIG'schen (?) Tabellen steht, die aber Vielleicht fuer 238 und nicht fuer 235 gilt, die ist 0,5 x 10^{-24}. Ich glaube auch, dass man nur diese Zahl fuer 238 wirklich gut wissen kann, weil ja bei schnellen Neutronen das 238 und das 235 gleichzeitig spaltet, und man also, wenn man Spaltungen misst, eben in den meisten Faellen die vom 238 misst, und ich nicht so recht sehe, wie man die beiden Spaltungen ueberhaupt unterscheiden kann.

HAHN: Weil das nur 0,7% davon waeren?

VON WEIZSACKER: Das wird man also nie messen koennen. Ich wuesste gar nicht, wie.

HEISENBERG: Man koennte vielleicht folgendes machen: man muesste Neutronen nehmen, die eine etwas kleinere Geschwindigkeit haben als die kritische Grenze beim 238, aber nicht thermisch; also sagen wir, Neutronen von 10^5 Volt. Mit denen musste man dann den Wirkungsquerschnitt messen und dann mit dem Verhaeltnis umrechnen.

VON WEIZSACKER: Ja, aber da weiss man auch nicht genau, wie dann der Wirkungsquerschnitt weiter von der Energie abhaengt.

HEISENBERG: Man koennte vielleicht sagen, so ungefaehr wird der Wirkungsquerschnitt in diesen grossen Energien doch vergleichbar sein mit der Kernoberflaeche. Von der Energie wird es wohl nicht so toll abhaengen. Die BOHR'sche Theorie gibt sogar an, dass er eigentlich im ganzen Energiegebiet konstant sein soll.

HARTECK: Wenn die so einen guten Spektrographen haben in U.S.A., dass sie es restlos trennen koennen, dann braucht man nichts anderes zu machen als so eine duenne Schicht zu haben und dann estsprechend zu multiplizieren. Denn wenn ein Kern aufbricht, gibt es ja viel mehr als ein alpha-Teilchen, die findet man immer mit dem Zaehlen, und dann gibt man das Verhaeltnis von thermischen zu schnellen. Da wuerde man es experimentell loesen. Die haben es sicherlich daher genau bestimmt.

HEISENBERG: Ja. Ich habe jetzt mal abgeschaetzt 0,5 bis 2,5 fuer den Wirkungsquerschnitt, indem ich mir gesagt habe, 0,5 ist der beim 238, das wird ja die untere Grenze sein, denn 235 wird leichter spalten. Und 2,5 das ist der Querschnitt des Kerns, der richtige Stossquerschnitt, und viel hoeher kann es auch nicht sein. Nun koennte es natuerlich ein bisschen groesser sein als πr^2 aber so in den Grenzen wird es liegen. Wenn man also die Grenzen zwischen 0,5 und 2,5 annimmt, dann ist die zugehoerige freie Weglaenge fuer Spaltung 9 cm bis 44 cm. Bei der Streunng dagegen weiss man, dass der Streuquerschnitt bei all diesen schweren Elementen ungefaehr 6 ist bei hoher Energie. Beim Uran kennen wir ihn ziemlich genau, da ist er, glaube ich, 6,2 x 10^{-24} beim 238. Beim Blei ist er ein bisschen niedriger, aber die sind alle ungefaehr in der Gegend. Also kann man sagen, fuer Streuung ist der Wirkungsquerschnitt ziemlich sicher in der Gegend von 6 x 10^{-24} und daher die freie Weglaenge etwa 3,7 cm. Dann ist eine wichtige Groesse, die man noch braucht, der Vermehrungsfaktor, also die Anzahl der Neutronen, die bei einem Stoss, bei einer Spaltung von 235 herauskommen. Da kennen wir aus unseren BERLINer Arbeiten sehr genau den Vermehrungsfaktor bei thermischen Neutronen, der aber nicht pro Spaltung sondern pro thermische Absorption gerechnet wird. Nun ist im thermischen der Spaltungsquerschnitt etwa drei und der Absorptionsquerschnitt etwa 6,2 d.h. wir kriegen eigentlich nur die Haelfte von dem wahren Vermehreungsfaktor, wenn wir unsere Zahl nehmen. Unsere Zahl ist 1,18, also kann man sagen, der Vermehrungsfaktor ist in Wirklichkeit 1,18 x 2, wird also, grob gesprochen, zwischen 2 und 2,5 liegen.

VON LAUE: Ich habe das nicht verstanden.

HEISENBERG: Wir haben immer experimentiert mit gewoehnlichem Uran, wo also nur etwas weniger als 1% 235 drin ist und sonst 238. In diesem gewoehnlichen Uran haben wir bei thermischen Neutronen folgende Verhaeltnisse: man hat einen Wirkungsquerschnitt von 6 x 10^{-24} dafuer, dass ein thermisches Neutron verschwindet, und zwar entweder einfach wegabsorbiert wird und damit den 239 Koerper macht, der in 23 Minuten zerfaellt, oder dass es spaltet, und zwar spaltet dann zwar das 235 - (int.)

HAHN: Ist das nur fuer thermische jetzt gedacht?

HEISENBERG: Ich spreche im Moment nur ueber thermische. In thermischen ist natuerlich her Wirkungsquerschnitt fuer Spaltung fuer das 235 furchtbar hoch, der ist in Wirklichkeit in der Gegend von ein paar Hundert, aber weil es nur mit 1% Haeufigkeit vorkommt, spielt das dann im Ganzen doch nur eine geringe Rolle; das heisst, der Wirkungsquerschnitt umgerechnet auf 238 ist dann auch nur etwa 3. Diese Zahl 3 ist, glaube ich, von DROSTE (phon.) mal bei Ihnen bestimmt worden.

HAHN: Ja.

HEISENBERG: DROSTE (phon.) hat, glaube ich, eine Messung von 3,1 oder so etwas 'rausgebracht. Wenn swei Neutronen absorbiert werden, hat man nur in einem Falle eine Spaltung, denn der gesamte Absorptionsquerschnitt ist 6, Spaltungsquerschnitt ist 3. D.h, wenn der Vermehrungsfaktor bei der Spaltung x ist, dann ist die effektive Vermehrung bloss $\frac{1}{2}x$ weil eben jedes sweite Neutron einfach vorher stirbt und eingefangen wird.

HARTECK: Wenn man die Resonanz...... hat, da hat man doch am Schluss $1/v$

fuer die geringen, und wenn das...... ist, gilt auch $1/v$. Wuerden Sie glauben, dass bei dieser BREIT (phon.) - WIEGNER Formel das ganz genau $1/v$ gibt fuer die niederen Geschwindigkeiten, oder koennte es, wenn man zu tiereren Temperaturen geht, ein kleiner Unterschied sein, wenn man die BREIT (phon.)- WIEGNER Formel extrapoliert und die direkte Fission(?)? Verstehen Sie, dass es sich da um 10% verschiebt relativ zueinander, oder ist da unten schon ganz haarscharf $1/v$?

HEISENBERG: Ich wuerde denken, dass es doch sehr genau $1/v$ ist.

VON WEIZSACKER: Diese Formel bedeutet doch gar nichts anderes als dass die Wahrscheinlichkeit des Prozesses pro Zeiteinheit konstant ist.

HARTECK: Das ist ganz klar, aber es ist ja noch die gewisse Einwirkung von der Resonanzstelle.

HEISENBERG: Ja, aber die ist so weit weg.

VON WEIZSACKER: Jedenfalls wuerde ich sagen, zu je kleinerem v man geht, desto besser - (int).

HEISENBERG: Also das spielt ja bei ber Bombe gar keine Rolle, die ganzen Resonanzgeschichten. Ich wollte ja diese ganze kleine Abschweifung in's Gebiet der thermischen Neutronen nur machen, um zu sagen, dass wir auf Grund unserer Experimente schliessen wuerden, dass der Vermehrungsfaktor bei der Spaltung des 235 etwa 2 bis 2,5 ist. Also, genauer wuerden wir sagen, 2 x 1,18. D.h., wenn ein 235 spaltet, dass dann 2 bis 2,5 Neutronen dabei 'rauskommen. Es scheint mir, dass die Amerikaner oder die Englaender der Ansicht sind, dass die Zahl ein bisschen hoeher ist, denn in der Zeitung war so ein Bild von dem Uran-Prozess, und da waren bei einer Spaltung immer drei Neutronen gezeichnet, die 'rauskamen. Also ich halte auch fuer durchaus denkbar, dass in Wirklichkeit die Zahl doch etwa 3 ist, und dass eben der Spaltungsquerschnitt, den DROSTE (phon.) seinerzeit bestimmt hat, doch noch ein bisschen hoch ist, dass der in Wirklichkeit ein bisschen niedriger ist. Die Amerikaner haben tatsaechlich frueher mal fuer den Spaltungsquerschnitt 2 bestimmt. und dann kaeme denau die Zahl 3 'raus.

DIEBNER: Die Werte wurden ja auch immer hoeher, die berechneten Werte. Bei unseren letzten Versuchen waren sie doch etwas hoeher geworden.

HEISENBERG: Ja, da handelt es sich aber nur um 1%, das schafft nicht viel. Ich habe also nachher gerechnet mit einem Vermehrungsfaktor, ich glaube, 2, 5. Damit haben wir also die wichtigsten Daten beieinander. Jetzt muessen wir fragen; Was passiert im Uran, wenn ein schnelles Neutron da auf die Reise geschickt wird? Dazu kann man nun folgendes sagen: Wenn wir ein beliebig grosses Stueck Uran 235 haetten und darin Neutronen, dann ist es ja klar, dass die Anzahl der Neutronen exponentiell anwaechst, weil ja die Kettenreaktion einfach beliebig weitergeht. Jetzt muss man sich noch ueberlegen, wie es exponentiell zuwaechst. Ich habe also da die Formel hingeschrieben, n ist proportional $e^{\nu t}$, wo eben ν eine charakteristische reziproke Zeit ist fuer dieses exponentielle Anwachsen. Diese charakteristische Zeit, die man nachher als Reaktionszeit braucht, kann man einfach so abschaetzen, dass man sagt, die Zeit, bis ein Neutron eine Spaltung macht, ist offenbar gegeben durch das Verhaeltnis der freien Weglaenge fuer Spaltung geteilt durch die Geschwindigkeit. Also ein Neutron laeuft etwa 9 cm bis zu der naechsten Spaltung. Nun muss ich einfach mir ausrechnen, wie lange laeuft es, bis es eben die Strecke 9 cm zurueckgelegt hat. Diese Zeit ist $\lambda_{Spaltung}$/Geschwindigkeit. In dieser Zeit ist die Vermehrung des Neutrons gerade dieser Faktor *x;* d.h., statt eines Neutrons habe ich nachher *x* Neutronen. Also bekomme ich fuer die Bestimmung dieser Groesse ν die Gleichung $e^{\frac{\lambda_{Spaltung}}{v}} = x$, Oder in anderen Worten, die charakteristische reziproke Zeit $\nu = \frac{v}{\lambda_{Spaltung}} \log x$. Die Geschwindigkeit kann man ungefaehr gleich 1/20 Lichtgeschwindigkeit annehmen, denn es sind Neutrone von etwa 1000000 Volt, so ganz voll gesprochen. Also jetzt wissen wir, wie eine Menge schneller Neutronen in einem beliebig grossen Uran-Block sich vermehrt. Aber wenn die Verteilung der Neutronen ungleichfoermig ist, und ich frage nicht, wie sich die gesamte Anzahl von Neutronen aendert, sondern wie sich die Anzahl von Neutronen in einem bestimmten kleinen Volumen aendert, dann koennen natuerlich aus diesem kleinen Volumen die Neutronen auch durch Diffusion wegkommen. Also muss ich so eine Diffusionsgleichung aufschreiben fuer die Neutronen und muss dazu die Diffusionskonstante der Neutronen wissen. Die Diffusionskonstante ist nach der kinetischen Gastheorie immer gleich der freien Weglaenge fuer Streuung mal der Geschwindigkeit mal dem Faktor 1/3. Also die freie Weglaenge fuer Streuung ist bekannt, die ist 3,7 cm. Die Geschwindigkeit ist auch bekannt, das ist 1/20 Lichtgeschwindigkeit. Daraus kriege ich die Diffusionskonstante. Schliesslich habe ich also fuer die zeitliche Veraenderung der Neutronendichte an irgendeinem Punkt die Gleichung ρ (das ist also die zeitliche Aenderung der Dichte) = Diffusionskonstante mal $\Delta\rho$ (Δ ist der LAPLACE Operator) + $\nu\rho$. Also man sieht ja, wenn das $\nu\rho$ nicht dastuende, also die Neutronenvermehrung, dann waere das einfach die Waermeleitungsgleichung, die man ja kennt. Dieses Glied $\nu\rho$ bedeutet eben, dass die Neutronen sich ausserdem hoch vermehren. Man sieht auch dimensionsmaessig - das ist noch wichtig - das Verhaeltnis d/ν ist offenbar das Quadrat einer Laenge, denn wenn ich mit ν durchdividiere, dan habe ich eben als Faktor, vom LAPLACE Operator angewandt auf ρ, d/ν, und das muss ja dieselbe Dimension haben wie ρ. Dieses d/ν nennt man seit den FERMI'schen Arbeiten das Quadrat der Diffusionslaenge 1^2. Dieses 1 kommt also jetzt hier numerisch 'raus zu 6,2 cm. geteilt durch Wurzel aus dem Spaltungsquerschnitt, wobei der letztere wieder in 10^{-24} gerechnet ist. Wenn der Spaltungsquerschnitt also gerade 1 x 10^{-24} ist, dann ist diese Diffusionslaenge 6,2 cm. Vielleicht kann man an dieser Stelle jetzt einen Moment den Vergleich machen mit unseren Maschinen. Bei unserer Maschine war die Diffusionslaenge immer viel groesser. Ich habe die Zahlen jetzt nicht mehr genau im Kopf, ich glaube aber, mich erinnern zu koennen, dass die Diffusionslaenge bei unserer besten Maschine, die wir zuletzt gehabt haben, etwa 35 cm war. Deswegen werden ja auch die Maschinen immer so furchtbar gross. Das liegt an folgenden zwei Punkten: erstens ist der Vermehrungsfaktor viel kleiner. Also hier ist er etwa 2,5, dort ist er nur 1,18, weil das 238 eben so viel wegnimmt. Zweitens - und dieser Einfluss ist noch wichtiger - laeuft eben ein Neutron sehr weit von der Stelle, wo es ausgesender wird, bis es thermisch wird und wieder absorbiert wird. Also in den Maschinen, mit denen wir zu tun haben, wird es durch D_2O abgebremst und dann als thermisches Neutron eingefangen. Es laeuft eben immer schon ungefaehr 20 cm weit, bis es schliesslich eingefangen wird. Diese Laenge addiert sich natuerlich zu dem 1, sie wird in das 1 verarbeitet. Also hier ist die Diffusionslaenge eben nur etwa 6,2 cm, waehrend sie dort in der Gegend von 35 cm war. Jetzt haben wir also die Diffusionsgleichung, nach der die Neutronen sich ausbreiten. Jetzt will ich diese Gleichung loesen fuer eine Uran-Kugel von einem vorgegebenen Radius R, und ich will annehmen, dass diese Uran-Kugel R wieder umgeben ist von irgendeiner anderen Substanz. Was fuer eine Substanz, koennen wir uns nachher ueberlegen. Wir koennen natuerlich auch Vakuum annehmen, aber jedenfalls nehmen wir eine Kugel an, dann kann man die Gleichung bequem loesen. Dann wird es sicher Loesungen geben, bei denen die Neutronendichte auch im Ganzen exponentiell zunimmt oder auch exponentiell abnimmt. Und zwar wird sich doch offenbar folgendes herausstellen: Wenn die Uran-Kugel klein ist, laufen immer mehr Neutronen nach aussen ab als innen produziert werden. Dann nimmt die Neutronendichte exponentiell ab. Wenn ich dagegen die Kugel hinreichend gross mache, dann kommt der Neutronenabfluss nach aussen nicht auf gegen die Neutronenvermehrung im Inneren, und dann nimmt die Neutronendichte exponentiell zu. Ich setze also willkuerlich an, dass die Neutronendichte gleich $\rho = \rho_0 e^{\mu t}$ sein soll, und suche zu bestimmen diesen charakteristischen Faktor μ im Exponenten. Wenn das μ positiv ist, dann geht die Maschine, und die Bombe explodiert. Wenn μ negativ ist, dann heisst es, dass irgendwelche Neutronen, die in dem Stueck sind, sofort herauslaufen und weiter nicht mehr interessant sind. Also wenn man das einsetzt, $\rho = \rho_0 e^{\mu t}$ dann kommt die Gleichung 'raus: diffusionskonstante mall $\Delta\rho + \rho(\nu - \mu) = 0$. Diese Gleichung muss man loesen. Dazu fuehrt man wieder zweckmaessig eine charakteristische Laenge, naemlich $d/(\nu - \mu) = l_1^2$. Die Loesung dieser Gleichung kennt man. Das ist einfach die Gleichung von dem bekannten Typus $\Delta\phi + K^2\phi = 0$. Die Kugelsymmetrische Loesung lautet sin x/x (x ist also in dem Fall die Variable) oder $1/1_1$. Die Loesung im Liesung im Inneren dieser Kugel heisst also $1/r$ x sin $r/1_1$. Also die Neutronendichte muss sich im Inneren verteilen nach dieser Formel $1/r$ x sin $r/1_1$. Und zwar muss das dann der Fall sein, wenn ich nicht annehme, dass innen irgendeine Neutronenquelle ist. Wenn ich innen eine Neutronenquelle haette, dann koennte es natuerlich sein, dass ich eine Liesung brauchen kann, die an der Stelle $r = 0$ z.B. singulaer ist, weil dort eine Quelle liegt. Aber ich will annehmen, dass gar keine Quelle da ist, so dass die Neutronen sich alle irgendwie gleichmaessig ueber die Kugel verteilen, und das geht nur nach der Formel $1/r \; x \sin r/1_1$.

VON LAUE: Ist das eine vom Uran unabhaengige Quelle?

HEISENBERG: Ja. Also im Inneren der Kugel haben wir sicher diese Abhaengigkeit $1/r \; x \sin r/1_1$. Wie ist es aussen? Da kann man folgendes sagen: Wir haben es mit schnellen Neutronen zu tun, die werden in der aeusseren umgebenden Substanz irgendwie gestreut und dabei verlangsamt. Das Verlangsamen aber schadet gar nichts, denn auch ein langsames Neutron kann spalten. Also kann ich so sagen: die Neutronenanzahl aendert sich aussen ueberhaupt nicht. Also aussen habe ich keinerlei Absorption, weder positive noch negative, und ich habe aussen reine Diffusion von schnellen Neutronen. Ich kann also sagen, aussen habe ich ein Medium, in dem die Neutronen nur diffundieren, aber nicht absorbiert oder vermehrt werden. In diesem Medium aussen gilt also auch so eine Diffusionsgleichung, wie sie da oben steht, bloss mit der Vereinfachung, dass $\nu = 0$ ist also keinerlei Vermehrung stattfindet. Die Loesung dieser einfachsten Diffusionsgleichung heisst natuerlich, ρ nimmt ab wie $1/r$. Also im Aussenraum muss ρ, die Dichte, abnehmen wie $1/r$, weil es sich um gewoehnliche Diffusion handelt ohne zugehoerige Absorption. Ich habe es mir da etwas erleichtert, indem ich das μ weggelassen habe, das μ ist doch gering gegen das ν.

HAHN: Die kommen doch von aussen wieder zurueck, denke ich.

HEISENBERG: Ja. Ein erheblicher Teil der Neutronen fliesst nach aussen ab, und ein Teil wird immer nur nach innen zurueckgestreut. Also ich muss dann die Grenzbedingungen ansetzen. Die heissen so: An der Trennungsflaeche zwischen Uran-Kugel und umgebendem Mantel muss die Dichte stetig uebergehen, und muss der Strom stetig uebergehen. Die Dichte innen habe ich schon hingeschrieben, die ist $1/r \; x \sin r/1_1$. Die Dichte aussen α/r. Ich muss also das α so waehlen, dass fuer $r = R$ am Kugelrand die beiden Ausdruecke gleich sind.

VON LAUE: Warum soll die Dichte stetig uebergehen?

HEISENBERG: Die Neutronendichte kann ja nicht springen. Die Neutronen bewegen sich ja doch sehr schnell 'rueber und hinueber ueber die Grenzflaeche. Also ich kann nicht einen Dichtesprung aufrecht erhalten. Also dann habe ich unter diese beiden Glieder mit der Dichte die Stroeme geschrieben. Den Strom bekommt man ja einfach, indem man die Dichte nach r differenziert. Der Strom ist ja gegeben durch den Gradienten der Dichte, multipliziert mit dem Diffusions-Koeffizienten. Also steht auf der linken Seite der Differential-Quotient nach r, multipliziert mit der Diffusions-Konstante d_i, und zwar habe ich jetzt im Index 'i' dazugeschrieben, weil es die Diffusionskonstante im Inneren bedeutet, und aussen steht der Differentials-Quotient nach r von α/r, das ist $-\alpha/r^2$ multipliziert mit der Diffusions-Konstante aussen. Und nun schreibe ich einfach die Bedingung auf, dass die beiden gleich sein sollen, und das werden Sie mir glauben, wenn man das also hinschreibt, dann faellt das α heraus, das kann man eliminieren. Es bleibt stehen, das Verhaeltnis der beiden Diffusions-Konstanten d_a/d_i muss sein 1 - Kugelradius/1_1 mal ctg $r/1_1$.

BAGGE: Wenn aussen freies Vakuum oder einfach Luft ist, wie ist es da?

HEISENBERG: Dann koennen Sie diese Formel auch anwenden, bloss muessen Sie dann da gleich unendlich setzen. Sie muessen sagen, dass jedes Neutron, was nach aussen kommt, sofort abtransportiert wird. Das laeuft darauf hinaus, dass am Kugel-rand die Neutronendichte Null wird (sic!), weil sie aussen mit beliebiger Geschwindigkeit abtransportiert wird. Jetzt sieht man also, damit man zu einem klaren Resultat kommt, muss man kennen nicht nur die Diffusions-Konstante im Inneren, die kennen wur ja, sondern auch die Diffusions-Konstante aussen. Das heisst, man muss nun irgend etwas annehmen ueber das Material, was man um die Uran-Kugel drumtut. Man sieht auch, es ist offenbar zweckmaessig, um die Uran-Kugel etwas drumzutun, was moeglichst viel Neutronen zurueckstreut, also, in anderen Worten, etwas drumzutun, was einen moeglichst grossen Streuquerschnitt hat. Andererseits handelt es sich hier um Streuung von schnellen Neutronen. Bei schnellen Neutronen hat man keine grossen Variationen der Wirkungsquerschnitte. Man kann sagen, wenn ich etwas Blei drumtue oder Uran 238 oder so etwas, dann habe ich immer Streuquerschnitte von der Groessenordnung 6 x 10^{-24}. Wenn ich Kohle drumtue, um ein ganz leichtes zu nehmen, dann habe ich etwa 4 x 10^{-24}, dafuer sind die Kohle-Atome etwas dichter gepackt als im Uran. Man kann es eigentlich machen wie man will, man kriegt immer freie Weglaengen in der Gegend von 4 cm. Also ich glaube eigentlich nicht, dass man irgend etwas damit profitiert, wenn man sich da besonders raffinierte Elemente 'raussucht, sondern wahrscheinlich liegt es halt immer in der Gegend von 4 cm.

VON WEIZSACKER: Der Wortlaut der Meldung war, man mische das Uran mit Graphit, um die Neutronen zu verlangsamen.

HEISENBERG: Ich koennte mir denken, dass man dadurch, dass man es mit Graphit vermischt, den Streuquerschnitt noch heraufsetzt, das heisst, also das di verringert und dadurch die ganze Maschine noch ein bisschen enger kriegen kann. Aber viel profitiert man damit nicht, denn fuer schnelle Neutronen streut die Kohle eigentlich genau so gut wie 235 - (int.)

HARTECK: Das Volumen nimmt kubisch zu, das ist immer schlecht.

HEISENBERG: Sie muessen doch immer ein 235 'rausnehmen, un ein Kohle-Atom 'reinzustopfen.

VON WEIZSACKER: Ich kriege doch im Ganzen von einer Substanz mit derselben Streuweglaenge mehr Material.

HARTECK: Aber es kommt uns doch nicht nur auf die Streuweglaenge sondern auch auf die Einfangweglaenge an. Wenn ich, sagen wir, die Haelfte dazugebe, dann habe ich in der Volumeneinheit die halbe Menge an Uran; ich muss also einen doppelt so langen Weg zuruecklegen bis zur Spaltung, und dadurch bekomme ich die vierfache Menge an Uran. Ich gewinne zwar ein bissel etwas dadurch, dass ich mehr dann drin herumstreue. Es wird immer schlechter.

VON WEIZSACKER: Aber dafuer werden die Neutronen erheblich langsamer und haben dann einen wesentlich groesseren Streuquerschnitt.

HEISENBERG: Das macht fast nichts aus. Das $1/v$ Glied geht ja erst viel tiefer los. Hier im Gegenteil ist doch der Wirkungsquerschnitt immer unabhaengig von der Geschwindigkeit, einfach immer der Kernquerschnitt. Aber aussen 'rum, wie gesagt, muss man 'was tun und zwar moeglichst etwas, was eine kleine Streuweglaenge hat, und ich habe jetzt mal versuchsweise angenommen, dass diese Streuweglaenge aussen genau so gross ist wie die innen. Also z.B. Kohle aussen tut das schon ungefaehr. Ich habe wegen der Ungenauigkeit, die hier hereinkommt, dann auch unterlassen, dieses μ - Glied aussen noch zu beruecksichtigen, denn es ist ja ganz willkuerlich, ob ich aussen da gleich 4 cm oder gleich 6 cm annehme.

GERLACH: Ob die den Kohlenstoff zugesetzt haben, um das Schmelzen zu sparen? Wenn man naemlich Uran-Karbid-Pulver nimmt und das reduziert, dann sind das zusammen schon eine sehr hohe Dichte.

HEISENBERG: So 'was koennte es sein, ja. Das koennte ich mir viel eher vorstellen, so eine technische Geschichte(?). Vielleicht koennen wir das nachher noch weiter besprechen. Also wenn die Diffusions-Koeffizienten aussen und innen gleich sind, dann ist $d_a/d_i = 1$, dann muss also einfach der ctgs = 0 sein, und der ctgs ist null, wenn das Argument $\frac{1}{2}\mu$ ist, also muss dann der Radius gleich $\frac{1}{2}\pi\ 1_1$ sein. Man sieht auch, dass es unguenstiger waere, etwa kein Streumaterial drumzutun, denn wenn ich etwa leeren Raum drum haette, um die Uran-Kugel, dann waere d_a gleich unendlich, d.h. der ctgs auf der rechten Seite muesste gleich minus unendlich sein, und das ist er dann, wenn das Argument π ist, und dann muesste also der Radius doppelt so gross sein. Also ganz grob kann man sagen, wenn ich gar nichts drumtue, um die Uran-Kugel, dann ist der kritische Radius, bei dem es los geht, doppelt so gross als wenn ich die Uran-Kugel mit Kohle oder Blei oder Uran 238 umgebe. Ich habe also jetzt angenommen, dass das Verhaeltnis der Streuquerschnitte innen und aussen eins ist; dann ist Kugelradius/$1_1 = \frac{1}{2}\pi$, und jetzt sehe ich den Wert von 1_1 aus der Gleichung ein und bekomme damit jetzt eine Gleichung fuer μ. Es kommt heraus

$$\mu = \nu - \frac{d\pi^2}{4R^2}, \text{ oder } \mu = \nu\,(1 - R_K^2 R^{-2}),$$

wo R_k eine kritische Laenge ist, $R_k = \frac{1}{2}\pi\ l$ und ungefaehr $\mu_K = 9{,}7$ cm/ Spaltungsquerschnitt hoch - ½. Daraus sieht man, das μ ist tatsaechlich negativ, wenn R sehr klein ist. Wenn ich also eine sehr kleine Kugel nehme, dann ueberwiegt das Glied $\mu_k^2 R^{-2}$ gegenueber der 1, und die Neutronendichte nimmt exponentiell ab. Wenn dagegen R groesser ist als R_k, dann wird μ positiv, und dann nehmen die Neutronen exponentiell zu, d.h., die ganze Geschichte explodiert. Man sieht also, dieses R_k ist die kritische Groesse, die die Uran-Kugel mindestens haben muss, damit die Sache explodiert. Die Explosion erfolgt dann exponentiell mit diesem Zeitfaktor μ. Nun sieht man, dieser kritische Radius haengt noch ziemlich empfindlich vom Spaltungsquerschnitt ab. Wenn ich den kleinsten Wert annehme, naemlich o,5, dann bekomme ich einen kritischen Radius von 13,7 cm, und wenn ich den groessten annehme, naemlich 2,5, dann komme ich auf 6,2 cm.

HARTECK: Ist das mit Rueckstreumantel?

HEISENBERG: Mit einem Rueckstreumantel, fuer den eben aussen die freie Weglaenge so gross ist wie innen. Nun sind diese 6 cm. eigentlich noch erheblich groesser als das, was die anderen behaupten. Die anderen behaupten ja, dass die ganze Explosivmasse nur 4 kg gewogen habe, und diese Kugel von 6,2 cm. wiegt ungefaehr 16 kg. Nun gibt es zwei Moeglichkeiten: das eine ist dass der Spaltungsquerschnitt in Wirklichkeit doch noch hoeher ist, und es ist ja nicht ganz undenkbar, dass er praktisch so gross ist wie der Streuquerschnitt, also etwa 6×10^{-24}, und auch dann reicht es nicht ganz. Also ich weiss nicht, was da noch viel gaendert werden koennte in den Daten. Sie koennen also noch einen besseren Rueckstreumantel haben, aber da koennen sie auch nicht viel dabei verdienen.

VON WEIZSACKER: Du hast vorhin gesagt, das $1/v$ Glied faengt erst viel weiter unten an. Das faengt doch sofort dann an, wenn die Wellenlaenge der Neutronen groesser wird als der geometrische Kernquerschnitt.

HEISENBERG: Ja.

VON WEIZSACKER: Vielleicht haben sie eben doch deshalb Graphit dazugegeben, damit sie langsamere Neutronen haben und einen groesseren Spaltungsquerschnitt.

HEISENBERG: Aber erstens bremst doch Kohle furchtbar schlecht - (int.)

WIRTZ: Aber wenn sie etwas schweren Wasserstoff - nehmen wir an so viele schwere Wasserstoff-Atome, sagen wir mal, in Form von schwerem Paraffin .... das zusammenzupressen, wie sie Uran-Atone haben.

HEISENBERG: Jedenfalls leichter waere es noch besser als schwerer.

VON WEIZSACKER: Bei welcher Energie der Neutronen ist die Wellenlaenge gleich dem Kernquerschnitt?

HEISENBERG: Bei 100000 Volt ist sie 10^{-11} vm.

VON WEIZSACKER: Beim Kern darf man doch ansetzen etwa 10^{-12}; das heisst, bei einigen Millionen Volt sind die Kernquerschnitte und die Wellenlaengen der Neutronen vergleichbar. Wenn es also nur gelingt, die Neutronen herunterzubringen auf, sagen wir, eine halbe Million Volt, dann wird sofort der Wirkungsquerschnitt um ein ganzes Stueck groesser sein.

HEISENBERG: Das glaube ich gar nicht, weil es ja nicht so ist, dass wir hier an der Resonanzstelle sind. Natuerlich bei der Resonanzstelle kann der Wirkungsquerschnitt einfach Quadrat der Wellenlaenge sein.

VON WEIZSACKER: Bei thermischen Neutronen ist dieser Querschnitt doch eben 50000 oder so 'was.

HEISENBERG: Wir wissen, dass in thermischen der Spaltungsquerschnitt 400 x 10^{-24} ist; wenn du nun auf 100000 Volt gehst, ist nach dem $1/v$ Gesetz, wenn es gilt, der Wirkungsquerschnitt bereits 0,4 geworden. Also kann man sagen, bei 100000 Volt wuerde nach dem $1/v$ Gesetz schon etwas Kleineres 'rauskommen als sicher da ist.

VON WEIZSACKER: Nun kommt die weitere Frage, wie stark werden die Neutronen gebremst, denn es ist ja so, dass praktisch jede Streuung unelastisch ist?

HEISENBERG: Ich wuerde sagen, nach drei bis vier Stoessen wird es vielleicht schon so auf den 10ten Teil, vielleicht sogar auf weniger 'runtergegangen sein.

VON WEIZSACKER: Wenn man einfach so die WEISSKOPF'sche (phon.) Formel anwendet, dann wird man doch immer sagen, ein Neutron, das mit einer bestimmten Energie E in den Kern hineingeht, kommt mit einer Energie der Groessenordnung $E/4$ vielleicht wieder 'raus.

HEISENBERG: Andererseits, wenn du einen Streuquerschnitt von 6 hast, dann sind mindestens 50% dieses Streuquerschnitts rein elastische Streuung, die so da aussen dran vorbeigeht und abgelenkt wird.

HAHN: Wenn man einen Uran-Mantel hat, dann gibt es ja doch noch das Auseinanderkrachen durch schnelle Neutronen; das gibt zwar nicht viel, aber doch eine kleine Vermehrung.

HEISENBERG: Also ich wollte ueberhaupt eigentlich folgende Deutung vorschlagen: ich wuerde glauben, dass das, was die im Zentrum haben, eben eine Mischung aus 238 und 235 ist; etwa so, dass es 50% 235 und 50% 238 ist, denn wahrscheinlich werden sie das 235 auch nicht gerade beliebig 'reingemacht haben. Von dieser Mischung, die sicher nicht viel schlechter ist als das reine 235, werden sie schon eine Kugel haben, die einen Radius von, sagen wir, 8 cm hat, aber in dieser Kugel von Radius 8 cm ist dann entsprechend wenig Uran 235 drin, und das haben sie angegeben als Gewicht.

HAHN: Die Liese MEITNER sagt doch aber hier vom Schweren Wasser etwas.

HEISENBERG: Die weiss ja sicher nichts. Schweres Wasser hat sicher ueberhaupt keinen Sinn, denn es handelt sich ja nur um Prozesse mit schnellen Neutronen, und fuer schnelle Neutronen ist Leichtes Wasser genau so gut wie Schweres.

HAHN: Wenn man ein schnelles Neutron hineinschiesst in Uran-Metall, dann vermehrt sich das auf 1,4 oder 1,35, bis es abgebremst ist.

HEISENBERG: Das haengt natuerlich ganz von der Groesse ab, also das ist auch wieder so ein Diffusuionsprozess, aber als Guenstigstes fuer eine unendlich grosse Kugel aus gutem 238 vermehrt es sich um 40%. Wahrscheinlich koennte man schon eine Mischung verwenden in der bloss, sagen wir, 10% 235 und 90% 238 sind. Das 238 hat immerhin doch den Vorteil, dass, wenn es mal spaltet, es dann gleich kolossal viele Neutronen abgibt, ungefaehr 5 Stueck. Es kann also, wie wir gerade festgestellt haben, sowohl am Spaltungsquerschnitt liegen als auch daran, dass sie solche Mischungen verwenden. Jetzt muss man also fragen, wenn man nun wirklich eine solche Kugel hat, wie geht die Explosion vor sich? Da habe ich jetzt mal angenommen eine Kugel, deren Radius nur um 5% groesser ist als der kritische Radius. Ich muss dazu sagen, ich glaube nicht, dass sie viel mehr gemacht haben als etwa 5% mehr, denn sie muessen ja sicher sein, dass die Haelfte der Kugel sicher unter der kritischen Grenze ist. Also sie setzen die Kugel zusammen aus zwei Halbkugeln offenbar.

KORSCHING: Koennen das nicht einfache Zylinder sein?

HEISENBERG: Das koennen auch Zylinder sein, aber ich wuerde sagen, Kugeln ist eigentlich immer der einfachste Fall, denn man will ja doch am Schluss die Neutronen mit einer kleinen Menge moeglichst in einem Raum drinhalten.

KORSCHING: Halbkugeln kann man schlecht so aufeinander schieben.

HEISENBERG: Da ist ja sowieso ein Eisenzylinder dahinter und so. Eine moegliche Konstruktion waere, man hat die Halbkugeln an einen Eisenzylinder drangemacht, und dieser Eisenzylinder laeuft in einem Kanonenrohr und wird in dieser Kanone auf die andere Halbkugel geschossen, die dann auch so gehalten wird. Wenn man annimmt, dass die Kugel, nachdem sie zusammengesetzt ist, einem um 5% groesseren Radius hat als den kritischen Radius, dann haette die halbe Kugel, wenn ich das als Kugel betrachte - in Wirklichkeit hat sie eine andere Form - einen Radius von etwa 85% des kritischen Radius. In anderen Worten, man muss schon auch mit den halben Kugeln ziemlicht dicht 'rangehen, an die kritischen Daten, und das muessten die also sehr genau gemessen haben dass ihnen da nichts passiert. Immerhin haben sie natuerlich auch noch folgenden Sicherheitsfaktor: sie haben diese Halbkugel ja auf der einen Seite mit

Luft umgeben, wenn sie leer noch da steht, und dadurch ist die Rueckstreuung dort viel schlechter.

VON WEIZSACKER: Ich finde diese geometrische Form doch unguenstig. Eigentlich muesste man aus dieser Halbkugel eine kleinere Figur mit runden Ecken ausschneiden, die wirklich wirksam ist, denn das Zeug mit den spitzen Kanten hilft ja nichts.

HEISENBERG: Eben. Also eine moegliche Konstruktion waere, sie haben immer diese Halbkugel in einem Eisenzylinder drin. Nun sagen wir, unten steht ein Eisenzylinder, die Halbkugel nach oben, nun kommt ein Kanonenrohr, und oben drueber ist wieder ein Eisenzylinder mit der Halbkugel nach unten, so dass sie flachen Seiten der Halbkugel sich zugewandt sind. Nun habe ich oben eine kleine Sprengladung, so dass zu einer bestimmten Zeit dieser obere Zylinder 'runterstoesst auf das untere, und das Ganze nun zu einer ganzen Kugel zusammengesetzt wird. Warum das so schnell gehen muss, will ich gleich sagen. Der Prozess laeuft nun ab nach der Gleichung $\rho = e^{\mu t}$, wo μ diese reziproke Zeit ist, die jetzt da oben in der Formel ausgerechnet ist. Wenn man mal annimmt, dass eben der Radius nur 5% groesser ist als der kritische Radius, dann kommt also 'raus: μ - 0,5 x $(10^7$ Sekunden$)^{-1}$. Also in 10^{-7} Sekunden nimmt die Intensitaet jeweils auf das e-fache zu. Das heisst in anderen Worten, die ganze Reaktion laeuft ab in etwa 10^{-5} Sekunden. Nun ist die Frage: erstens, warum muss man die beiden Halbkugeln so schnell aufeinandersetzen? Koennte man nicht die ganz langsam aufeinandersetzen und sagen: "Na gut, wenn ich sie eben langsam aufeinandersetze, irgendwann mal geht eben dann die Geschichte los, wenn ich sie nahe genug beieinander habe, und dann ist ja alles gut"? Das scheint mir aus folgendem Grund schlecht zu sein. Wenn ich es langsam zusammensetze, faengt die Sache an zu einem Zeitpunkt, wo ich nur viel weniger ueber dem kritischen Radius bin als diese 5%, also sagen wir, bloss 1/10 pro mille ueber dem kritischen Radius. Zu dieser Zeit ist der Zeitfaktor, mit dem die Geschichte geht sehr viel langsamer, also nicht $(10^7$ Sekunden$)^{-1}$, sondern eben nur 10^{-4} oder so 'was. Dan wuerde die ganze Explosion eben nicht 10^{-7} sondern 10^{-4} Sekunden brauchen. In dieser Zeit wuerde also die Sache sich so schnell so weit erwaermen, dass man auf 5000° oder 10000° oder 20000° kommt. Sobald ich aber diese hohen Temperaturen habe, dann platzt die ganze Geschichte ja auseinander. Dieses Auseinanderplatzen erfolgt dann auch in Zeiten vergleichbar mit 10^{-4} Sekunden. Das heisst, die Reaktion laeuft gar nicht bis zu Ende ab, sondern sie laeuft eben praktisch nur ab, bis sie etwa 10000° erreicht hat, und dann ist sie schon zu Ende. Also ich muss die Zeit schon hinreichend kurz waehlen, damit das nicht vorher auseinanderfliegt. Das habe ich nun mal so ueberschlagen, und da kommt eben 'raus, die Zeit darf eigentlich nicht laenger als 10^{-6} Sekunden sein, vielleicht noch $\frac{1}{2}10^{-6}$, aber das ist dann so etwa die Grenze. Das kann man erreichen, wenn man die beiden Halbkugeln aufeinanderschiesst. Man kann es aber nicht erreichen, wenn man sie etwa bloss langsam aufeinander legt. Ich kann sagen, es gibt einen kritischen Abstand, sagen wir mal, 1 cm. Wenn ich sie mahr als 1 cm auseinander habe, passiert gar nichts, geht es ueberhaupt nicht los. Wenn ich sie naeher als einen Zentimeter zusammen habe, dann geht die Sache explosiv los. Dieser letzte Zentimeter muss in 10^{-6} Sekunden zurueckgelegt werden.

HARTECK: Das geht ja nicht, das waeren ja 10000 m pro Sekunde.

HEISENBERG: Na ja, 10^{5} (means 10^{-5} Sekunden?) ist vielleicht gerade noch moeglich. Darum werden sie es auch nicht ganz ausnutzen koennen.

HARTECK: Es genuegt nicht nur 'moeglichst schnell', sondern es kommt sehr darauf an.

HEISENBERG: Es kommt sehr drauf an. Es gibt noch eine Moeglichkeit, da anders zu argumentieren. Man koennte naemlich folgendes sagen: wenn man gar keine Neutronenquelle in die Naehe bringt und es sozusagen der Hoehenstrahlung ueberlaesst, dann koennte ich ja so sagen: "Ich bringe es verhaeltnismaessig langsam zisammen, dann wird zunaechst gar kein Hoehenstrahlungsneutron da sein". Aber das ist ein Spiel des Zufalls. Da kann es eben passieren, dass doch gerade in dem Moment, wo der kritische Radius ueberschritten ist, schon das Neutron da ist, und dann ist der Wirkungsgrad der Bombe viel schlechter.

VON WEIZSACKER: Wie oft entstehen in so einem kleinen Gebiet Hoehenstrahlen-Neutronen?

HEISENBERG: Das ist jetzt die Frage der Zuendung. Da habe ich folgendes abgeschaetzt: Erstens kann die Zuendung etwas durch die spontane Spaltung erfolgen. Das ist ja eine Moeglichkeit. Da sieht man aber, das reicht nicht. Die Lebensdauer fuer spontane Spaltung ist etwa 10^{14} Jahre, wenn ich mich recht erinnere.

HARTECK: 10^{14} bis 10^{16}.

HAHN: Woher weiss man, dass 235 spontan spaltet?

HEISENBERG: Ach so, da ist 238 wahrscheinlich auch dabei. Die beiden sind auch ungefaehr vergleichbar. Das waeren also in Sekunden etwa 10^{22} Sekunden, und wenn ich nun in 10^{-6} Sekunden ein Neutron haben will, dann muesste ich 10^{38} Atome haben; die habe ich natuerlich nicht.

VON WEIZSACKER: Man koennte natuerlich umgekehrt sagen, wenn bestimmt kein andere Zuendungsmechanismus da ist, dann fuehrt man es ganz langsam zusammen, und dann geht es los.

HEISENBERG: Ja, ich weiss auch nicht. Vielleicht haben sie so einen Trick gemacht. Dann koennen sie natuerlich Pech haben, sie muessen dann riskieren, dass ihnen die Bombe einen sehr schlechten Wirkungsgrad gibt.

VON WEIZSACKER: Wenn sie es dann noch relativ schnell zusammenfuehren, ist die Wahrscheinlichkeit, dass es schief geht, doch sehr gering.

HEISENBERG: Bloss wenn sie es da schnell aufeinanderschiessen, muessen sie dafuer sorgen, dass es nicht wieder durch Reflektion auseinandergeht.

HAHN: Was wuerde passieren, wenn die 1/1000 mg Radium da 'reinnaehmen?

HEISENBERG: Das wuerde sicher reichen.

GERLACH: Wenn die eine Haelfte so 85% von der kritischen Groesse ist, wie lange lebt dann ein thermisches Neutron drin?

HEISENBERG: Sie meinen, wenn man mal also die thermischen Neutronen von der Hoehenstrahlung betrachtet?

GERLACH: Ja, denn die machen doch wieder neue Neutronen in der Haelfte drin.

HEISENBERG: Das lebt nicht lange.

HARTECK: Das hat doch die Geschwindigkeit von 10^5 cm, ein thermisches. Und wenn wir reines 235 haben, dann ist die freie Weglaenge nur einige Hunderdstel Zentimeter. Also wird die Groessenordnung sein etwa 10^{-7} Sekunden.

HEISENBERG: Ja, stimmt, das ist viel zu klein.

GERLACH: Dafuer ist aber wieder ein Neutron da.

HEISENBERG: Das laeuft weg, das ist zu schnell.

GERLACH: (Fuer) solche Abfallszeiten war doch mal eine Formel angegeben.

VON WEIZSACKER: Hier ist es ja ausgerechnet. Ja. in unseren grossen Maschinen sind die Abfallszeiten viel groesser. (Pause).

HEISENBERG: Es kann ja sein, dass sie es doch so machen: sie fuehren die beiden Teile moeglichst schnell zusammen und sagen: "Wir hoffen eben, dass wir nicht das Pech haben, dass gerade, wenn die kritische Groesse ueberschritten ist, schon das erste Neutron da ist".

KORSCHING: Wenn die Zerfallsseite so ungeheuer abhaengig ist, dann muesste man, wenn man ganz langsam mit einer Mikrometerschraube die Dinger zusammenschraubt, dann erreichen koennen, dass man die beliebig langsam abbrennen lassen kann.

HEISENBERG: Da kommen Schwankungserscheinungen. Natuerlich, im Prinzip waere das moeglich. Dann habe ich also noch eine energetische Betrachtung drangefuegt. Nehmen wir an, es sind 4 kg 235, was wirklich 'verbrennt'. Dan werden dabei frei 3×10^{21} erg. Das ist ganz trivial. Pro Spaltung werden ungefaehr 3×10^{-4} erg. frei, das muss ich dann mit irgendwelchen Faktoren multiplizieren. Nun kann ich fragen, wie hoch wird die

Temperatur in diesem Uran-Block? Da kann man nun einfach so rechnen, dass diese ganze Energie praktisch in Strahlung umgesetzt wird, denn die spezifische Waerme von dem Metall ist von T unabhaengig, also die Energie geht etwa mit T.

VON LAUE: Es handelt sich um Uran-Gas.

HEISENBERG: Ja. Dagegen die Strahlung, die auch in diesem Raum gleichzeitig entsteht, die hat einen Energie-Inhalt, der nach dem BOLTZMANN'schen Gesetz mit T^4 geht. Also bei hinreichend hoher Temperatur ueberwiegt immer die Strahlung ueber alles andere. Also praktisch wird die ganze Energie in Strahlung imgesetzt, und ich kann die Temperatur so ausrechnen, dass ich sage: die Temperatur wird ebenso wie die Neutronendichte exponentiell zunehmen, wieder nach einem Gesetz $e^{\mu t}$. Sie wird bis zu einem Maximum zneuhmen, bis naemlich die gesamte Energie verbraucht ist. Die gesamte Energie muss bis zu diesem Zeitpunkt dann durch die Oberflaeche auch abgeflossen sein, denn die Strahlung geht ja mit Lichtgeschwindigkeit, also praktisch braucht sie keine Zeit im Inneren des Urans. Nun geht durch eine Oberflaeche nach dem STEPHAN-BOLTZMANN'schen Gesetz pro cm^2 und Sekunde die Energie σ, wobei σ die Konstante ist 5,8 x 10^{-5} Erg pro Sekunde, cm^2 und $Grad^4$. Es muss also das Zeitintegral von Oberflaeche F des Uranstuecks mal σT^4 gleich der gesamten Energie sein, und dieses Zeitintegral ist bei expansieller Zunahme einfach gleich

$$\frac{F \text{ x } \sigma \text{ x Maximaltemperatur}^4}{4 \text{ x expansiellen Zunahmefaktor}}$$

und wenn ich nun die Daten einsetze, die wir da oben angenommen haben, dann kommt man auf eine Maximaltemperatur von 3 x 10^7 Grad. Das ist etwas hoeher als die Zeitungen angeben; in den Zeitungen habe ich mal die Zahl 3,6 x 10^6 Grad gelesen. Es scheint, dass nicht die ganze Energie umgesetzt wird. Es kommt ja vielleicht auch nicht so auf die letzte zehner Potenz an, denn die Bedingungen kennen wir ja nicht genau von der Bombe, und es ist ja auch wieder die Frage, ob diese Zeitungszahl richtig ist. Wenn man wirklich die Kugel zusammen hat, bevor die Sache losgeht, dann dehnt sie sich in den 10^{-7} Sekunden auch faktisch nicht aus. Ich habe ausgerechnet, um wieviel die Kugel sich ausdehnt, und da kommt man eben bei 10^{-7} Sekunden erst auf Ausdehnung von Groessenordnung von wenigen Millimetern.

VON LAUE: Vor allen Dingen kommt ja doch der Strahlungsdruck in Frage fuer den Streumantel.

HEISENBERG: Der Strahlungsdruck ist bei der Maximaltemperatur etwa 10^9 Atmosphaeren; das ist an sich irrsinnig viel, aber es reicht eben nicht, um in dieser ungeheuer kurzen Zeit schon merklich die Geschichte auszudehnen.

HARTECK: Also wie ist denn die Temperatur ausgerechnet worden?

HEISENBERG: Also ich habe einfach so ausgerechnet: es soll praktisch bei Ende der Reaktion die gesamte Energie, die frei geworden, ist, durch Strahlung weggegangen sein. Das ist natuerlich deswegen nicht ganz richtig, weil ja auch ein gewisser Bruchteil der Energie dazu verwendet wird, um wirklich die Teilchen zu beschleunigen, also die Molekule zu bewegen. Aber praktisch spielt eben bei diesen ganz hohen Temperaturen eigentlich die Strahlung die Hauptrolle. Also das meiste von Energie geht in Strahlung, und nur relativ wenig Energie geht in Bewegung von Teilchen. Diese Strahlungsenergie kommt mit Lichtgeschwindigkeit 'raus. Also habe ich gesagt, es muss bis zum Ende der Reaktion praktisch die gesamte Energie, die ueberhaupt frei wird, durch die Oberflaeche nach aussen abgestroemt sein. Die Strahlung von 3×10^7 Grad ist, weil sie PLANCK'sche Strahlung ist, eine Roentgenstrahlung, deren Maximum bei etwa 1 Angstroem Wellenlaenge liegt, also die genauere Zahl ist 0,86 Angstroem. Der Hauptanteil der Strahlung kommt 'raus als Roentgenstrahlung, und wird nun in der umgebenden Luft absorbiert. Das erste, was dann passiert, wird sein, dass die ersten 10 m Luft zur Weissglut gebracht werden. Was man zu allererst sehen wird, wird ein gluehender Ball von etwa 20 m Durchmesser sein, der weiss glueht infolge der absorbierten Roentgenstrahlung. Diese Strahlung wird natuerlich wieder weitere Strahlungen nach aussen schicken, und so wird sich die Geschichte allmaehlich ausbreiten.

VON WEIZSACKER: Wobei dann allerdings wohl nur noch ein kleiner Teil Strahlung ist von der Energie, denn, sowie das Ding auf vernuenftigere Temperaturen gekommen ist, dann ist der groessere Teil des Drucks Gasdruck, und dann geht det Rest als Druckwelle weiter.

HEISENBERG: Dann wird die Geschichte am Anfang wunderbar kugelsymmetrisch sein, und dann allmaehlich erst kommen die Konvektionserscheinungen dazu, und dann wird es kompliziert.

VON WEIZSACKER: Ich wuerde drei Phasen sagen: eine Strahlungsphase, dann eine Phase, wo zunaechst einmal eine Druckwelle erzeugt wird, und nun dehnt es sich mit dieser Druckwelle gleichzeitig aus, und dann wird eine Konvektion einsetzen und das Ganze nach oben getragen.

VON LAUE: Wieviel Energie geht von vornherein als sichtbare Strahlung weg?

HEISENBERG: Ich vergleiche etwa diese innere Uran-Kugel mit einem gluehenden Koerper, sagen wir einmal, auf 1000°. Dann wuerde es um 10^4 mal mehr strahlen im sichtbaren als dieser gluehende Koerper. Die Oberflaeche dieser Uran-Kugel strahlt also um etwa 2000 mal mehr als die Sonnenoberflaeche.

VON WEIZSACKER: D.h., wenn man es aus einer Entfernung sieht, in der es so gross ist wie die Sonne, dann strahlt es eben 1000 mal mehr als die Sonne.

HEISENBERG: Es waere interessant, ob z.B. Gegenstaende von dem sichtbaren

Strahlungsdruck schon umgeworfen werden koennen. Das ist durchaus denkbar. (Pause). Jedes Neutron hat in unserer Maschine die Chance 1/20 eingefangen zu werden im Resonanzkoerper, also auf jede Spaltung ist praktisch 1/20 '94' gemacht.

HAHN: Das habe ich jetzt nicht verstanden.

HEISENBERG: In unserer Maschine is es immer so eingerichtet gewesen, dass jedes Neutron mit der Chance 1/20 weggefangen wurde durch die Resonanzstellen, bevor es thermische geworden ist.

HAHN: Ach so.

13 September 1945

Subject: Farm Hall Report No. 6.

To: Major Francis J. Smith, Room 5004, New War Dept. Bldg., Washington, D. C.

1. There is inclosed herewith Farm Hall Report No. 6 of operation "Epsilon", dated 23 August to 6 September 1945.

2. As you will note in paragraph 3 of the attached report there is mention of a quantity of radium hidden at Hechingen. This matter has been discussed by Major Smith, Lt. Cmdr. Welsh and myself and it was decided that as far as this material is of no direct use for TA, and, as the operation of securing of this material would likely be detected by the French, it was thought best to take no action at the present in regard to securing it. If your office is not in accord with this, please advise.

For the Military Attache:

H. K. CALVERT,
Major, FA,
Assistant to the Military Attache.

Incl. - 1

STANDARD FORM NO. 64

Office Memorandum · UNITED STATES GOVERNMENT

DATE:

TO : Gen Groves

FROM : Maj Britt.

SUBJECT :

WAC

I do not see any need for our trying to get the radium —

G

write & tell Calvert

OK.

Subject: Farm Hall Report No. 6.

1st Ind.

P. O. Box 2610, Washington, D. C., 26 September 1945. To: Lt. Col. H. K. Calvert, Office of Military Attache, American Embassy, London, England.

Reference paragraph 2 of basic correspondence, this office concurs in your decision not to make any attempt to secure the radium reported hidden at Hechingen.

AMOS E. BRITT,
Major, Corps of Engineers.

Incl. w/d

Copy No. 1.

Capt. Davis for Gen. Groves.

Ref. F.H.6.

To: Mr. M. PERRIN and Lt. Cdr. WELSH.
From: Major T.H. RITTNER

OPERATION "EPSILON"

(23 August - 6 September, 1945)

I. General

1. The principle incident during the period under review has been the receipt of replies to the letters written by the guests to their families. This is dealt with in detail later in this report. Although the letters show that the families are apparently well, the guests are picking on quite trivial remarks in the letters as giving cause for anxiety. They have already asked that further letters should be sent in the near future. In this, as in many other ways, they show complete lack of appreciation of the fact that they are nationals of a defeated nation and seem to think that the Allied military authorities have nothing better to do than to send couriers round GERMANY for their benefit.

2. As the weather deteriorates the guests are unable to spend so much time out of doors. This is having the effect of making them bored and querulous and will tend to get worse as time goes on. HEISENBERG has threatened to withdraw his parole unless some decision is come to regarding his future and that of the rest of the guests.

3. Professor HAHN was lent a copy of the American magazine "Life" of 20 August containing articles on the atomic bomb and a number of photographs of personalities, including himself, connected with atomic research going back some hundreds of years. Von WEIZSACKER was looking at this journal when he was heard to remark: 'Of course they are mostly Germans'. The fact that this statement was untrue merely emphasises the inborn conceit of these people, who still believe in the '<u>Herrenvolk</u>'. This applies to every one of the guests with the possible exception of von LAUE.

4. The usual bi-weekly lectures have been given, but they have been confined to general physics. There have been no other technical discussions worth reporting.

5. Major CALVERT, Major SMITH, Captain DAVIS and Lieutenant VOLPE, all of the U.S. Army, visited FARM HALL on Sunday, August 26th, and dined with the guests.

II. News From Home.

1. The American Military Authorities sent Lieutenant WARNER, U.S. Army, to GERMANY with the letters the guests had written to their families, with instructions to bring back replies if possible. This officer returned with a numbers of replies and visited the guests at FARM HALL on 28 August.

2. Replies were received from the wives of von LAUE, HAHN, GERLACH, HEISENBERG, WIRTZ and BAGGE showing that these families were well and living under reasonable conditions.

3. Von WEIZSACKER's wife had gone to her parents in SWITZERLAND. HARTECK and KORSCHING are batchelors and had not written.

4. The only unfortunate case was that of DIEBNER. It will be remembered that arrangements were made in June, at DIEBNER's request, to move his wife and child from STADTILM in THURINGIA which was about to be occupied by the Russians. They were taken to BAGGE's parents at NEUSTADT nr. COBURG. When the American officer arrived at the home of the old BAGGES to deliver DIEBNER's letter he found that DIEBNER's wife and child had left for an unknown destination accompanied by a certain Herr RACKWITZ. Old Mrs. BAGGE made it quite clear to the American officer and also in a letter to her son that she had disapproved of the 'goings on' between Frau DIEBNER and RACKWITZ. DIEBNER has merely been told that it was impossible to deliver his letter and he appears to be satisfied that his friend RACKWITZ is protecting his interests in looking after his wife and child.

5. The following points of interest emerged from the letters:

(a) Frau WIRTZ wrote on 30 June that there was a rumour that HEISENBERG had been brought to ENGLAND by plane and there was also a rumour that the rest of them were in U.S.A. She also stated that Professor JOLIOT had been in HECHINGEN. He told her that they would soon be home.

(b) On 2 July, Frau WIRTZ mentioned that there had been a lot of Russians in HECHINGEN. She also mentioned that BOPP had been dismissed from his post as Deputy Director of the Institute and had been imprisoned for 5 days. SCHUELER had persuaded the French to appoint him Deputy Director in place of BOPP.

(c) In a letter, Frau WIRTZ stated that the Institute was still working, and that she and the other wives were being paid 60% of their husband's salaries.

(d) Frau von WEIZSACKER has gone to SWITZERLAND. She wrote in a letter to Frau WIRTZ that she had heard a rumour that their husbands were in England.

(e) Frau von LAUE also mentioned the presence of Russians in HECHINGEN, and stated that she had heard that the entire staff of the KAISER WILHELM INSTITUTE in BERLIN , together with all the apparatus and documents, had been taken to RUSSIA.

(f) Frau von LAUE also stated that she had heard that Professor HAHN had been elected President of the KAISER WILHELM GESELLSCHAFT.

III. WIRTZ's Radium

1. Reference has been made in previous reports to a quantity of radium which WIRTZ was alledged to have hidden.

2. The following conversation took place between WIRTZ, von WEIZSACKER and HEISENBERG on 27 August. A copy of the original text was immediately sent to Lt. Cdr. WELSH.

WIRTZ: I hope the radium is still at HECHINGEN. I only discovered afterwards that it had been buried under the case containing the uranium cubes (?).

HEISENBERG: Oh there!

WIRTZ: Yes, there. The Americans had taken these cubes(?) out and then someone shouted to GEISSMAN(?): "Must I take the case out?" GEISSMAN(?) said he didn't care so they didn't take the case out and threw the rubble back into the hole on to the case - it was quite a job as it was a big case and they had only taken away about a third of the rubble, just enough to get the lid open and it would have been quite a job to have got it right out. The radium was underneath it, it is still there. GEISSMAN(?) succeeded in putting one of the cubes(?) on one side whilst those 14 or 15 men where searching. We made a mistake, we ought not to have handed it over.

IV. The Future.

1. A number of the guests have discussed their future with me. In the main their ideas expressed to me correspond to those expressed in private conversation with their colleagues, which have already been reported.

2. HEISENBERG made it quite clear to me that he wishes to continue work on uranium, although he realises that this could only be done under Allied control. His main interest at the moment is to get back to

GERMANY to look after his family, who appear to be in some difficulty as they live in the mountains near Munich, and his wife has no one to help her with her seven children. He is very distressed to hear from his wife that his mother died two months ago, and that a women friend of his wife, who had been helping her had also died. He is perfectly prepared to give an undertaking on oath not to work on uranium, except under Allied control, if he is allowed to return to his family. HEISENBERG has threatened to withdraw his parole, which he gave me in writing some time ago, unless some arrangement is made regarding the future of himself and his family.

3. BAGGE had a long conversation with me regarding his future. He professes to be no longer interested in uranium, and would prefer to concentrate on work on cosmic rays. He says he only took up research on uranium because of the war. BAGGE appears to be more concerned about his family than any of the other guests. It will be remembered that he had pictured his wife being raped day and night by Moroccan troops, and when it was clear from his wife's letter, that this was not the case, he did not show the relief and joy that I expected. He now complains that his wife is expected to cook for the French troops that are billetted in the house. Like many others he just fails to realise that GERMANY has lost the war. He is working himself up into a possibly dangerous state of nerves.

4. The following conversation took place on 25 August between HEISENBERG, HARTECK and WIRTZ:

HEISENBERG: I would think that one could come to an arrangement with the Anglo-Americans, so that they would on the one hand superintend the work - which they will do in any case - and that at the same time they would see that we got sufficient material.

HARTECK: I can tell you that as soon as that happens we will be looked upon as traitors in the eyes of the masses.

HEISENBERG: No. One must do that cleverly. As far as the masses are concerned it will look as though we unfortunately have to continue our scientific work under the wicked Anglo-Saxon control, and that we can do nothing about it. We will have to appear to accept this control with fury and gnashing of teeth.

HARTECK: One couldn't get away with that at HAMBURG.

HEISENBERG: One could get away with it at HECHINGEN.

WIRTZ: I don't understand why you are so optimistic about the Anglo-Saxons; HAHN and the others are the same. Surely the last five months have shown you that there is not much ground for optimism.

HEISENBERG: It is true that they have now held us prisoners for five (sic) months, but when you consider the position the Allies are in, then it is understandable. Now that they have got the atomic bomb, and realise what a terrifically important thing they have in their hands, it is not very easy for the politicians to allow German uranium specialists to be at large. You can't expect that.

5. The following conversation took place between HEISENBERG and KORSCHING on 6 September:

HEISENBERG: I have the feeling that the only thing they really fear, is that we might go over to the Russians. The English say to themselves quite rightly: "They don't know our technical details, but they know so much about the whole business, that they could be a very considerable help in speeding the thing up in RUSSIA."

KORSCHING: But these people know us so well by now and know that we are not pro-Russian.

HEISENBERG: But if in a year or six month's time we find that we are only able to eke out a meagre existence under the Anglo-Saxons, whereas the Russians offer us a job for say fifty thousand roubles, what then? Can they expect us to say: "No, we will refuse these fifty thousand roubles as we are so pleased and grateful to be allowed to remain on the English side".

KORSCHING: No. We must simply say that if the English expect us not to work for the Russians, then we expect them not to let us starve, but to let us live decently.

FARM HALL, (Sgd) T.H. Rittner
GODMANCHESTER. Major.
8 September, 1945.

AMERICAN EMBASSY
OFFICE OF THE MILITARY ATTACHE
1. GROSVENOR SQUARE. W. 1.
LONDON. ENGLAND

26 September 1945

Subject: Farm Hall Reports, 7 & 8.

To : Major F. J. Smith, Room 5004, New War Dept. Bldg., Washington, D. C.

1. Enclosed herewith are Farm Hall Reports, 7 & 8.

2. Inasmuch as these reports contain information of an extremely valuable nature which should aid in making the ultimate decision disposition of our prisoners, it is being forwarded without taking time to add comments however, our comments will follow shortly.

For the Military Attache:

HKCalvert
H. K. CALVERT
Lt. Col., F.A.
Assistant to the Military Attache.

2 Incls:
As listed above.

Copy No. 6.
Lt. Cdr. Welsh.
Ref. F.H.7.

To: Mr. M. PERRIN and Lt. Cdr. WELSH.
From: Major T.H. RITTNER

Distribution:
Copy No. 1 – Capt. Davis for Gen. Groves.
" " 2 – Capt. Davis.
" " 3 – Pro. Blackett.
" " 4 – Mr. Perrin.
" " 5 – C.P.A. for C.S.S.
" " 6 – Lt. Cdr. Welsh.
" " 7 – Major Rittner.

OPERATION "EPSILON"

(7th-13th September, 1945)

I. GENERAL

1. The guests are now in a cheerful and friendly mood expecting, since the visit of Professor BLACKETT, an early return to GERMANY.

2. Indeed, on Tuesday, 11th September, Professor HAHN came to ask about the possibility of using two Swiss 20-Franc notes, which he has with him, to buy certain little luxuries for his wife.

3. An easing of the letter difficulty is also expected, Professor HAHN having asked for a supply of plain envelopes for possible letters to Professor BLACKETT, Sir Charles DARWIN, Herr SCHUHMACHER and so on.

II. VISIT OF PROFESSOR BLACKETT

1. Professor BLACKETT and Lt. Cdr. WELSH arrived at FARM HALL on Saturday, 8th September, in the early evening, staying until after lunch on Sunday.

2. Before and after dinner on Saturday Professor BLACKETT had long conversations in English with HEISENBERG, who was an old friend, about the future of German science, about the guests at FARM HALL and their immediate prospects, and about HEISENBERG's family troubles.

a) Before dinner:

HEISENBERG: We had some discussions with DARWIN and he told us that BOHR held very strongly the opinion that everything (concerning the bomb) should be made quite public and not be kept secret; on the other hand, of course, it is such an important thing for politics that it is almost impossible to do away with all secrecy.

BLACKETT: A great many serious people are in favour of it. I think, the arguments are very strong. Of course, if you do, it is a little difficult logically, because why don't you do it with all other armaments? After all, the bomb is only one. DALE, in his letter to the "Times", said, he believes, one should have no secrets about armaments at all.

HEISENBERG: That is probably not possible to achieve. You will not be able to force the different governments to give their armaments away unless you can actually get a total disarmament of the world which would obviously be the best thing to do.

BLACKETT: How does one start?

HEISENBERG: Yes, I am afraid that the Russians for instance simply won't.

BLACKETT: The Americans are the only people who have got it and what they do is according to what the Russians do and, clearly, the Russians are behind the Americans. How many years, I do not know.

HEISENBERG: At least three.

BLACKETT: Yes, at least three. I have not had anything to do with it. I was a little connected with it in '41 and I have only recently come in through this Committee – you saw it in the paper – which is partly a result of the change of government and things. You have seen the British White Paper?

HEISENBERG: Yes.

BLACKETT: There is a very much more detailed one coming out in AMERICA. It has not yet been published in ENGLAND.

HEISENBERG: From the things I have seen in the papers and in the White Paper which DARWIN gave us, I think that I can imagine all the details of what they have done. The physics of it is, as a matter of fact, very simple, it is an industrial problem. It would never have been possible for GERMANY at all to do anything on that scale. In some way, I am glad that it has not been possible because it would have been terrible for us all. We have started on a very small scale. We were interested in a kind of machine, but not a bomb, the idea being first of all that we knew that there was no chance to do anything on that scale and we knew that, in order to separate the isotopes, we would have to do it on that scale. Then we thought, we could, with much smaller industrial effort, actually build a small machine which gives us energy.

BLACKETT: Which is a sensible thing; that is what we want to do.

HEISENBERG: Yes, and we knew also, if we did that, then our government will be satisfied with it and we have had a good time in working on it. Actually, I think, we have almost succeeded in doing it. However, all the plants producing Heavy Water were destroyed by the R.A.F. and that was really why it was not completed. But still, from the scientific side, one knows all the things. The Russians certainly also know it, KAPITZA and LANDAU.

BLACKETT: Frankly, there aren't any secrets, there are some tricks of the trade. You know the recipe for making an omelette, but you can't necessarily cook a nice one. There aren't any scientific secrets. In fact, BOHR told me, in '39 there was a discussion on the whole thing in COPENHAGEN, it was extraordinarily complete. Tell me about yourself.

HEISENBERG: There are many subjects, I don't know which we can discuss, because, of course, there are some limits. What do you think will happen in GERMANY at all, for science etc? Of course, we are interested whether one can make a living in GERMANY or not. Then the special question: If there is a possibility to have science re-established in GERMANY, which would be a practical scheme for, say, our Group? I thought that you probably would know the conditions which your government has in mind in this respect and I thought that perhaps within the frame of these conditions we could have a discussion, that we both would agree – within this frame – what we both think would be the most reasonable scheme.

BLACKETT: That we certainly can do.

HEISENBERG: Of course, it is another question whether your advice would be taken by the government.

BLACKETT: I have only been on the fringe of the question of the Control Commission and things. Frankly speaking, it is in a pretty bad mess. We were not very well prepared for the war, but we were probably better prepared for war than we were for the peace.

HEISENBERG: Exactly what DARWIN has told me!

BLACKETT: You must remember that we just were not prepared for winning the war and the people did not see far enough. There obviously are, from what I have heard, a lot of different opinions, you get the military against the other people. In fact, generally in AMERICA and ENGLAND there is a formative period going on in which we are really trying to think out what is practical politics. And this question of science; well, all kinds of different ideas have been put about by different people. I have got a perfectly clear idea of my own and I think, I can say that a great majority of English scientists agree. Whether our views will be accepted, I don't know. Only a few points are certain. In my view, all universities should be re-started and encouraged with their teaching and their research. It seems to me that the universities are, to a great extent, the key to what is commonly called 're-education'. I believe that ordinary research should go on quite freely. I think it will come – this is only my own personal view – in not too long a time that ordinary academical research and private research in the smaller industries, in the light industries, would be fairly uncontrolled, fairly free. There may be some inspectorate.

HEISENBERG: The first question would be: "What do you think will happen to EUROPE as a whole?" Then I would probably say: In some ways, I don't fear the controls, because, I think, the world is anyway going into a state in which everything is more controlled than it has been before. All industry will be controlled; it is partly socialism, I should say. Socialism always means that you can get things more effectively done by the government taking over the organisation. As a matter of fact, we see in RUSSIA and, partly, we have also seen in GERMANY perhaps how effectively such a thing can work. In so far I wouldn't mind if there were control in many respects, even to some degree at the universities.

BLACKETT: In a way, we have oversold science. A lot of people are thinking that scientists are magicians who, by themselves, are dangerous; what is dangerous is the great machine which makes use of these things and produces them. We have to prevent people from thinking that scientists as such are dangerous people, that is rather current. Whether nuclear physics will ever become clean open subjects again, we are as much in the dark as you are there.

HEISENBERG: That is what I have felt. And also, necessarily, after the atomic bomb had been dropped, I knew, it must be an extremely difficult subject.

BLACKETT: Many of us, who knew about it being done, just really prayed that it would not come off because it is a great complicating factor. It is going to need a very very great deal of political and technical courage to get the world . . . under control.

HEISENBERG: If your people would agree that science in general can be encouraged in GERMANY and also that, say, nuclear physics is allowed again – of course not preparing a bomb, but just ordinary physics, high tension and all this old game – what scheme, do you think, could be applied in our case, because our case is difficult? Our institute is really in BERLIN. Now, BERLIN is almost Russian and, therefore, I would understand that your politicians would not like us just to work in BERLIN, because they would be afraid, that we could in some way combine with the Russians and so on. We have transferred our institute from BERLIN to HECHINGEN, as there was so much bombing in BERLIN. Now HECHINGEN is in the hands of the French. (Laughs) We moved as far west as possible because we preferred to be occupied by you or the Americans instead of by the Russians, but then we had the bad luck that the French came. And now they probably would not like us to work in the French zone. On the other hand, in this respect you have managed it very badly because they have just imprisoned a part of us and the other half just stays in HECHINGEN and can tell the French everything they want. LAUE, for instance, he had never heard anything about uranium during the whole war, he never knew anything about the machine we were building and Mr. BOPP, for instance – he is a theoretical physicist working in my laboratory – was acquainted with every small detail in the whole business and he stayed in HECHINGEN. And JOLIOT visited him and (JOLIOT) told me, he had told JOLIOT everything about it. Well, I could not help it, it was not my business to prevent him.

BLACKETT: My own personal views are that the majority of you should go back reasonably soon. Whether you should leave nuclear physics for the time being and work in other things – I don't know whether it will be necessary or not. I don't think, there is any danger in working in nuclear physics, provided that no great uranium or . . . plants are built. Obviously, the subject is perfectly right. I am quite sure, personally, that the rest of physics and science generally will be clear for you within a fairly reasonable time in GERMANY. That is my own view and nothing will shake it. Whether the people agree with me, I am not quite sure. I know, most scientists do. It is obviously ludicrous to take any other view on the long run.

HEISENBERG: There is still one other question in this connection, this control we spoke about. This is almost a necessary thing in GERMANY now in order that anything should work in science. We have not got any German government. In ordinary times, I would have gone to my government and said: "Well, this is the situation, I would like you to give me some money" and so on. And this is the difficult situation in GERMANY as a whole. Since you don't allow a German government, you must rule GERMANY yourselves. You can't have the cake and eat it and, therefore, if it is done as it seems to be done now, then really it ought to be done and then there should be a control to every institute of physics, either American or English. In any case, there should be some provision made that the English adviser or the local government can see that we get money or anything like that.

BLACKETT: The trouble has been that we were not really prepared for the end of the war and the whole conception of what is going to happen to GERMANY was very vague, but is now a little clearer since the BERLIN Conference. The rough outlines are settled and what is clear and, I think, fundamentally sensible is that the government starts locally, building up, rather than to start right at the top.

HEISENBERG: Yes, obviously.

BLACKETT: I don't know what the group is going to be, whether there will be, roughly speaking, the English, the French and the Russian groups, but, as regards getting resources, there will have to be an economic and financial administration. Presumably there will be largely German personnel with a few Allied people at the top. I should have thought, supposing you and some of your colleagues went back to the English zone, GOETTINGEN or somewhere (int.)

HEISENBERG: There are only two universities which are still intact in the English zone, GOETTINGEN and HEIDELBERG. Of course, HEIDELBERG is in the American zone. But these institutes are filled with people already, they would not need any more.

BLACKETT: It is a question of mere accommodation?

HEISENBERG: It is a question of accommodation and also the following: I am Professor at the University of BERLIN and hitherto I got the money from there, but since the end of the war I have not got any money at all and I don't know whether my wife can live or not. Of course, the same problem arises for everybody and this is a matter which must be settled in some way.

BLACKETT: Those domestic things, there I would be in a very good position to get something done. I think we can get a reasonable policy adopted. I don't think, I should be frightfully worried about the bread and butter side of it. Things are difficult, well, they are difficult for the whole of EUROPE. I don't think, you need be frightfully worried.

HEISENBERG: There is really very much trouble in my family. You have probably heard that two members of my family have died during the last months when I was away. My mother has died and then a friend of my wife. She did not actually belong to the family, but she used to live with us. These things happen, but I was a little angry with the Americans because, when they took me, they told me they would take care of my family. My wife was not even allowed to write to me that my mother was ill. So I could not see my mother before she died. Also, after she had died, I did not get any letter from my wife. I only got it two months later. I think, something must be done to see whether the people have anything to eat or not.

BLACKETT: Those are the things that will right themselves in the end.

HEISENBERG: The situation was this: I had lived in LEIPZIG. Our house was destroyed in 1943 and I moved the family to a cottage I owned before the war in the BAVARIAN ALPS. This cottage was meant for a very few people and at the end of the war there were about 13 people in this very small house. My family is still there. The last day, when this Colonel PASH came to take me to HEIDELBERG, I told him about the extreme difficulty of the food situation and he was very kind and allowed me to go with an American jeep to some place where I could buy bread for about a week.

BLACKETT: When have you last heard from your family?

HEISENBERG: That was about three weeks ago. That was the first letter I had since we were taken prisoner. Since that time I have felt that something must be done about the family and I told our Major about that. Let us speak about the general questions. If we go back, say, to GOETTINGEN, then the terrible question arises where people should live in GOETTINGEN, because all the houses in GOETTINGEN are certainly overcrowded to the last. That would be so in practically every German city. Have you been to GERMANY?

BLACKETT: Only in FRANKFURT. That is pretty bad. What universities are in the American sector?

HEISENBERG: HEIDELBERG. I should say that HEIDELBERG would be the best, in so far as the city of HEIDELBERG is still intact.

BLACKETT: MUNICH University is gone, is it?

HEISENBERG: That is completely burnt out. Well, it may be that in the outskirts of MUNICH you would perhaps be able to find some room. HEIDELBERG would be a good place in so far as there are several institutes which are empty now. Then BOTHE is in HEIDELBERG, so there would be co-operation with BOTHE. From the accommodation point of view, it would be best if we could simply stay in HECHINGEN until something is found. I was the only one who had no accommodation for my family in HECHINGEN, but probably I would be able to find a flat.

BLACKETT: All the people except LAUE and HAHN worked in HECHINGEN?

HEISENBERG: Even LAUE and HAHN were in HECHINGEN. We were practically all in HECHINGEN except for DIEBNER.

BLACKETT: The institute is running?

HEISENBERG: Yes, the institute is running, as far as I know. If it is simply the problem that it should be avoided that we discuss matters of nuclear physics with the French, of course, we would be able to give our word of honour not to do it or something like that.

BLACKETT: I don't personally believe – that is, when people wake up to the facts – that there is anything in it. There is more published in the American document than we know, at least more than I know. It is published, it is not secret. There are only about a dozen copies in ENGLAND yet. It is extremely detailed. It is all about plutonium, separation processes etc. I don't think, there are any secrets.

HEISENBERG: No, from a reasonable point of view there is, of course, not the slightest reason to keep us here whilst Mr. BOPP is in HECHINGEN.

b) At this point the conversation was interrupted and was resumed again after dinner, as follows:

HEISENBERG: Apparently there will only be a kind of provisional arrangement now and the final decision about what shall happen to German science will be made later.

BLACKETT: Yes, I think it has been formulated now, but it will take some time to get finally settled. As to your own future, action is going on very energetically at the moment.

HEISENBERG: When do you think that that will be decided so that we could perhaps leave here?

BLACKETT: I am a little bit new here to the position, only having heard of your position here three days ago, but, I think, I am now in a position to put my weight about. It is the change of government and this new Committee. We are very much junior partners in the collaboration with AMERICA, and your position here is really American. It is not a British responsibility. They have said what they want done and we have done it. We have taken part in this bomb project only in a very small way, and we are not, in fact, free agents. One can see, till the bomb was dropped, that there was an argument for our extreme secrecy, but after that there clearly is not. But now the bomb has been dropped and this committee has been formed, it is quite possible to outline a reasonable policy and push for it. I think, the circumstances are quite favourable for pushing for a decision. The decision may be made in WASHINGTON, not here, but, obviously, we will have a say in the decision. The decision should be – one extreme would be that you should all be sent back to GERMANY, just carry on your ordinary work with some restrictions on nuclear physics. And then I can say to you, there is a possibility of some of you being invited to work in ENGLAND or AMERICA. I don't know, and I don't know whether it will be made, and I don't know what the reactions would be. I am quite sure there wouldn't be any compulsion about it. You will be given the option. At least, in my own view. Remember, I am talking very much as a private individual. You know, no one can talk, in a sense, anything but as a private individual, because the decisions are made by an enormous number of people, and you never know what is going to happen. But I suggest this as a possibility. I don't think, it will be many of you, frankly. You and HAHN or somebody like that. What your reactions would be, and what my advice to you would be, I don't even quite know.

HEISENBERG: I can tell you what my feelings would be. I think that LAUE, if he were asked, would always say "yes", because he has his son living in AMERICA. That is what I could imagine. But for the rest of us, in some way, we feel that it is our duty to stay in GERMANY. Now, of course, if the situation in GERMANY becomes so bad that one should say: "We'll never have the opportunity of doing any reasonable research work", that is a different matter again because finally my life is bound to physics. Therefore, I really don't know, but I feel more inclined to say: "If possible, I should like to be in GERMANY and re-build the thing".

BLACKETT: I think, you are right. I think, the future of GERMANY is extremely important for EUROPE. You are one of the people who can help to get things going again.

HEISENBERG: There is one thing which we haven't mentioned: what shall HAHN do with his institute? First of all, he is the discoverer of the fission. All the experiments which he carried on in his laboratory are pure chemistry of radio-active substances and certainly are not of the slightest danger to anybody.

BLACKETT: No, I agree entirely.

HEISENBERG: The third is that it is difficult to convince a politician that HAHN is not a dangerous person. But just in the discussions which you now have with Mr. ANDERSON, I think, it would perhaps be convenient to say, it is good that there should be a control in every German scientific institute because, if they make such a control, then perhaps they are less afraid of, say, Mr. HAHN continuing his experiments.

BLACKETT: If I had my way, I would say, that, subject to a very rigorous, overriding control on heavy industry and production and prototypes, I would leave you free to get on with some work. I would allow you to have uranium in kilograms or so. Whether that will happen, I don't know.

HEISENBERG: Personally, I would go entirely away from experimental physics. But I have had this institute for about five years and I think that these people in some way depend on me and it is nice to go on. The idea always was, I would like to have some small uranium machine going, just so much that one can get very strong radio-active preparations so that HAHN could get (int.)

BLACKETT: I agree. And all the biological aspects.

HEISENBERG: Well, we were interested in having a strong activity of neutrons. I thought, I can make some nice experiments with high intensity neutron beams and things like that. This will be allowed?

BLACKETT: It may happen, but I don't even know whether I can get back. We don't know ourselves what the status of nuclear physics in ENGLAND is going to be. Are we going to have a Shangri-la? We don't know. To you generally as a group I would say: "In general, go back and study the harmless part of physics and be as good as you can and do your very very best to build up a new GERMANY".

HEISENBERG: If I am going to re-build the institute, then, of course, I would be glad if I could re-build it in this place where it really is going to be. The difficulty is, that, at present, people are in HECHINGEN. Now I don't know whether HECHINGEN will, in the long run, be the best place. In some way, of course, it would be best to say: "Let us go back to BERLIN where we have an institute which is still entirely intact". I think, BERLIN would be the best solution in the long run or, may be, one of the big cities.

BLACKETT: I entirely sympathize with your concern with these practical difficulties. But the real difficulty is on a broader level. There are differences of opinion about this question. There is the view that all scientists are dangerous, and, if we can get over that and substitute for "the scientists are dangerous", "the application is dangerous and wants control", then one can find individual ways of sorting the practical difficulties. The biggest difficulty perhaps is the question of the different zones and the relations with RUSSIA. That is a very serious one. I simply don't know. I don't believe, there is a policy about it yet. But it seems quite clear that there is a very, very strong feeling that one must not risk any of you brilliant key people working for RUSSIA. There is an interim period when AMERICA alone has it. And in that case it may be sensible to prevent you working for the Russians.

HEISENBERG: Obviously. Of course, I would not mind at all if we should give a kind of word of honour never to join the Russians. But, if in the long run it turns out that the living conditions in GERMANY are just impossible and if then, say, the Russians would make very good offers to some of the physicists, it would be difficult for those to say: "I rather starve in GERMANY than go to RUSSIA". – Then the situation is this, that you will get some decision on the principal questions first, and then all private problems will be discussed.

BLACKETT: I think, that, if they can get the right decisions on the main issues, it will be quite easy to take steps to get private things right. But I can't be certain about how the primary decision will go. I don't think it will go adversely in the long run, it may easily be delayed. I just don't know.

HEISENBERG: If it is delayed, that would mean that we would be detained still longer.

BLACKETT: Well, I don't think so. I suspect that you wouldn't be detained (longer). I mean, this is an absolutely pure guess.

HEISENBERG: With regard to my family, I think that something must be done. When I got this letter from home, I felt that, if I only told the Major that I am worried about the family, then exactly nothing will happen. He, perhaps, tells it to the next one, but the next one certainly does no . . . , so I decided to tell him that I am taking back my word of honour that I would not go out of this house. This, apparently, has had the consequence that they told you about the whole thing, and, in so far, already something very good has come out of it, but I rather think that something must be done about my family, and I suggested, which the Major approved, that one should perhaps try to get in connection with my wife's brother.

BLACKETT: I know him well, he stayed with us a few days ago.

HEISENBERG: Did he tell you about my family?

BLACKETT: Oh yes, he tried to get in touch with you, but he was not allowed to.

HEISENBERG: Do you think there is a chance that I could get in touch with him? Or do you think that would not be allowed?

BLACKETT: There is absolutely no conceivable reason, in my view, for this complete black-out on information about your being here, and the fact that you have been kept incommunicado from everybody is just silly now. Up to the bomb being dropped there was a black-out on everything, now I don't see the slightest reason from any point of view why SCHUHMACHER should not come to see you, but this is not, in fact our responsibility – (int.)

HEISENBERG: Who is going to decide, WASHINGTON?

BLACKETT: Well, it will have to be settled in WASHINGTON and I think we are in a very strong position to be pretty firm about it and take steps. I can see no conceivable objections on any grounds and I cannot understand exactly why there is this extreme blackout about the whole thing at all.

HEISENBERG: Did my brother-in-law tell you what he thinks should be done about the family, because he has been there?

BLACKETT: He stayed with us one night before he went to GERMANY the last time.

HEISENBERG: Did he know at that time that I was taken away from GERMANY?

BLACKETT: Yes, he knew you were not in GERMANY. He did not know where you were. I see no objective reasons at all why SCHUHMACHER should not come and see you and should not, if necessary, go over there and get your family going. The difficulty is that it is not, in fact, at the moment our responsibility, but I think, we are in a position now to do something about it.

HEISENBERG: How long ago is it that my brother-in-law has been with you?

BLACKETT: I should guess, about the end of July.

HEISENBERG: That was before he went to GERMANY?

BLACKETT: He had been before. He is in American uniform on the Bombing Research Mission. You know, he has become rather an extremely successful economist here. He is very friendly with MORRISON, the Lord President of the Council, and is very much in with the Labour Party. He is rather the sort of private adviser to the Labour Party on economics. He is very well thought of indeed. I met him two years ago with CRIPPS. He was one of this new, what I call, neo-KEYNESian school of economists.

HEISENBERG: Of course, I would be very glad to see him also. But if you perhaps could just telephone to him and ask him what he thinks one could do perhaps – (int.)

BLACKETT: I think, the chief thing to do, is for me to try and get some contact to be allowed in general between you people and other people. And later that will come too. I don't see any reason why other scientists should not come and see you here too. There was laid down a law that no one should see you. We didn't lay it down.

HEISENBERG: That is clear. Only the problem is, it may take again weeks until you get this permission. If you could just telephone (int.)

BLACKETT: I don't think, SCHUHMACHER is in ENGLAND at the moment actually.

HEISENBERG: The important thing would be that somebody takes care of my family in some way, the best thing would be, if SCHUHMACHER could go there.

BLACKETT: I think that can be arranged quite easily. Tell me, on the whole, everybody here would be glad to get back to GERMANY as soon as possible, roughly speaking?

HEISENBERG: Yes.

(Pause)

I have a few more questions. One is that BOHR has his 60th birthday on October 5th. I would have liked to write to him and to send him a paper which I had intended as a kind of 'Festschrift' for him. I don't know whether you think that this would be possible.

BLACKETT: Send it to me and I see what can be done. If the decision is made that you should be allowed to go back, it will want a visit to GERMANY from myself or someone to the Control Commission to go round and explore the possibilities of . . . But we got quite a good, very sympathetic scientist on the Control Commission now who will be in charge of that sort of arrangement. I don't think there is much doubt about the majority of your junior colleagues here. They will be allowed to go back to GERMANY fairly soon. But I think there is a possibility of you key people being asked whether you would like to work in this country. I don't know.

HEISENBERG: There is one thing I would also ask in this connection. Could you just keep us in touch with things. Of course, you are not allowed to tell us about all the discussions. That is obvious. Only it is very difficult just to have to sit here and know nothing about what is happening.

BLACKETT: I cannot make any promises, because I am a bit new in the job. I am fairly optimistic that things will clarify fairly quickly now.

HEISENBERG: This question of my family, do you think that in a short time I could get some statement on the business?

BLACKETT: Yes, you will get a statement quickly.

HEISENBERG: What do you mean by "quickly"?

BLACKETT: A fortnight.

In connection with the above, it is worth noting the following remarks made by HEISENBERG to HAHN on 5th September, 1945:

HEISENBERG: I have the feeling that we shall be nearer getting out of this detention once we do get permission to contact my brother-in-law. If I could talk to him, we could achieve a good deal more because, after all, he has got various connections here. And he, on his part, could contact BLACKETT for instance or somebody like that. Besides, he could perhaps send some food to my family.

3. On the morning of 9th September, HAHN told Professor BLACKETT of his anti-National-Socialist feelings and willingness to co-operate with the Allies in the following conversation in English:

HAHN: As I know from the letters from HECHINGEN and TAILFINGEN, the institutes seemed to be allowed to work now as before, but you think, HEISENBERG told me, it will not be allowed – (int.)

BLACKETT: No, I did not say that. I said that there is no policy settled yet. There are certain differences of opinion in the Allied Control Commission about German science, but I have no doubt at all that the view of the majority of English scientists and the sensible people is that ordinary academic and pure and fundamental physics will be completely uncontrolled. There may conceivably be some bar on nuclear physics. Already there is a bar on nuclear physics in ENGLAND. The trouble is that nuclear physics in a big way can only be done with big government support. You have got to have cyclotrons and things like that and so it ceased to be, even in ENGLAND, a free subject. If I had my way, I would say that nuclear physics in a small way will be as open to you as it will be to us.

HAHN: It was really a more secret subject in your country than in ours, because I published my papers. I told the people that, if I do not publish our harmless things, we make the American and English people think, we are making bombs etc. Therefore, we show them that we do quite harmless things, but I told them, you never can tell what will happen to-morrow. Therefore, I saved the people in my laboratory. They did not go to war because I told them, it was awfully important; of course, we knew that it was of no importance at all. We really were cheating our government. The same as my friend MATTAUCH he was just as much a fanatical anti-National-Socialist as I was.

BLACKETT: I don't think there will be restrictions.

HAHN: I have some people – for instance my friend STRASSMANN and there are other young people – who have really good experience in separating isotopes or, say, indication method and I think it would be interesting even for the Americans, because I don't know whether they have so very many chemical people with that experience and therefore, it is a waste of manpower, even in the interest of the Americans. Of course, we would welcome any amount of English and American friends or people who look at our work, perhaps they could even learn something chemically.

BLACKETT: I think that you need not worry very much, that is coming out quite all right; what is annoying to you is the delay. People have been working here very hard to help you and to get things right and my own view is that their effort will succeed quite soon.

HAHN: I am not speaking for myself but for the young people. What shall they live on if they are not allowed to work?

BLACKETT: Are they working?

HAHN: I suppose so.

BLACKETT: That I can find out.

(Pause)

HAHN: My great trouble is really my institute. What is the future of people who really behaved very well during the war? I hardly had a man who was in the Party. If I had not discovered this uranium fission, so that the people said: "Oh, that is very important", I think, I should have lost my place.

BLACKETT: I may say that your reputation is very well known over here because of your very fine record as anti-Nazi. It is very much appreciated, so don't you worry.

HAHN: I always think that one might persuade the people that we can be of use and help to the Americans and, of course, English scientists, if they really allow us to continue.

BLACKETT: We are very sympathetic. You have been stopped, through the circumstances of the war, for some months. Heaps of us have been stopped in our own subjects for six years.

4. In the following conversation in English Professor BLACKETT advised Von LAUE to return to the re-building of GERMANY and offered to try to arrange meetings with Professor BORN and Sir William BRAGG:

BLACKETT: You are very keen to get back to GERMANY?

LAUE: Yes.

BLACKETT: It ought to happen soon. It is all very unfortunate. Great efforts have been made over here by the people concerned to get everything cleared up, but, of course, we have had a very difficult position. We have not, in fact, been able to do what we want and I think, they tried to make it as comfortable as you can be under conditions which we did not invent.

LAUE: Some of our younger colleagues have the wish to go for one or two years to AMERICA or ENGLAND. Is that perhaps possible?

BLACKETT: Who?

LAUE: BAGGE.

BLACKETT: You mean just for ordinary pure science?

LAUE: Yes.

BLACKETT: I don't think that it is any use trying to do that immediately. I think the sensible thing is for you to be sent back as soon as possible and to get going under the best conditions and then perhaps, in a year's time, visits may be possible. But I cannot tell at all about that. I don't think it is worth looking ahead to ask when you will be able to move about. We don't know ourselves when we will be able to move about EUROPE freely.

LAUE: Do you think it is possible for me and my wife to go to our son at PRINCETON.

BLACKETT: When I don't know, but there is no doubt the time will come. The strain on transport is extremely great and it will be quite impossible to make a case for you to have priority of transport. That I think is not reasonable to expect, but what I think is reasonable, is that we shall try and do our best to make sure, when you are back in GERMANY, that you have reasonable conditions for working. It is easier for you as a theorist. I can assure you that we scientists in ENGLAND are very sympathetic. I am very keen that publications will start properly again and also the exchange of periodicals. If you went back to GERMANY, what would you do?

LAUE: I would go to the seat of the 'KAISER WILHELM INSTITUTE'.

BLACKETT: That is now moved to HECHINGEN.

LAUE: Only a part of it has moved to HECHINGEN and I don't think it is possible to go to BERLIN yet.

BLACKETT: I feel that you are one of the people who should, in my own personal view, be back there helping to re-build academic and scientific life. Your reputation is very high as a very wise man who has taken a very good line and you are respected enormously. I remember talking about your attitude to the

Nazi-movement – do you remember – right back in '38. I feel, you ought to be one of the people back there, trying to get the right views across and re-building things. That you would like to do?

LAUE: Yes.

BLACKETT: I would like it to be possible to have wider discussions between you and other people about the future of German science. I do not know why you should not see BRAGG. As I say, we have not been in control of this, but I will certainly remember it and see what I can do. BORN and BRAGG are the two people whom you would like to see again?

LAUE: Yes, and G.P. THOMSON.

5. Later on Sunday morning, Professor BLACKETT had another talk with HEISENBERG in which they discussed the guests at FARM HALL, as follows:

HEISENBERG: WEIZSACKER is not interested much in going back to the uranium business, he is more interested in astro-physics at present. He was professor at the University of STRASSBOURG, but that of course is finished. But every German government will always be glad to give – (int.)

BLACKETT: He is extremely good?

HEISENBERG: He is a very good man.

BLACKETT: He wants to go back to GERMANY?

HEISENBERG: Yes, I think so. He feels the same way as I do we are needed in GERMANY and that we are not needed in the UNITED STATES.

BLACKETT: I agree with you about that. What about KORSCHING?

HEISENBERG: He is not a brilliant physicist, but he is a very good experimenter and he had a nice idea on separation of isotopes. He had an apparatus going which was in some way a combination of the idea of CLUSIUS and DICKEL and the . . . diffusion.

BLACKETT: He looks a little unhappy.

HEISENBERG: He is the type of man who has never been abroad. He is German and he has never come out of his German cities. He thinks it is terrible to stay here, he has no work to do.

BLACKETT: BAGGE?

HEISENBERG: His primary quality is great energy. He is a very active man and in so far he has done good work. He has done partly theoretical, partly . . .

work. I don't know whether you have seen his papers on cosmic radiations. In some ways he is a proletarian type, he comes from a proletarian family and that is one of the reasons why he went into the Party, but he never was what one would call a fanatical Nazi. In some way I like him quite well. He has his great ability to work . . . If I tell him: "You try to build up this apparatus", then I can be sure that in a short time he actually has done it.

BLACKETT: He is mainly an experimenter but does a little theory?

HEISENBERG: Yes.

BLACKETT: WIRTZ?

HEISENBERG: He is a much better physicist, he really knows the whole game of physics. He is an experimenter and he has been very helpful to the organisation. He is the type of man who knows how to organise an institute. If WIRTZ had not been in our institute, we certainly would not have come so far in the uranium business and all the other things; especially when we had to try to get some apparatus, he was the man to organize it. In some way he is perhaps the most important of the younger people in my institute. He really managed the whole thing, I could not manage an experimental institute at all. WIRTZ is very good. He has never made a real mistake in his experiments.

BLACKETT: What was he working on before?

HEISENBERG: He has been working on thermo-diffusion.

BLACKETT: What would he go back to?

HEISENBERG: For instance, if we re-started the institute for cosmic rays, that would be nice. Then he certainly would be able to do any kind of work connected with cosmic rays.

BLACKETT: There is nothing special about him, politically or anything else?

HEISENBERG: No, he was politically always on the good side, on our side.

BLACKETT: KORSCHING is the same, has he always been neutral politically?

HEISENBERG: Yes, entirely neutral. The only ones of our party who have been in the national-socialist Party are DIEBNER and BAGGE. DIEBNER is of course, politically perhaps the most difficult case of all.

BLACKETT: He is a physicist?

HEISENBERG: He has got his degree in physics, but he is not really a physicist, he is more a kind of 'Verwaltungsmann'. He was connected with the 'Heeres-Waffenamt'. He was the organiser of this whole uranium business on a very small scale from the beginning.

BLACKETT: Under GERLACH:

HEISENBERG: He was not under GERLACH. In the beginning, it was wholly a matter of this department of ordnance; his man was a general General SCHUHMANN (?). Then SCHUHMANN (?) was replaced by ESAU.

BLACKETT: What would he do, if he went back, what would DIEBNER do?

HEISENBERG: That is one of the difficulties. He suggested several plans. The one plan said, in some way he could be connected with administration of scientific work; the other possibility would be that he could be connected with industry.

BLACKETT: There are innumerable thousands of people in GERMANY who have got to find a new occupation. We have a sort of scientific responsibility for the good people, the not-so-good people will have to find their own way. That is the lot? We have been through all?

HEISENBERG: HARTECK is professor at HAMBURG.

BLACKETT: He just goes back, I suppose?

HEISENBERG: He just goes back.

III. REACTION TO PROFESSOR BLACKETT's VISIT

The only detailed reaction to Professor BLACKETT's visit was a conversation between HEISENBERG and von WEIZSACKER, showing concern for the future of the KAISER WILHELM Institute in particular and of science in general. VON WEIZSACKER, whilst admitting the necessity – in their own interest – to work at present for science on an international basis, admits doubts whether this would be a paying proposition to any nation while national sovereignty still exists:

HEISENBERG: On the whole, I find the situation satisfying.

WEIZSACKER: I should say, BLACKETT is a sensible man with whom one can get down to brass tacks.

HEISENBERG: Yes.

WEIZSACKER: Is it your impression that the rest of us might get back to HECHINGEN long before you and HAHN?

HEISENBERG: That is conceivable, although I think that once the first change in the manner of our detention is made, the whole thing must soon come to an end.

WEIZSACKER: There is just one snag; if you don't get to HECHINGEN while LAUE does, the institute will have the most incompetent leadership imaginable.

HEISENBERG: In fact, WIRTZ would have the actual control. We would have to think over carefully how to work that. It will be difficult officially to take away LAUE's authority. I might well do it in a friendly manner, saying: "You are the official head, but I recommend that, for the time being, you let WIRTZ do everything".

WEIZSACKER: I would even undertake a little intrigue to see that LAUE doesn't get back either until you get back, but I should say that is rather difficult to bring about, as one should not do too much in that direction.

HEISENBERG: As LAUE is the only one who would gladly go to AMERICA, it might be possible to do it in that way.

WEIZSACKER: I don't mean to be unkind to LAUE, but one must think of the institute.

HEISENBERG: My impression is, even if they are still doubtful about us, that it is most likely they will say: "We don't want you to go to the French sector for any length of time, but by all means go there with BLACKETT and have a look round, but actually we should like to set up the final show in GOETTINGEN".

WEIZSACKER: On the other hand, I don't want to give too much power to WIRTZ which he, himself, doesn't want anyway. If for instance WIRTZ gets back, then the question arises what the attitude will be towards SCHUELER. You are the only one who can deal with him. It is obvious that you will take SCHUELER's place when you get back and, if WIRTZ gets there before you, then he would have to be the nominal head. But the whole problem of subsequent relations with SCHUELER is so unpleasant that it will be pointless for anyone else but you to deal with it.

HEISENBERG: I am actually not pessimistic in this respect as, even if the Americans did offer me an opportunity to go to AMERICA, I could still say that I should have to return to GERMANY first in order to see how I stand or something like that. BLACKETT repeated again and again that he was firmly convinced that it would be senseless to control science in the future as science as such is not at all dangerous. Only the industrial use of science presents a danger. He said that one must make a clear difference between science as such and its application.

WEIZSACKER: It is not in our interst to say so, but – and even BLACKETT must realize that – if we, in GERMANY, had withheld the publication of HAHN's discovery for only about a year, then the Americans would not have been able to complete the atomic bomb in this war. From the point of view of pure science, it is clear that every limitation is primarily a hindrance to progress. From the point of view of the politician, however, I should say that the discovery of a principle such as a dynamo or uranium fission is of the utmost interst just at the stage before it has been developed technically. Because, if you get it at such an early stage, you have a chance to get ahead of your competitors.

On the other hand, once the principle has become common knowledge, then every country has enough competent people to develope it. However, . . . not say that now, as it might spoil our own chances but it is true nevertheless.

HEISENBERG: The consequence of what you say would be that <u>all</u> science would be kept a close secret everywhere, but I do not think that things will turn out in that way, because for instance, it appears that the Americans have in fact published pretty well all the details of their atomic bomb.

WEIZSACKER: I am by no means in favour of that happening either. It depends on what one wants. One can say: "We want politics to be international and so we also want science to be international". But if for instance the Americans are afraid of the Russians or the Russians of the Americans, then you cannot say with a clear conscience: "We want science to be international". If the theory I am now propounding were to be accepted, then it could only mean that we should be forbidden to do scientific work in GERMANY. Thus, obviously, we must work for international science.

HEISENBERG: Yes, I am convinced that the only possible thing for us in GERMANY is to adopt the international theory.

WEIZSACKER: Obviously.

HEISENBERG: The international theory will win in the end, even though there may be five more wars fought with atomic bombs in the meantime.

WEIZSACKER: Actually, this would necessitate the abolition of national power. As long as there is anybody who is interested in power, he must try, under present circumstances, to monopolize the results of scientific research. Only when all national power has been abolished, will it be possible to have international science.

HEISENBERG: Yes, and that is exactly the theory the Americans have adopted and therefore they are gaining practically all the power.

WEIZSACKER: But I must say that, what they have really achieved, they have achieved through secrecy.

(Laughter)

HEISENBERG: BLACKETT thought nevertheless that nuclear physics and uranium physics would be impossible for us for some time so that one would perhaps have to see for the time being that what HAHN does is not so defined. It seems to me that one must wangle one's way around these things by clever definitions. We shall have to see what can be done in my institute, but I would just as soon work on cosmic rays.

WEIZSACKER: Have you the impression that BLACKETT has influence?

HEISENBERG: His influence is certainly increasing at present, because he has been a Labour-man for a long time and is now slipping into politics. Whether or not he is clever enough to retain his influence later on is unforeseeable.

WEIZSACKER: Do you think that he is now working for political influence? Is he at heart a scientist who is glad to get back to science or is he prepared to sacrifice science for politics.

HEISENBERG: Certainly partly the latter. Although, of course, he wants to do scientific work, he realizes that he has an influential position and wants to do something in that sphere as well.

WEIZSACKER: Yesterday I overheard him saying something which prejudiced me enormously in his favour. I do not know exactly what question he was asked, but he replied that, in his experience, when physicists start meddling in politics, then they become just like other politicians. That seems to me so realistic compared to all other physicists' opinions that I feel that he may be the only physicist who could really be of any use as a politician.

IV. LECTURES

The usual bi-weekly lectures have been given but have been confined to general physics and there has been no other technical discussion worth reporting.

FARM HALL,
GODMANCHESTER.
15th September, 1945.

(Sgd) P.L.C.Brodie
Capt. I.C.
for Major T.H. Rittner

Copy No. 6.

Lt. Cdr. Welsh.

Ref. F.H.8.

To: Mr. M. PERRIN and Lt. Cdr. WELSH.

From: Major T.H. RITTNER

Distribution:
Copy No. 1 – Capt. Davis for Gen. Groves.
" " 2 – Capt. Davis.
" " 3 – Pro. Blackett.
" " 4 – Mr. Perrin.
" " 5 – C.P.A. for C.S.S.
" " 6 – Lt. Cdr. Welsh.
" " 7 – Major Rittner.

OPERATION "EPSILON"

(14th Sept.-15th Sept. 45)

1. This report deals only with a letter written by HEISENBERG to Professor BLACKETT and the discussions attending the writing of the letter.

2. Before mentioning the letter, HEISENBERG discussed the future with GERLACH in the following conversation on the afternoon of Sept. 14th.

HEISENBERG proposes the "Deutsche Physikalische Gesellschaft" as a central authority, suitable to control physics in GERMANY and, he hopes, able to overcome the difficulties of the several separate zones of occupation.

There is also mention of certain funds held by 'Kaiser-Wilhelm Institut' and by GERLACH.

HEISENBERG: Die 'Deutsche Physikalische Gesellschaft' (German Physical Society) is a useful organisation, as it would see to it from the beginning that the various zones will not be isolated from each other.

GERLACH: Yes, it has already occurred to me that the 'Deutsche Physikalische Gesellschaft' is an organisation which is politically harmless.

HEISENBERG: Quite, it has been in existence since before the Nazis and it has not been nazified.

GERLACH: It hasn't become a member of these various Nazi organisations but last year or the year before those two Nazi leaders were put in.

HEISENBERG: But that wasn't the Society's fault. Of whom were you thinking?

GERLACH: Of FINKELNBURG in the first place.

HEISENBERG: Yes, FINKELNBURG, of course, is slightly Nazi – (int.)

GERLACH: – whom RAMSAUER wanted to have for purely political reasons.

HEISENBERG: Although I believe, FINKELNBURG is not of ESAU's type.

GERLACH: No, he has conducted himself in this discussion on religion – (int.)

HEISENBERG: – very decently.

GERLACH: Also, when this discussion was being prepared, he always was on the right side.

HEISENBERG: As far as I know him, I never had the impression at all that he was very wildly Nazi. I don't think that he would have joined any Jew-baiting. I think, he was at one time a member of the S.S. I suppose that is true. I would say, the Deutsche Physikalische Institut has now got to choose clever people who can cope with the others. Indeed, I think, one should have people on the executive who can cope with the Russians and also people who can cope with the Americans. They will have to negotiate with the Inter-Allied Control Commission from whom they will have to get fundamental decisions, so that they can carry on and do any necessary wangling. I would very much like to go to BERLIN now, for instance, and have a talk with SAUERBRUCH. I am sure, he could do something for me. By now, he must have one or two things running and then one could probably get quite a decent position in BERLIN. My feeling is that the solution which has to come and probably will come is that the Anglo-Saxons will see to it that we shall have again a unified administration for all zones.

GERLACH: With the 'Deutsche Physikalische Gesellschaft', there were also the 'Gauvereine'. I should think, it is quite likely that these 'Gauvereine' as such will carry on independently for the time being, until there is a central controlling administration.

HEISENBERG: Besides, there must be a chief whom we can choose ourselves, as in the past, and this chief can then direct affairs all round.

GERLACH: And he would have to try also in the Russian zone – (int.)

HEISENBERG: Yes, especially there. I would say that the BERLIN 'Gauverein' in particular must work with all three at the same time. We still had money from the 'Reichsforschungsrat' left in HAIGERLOCH – 100,000 marks.

I put it in my institute's account for safety. That, of course, is at our disposal again when it is wanted.

GERLACH: Officially, I have half a million marks in a special account in MUNICH. Whether that is still there, I don't know, of course. That was money which had got lost on its way from BERLIN: so I went to ETTAL, where the 'Kultusministerium' was and asked them to sign a letter for me to the 'Reichsbankstelle' in MUNICH, asking them to advance half a million marks to the 'Bayrische Staatsbank' and to debit the amount to the 'Kultusministerium'. The President of the 'Reichsbank' did it and the President of the 'Bayrische Staatsbank' acknowledged it. From there, it was transferred to my bank, the 'Hypotheken- und Wechsel-bank'.

HEISENBERG: If we can save that money somehow, we should do so at all costs and use it for research in some form.

3. Early in the morning of 15 September, HEISENBERG proposed to HARTECK, VON WEIZSACKER and WIRTZ that he should write a letter to Professor BLACKETT setting out their own wishes for their future which he could bring up at a supposed conference on Thursday, 20th September.

All believe that their best hope for the future is to work under Allied, including Russian, control, as they will then be given more ample facilities.

There is an undercurrent of German nationalism in the minds of VON WEIZSACKER and WIRTZ.

This was the discussion:

HEISENBERG: What do you think of my once more putting the different points of view in writing for BLACKETT? The things we have discussed may well have gone in at one ear and out at the other, but, if they have been put down in writing, they can look them up again during their discussion and say: "It's like this."

WEIZSACKER: What would you want to write?

HEISENBERG: With regard to our institute I would say: I would prefer a wider field with control to a very restricted one without control. If there is no control, we would certainly not be able to do any work on uranium and certainly nuclear physics only on a very restricted scale. That leaves only work on cosmic rays. If there were no control, we would certainly do no work on cosmic rays that needed aeroplanes or balloons either. Perhaps, if we were lucky, we might just do something in connection with the JUNGFRAUJOCH. That means, all that will really be left to us would be a kind of tuppenny-ha'penny physics with cosmic rays. The Americans have worked in that field so thoroughly that it would take many years' work to catch up technically. That is not really worth while. In fact, it would be better for the institute, if they allowed a wider field of research and controlled it. If they don't want to do that under any circumstances, then it would be better for the institute, if, say, DEBYE came and worked on quite a different branch of physics. We ought to consider whether we should not simply put that in as another suggestion. Then the institute would have to be changed completely to something where there are no problems of secrecy and so on. I would then go back to theoretical physics and we'll have to see how things develop. In favour of control we might also mention that a reasonable control is the beginning of collaboration which, after all, should be desirable for them as well. If we write that to BLACKETT in a private letter, I don't believe that there is any danger of being hanged by the Werewolves later on. We might also consider whether we should not actually suggest that we shall work only on cosmic rays and not on nuclear physics but that we should at least work in BERLIN under a control which will make it possible for us to get aeroplanes for our work. There will always be English pilots who can fly them, and, if there is an English or American supervisor in our institute, then he would have the right connections in that quarter.

WEIZSACKER: (gloomily) This, of course, could even more easily be interpreted as 'collaboration'. In any case, I think, it is an excellent thing for BLACKETT to know exactly what we would like. Didn't you tell him?

HEISENBERG: Well, I told him more or less what we have discussed tonight. (Pause) If we had to work in a field that has been exhausted, then we would have to interest ourselves in experimental precision methods. I am a completely unsuitable person to organize that. But DEBYE could find solutions and entirely different ways to be able to work in some field not considered dangerous.

HARTECK: Then comes the question: Do they want us to start highly qualified work in GERMANY again?

HEISENBERG: BLACKETT certainly does.

HARTECK: One could imagine some physical institute in HAMBURG on the lines of the 'Reichsanstalt'. I don't believe that the English would allow that.

HEISENBERG: I am not so sure. The 'Deutsche Physikalische Gesellschaft' ought now to do this: Firstly, they ought to tell the 'Gauvereine', each 'Gauverein' should get on good terms with its local government. But, secondly, the 'Physikalische Gesellschaft', as a whole, continues to exist. The others are not likely to forbid that, because it is in a very good position politically. And now, if they are clever, they'll get people onto their board who know how to deal with the various people. There will be enough to deal with the Americans, but there ought to be somebody who can somehow deal with the Russians. In that way, we'll have to see to it that there is a central scientific organisation in GERMANY. I don't believe, the others will prevent that.

WEIZSACKER: We have been told that the future of German research is still under discussion. We don't know what the results of these discussions will be. If they decide that research in GERMANY is not to be encouraged, there is no future for our institute. Apart from that, it is, of course, as well if BLACKETT knows in time what we would like to do in case we get permission to work.

HARTECK: Will they, for instance, allow us to continue work on the benzine syntheses (Benzinsynthesen) in the chemistry of the carbohydrates? If there is anything that interferes in the slightest with their own business, are they likely to prohibit it?

HEISENBERG: They will prevent all types of organisation which will again make GERMANY an independent industrial power.

WEIZSACKER: These things will change. If they now, for instance, forbid research, it is important that we do university work, that is train young people who actually can't do anything now. But in five years' time, when they have learnt something, research may perhaps be allowed again after all, and then we shall need these people. I am quite convinced that after a few decades research will be done again in GERMANY on a large scale. So now the main thing is to bridge the gap during which, perhaps, it won't be possible.

WIRTZ: Well, I think, we might just as well ask for as much as we can hope to get done for us. (Pause) To return to the letter to BLACKETT, I am all for telling him, we would like to work on as many different subjects as possible and we would have no objections against some sort of control. Perhaps you could add that he should keep this letter as private as possible.

HEISENBERG: Naturally, the Commander, and all these people will see it.

WEIZSACKER: It does not matter if they see it, only the Werewolves shouldn't. I would like to know if any of you has any doubts about this letter because that is a matter on which we can decide now.

HEISENBERG: I wanted him to have the letter before Thursday's meeting, because it is important that he should by then thoroughly understand the position. I am a little afraid that the easiest solution for the English would be to say: "They must not do nuclear physics. They may work on cosmic rays." After all, that is the easiest way out for them and one should not make it so easy for them.

WIRTZ: I think, if we can keep the things going now, then later on perhaps we might achieve more instead of being tied to the best which we can get for ourselves at present.

HEISENBERG: I don't know about that.

WEIZSACKER: I think that, with people like BLACKETT, one should ask for things on a grand scale and, above all, make proposals that will keep us in permanent contact with BLACKETT or some of their people. After all, that is quite a good thing.

HEISENBERG: Yes.

WEIZSACKER: I would say, it is fundamentally very good for us, if an Englishman is obliged to be personally present in our institute all the year round.

HEISENBERG: Yes.

WEIZSACKER: I would, therefore, say, the control is not at all unpleasant but rather it is desirable, just as I think it desirable that the Allied Government of GERMANY lasts as long as possible so that the Anglo-Americans have full responsibility for whether the people starve or not. Of course, we must not get a man who is too awful. But perhaps BLACKETT can see to that. Naturally, if we get a man like DIEBNER, that would be intolerable. I would not like to have the Commander, of course; he again is too high up for that, nor would I even like to have his son. The only problem is whether that will later be held against us as 'collaboration'.

HEISENBERG: Actually, I think so.

WIRTZ: If the whole thing is properly worded, I see no danger.

WEIZSACKER: Of course, one could consider starting a big low-temperature laboratory, if everything else fails.

HEISENBERG: That would have to be in BERLIN as well and I don't know at all whether one shouldn't perhaps say, the best solution is to go back to BERLIN in a year's time or so. Wouldn't it really be the best solution if we were to work in BERLIN on cosmic rays and low temperature physics and be controlled by an American as well as by a Russian?

WEIZSACKER: Now, that's quite ideal. Probably that won't be possible, that would be too much to expect. Because, once they let us go to the Russians, then they'll also let us work on uranium. That is the same thing.

HEISENBERG: Through my letter to BLACKETT, I want to prevent him from adopting the solution which is easiest for the English. Such a solution annoys me as it would merely mean that we may go home and they would take no further interest in us. That is just what I want to prevent. I would say that in that case collaboration with HAHN's institute is also a point in favour of that as it would facilitate a more favourable solution insofar as the English will say, from the point of view of prestige, it is hardly possible to stop HAHN from working on uranium. That would be so niggardly that they could hardly bring themselves to do it. If we could now continue to work together with HAHN on such things, under control, then they'll say to themselves: "We'll have to give permission to HAHN, we shall have to control him anyway; we might as well allow the whole group HAHN – 'Kaiser-Wilhelm Institut' to continue to work under control. That does not apply to the rest of German science which shall be free, but this is a special case. That is perhaps after all the most reasonable solution."

WIRTZ: I am pessimistic on that point, because they are so terribly afraid of German research.

HEISENBERG: In any case, if they discuss whether HAHN should go on working on uranium, all physicists on the American commission, on which OPPENHEIMER is, or the people here will say: "Yes, we'll have to give him permission, we can't do anything else."

HARTECK: Are you sure?

HEISENBERG: I am certain about that.

WEIZSACKER: There is a certain trend in the world which is now beginning to appear; let us call it 'internationalism'. There are quite a number of people, especially in ENGLAND and AMERICA who think that way and I don't know at all whether they're doing their countries any good. But they are the people to whom it is best for us to attach ourselves and we'll have to support that. Those people who don't want to keep any secrets about the atomic bomb are the people who are useful to us.

4. Later on the morning of the 15th HEISENBERG, having read his letter to HAHN, discusses it with him. He says that the letter, in fact, implies that, unless he is generously treated by the Western Allies, he will seriously consider working for the Russians.

HEISENBERG: How do you feel about the whole thing?

HAHN: I am afraid that they want to stop us doing nuclear physics.

HEISENBERG: In your case, there is a good chance that they won't do that. It would be very hard for them, also from the point of view of prestige, to stop you from working, as without any doubt you discovered the whole thing.

HAHN: I think that what you have written there is excellent. Do you think, it would perhaps be wiser if DEBYE joined us?

HEISENBERG: I would say that, if they are determined to curtail nuclear physics, then DEBYE is certainly a good solution. DEBYE would then re-organize the institute completely to suit his own work – (int.)

HAHN: No, I mean that you should be there, only that a man like DEBYE, being definitely 'Non-Nazi', should be there as well.

HEISENBERG: Yes.

HAHN: I don't believe DEBYE would ever go to HECHINGEN.

HEISENBERG: Certainly not to HECHINGEN, but perhaps, say, to HEIDELBERG. Actually, he would perhaps prefer above all to go back to BERLIN. In some ways, DEBYE's return would, of course, be the simplest way to get over the difficulties. If DEBYE takes over, then he has got to do everything and then it is a question whether I should not simply go to MUNICH or do something else. After all, why shouldn't I simply do theoretical work? I would say, if nuclear physics can be done in a big enough way, or cosmic rays with sufficient means, then I should, of course, like that. If, however, one has to work with a few shabby measuring tubes (Zaehlrohr) on subjects which have already been exhausted by the Americans, that just is not worth while.

HAHN: Can you do work on cosmic rays in GERMANY at all? Have we got mountains 3500 and 4000 m high?

HEISENBERG: What I should like would be firstly: I should like to put apparatus into an aeroplane. There is an immense number of aeroplanes in GERMANY which, of course, cannot be used by Germans. But I don't at all see why any American pilots, who have to practice flying anyway, should not take up with them some apparatus or, for instance, one could send people with apparatus up the JUNGFRAUJOCH. I think that this slightly more international co-operation is definitely the right thing. I had imagined doing nuclear physics and cosmic ray work in greater style in peace time. To do modern physics in a small way is of no use at all. There is a great temptation now for the Anglo-Saxons to say: "These people shall, of course, continue to work. We don't want to take away all their chances, but these big technical things are quite out of the question for them." This temptation is great. That means, we are permanently put on ice and can do physics on the Roumanian or Bulgarian scale. But that is exactly what we must endeavour to prevent these people from doing. Then it is greatly preferable that they, themselves, should take the responsibility for what is done in GERMANY, and then it will be in their own interest to see that it is done properly.

HAHN: Actually, it ought to be quite agreeable to the English, at least, if not everything is done in AMERICA.

HEISENBERG: I would also say, the English ought really to be interested in re-building EUROPE, and they are bound to say to themselves – I am not going to tell them that, because then they would detain us still longer – if the English tell me: "At the very best, you will be allowed to work with tupenny-ha'penny apparatus" and the Russians say: "You will get an institute with a yearly budget of half a million." Then I would consider if I shouldn't go to the Russians <u>after all</u>. Because that is really simply idiotic. After all, we are there to work sensibly. Up to now, it is certainly so that in these scientific matters the Russians are much more generous than the English. I am firmly convinced of that, because the Russians have already learnt the modern way of thinking which the English, of course, have not up to now. I don't know at all that the future of EUROPE does not lie with the Russians after all. What impressed me again so very much in that direction recently was the reaction of the very nice American Major who said: "Yes, we Americans want to go home again. We don't want to see those destroyed towns. We are going to send our boys home and we are going to leave here just as many as are absolutely necessary, but we don't want to have anything to do with the whole business." Well, my God, if that is what they want, then we are bound to work with the Russians.

HAHN: Unless the English take the initiative.

HEISENBERG: They can't, they are much too weak. They can't do anything against the Russians.

HAHN: If we could achieve being allowed to carry on under English supervision, that would, of course, be by far the best.

HEISENBERG: Yes. I find, there are only these two possibilities: either the Anglo-Saxons take so much interest in EUROPE that they really want to hold it, and holding in that case means that they take part in nearly everything which is being done and see to it that it is being done properly. But if they don't want to and say: "No, for all we care, EUROPE can go bankrupt." Then there is nothing left for EUROPE but to turn to RUSSIA. Because the Russians wouldn't let it go that way, they would so something. I think, we ought to make it a little more difficult for them to take the line of least resistance, and naturally the line of least resistance is: "Well, yes, we will allow you Germans to do a little physics again on a small scale." That is exactly what we don't want.

HAHN: But, of course, it is dangerous to threaten them with RUSSIA.

HEISENBERG: We can't do that. After all, that is implied when I say, I don't want to do petty physics. Either, I want to do proper physics or none at all. If the final decision is that I can't do any proper physics and I go back to GERMANY again, naturally they, too, will realize that I am then going to consider doing physics with the Russians after all.

5. The letter, the original of which was passed on on September 17th to Lt. Com. WELSH:

"Dear BLACKETT, After our recent talk, I thought matters over and came to the conclusion that it might be as well to tell you about the special situation of my institute before we discuss the future of our institutes in general. The 'Kaiser-Wilhelm Institut' for physics in BERLIN was built by DEBYE and was under his direction until January 1940. After DEBYE's departure, DIEBNER was in charge of the administration for some time. He was responsible for the conversion of the institute to nuclear physics. From spring 1941 onwards, I was practically in charge and later also officially. Since that time, the bulk of our work was done on uranium and on nuclear physics in general (high tension apparatus), besides that, work was done on cosmic rays, as you will know from our book, and also work in the various departments (X-ray laboratory, low-temperature laboratory and optics) was continued on a reduced scale. As the first aim of our scientific work we had intended to build a 'Brenner' (burner) with D_2O, graphite and uranium metal. This 'burner' was to be a strong source of neutrons. In war-time, naturally, these results would have been followed by technical developments which would have aimed at a practical use of the energy. We had, however, hoped that the 'burner' could in peace-time be used for the preparation of powerful radioactive substances for chemical research at HAHN's institute and also perhaps for experiments at my institute with neutron-beams of great intensity. The quantities of D_2O, graphite and uranium at HAIGERLOCH and STADTILM would probably have been just sufficient for the construction of the 'burner'.

The first question regarding the future of the institute is whether DEBYE will return and take charge once more. We managed for DEBYE to be still in charge officially, therefore it depends only on him whether he is willing to take charge again and, if so, there would probably be no political difficulties since he is of Dutch nationality. If, on the other hand, DEBYE does not return, the question of the future subject of our research will arise. If we are to choose between the following two possibilities, either a very much reduced field of work without control or a wider one but controlled by Allied physicists, then I should certainly prefer the latter possibility for the following reasons: If there were no control, policy would probably prohibit work on nuclear physics in general. In that case, provided the line of research is not completely changed, the institute could only work on cosmic rays. And in that case any collaboration with HAHN's institute would stop. Besides, in the field of cosmic rays all possibilities would be closed to us if they involve considerable technical means, i.e. experiments with aircraft and such like. Then the institute would have to rely on working with the most modest technical means in a field in which others have already done such thorough work that the result would not justify such an effort. In that case, it might be better to switch over the institute to quite different subjects with which I have nothing to do. If, however, work on nuclear physics could be continued, at least with high tension apparatus, so that the collaboration with HAHN's institute would continue, I would certainly be prepared to put up with control. Then perhaps one might also be able to get the support of the authorities for experiments on cosmic rays so that a considerably greater technical effort becomes possible. After all, in the long run, it will be possible for physicists who are themselves interested in science to exercise a control.

The question of the location of the institute can probably be decided only when the future of the 'Kaiser-Wilhelm Gesellschaft' as a whole is being discussed. Perhaps HECHINGEN is a suitable interim solution which, however, should be replaced as soon as possible by a permanent solution. Probably there would be empty institutes in HEIDELBERG and GOETTINGEN which could be used, but no living accommodation. The institute in BERLIN is undamaged and is occupied by American troops. It is possible that the Russians had previously removed apparatus and equipment. Perhaps it would be easier in BERLIN to get our old flats back, but the specific difficulties of BERLIN are not in favour of that town.

I hope, we shall be able to discuss all these problems once more when the fundamental questions have been settled.

With kindest regards

Yours (sgd) W. HEISENBERG"

Farm Hall
18 Sep 45

(Sgd) P.L.C. Brodie
Captain
for Major T.H. RITTNER

Lt. Col. Calvert for General Groves.

Ref. F.H.9.

To: Mr M. PERRIN and Lt. Cdr. WELSH.

From: Major T.H. RITTNER

OPERATION "EPSILON"

(16th-23rd Sept. '45)

I. GENERAL

1. This has been a difficult week, chiefly spent in placating HEISENBERG, who had expected rather more from Professor BLACKETT than was perhaps justifiable.

2. HEISENBERG's letter to Professor BLACKETT – mentioned in F.H.8 was delivered to Lt. Cdr. WELSH on September 17th.

3. On 20th September, I had an involuntary conversation with BAGGE, in which I tried to prepare the ground should there be, as I expected, no immediate result of the Conference supposed to take place on that day.

II. THE MEETING ON 20th SEPTEMBER

4. On 21st September, I went to LONDON at the request of the Professors to find out the result of the meeting.

5. On my return, without immediate permission to return everyone to GERMANY, a tense situation arose.

6. Professors HAHN, HEISENBERG and GERLACH impressed upon me that the vital question was really the future of Frau HEISENBERG and his children. Unless they were looked after, HEISENBERG would feel morally bound to withdraw his parole and the rest would loyally do likewise.

7. I gained a breathing space by repeating that the meeting was still in progress and that a decision was expected shortly. It is always disastrous to give a definite date, as, when nothing happens on that date, the effect on our guests is unfortunate.

8. I also agreed to try to get in touch with Professor BLACKETT through Lt. Cdr. WELSH to see whether he had made any arrangements with HEISENBERG's brother-in-law, Herr SCHUMACHER at OXFORD, to look after HEISENBERG's family.

9. HEISENBERG discussed this conversation with VON

WEIZSACKER, WIRTZ, KORSCHING and DIEBNER that evening without, however, referring to Herr SCHUMACHER:

HEISENBERG: We were talking about my family and, generally speaking, the result of to-day's meeting was such that one still doesn't know what they want. This is certainly a plausible result, but not one, one might have wished for. There will be another meeting next week and it is hoped that a decision will be taken then.

DIEBNER: Nobody knows what, I suppose.

HEISENBERG: Everything up to now seems to point towards the one thing that, if anything will be decided at all, it will be decided that we should go home. But the point is that they cannot make up their minds to do that.

KORSCHING: Is this an all British Committee?

HEISENBERG: On the contrary, BLACKETT told me, I should realize that in these matters the British are very much playing the second fiddle.

WIRTZ: I think that one should turn completely to AMERICA.

HEISENBERG: That is why it is a pity that we have so little connection with leading Americans. To talk to a mere American Major is of no use.

WIRTZ: The tragedy is that we are in ENGLAND. We should either be flown to AMERICA or be sitting at American Headquarters in GERMANY. But here in ENGLAND we are right off the map.

HEISENBERG: Yes, there is something in that.

DIEBNER: What did the British hope for in bringing us here? If they don't have any influence, what does it all mean?

HEISENBERG: They obviously have some influence.

WEIZSACKER: One should nurse friendly relations with both of them. The Americans may be stronger, but Britain may have more say during the coming years in determining our fate. I have the feeling that the Americans are less interested.

HEISENBERG: At any rate, the British have a really vital interest in EUROPE, because, if they do not succeed now in making a lasting arrangement in EUROPE, they will be in for it next time. This is quite obvious.

10. On Saturday morning, HEISENBERG asked me whether I could not by-pass Lt. Cdr. WELSH and telephone to Professor BLACKETT direct. I refused, but explained that it was much better not to bother

Professor BLACKETT unduly who was doubtless doing all he could and that veiled threats such as a withdrawal of parole would undoubtedly have the effect they least desired. In a conversation later on, HEISENBERG expressed his annoyance to HAHN as follows: "I can wait a bit, but it is nevertheless annoying that the Captain may not telephone to BLACKETT."

III. <u>TECHNICAL</u>

11. On Tuesday, 18th September, HARTECK gave a lecture on 'Atmosphere', based on the work of FOWLER and MILNE.

12. There has been nothing else of technical interest during the week.

FARM HALL
GODMANCHESTER

(Sgd) P.L.C. Brodie
for Major T.H. RITTNER

27 Sep 45.

Copy No. 1.

Appendix to F.H.9.

To: Mr. M. PERRIN and Lt. Cdr. WELSH
From: Major T.H. RITTNER

OPERATION "EPSILON"

(Production of Heavy Water)

1. On 23rd September HARTECK described to HAHN a method of producing D_2O which he and SUESS had discovered and which is considered more economical than any other method known to them.

2. As adequate facilities for translation were not available at FARM HALL, this has been done in LONDON.

3. The, translated, conversation was as follows:

HARTECK: It is a question of an ammonia distillation. The most expensive part is the vessel for distillation. Then you have to take some substance which has as high a vapour pressure difference as possible at the highest possible pressure. Apparently that is ammonia. Moreover, that has three hydrogen atoms and will exchange with water when you want.

HAHN: Does the ammonia gas flow straight through the water?

HARTECK: No, you continually fractionate the ammonia and exchange the ammonia which has been once regenerated with water. You have a column in which there are 5 or 10 tons of ammonia.

HAHN: There is more light ammonia above -

HARTECK: And the heavy is at the bottom. Then you take the ammonia, the light upper fraction, allow it to exchange by flow against water until it has its old content. So all I need is really just a few tons of ammonia which is certainly not dear.

HAHN: Do you then impoverish the water?

HARTECK: No, it's any old water.

HAHN: And the ammonia that you get again from the water -

HARTECK: That is again normal.

HAHN: And the light ammonia just disappears?

HARTECK: No, the light ammonia is not lost.

HAHN: Well, then the hydrogen exchanges -

HARTECK: With the water and is driven out again later. It is as if for example you take 20 per cent out and 80 per cent is normal, then you carry out the exchange. It is just a question of economy.

HAHN: How much does it affect it if you have 5 tons of ammonia? How long must you distil this fractionally so as to get a reasonable enrichment at the bottom?

HARTECK: One step is 2.4 per cent. You would have to have a column about 80 metres high if you want to get twelve-fold enrichment.

HAHN: For one step -

HARTECK: No, the important thing is how many KW. hours must you expend in order to get 1 gm of heavy water.

HAHN: I believe yours is cheaper. I would like to know just how you work it. Would you put 12 columns side by side?

HARTECK: No, one giant column. Diameter about 2 metres, height 80 metres. It works like this: If I have cold water and hot water, say, at 5° and 20°, then I have got an evaporator at 20° and condenser at 5°.

HAHN: And in what form is the ammonia?

HARTECK: Liquid ammonia.

HAHN: Under a certain pressure then?

HARTECK: Yes, at 9 or 10 degrees it has its pressure of 4,000 mm. or so. That's the best.

HAHN: Then you get a few atmospheres on it?

HARTECK: Several Atmospheres. In a column with so many bends you have a pressure drop of about 1½ atmospheres. You can estimate that straight away. And this fall of pressure of 1½ atmos. corresponds to several degrees difference of temperature. And then it boils down to this: I can heat with hot water and cool with cold water. But I could however, have a heat pump; that is, a refrigerator, and if I have this, then I need continuously 2000 KW. To make 5 tons per annum. For one gramme one needs 3 to 4 KW. hours. That is, if you use a heat pump. If you use hot and cold water, it is very much less and of course costs very much less. I only got that worked out after we had designed the column.

The procedure is cheaper. It seemed so to everybody that I spoke to about it according to any means of estimating. Yet apparently these people work with graphite-blocks and not with heavy water. But perhaps heavy water has difficulties, if it picks up neutrons and becomes radioactive. And if one has in fact played about for far too long with the thing, like I have, and has a definite solution, it is not at all good, to be stopped in such a way. Don't you think so?

HAHN: Yes, I agree.

HARTECK: Now the Question is, in Germany apparently we cannot do any more.

HAHN: Certainly not with heavy water.

HARTECK: What I would really like, would be, if I could have a talk with one of the physical chemists, that is Rideal or H.S. Taylor. If they have not already got this process, then I certainly would not show up badly.

HAHN: Agreed. You talk to them then and they say to you, "Yes, we knew that long ago", and they go out and take out the patent afterwards, or they would use the process.

HARTECK: Oh, a patent application is all the same to me. But all the same one would at least like to be able to collaborate a bit or to come a bit more in contact with the scientific people. We became so discreet in Germany. It is a pity that we put so much time into the thing and this is the result. We have built several columns and the I.G. constructed -

HAHN: You would really like Taylor or Rideal to come here? They do not let anybody come in here, that is the shame of it.

HARTECK: But this you could say to the people, that of all that has been done so far you got the impression that the process that we have for heavy water is better than what they have by an appreciable factor. So it is rather an act of friendship on our part in the special case of heavy water. I do not want to press myself on these people but it is about half a year since we are here and you could not say we were running after them. The Major can say to the Commander that I would very much like to see RIDEAL on such and such a matter.

HAHN: That you had a method wereby you were convinced that you could prepare heavy water 2 to 3 times as cheaply as by previous processes.

HARTECK: I won't exaggerate but you can certainly say 3 times. The Director said if it was set up and you could sell heavy water at 20 pf. per gm. you could do quite good business. There is nothing in writing that you could show. In my notes I have only "According to the new process from the Institute for Physical Chemistry – Method for Heavy Water – an apparatus will be built at a cost of 1.2 million marks to be capable of producing 5 tons of D_2O." Otherwise there is no word about it. Right from the highest levels it was admitted that that was naturally quite unreasonable last February. The vapour pressure of ammonia had not even been measured and then you need an exchange apparatus; that is in fact the point, that one fractionates not only the ammonia but that one continuously exchanges it with water. These are in fact the two dodges.

HAHN: Now then at the bottom you have your ordinary ammonia which you gradually enrich.

HARTECK: Yes, indeed.

HAHN: What happens to it?

HARTECK: This ammonia flows into the second column and the second column is set vertically in the interior of the first column. Then in the second you set a third inside. The first is round about 2 metres diameter, the internal one is perhaps 40 cm. and the innermost 15 cm. Can I draw it for you? This is the first column the big one. That is, we will say, normal, that is perhaps 30% impoverished and here three-fold. It exchanges here with water and comes here again against the counter current. Then here comes the three-fold – here we have the column inside, then the lower material comes up here, that is three-fold. In reality it may be much more here, it is already ten-fold. So here it is ten-fold and here it is round about 80 fold.

HAHN: For what reason then?

HARTECK: Because it then goes on being fractionated. That runs down and the gas rises. Then they both go through again. Here the 80-fold and here the 800-fold.

HAHN: What happens to the light fractions? Are they always carried off by the water?

HARTECK: Yes, they act on each other and here the light one, that is simply the counter current, the ammonia flows in here, water in the opposite direction and it is only a few seconds before they have come into equilibrium and then the water which flows away, with of course all the ammonia in it, has to be boiled. And the remainder which I have, that is already let us say 10%, that is simply burnt, it plays no more part. Now we have 20% water with which I can do what I like.

(Sgd) PLc Brodie
Capt.
24 XI 45

DEPARTMENT OF SCIENTIFIC & INDUSTRIAL RESEARCH

DIRECTORATE OF TUBE ALLOYS,
16 OLD QUEEN STREET,
WESTMINSTER, S.W.1.

Telephone: Whitehall 1632.
(Extension 202).

8th October 1945.

Sir James Chadwick, F.R.S.,
Technical Section, B.S.C.,
P.O. Box 680,
Benjamin Franklin Station,
Washington 4, D.C.

TOP SECRET

Dear Chadwick,

I am sending you herewith two further reports by Rittner on Operation Epsilon (F.H.10 and F.H. 11).

Yours sincerely,

(M.W. Perrin)

Copy No. 1

Lt. Col. Calvert for Gen. Groves

Ref. F.H.10

To: Mr. M. PERRIN and Lt. Cdr. WELSH
From: Major T.H. RITTNER

OPERATION "EPSILON"

(24th–30 Sept. 1945)

I. GENERAL

1. This has been a quiet week, the guests being largely engaged in writing a memorandum concerning their own future. They all seem quite happy at present.

II. THE MEMORANDUM

2. On 24th September, Professor HAHN discussed the situation with Major RITTNER, Professor HEISENBERG, in a conversation with Professors VON WEIZSACKER and WIRTZ, having nevertheless already made it clear that he no longer really wished to withdraw his parole:

HEISENBERG: I have had another talk with HAHN. He will tell the Major that we are, on the whole, really very depressed and that it has become increasingly clear to us that nothing will be decided about us as soon as we had hoped, as it won't be allowed on account of higher politics, and that perhaps it would be a good thing if some kind of a limit could be fixed by which at least something could be decided about our families. Besides, he shall make it quite clear to the Major that he (HAHN) had only succeeded after tremendous trouble in persuading me to wait at least for a few days before proceeding with this parole business. I think, it is quite a good thing to continue exerting pressure in this way, although my personal feeling is that I should not withdraw my parole in the immediate future.

WEIZSACKER: Actually, I think that, to-day being the date on which they had expressly told you that you would be given some news, you are perfectly justified, in fact, it is actually your duty, to apply pressure in this direction if you want to have a basis for further negotiations.

HEISENBERG: Yes.

WIRTZ: However, we must not push matters so far that we have suddenly no alternative left.

WEIZSACKER: We must not push matters so far that we have to carry out our threat.

3. HAHN reported his discussion with Major RITTNER to HEISENBERG, Major RITTNER having suggested, as they were getting restive, that they should write a memorandum on their political convictions and hopes for the future, but HEISENBERG did not wish to be associated politically with DIEBNER:

HAHN: The Major thought that we should put something about our political convictions into this memorandum, that is, something about the anti-Nazi convictions of our whole group.

HEISENBERG: Yes, but if we put down something in this memorandum about a general anti-Nazi attitude and DIEBNER signs it, then I could not conscientiously sign it as well. You know as well as I do that DIEBNER has joined in the Nazi-game.

4. GERLACH, in a conversation with HAHN, expressed similar doubts:

GERLACH: DIEBNER could never sign with a clear conscience that he has never taken part in war work of any kind.

5. KORSCHING, who is unmarried, in a conversation with BAGGE and DIEBNER, thought that this was not the time for making demands, but felt that, once back in GERMANY, it would be easier to get permission to work on an uranium engine:

BAGGE: I do think that this memorandum must be compiled with great care.

KORSCHING: The main thing is to make them send us home. Later on, things will arrange themselves. I mean, in a year's time, we can certainly build a uranium engine again, nobody will care a damn. But if we say now that we won't go unless we can build a uranium engine, they will just keep us here. We must just renounce our wishes in that respect, to start with at any rate. Later on, we must just ask these people for permission again, then they are bound to say "yes" after a while. That is also in the interest of the English. The English obviously want the Americans to keep their hands out of it. The Americans, of course, will only do that, if, for the time being, we promise that we will not work on uranium at all. If after a year we go to the English and say: "Give us permission", then they will certainly grant it, then they will have able assistance. In a year's time all this will be out of date, then it will simply be an English enterprise. Then you may have to work in ENGLAND, but there is no reason why one should not go there.

6. In a general discussion, the guests found some difficulty in composing the memorandum, as their principal desire is for something to be done for their families and not to protest their political convictions which they feel are well enough known. Nothing of interest transpired from this discussion which has not already come to light in previously monitored conversations.

7. A photostat copy of the memorandum is attached as Appendix I.

III. LETTERS

8. Letters arrived from TAILFINGEN on 25th September and, after translations had been made here – which translations were handed to Lt. Cdr. WELSH on 28th September – the letters themselves were handed to the guests on 26th September, together with letters from Prof. BLACKETT and FRANK.

9. The letters gave general pleasure except to BAGGE who is something of a masochist and immediately thought the worst of everyone when there was no letter for him. In a conversation with DIEBNER, he said: "That man was in HECHINGEN and I have had no letters. Have we really deserved that? We have told them everything and now they simply detain us. What have we done?"

10. On 28th September, letters arrived from various scientific friends which have further improved the temper of the household and given some anxiously awaited news of the 'Kaiser-Wilhelm-Gesellschaft'. These were not previously opened here.

IV. THE FUTURE

11. Von WEIZSACKER, a propos of nothing in particular, suggested in a conversation with HAHN and GERLACH that it would be unwise to return the party to GERMANY:

WEIZSACKER: If I were in charge of the whole question, I would certainly think it over most carefully before letting us go home.

GERLACH: I have also come to the conclusion that they probably will not let us go to GERMANY now.

V. <u>TECHNICAL</u>

12. On Friday, 28th September, HAHN gave a lecture on his first discoveries on uranium fission, as previously published.

13. There was nothing else of technical interest during the week.

FARM HALL,
GODMANCHESTER
3 Oct 45

(Sgd) PLc Brodie
Captain
for Major T.H. RITTNER

Copy No. 1.

Ref. F.H.10.

APPENDIX 1.

Die unterzeichneten deutschen Wissenschaftler befinden sich seit etwa fünf Monaten in Haft, offenbar weil ein Teil von ihnen zu der Arbeitsgruppe gehört hat, die in Deutschland während des Krieges das Uranproblem bearbeitet hat[1)]. Da die alliierten Behörden bisher noch keinen endgültigen Beschluss über das weitere Schicksal der Unterzeichneten haben fassen können, möchten wir darum bitten, die folgenden Gesichtspunkte in Erwägung zu ziehen.

Die unterzeichneten Wissenschaftler haben vor dem Krieg, insbesondere soweit sie auf dem Gebiet der Kernphysik gearbeitet haben, stets reine Grundlagenforschung betrieben, und zwar in der gleichen Weise, wie dies international in der Wissenschaft überall üblich war. Sie haben die brauchbaren Ergebnisse ihrer Arbeiten veröffentlicht, haben sie auf internationalen Tagungen vorgetragen, und ihre Institute waren regelmässig von ausländischen Physikern oder Chemikern besucht[2)]. Erst im Verlauf des Krieges sind in den Instituten auch Kriegsaufträge bearbeitet worden; trotzdem ist die rein wissenschaftliche Arbeit weitergegangen[3)] und es ist uns gelungen, einen wertvollen Teil der wissenschaftlichen Tradition und des wissenschaftlichen Nachwuchses durch den Krieg zu retten. Die Unterzeichneten glauben, dass der Wiederaufbau der Wissenschaft und der Universitäten in Deutschland auch im internationalen Interesse liegt, und dass man dafür sorgen soll, dass das durch den Krieg Gerettete nicht nachträglich noch verloren gehe. Die Unterzeichneten werden für diesen Wiederaufbau dringend gebraucht; sie haben den Wunsch, die Arbeit in der gleichen internationalen Weise wie vor dem Krieg fortzusetzen. Wenn die Politik zur Zeit hierin gewisse Einschränkungen verlangt, sind sie bereit, diese in Kauf zu nehmen.

Schliesslich möchten die Unterzeichneten darauf hinweisen, dass die völlige Trennung von ihren Familien bei den äusserst schweren Lebensbedingungen in Deutschland für diese Familien und damit für sie selbst eine sehr grosse Härte bedeutet[4)].

Aus diesen Gründen möchten die Unterzeichneten bitten zu erwägen, ob nicht die folgende Regelung möglich wäre:

1. Es wird den Unterzeichneten gestattet, nach Deutschland zurückzukehren und dort für ihre Familien zu sorgen.

2. Sie nehmen ihre Lehr- und Forschungstätigkeit wieder auf, wobei der Umfang ihres Arbeitsgebiets und der Ort ihrer Arbeit noch im Einzelnen (etwa im Einvernehmen mit den alliierten Behörden in Frankfurt a. M.) festzulegen ist.

3. Umfang des Arbeitsgebiets. Wenn die Wahl besteht zwischen einer sehr starken Einschränkung des Arbeitsgebiets ohne Kontrolle einerseits und

weiteren Arbeitsmöglichkeiten mit Kontrolle andererseits, so ist die zweite Möglichkeit vorzuziehen. Denn erstens würde eine starke Einschränkung, zusammen mit den schon durch die Wirtschaftslage gegebenen Schwierigkeiten den Wiederaufbau wertvoller Forschungsarbeit fast unmöglich machen, zweitens würde eine Kontrolle jedes Misstrauen beseitigen und vielleicht von Wissenschaftlern ausgeübt werden können, die an der Arbeit selbst Interesse Haben.

4. Arbeitsort. Die Kaiser Wilhelm-Institute für Chemie und Physik haben ihren eigentlichen Sitz in Berlin (amerikanische Besatzungszone), sie sind jetzt in Verlagerungsstätten in Tailfingen und Hechingen (französische Besatzungszone) untergebracht. Wenn aus äusseren Gründen für die Weiterarbeit weder Tailfingen-Hechingen noch Berlin in Betracht kommt, so schlagen die Unterzeichneten vor, Heidelberg, München oder Göttingen in Erwägung zu ziehen, wo es wahrscheinlich leere Institute bzw. geeignete Anlagen gibt. Die Beschaffung von Wohnraum würde hier allerdings die grössten Schwierigkeiten machen.

Anmerkungen

1) Inzwischen sind Dr. Bopp von der gleichen Arbeitsgruppe und Prof. Schüler von den Franzosen verhaftet und nach Frankreich gebracht worden. Es wäre wohl zweckmässig, wenn auf diese Institutsmitglieder die gleichen Gesichtspunkte angewendet würden, die auf uns angewendet werden.

2) Einige der Unterzeichneten haben Institute geführt, in denen eine Reihe von englischen und amerikanischen Wissenschaftlern gearbeitet oder einen Teil ihrer Ausbildung erhalten hat. Z.B. haben wissenschaftlich gearbeitet

an dem von Prof. Hahn geleiteten Kaiser Wilhelm-Institut für Chemie: Dr. M. Francis, Prof. King.

an dem von Prof. Gerlach geleiteten Physikalischen Institut der Universität München: Dr. Bragg, Dr. Colby, Dr. Lane, Dr. Goudsmit, Dr. Little, Dr. R.W. Wood, Dr. Wulff.

an dem von Prof. Heisenberg geleiteten Institut für theoretische Physik an der Universität Leipzig: Dr. Eckart, Dr. Houston, Dr. Jahn, Dr. Feenberg, Dr. Podolski, Dr. Slater, Dr. Uehling, Dr. Nordsieck, ferner die später nach Amerika oder England ausgewanderten: Dr. Bloch, Dr. Weisskopf, Dr. Halpern, Dr. Placzek, Dr. Teller und Dr. Peierls.

3) Die Arbeiten des Kaiser Wilhelm-Instituts für Chemie über die Spaltungsprodukte des Urans sind auch während des Krieges fortgesetzt und in vollem Umfange veröffentlicht worden. Das Kaiser Wilhelm-Institut für Physik hat noch 1943 ein Buch über die Kosmische Strahlung herausgegeben. Ausserdem wurden verschie-dene Arbeiten aus dem Gebiet der theoretischen Physik, über magnetische, elektrische und optische Probleme veröffentlicht.

4) In diesem Zusammenhang ist zu berücksichtigen, dass die Familien der Unterzeichneten zum Teil ihre eigentliche Wohnung durch Luftangriffe verloren haben und in Notquartieren wohnen. Die derzeitigen Lebensbedingungen in

Deutschland, besonders in der französischen Zone, machen es einer Frau, die Kinder zu betreuen hat, fast physisch unmöglich, auch für Lebensmittel- und Brennholzvorräte für Winter zu sorgen. Wahrscheinlich sind die Familien wegen unserer Abwesenheit auf den Winter überhaupt nicht vorbereitet. Ausserdem erhalten sie zum Teil sicher keinerlei Zahlungen mehr. Die Unterzeichneten glauben auf Grund ihres persönlichen Verhaltens in den vergangenen Jahren die Erwartung aussprechen zu dürfen, dass man ihre Familien nicht im Winter in Deutschland verkommen lässt.

26.9.45.

Copy No. 1.

Capt. Davis for Gen. Groves.

Ref. F.H.11

To: Mr. M. PERRIN and Lt. Cdr. WELSH
From: Major T.H. RITTNER

OPERATION "EPSILON"

(1st-7th October 1945)

I. GENERAL

1) All interest this week centres on a meeting between Professors HAHN, VON LAUE and HEISENBERG with various British scientists at the Royal Institution on Tuesday, 2nd October, and a letter which HEISENBERG has written to Professor BLACKETT as a result.

II. THE MEETING

2) In addition to the various British scientists, HEISENBERG also met his brother-in-law, Herr SCHUHMACHER, who apparently gave him an encouraging report about conditions in the British zone, and whom HEISENBERG is most anxious to meet again in the near future. He has asked for this second interview as he thinks, SCHUHMACHER will be able materially to assist him in selecting a site for the future Kaiser-Wilhelm Gesellschaft.

3) On returning from the LONDON meeting, HEISENBERG gave the following account of what had taken place:

"BLACKETT has told me the following: the Americans have decided that we should return to GERMANY. They have made a condition, however, that we are not to return to the French zone. BLACKETT has expressly told me that he does not believe that we are to join our families at home in order to fetch them from HECHINGEN, but he thinks, it is more probable that the families will be brought to us from HECHINGEN. The Americans do not appear to raise objections against our going to the British sector. In accordance with a very detailed report, which my brother-in-law gave me, conditions seem to be best in the English sector at present. GOETTINGEN, HAMBURG and BONN came under discussion. The idea is that some central organisation is to be set up which will somehow arrange the reconstruction of German science. GOETTINGEN was mentioned in this connection, as, first of all, the Kaiser-Wilhelm Gesellschaft has been evacuated to GOETTINGEN and, secondly, the Academy is there and, thirdly, on account of the university. The only snag about GOETTINGEN is the proximity of the Russian sector and BLACKETT thought that the Americans would not consider

GOETTINGEN for that reason. The plan now seems to be that the Allies want to agree on a place where to send us. It will naturally be a tremendous task for any Allied organisation to endeavour to obtain an institute and accommodation for us, to get us there and to have our families and furniture brought along as well, but, fundamentally, that would be the only solution to make this plan attractive for us. The second alternative would be that the furniture will be left there and only our families brought along. In that case, only a few flats would be vacated for us. I have the impression that the Americans consider this latter alternative as the easier one and actually intend to carry it out. But I think that we can still exert some influence in this matter and I almost feel that we shall be able to choose between our families staying in HECHINGEN until a move on a larger scale will be possible and us remaining detained in the meantime on the one hand, and being united with our families as soon as possible but not being able to take our furniture along on the other hand. According to what BLACKETT told me, it will probably be fairly easy to arrange for those, whose homes are already in the American sector, simply to join their families now. When talking of my own family, I sort of asked whether there was a chance for me to return to my family. BLACKETT considered this very probable. As we stand at present, we can influence matters considerably as they themselves don't know yet what they shall do. All this, naturally, within the scope of what they have said."

III. LETTER TO PROFESSOR BLACKETT

4) Later that evening HEISENBERG suggested writing a letter to Professor BLACKETT outlining their wishes for the future on the strength of matters discussed during the meeting. The following is a report of the conversation which took place:

HAHN: Well, what do you think we should decide upon now?

HEISENBERG: I feel, we should do the following: I should simply write another letter to BLACKETT to-morrow or the day after in which I shall put on record firstly what has been discussed in the meeting this afternoon and secondly what we shall now have discussed amongst ourselves. We should decide upon a strategy in accordance with which we can make suggestions to these people, as I feel that at present we have a fair amount of influence on what will be done. For instance, I think we could get a decision through that within the next few days an Englishman, an American and one of us is to go to GERMANY in order to inspect the various places to which the institutes could be moved, and they, together, will then make a report on the possibilities. Perhaps even a programme could be worked out as to what this commission should do in GERMANY.

HARTECK: To judge from my experience, it is always better if several people go, say two or three.

HAHN: I doubt whether they will let three of us roam about, but perhaps we could get them to agree to two.

HEISENBERG: I should say that something could be achieved in this direction as this would not contravene the condition imposed by the Americans to the effect that we are not to leave here until something definite has been decided upon. It is, after all, an obvious and reasonable demand that somebody should travel about to find out things.

HAHN: But the French zone would be out of bounds during such trip.

HEISENBERG: I should say, the places to visit would be MUNICH, HEIDELBERG, GOETTINGEN, BONN and HAMBURG.

BAGGE: HEIDELBERG has not been mentioned anymore earlier on. Why not?

HEISENBERG: Actually only because we would rather steer clear of the American zone of occupation at present which, after all, is very much in the interests of the British just now.

WEIZSACKER: The only ones who have really got a vital interest in the well-being of GERMANY are the British apparently. By the way, I think that the results of the conference this afternoon have been unexpectedly positive.

WIRTZ: Yes, but I think that, if the institutes or even some of their members could not join us from the French zone, these positive results are rather doubtful.

WEIZSACKER: Of course, but these are difficulties of which we have known already.

LAUE: I should think that the English place more importance on general questions than on our institutes.

HEISENBERG: Yes, that is the difference between BRITAIN and AMERICA. The Americans place more importance on people and the uranium bomb and the English are more interested in science in general.

WEIZSACKER: That is all well and good, but, as we are pawns in this game, we must stress that to be able to work is the main thing for us.

GERLACH: It would be unthinkable to waste us on scientific education.

WEIZSACKER: The only thing we can do is to work ourselves in a place where we are in a position to talk with these people. But we cannot be made into just an official administrative body.

HEISENBERG: That is surely not the intention of the English.

5) Having slept on the matter, the drafting of the letter was discussed again on the morning of the 3rd October. In this discussion it appears that

the meeting at the Royal Institution has greatly encouraged the guests, even to the point of dictating their own terms. VON WEIZSACKER's diplomatic cunning is especially worthy of note:

HEISENBERG: The key to the whole problem seems to be that we cannot work in any place in GERMANY without our institutes. It is, therefore, vital that, if we are to be moved, all the institutes and at the same time all our furniture and everything belonging to the families must be moved as well. We must make it clear to BLACKETT, therefore, that the Allied authorities cannot avoid negotiations with the French under any circumstances. The most important point of the letter to BLACKETT must, therefore, be: "Negotiations with the French will have to be taken up to get their agreement on the institutes moving from HECHINGEN and TAILFINGEN to another place yet to be decided upon." Then it is to be hoped that, once these negotiations have been successfully concluded, we will be allowed to join our families in HECHINGEN and TAILFINGEN. The official reason for this could be that we are to prepare the move. But I don't know whether one should stress this point already now as this probably belongs to a later stage.

WEIZSACKER: I suggest, we keep on telling the English that we wish to leave here but, at the same time, keep on making suggestions which are difficult to comply with. And whilst these suggestions are pending, we are kept here and that will serve as a further moral pressure until our suggestions have been carried out. But if we now make suggestions which can be easily complied with but are not worth very much to us, then they will be accepted immediately and then these unsatisfactory conditions will last for years. BLACKETT will encounter tremendous difficulties now. Mr. BEVIN will say: "How can you expect me to do such a thing with the French at this stage, it is quite out of the question." Then BLACKETT will say: "The only famous German scientists who have never been Nazis are sitting about here and not even their families are cared for", and thus, again, BEVIN will be obliged to do something. But if they tell him: "They are in GODESBERG or URFELD and everything is taken care of", then BEVIN will say: "Good, then let us wait for five years."

LAUE: The fact that we are in ENGLAND, however, is favourable as the English simply will have to take an interest.

WEIZSACKER: The English are responsible for us now and we must make use of that.

HAHN: From month to month we have stiffened our attitude towards them and just through that our situation has improved.

WIRTZ: The firmer we are, the more results we will get.

WEIZSACKER: Strategically, it is a wonderful situation for us, if they have to refuse us something, as we can then say: "Alright, but then give us this or that." But, if from the start we suggest something which they don't have to turn down, as it is easy to carry out and yet is not of much use to us, then we shall have to do everything for ourselves.

6) The final letter to Professor BLACKETT was handed to me on the 4th October and was given to Lt. Cdr. WELSH the following day. The following is a translation of the letter:

"Dear BLACKETT,

We were very grateful for the manner in which our discussions in LONDON were carried on and the subjects discussed and, on the strength of these, we have once again amongst ourselves discussed the pending questions in detail. Perhaps it will be useful if I report to you on this.

After what we have been told by you and your colleagues, we appear to be bound to the following conditions: The Allied authorities agree on the whole to our return to GERMANY. However, for the time being, a return to the French zone of occupation is not intended, but rather the scene of our activities is to be shifted to a place within the Anglo-American zone of occupation, such place to make it possible to resume work in accordance with the former scientific tradition and thus be suitable as starting point for the re-building of science as a whole.

Bearing these conditions in mind, the following difficulties have to be pointed out: The resumption of our work is completely dependant on our institutes. Our removal to a larger town with a university will only prove useful if our institutes with all their equipment and their members – who are at present still working in HECHINGEN and TAILFINGEN – are also moved to that town. Such removal, under the conditions prevailing at present in the zone of occupation, will necessitate the authority of the French Government which, as you know, has taken a lively interest in the work and equipment of the institutes. I, therefore, believe that negotiations about the future of the institutes cannot be avoided between the Anglo-American and French authorities. It can be rightly added in favour of such a removal into the Anglo-American zone that these institutes are actually stationed in BERLIN-DAHLEM, i.e. the American sector, and that WUERTTEMBERG only represents a temporary place of evacuation of the DAHLEM institutes. Furthermore, it can be stressed that these institutes represent the private property of the Kaiser-Wilhelm Gesellschaft, which is at present administered from GOETTINGEN, and that these institutes are not the property of the German state.

Quite independent of the result of these negotiations with the French authorities, one should start immediately to look for a suitable place to which to move the institutes. In the discussions, which we had amongst ourselves on this question, various arguments were put forward which, if such place has to be in the English sector, were more in favour of HAMBURG than of BONN. The decision as to the sector itself will hardly depend on us. In any case, we were all agreed that it would be most desirable if two people of our lot here and perhaps also members of the institutes in HECHINGEN and TAILFINGEN were given an opportunity to inspect for themselves the respective places of work, the possibilities for accommodation etc., before the removal of the institutes is finally decided upon. In order to commence work in this direction immediately, we should like to ask you whether the following decisions cannot be brought about:

1) A commission comprising suitable persons of the Allied authorities and two members of our party here (perhaps WIRTZ and I) will be asked to find a suitable place for the institutes and to make a report on their findings. The following should be part of the task set this commission: Visit to the various likely towns (MUNICH, HEIDELBERG, GOETTINGEN, BONN, HAMBURG) and contact with the respective authorities. The headoffice of the Kaiser-Wilhelm Gesellschaft and several members of the HECHINGEN and TAILFINGEN institutes would also have to be contacted. The commission would also have to solve the problems of finance and supplies.

2) At the same time, the Anglo-American authorities will start negotiations with the French authorities on the subject of the intended move of the institutes.

3) As, for instance, GERLACH (MUNICH) and HARTECK (HAMBURG) could give valuable assistance to the Commission in their work in MUNICH and HAMBURG, there should be no objections if such members of our circle who have their families or institutes in the Anglo-American sector were to return to GERMANY immediately – perhaps under the condition that they don't leave their respective sectors.

If this scheme would prove useful in its main points, I should imagine my own activities to be as follows: I would return to GERMANY at the earliest possible date with the group mentioned under 3), that is, to start with, I would return to URFELD in order to try and attend there to the most dire needs. Then, perhaps a few days later, I would be at the disposal of the commission who could perhaps start their work in MUNICH and fetch me there. After completion of our trip, I could return to ENGLAND for a further conference or I could await the final decision in URFELD and then commence with the removal myself. On the whole, it would be – probably also for deciding on a starting point – advisable for me to have frequent opportunities to discuss these questions with you, but

I don't know, of course, to what extent transport could be made available for such purpose.

As regards the question for a starting point, we are all agreed here that, as far as natural science is concerned, the Kaiser-Wilhelm Gesellschaft would be a suitably central organisation and that, for historical and philosophical subjects, it should be the Kartell Der Deutschen Akademien. Connection with the universities could be maintained from both these places.

Once again, many thanks and kindest regards,

Your (sgd) W. HEISENBERG"

IV. OTHER LETTERS

7) Letters, from HAHN to Geheimrat PLANCK, Professor WESTGREN, Professor Liese MEITNER, from VON LAUE to Geheimrat PLANCK and Professor Liese MEITNER and from VON WEIZSACKER to his wife, were handed to Lt. Cdr. WELSH on the 6th October, translations having been made.

V. THE FLY IN THE OINTMENT

8) BAGGE, true to his character, made the following comment to HAHN: "If we go to BONN now, what should I do there alone? Then I shall be sitting alone in a furnished room and have all the trouble of getting my own food."

VI. TECHNICAL

9) There were no lectures and no matters of technical interest during the week.

FARM HALL,
GODMANCHESTER
8th October 1945.

(Sgd) P.L.c. Brodie
Captain;
for Major T.H. RITTNER

AMERICAN EMBASSY
OFFICE OF THE MILITARY ATTACHE
1. GROSVENOR SQUARE, W. 1.
LONDON, ENGLAND

31 October 1945

Subject: Farm Hall Reports No. 12 and 14 - HEISENBERG and HAHN.

To: Maj. A. E. Britt, Room 5004, New War Dept. Bldg., Washington, D. C.

1. Inclosed herewith is copy No. 1 of Farm Hall Reports No. 12 and 14 and Appendix to Farm Hall Report No. 9.

2. In regard to paragraph 3 of page 1, Farm Hall Report No. 12 wherein HEISENBERG and HAHN complained that nothing was being done about their families, you are advised, the following action has been taken: On 30 October 1945, a cable was dispatched from this office to General SIBERT advising him that General MONTGOMERY was reluctant to accept the guests under the conditions first laid down. These conditions, as you know, were that the guests would be returned to some location in the British sector, preferably a university town; that, while they would not be physically confined, it would be necessary for them to obtain special permission for leaving such location. General MONTGOMERY stated that if these men were to be under his custody, the only conditions he would accept them under were that they would be confined and under close surveillance. General SIBERT was asked, in the above mentioned cable, that until the decision came down on these men, if the theater would assume responsibility for the protection of dependent members of their immediate families. This is similar to Project Overcast which was described in our cable 65971 of 11 October. The only two scientists General SIBERT would have to provide for are those in the American sector who are Prof. Werner HEISENBERG and Prof. W. GERLACH. As of this writing we have received no reply from General SIBERT, however, it is believed that the theater will assume responsibility of these two families.

For the Military Attache:

HK Calvert

H. K. CALVERT,
Lt. Col, CE,
Assistant to the Military Attache.

Inclosures - 3
as above

Copy No. 1.

Ref: F.H.12.

To: Mr. M. PERRIN and Lt. Cdr. WELSH
From: Major T.H. RITTNER

OPERATION "EPSILON"

(8th-14th October '45)

I. GENERAL

1. This has been a quiet week, but the guests are showing increasing signs of restiveness.

2. Captain SPEER, U.S. Army, arrived on Monday, 8th October, a relief as welcome as it was necessary.

3. On Tuesday, 9th October, first HEISENBERG and then HAHN approached us, complaining that nothing was being done about their families and that their own detention was altogether immoral. The usual threat of a withdrawal of parole was brought up and, as usual, we were able to prevent this for the time being.

II. TECHNICAL

4. On Tuesday, 9th October, HAHN gave a lecture on previously published experiments in chemistry and fission, carried out by himself, STRASSMANN and GOETTE.

5. There was nothing else of technical interest.

FARM HALL
GODMANCHESTER
18th October 1945

(Sgd) P.L. Brodie
Captain,
for Major T.H. RITTNER

Copy No. 1.

Ref: F.H.14

To: Lt. Cdr. WELSH and Mr. M. PERRIN

From: Major T.H. RITTNER

OPERATION "EPSILON"

(14th-21st October 1945)

I. GENERAL

1. Another quiet and uninteresting week, in which the main emphasis has been on our guests' worries about their families.

2. WIRTZ suggested to me on Sunday, 21st October, that, as he and some of the others did not expect that they would be allowed to return to GERMANY for some considerable time, it would be a good idea to bring their families over here.

II. FOREIGN RELATIONS

3. There is at meal times an increasing tendency among the guests to cavil at the behaviour of the occupying powers in GERMANY, the Russians and French being considered particularly wicked.

4. The general attitude seems to be that the German war was a misfortune forced on the Germans by the malignancy of the Western Powers, who should by now have forgotten that it had taken place (the guests seem to have done so) and that the United Nations should all be largely concerned to set GERMANY on her feet again.

5. Both WIRTZ and WEIZSACKER have argued that the Japanese war was engineered by President Roosevelt who deliberately allowed the attack on PEARL HARBOUR without giving the due warning he was in a position to give. In any case, Commodore PERRY's first expedition to JAPAN was the prime cause of the war which was, therefore, the responsibility of the Americans.

III. LITTLE HITLERS

6. DIEBNER is considered something of a Nazi by his hosts. The compliment is apparently returned as can be seen in the following conversation between BAGGE and KORSCHING:

BAGGE: Tell me, what is DIEBNER so excited about?

KORSCHING: He is quite rightly annoyed about our Nazi-nationalist English guards.

IV. TECHNICAL

7. There was nothing of technical interest during the week.

FARM HALL
25th October 1945

(Sgd) P.Lc. Brodie
Captain,
for Major T.H. RITTNER.

Copy No. 1.

Ref: F.H.16

To: Mr. M. PERRIN and Lt. Cdr. WELSH

From: Major T.H. RITTNER

OPERATION "EPSILON"

(22nd-28th October '45)

I. GENERAL

1. Yet another quiet week. The guests are becoming apathetic, no longer bothering to hold colloquia and, according to WIRTZ, none of them now troubles to do any scientific work.

II. CONSERVATISM

2. The guests do not easily accept new ideas. On Thursday, 25th October, I mentioned that Professor OLIPHANT had said that uranium was not the only possible source of atomic energy. This suggestion was greeted with derision, but, on showing them the relevant article in "Picture Post" for Saturday 27th October, it was generally agreed that there was something in the suggestion.

III. LETTERS

3. All the guests and orderlies have written letters for eventual delivery to their next-of-kin. These, together with duplicate translations, were delivered to Lt. Cdr. WELSH on Tuesday, 23rd October. A letter from von LAUE to his son in AMERICA was handed to Lt. Cdr. WELSH on Friday, 26th October.

IV. MONITORING

4. The monitoring service, which is continuous during the guests' normal waking hours, has produced nothing at all this week.

FARM HALL
GODMANCHESTER.
31st October 1945

(Sgd) P.L.C. Brodie
Captain,
for Major T.H. RITTNER

AMERICAN EMBASSY
OFFICE OF THE MILITARY ATTACHE
1, GROSVENOR SQUARE, W. 1,
LONDON, ENGLAND

14 November 1945

SUBJECT: Transmittal of Document

TO: Major A.E. Britt, Room 5004, New War Department Building, Washington, D.C.

1. Enclosed is Copy No. 1 of Farm Hall Report No. 16, covering the period of 29th October - 4th November.

2. For your information there has been some muddle with the numbering of these Farm Hall reports. There is no No. 13 and two 16's.†

For the Military Attache:

1 Incl: As above

H.K. Calvert

H.K. CALVERT
Lt.Col., C.E.
Assistant Military Attache to the

† The two reports FH16 are listed in the contents to this book as FH16(a) and FH16(b).

Copy No. 1.

Ref. FH16.

To: Mr. M. Perrin and Lt. Cdr. Welsh

From: Major T.H. Rittner.

OPERATION "EPSILON"

(29th Oct.-4th Nov.)

I. GENERAL

1. Capt. SPEER, U.S. Army, left on October 29th.

2. A visit from Dr. FRANK, on Friday 2nd November was the highlight of the week.

II. WEREWOLVES

3. A paragraph in the "Daily Express" for Tuesday 30th October, suggesting that HAHN might work at DIDCOT provoked him to say that it was monstrous to write so, and so to provoke the WEREWOLVES.

III. THE FUTURE

4. In the late evening of 31st October HEISENBERG, HARTECK and KORSCHING discussed their future, in the following conversation; they thought the BRITISH more likely to find a use for them than the AMERICANS.

5. HARTECK: Do you think that the offer of the RUSSIANS to the GERMAN scientists has any influence on our remaining here?

HEISENBERG: I would say: There is a party quite high up near Mr. TRUMAN and this party says: "One must lock up those GERMANS for years, because wherever they are, they are a danger. If they are just anywhere, they can go over to the RUSSIANS, and even if they don't want to, the RUSSIANS can kidnap them." And then there is another party who says: "That is a quite impossible approach. Firstly they are Anti-Nazis. Secondly they are upright scientists. Thirdly this whole secrecy is nonsense in any case. We must make use of these people, in order to build up science again in GERMANY." This second party has now obviously achieved a principal concession from the opposition-party: The fact that we are to go back to GERMANY. But now the opposition-party says: "If you are so insistent on that point, then you must arrange everything beforehand that they can on no account get somehow kidnapped by the RUSSIANS." I should think that if we get back to GERMANY at all, this factor will perhaps even get us into a favourable position in GERMANY. For they will say to themselves: "Firstly we have got to put up guards, of course, so that they don't get kidnapped. Secondly, however, we also have to offer them such good terms, that they will willingly stay with us."

HARTECK: But it would be too much trouble for them to have the main party messing around in MUNICH and me in HAMBURG.

HEISENBERG: Yes, that's just it. That's the great difficulty. I can't see yet how they want to do it. Of course, it is possible that at some point this opposition party up there will also say that they are no longer interested and will then say: "After all they don't know very much about the secrets, and fundamentally, the RUSSIANS already know everything." Nevertheless I would still think that in any case it would be unpleasant for them if we went to the RUSSIANS.

HARTECK: Anyway, have they got enough people?

HEISENBERG: The ENGLISH or the AMERICANS?

HARTECK: The ENGLISH. I have an idea that the AMERICANS have an abundant supply of people, and the ENGLISH simply have too few.

HEISENBERG: Yes, I can imagine that the ENGLISH would say: "Well, that's quite good. Let's take these people to ENGLAND." I should think that there is perhaps one party here who want that. But I believe that the AMERICANS wouldn't allow that either because they say, and they are justified to some extent: "The ENGLISH won't give them a huge salary, and especially they won't employ all of them. Then, let's say, there are Nazis among them and they certainly would incline more towards RUSSIA than towards ENGLAND. And if they are at large, who can stop them from simply going to the RUSSIAN Embassy in LONDON and saying: "We put ourselves at the disposal of RUSSIA." I don't know whether the AMERICANS would allow that. Now the fact is that the AMERICANS are not so greatly interested in the ENGLISH starting a gigantic atom factory over here in the near future.

HARTECK: But can they prevent that at all?

HEISENBERG: No, they can't stop them from doing any amount of atomic research over here. But they can stop them from employing us over here by saying: "This is an inter-allied affair. You are looking after these GERMANS but we want to have a say in what is to be done with them."

HARTECK: The AMERICANS can draw from a huge number of institutes. This is not the case in ENGLAND.

HEISENBERG: They have not such an abundance of people.

HARTECK: And if they are joined by a dozen or so people, that would be quite nice.

KORSCHING: But I too believe the AMERICANS will be against it. People of whom one doesn't know how far one can trust them – I mean there are certainly enough people who are going to say: "One might trust Professor HAHN and perhaps Professor HEISENBERG, but one doesn't know exactly what the others are like."

HEISENBERG: Yes, a couple of us are Party Members. That will naturally cause a certain amount of difficulties. One must realize that there are certainly some high-ups in AMERICA who will say, and from their point of view quite rightly so: "The easiest way out would be if these ten GERMANS were dead." The AMERICANS don't need us at all. There is no doubt that they can do it better than we can. They are not at all interested in seeing that the ENGLISH are assisted by us. It is quite nice if the ENGLISH are some years behind. And under no circumstances must must they go to the RUSSIANS. But naturally these people aren't the only ones who have a say in these matters. Particularly as these people are being strongly attacked by the AMERICAN scientists for other reasons.

HARTECK: After all, I must say if one knows the RUSSIANS and the Nazis, these people here are really very decent to take so much trouble with us.

HEISENBERG: The ENGLISH say: from their point of view, the AMERICAN policy to smash Europe to pieces is just as logical as the corresponding RUSSIAN policy, but from the ENGLISH point of view it is the worst thing that could happen, because the ENGLISH must still control Europe. And therefore for the ENGLISH it is of the utmost importance to reconstruct Europe, and for that purpose we are useful to the ENGLISH. It is quite clear that they could use us, because up to now they haven't got a large number of trained young people. They could even use us in GERMANY. Again they might say: "Let these people work in GERMANY on this problem, and let them promise to let us have any new results first, and only produce what we tell them to." This control plan might well be quite acceptable to them.

KORSCHING: That would mean that the people concerned would not publish anything.

HEISENBERG: No, it is like that. It would mean that one could, of course, not publish these technical details any more, but one could publish the purely scientific matters on which one could work just as well. I would prefer to run the Institute on a purely scientific basis. One could easily tackle problems which, even from the point of view of nuclear physics, still tend to the side of pure science, for instance cosmic rays. In this respect I am for waiting until these people have decided what to do with us. The trouble is that up to now they are completely at loggerheads. There are some who are extremely friendly towards us, and on the other hand there are these obstinate people, these AMERICAN HEYDRICHS AND KALTENBRUNNERs who say: "What? The best thing these GERMAN scientists can expect from us is to stay locked up." I think the ENGLISH have got every reason to wish for the help of GERMAN science†, because in a manner of speaking this is going to be their sphere of influence and they'll get something out of it. On the other hand it doesn't matter to the AMERICANS.

IV. Dr. FRANK's VISIT

6. The contemplation of Dr. FRANK's visit aroused the usual pleasure that is given by contacts with the outside world.

7. It was not thought to be anything but a social call, and HEISENBERG answering a question of KORSCHING, did not propose any special plan of campaign for the day:

8. KORSCHING: How shall we behave when FRANK comes?

HEISENBERG: The best thing is just as usual.

9. Dr. FRANK had long talks with WIRTZ, who was the hero of the day, but nothing new was said, at least in the House.

10. There was little reaction to the visit, but in conversation with HAHN WIRTZ thought Dr. FRANK had worked hard to be able to visit them at all. HEISENBERG detected the bogey of Party Politics:

11. WIRTZ: FRANK said the whole business of our detention is purely political.

HAHN: If it had nothing to do with physics we would not be here.

WIRTZ: He himself says what is being done is, of course, all quite nonsensical. The wrong people are detained, nearly everything that can be known has already been published etc and he laughs about TRUMAN and these people, but he says there is nothing one can do at the moment. I mean actually he has achieved quite a lot by the fact that he comes here, after BLACKETT has apparently been excluded for the time being. FRANK, as he says himself, is conservative, while BLACKETT is not.

HEISENBERG: There you are, it is, after all, that party-business.

12. VON LAUE's reaction, to HAHN, is to show the first signs of impatience, but HAHN recommends waiting until Mr. ATTLEE returns from the U.S.A.

---

† The following marginal note appears in the transcript: 'that is to assist German science'.

13. LAUE: But now at last something must happen, after all we can't sit here for ever.

HAHN: We must wait until ATTLEE comes back from the conference with TRUMAN before doing anything further.

14. KORSCHING suggested to BAGGE a secret separatist movement in which the younger scientists should dissociate themselves from the older, as the former have nothing to do with the Atom bomb. KORSCHING would write about this to Lt. Cdr. WELSH offering to give their word to abandon nuclear physics, provided they may continue to work in some other sphere. They also decided to await Mr. ATTLEE's return from the U.S.A.

V. FRENCH SCIENCE

15. The article in 'The Times' (London) of Saturday 3rd aroused no comment other than angry resentment at the French having removed scientific equipment from Germany, and some scorn for the reported french achievements during the occupation.

Farm Hall,
Godmanchester.
5th Nov. 45.

(Sgd) P.L.c. Brodie
Capt.,
for Major T.H. Rittner

Copy No. 1.

Ref: F.H.17.

To: Mr. M. PERRIN and Lt. Cdr. WELSH

From: Major T.H. RITTNER

OPERATION "EPSILON"

(5th-11th Nov. 1945)

I. GENERAL

1. The week began with HEISENBERG proposing drastic action and ended with a very successful and pacifying visit from Lt. Cdr. WELSH.

II. HEISENBERG's PLANS

2. In conversation with HAHN on 5th November, HEISENBERG discussed the possibility of taking some action to clarify their own position and future. HAHN agreed that things were becoming impossible:

3. HEISENBERG: I don't know whether it would not be a good thing after all for one of us to return his parole and scram.

HAHN: I consider that to be a dangerous undertaking, because then people would have good reason for saying: "We have treated them well and now they are obstinate. Therefore, we shall now treat them badly." Perhaps one should speak with the Captain again, at any rate about the sinking spirits of several of us, and then I should say to him: "Listen, some of our people have just about reached the end of their tether –" he can see that for himself. LAUE staying in bed for a day and that sort of thing and that GERLACH gets fits of hysteria; he can hear that when GERLACH suddenly gets a fit in the kitchen and starts shouting at the staff. I can tell him that, in the long run, it will not be possible anymore to calm people down again and again.

HEISENBERG: I think, what they are doing is so very unnecessary. I simply cannot understand why they don't tell us what is going on.

HAHN: Shall we go there together, perhaps the day after to-morrow?

HEISENBERG: I just think that talks with the Captain are somehow futile, they don't help. He listens to us and then passes it on to the Commander, already with certain reserves. Then they have a talk about it, air their feelings a bit as to how unpleasant the whole thing is and with that things have really come to an end as already the Commander does not pass this on to a higher authority anymore. Perhaps, if we are very lucky, the Commander tells the competent Colonel or General who is sitting here in LONDON. It is impossible (?) that it will ever reach AMERICA where a decision could be made.

HAHN: Then let us tackle the problem in such way that we really ask for a visit of the American Colonel. We shall never see the superior of the Commander, he keeps him away from us.

> 4. Speaking to WIRTZ on November 6th, HEISENBERG elaborated three possible means of forcing a decision. WIRTZ, as usual, tried to exert a restraining influence:

5. HEISENBERG: I should like to have your opinion about three possibilities I have thought of. The first one is that we should make an application for an interview with Sir John ANDERSON. The second plan would be that I withdraw my parole. I have thought of the following rather daring plan: Let us assume, I had no further obligation on account of my parole. Then I could run away one day and could go to the Danish Embassy. I would ask them for shelter and could then establish contact with BOHR, or the Danish Embassy could do that, and from there I could then somehow negotiate independently through BOHR. I would tell them that I am quite prepared to return into English or American captivity, but only under certain conditions. These conditions we would have to think out. To start with, we would want to know how the negotiations about our future are progressing, we don't want to be treated as stupid schoolboys. Secondly, we would want to have contact with our families or something of that nature. The third plan would be that we have a general discussion here in which we would decide that we shall all withdraw our parole on, say, the 1st December. In that case, HAHN could go to the Captain and tell him that the utmost he was able to achieve with us was that we would wait until the 1st December, but it had already been decided to withdraw the parole on that date and he should ask at the same time that a note to that effect should be made on our written parole right now: "terminates on 1st December" or something like that. Those are the three possibilities. How would you feel about them?

WIRTZ: Well, to start with at any rate, I would not be against making an application to be received by Sir John ANDERSON.

HEISENBERG: In my opinion, I don't think, this has any chance for success.

WIRTZ: Well, I don't think much of anything connected with escape. I can't tell you exactly why, but – (int.)

HEISENBERG: Too forceful, you think. My impression would be that at best we could force an issue now, but we would have to pay for it later on. In any case, those circles who now detain us here would be so badly disposed towards us for all time that, in the end, the result would be detrimental after all. I should say that the only chance for lasting success would be, after we have achieved something temporarily, that those people who are badly disposed towards us will definitely disappear from American politics and will be replaced by people like

OPPENHEIMER who mean well. But it is a very risky game and there is a very good chance that, if we do make such daring coup, the opposition will bring forth all its hatred of the Germans and they will say: "Well, on the whole, they have, after all, all tried to help the Nazis; they may not have achieved an atomic bomb, but had they done so, they would have naturally given it to HITLER" and so forth.

6. Discussing the matter with WIRTZ, VON WEIZSACKER showed the same caution:

7. WEIZSACKER: If we really must have an outlet for our bad humour and our worries, then I would much rather insult the Captain (BRODIE) three times during the course of a meal than compose just one document which might do us harm for years to come.

WIRTZ: Yes, one must not shoulder responsibilities – as HEISENBERG apparently is willing to do – only to vent one's own feelings and thereby causing great harm.

WEIZSACKER: HAHN is also in the mood for action now, but when he does something, it will be sensible.

III. Lt. Cdr. WELSH's VISIT

8. Lt. Cdr. WELSH came down on 10th November and had long and satisfying talks with various members of the company, particularly WIRTZ.

9. WIRTZ repeated his conversation with Lt. Cdr. WELSH to HEISENBERG and then said:

10. WIRTZ: My object this evening was to let him know that we understand his difficulties, but it is very valuable to us to have regular contact with him.

HEISENBERG: I quite agree.

11. On 11th November, HAHN also expressed his satisfaction with the visit:

12. HAHN: I am very glad that the Commander was here again and that we could put our views before him.

WIRTZ: I think, if the Commander were to discuss things with us often and as thoroughly as yesterday he did with me, it would be much easier for us to put up with all this.

HAHN: I quite agree.

IV. TECHNICAL

13. Colloquia are again in fashion which indicates an improvement in morale.

14. On 6th November, BAGGE gave a lecture about the chemical properties of the transuranic elements, based on work by FERMI.

15. On 9th November, WIRTZ lectured on the subject of chemical reactions of elements.

16. There was nothing else of technical interest during the week.

V. ADMINISTRATION

17. During my absence from FARM HALL on 8th and 9th November, Mr. OATES was very kindly brought over from PARIS to take my place. He stayed until the 11 November.

FARM HALL
15th November 1945

(Sdg) P.L.c Brodie
Captain
for Major T.H. RITTNER

Copy No. 1.

Ref. F.H.18

To: Mr. M. PERRIN and Lt. Cdr. WELSH
From: Major T. H. RITTNER

OPERATION "EPSILON"

(12th November-18 November 1945)

I. GENERAL

1. This has been a quiet week with no noticeable signs of discontent.

2. The announcement of an award of a NOBEL prize to HAHN in the 'Daily Telegraph' for Friday 16th November was the highlight of the week.

II. HEALTH

3. On Monday, 12th November, LAUE, HEISENBERG and GERLACH visited the doctor, who considered that their various troubles were principally nervous. The subsequent improvement tends to confirm this.

III. HEISENBERG's FINANCIAL AFFAIRS

4. At the request of Colonel CALVERT, I took to LONDON on Saturday 17th November forms of power of attorney for Mrs. HEISENBERG, one in English and one in French; the former being in the normal English form, the latter home-made. These were handed to Lt. WARNER.

IV. COMING MEN

5. On the 13th November I delivered a list of promising young German physicists, compiled by HEISENBERG, to Lt. Cdr. WELSH.

6. WIRTZ expressed disapproval of the compilation of the list in a conversation with DIEBNER:

7. WIRTZ: To be quite honest, I would have nothing to do with that business. It's wrong, I wouldn't give any names, I should have time enough to see that these people did not suffer when I got back to GERMANY. I do not mind dealing with the Commander, but I don't trust him. I have another reason for not doing it: After the coming war with the Russians, if GERMANY is occupied by RUSSIA, they will say: "Who gave lists to the English?" They will find the lists somewhere in GERMANY and then one would be in a fearful hole.

DIEBNER: Have you discussed this with HEISENBERG?

WIRTZ: No. It makes me uncomfortable when our fellows give them such lists.

8. Both HAHN and HEISENBERG also had scruples, but readily believed that no harm was intended.

V. NOBEL PRIZE

9. The 'Daily Telegraph's' announcement of the award to HAHN of the Nobel Prize for Chemistry caused general pleasure and also deep misgivings as no official confirmation was forthcoming. It was even thought that some unaccountable malice was responsible for our withholding the news. However, great efforts were made in LONDON to try to verify the report and, as the source seemed reasonably reliable, the award was duly celebrated with songs, speeches, baked meats and some alcohol. Proceedings started very badly with an unfortunate speech by von LAUE, at the end of which, both he and HAHN, were in tears to everybody's great discomfort, particularly mine, as I was sitting between them. However, the united efforts of the rest of the party restored our normal good spirits.

10. I hope to obtain copies of the better songs and speeches and to publish them later in the form of an appendix to this report. They will, of course, be of no operational interest.

11. VON LAUE told HAHN that there was a good reason for the award of the prize.

12. LAUE: You have not got the NOBEL prize as a consolation for we ten German scientists who are shut up here, but I think that is their reason for giving it to you now instead of waiting till next year which was probably their original intention.

13. HAHN said that, if he were allowed to go to SWEDEN to receive the prize, he would not be able to give his word to say nothing about his detention here.

14. HARTECK: They (Allies) cannot make you give your word not to say where you have been and with whom.

HAHN: Out of the question!

HARTECK: They won't make you give your word, but they will say: "Of course, we are delighted to send you to SWEDEN, but please don't say with whom you are."

HAHN: I should tell them: "I have not for years been in a country at peace. I used to like my glass of wine. I have friends in SWEDEN, such as Professor QUENZEL (?) who has a very good cellar. I cannot guarantee, you know what old German students are, not to get a bit tight in a friendly way. I obviously can't give you my word for what might happen then."

VI. THE FUTURE

15. GERLACH told HAHN that he would on no account go to AMERICA.

16. GERLACH: I would not got to AMERICA for all the tea in CHINA. I considered it thoroughly, because it is so often said that that might be a possibility.

HAHN: As soon as one can do anything in GERMANY, of course, it is our duty to be there. Personally, I shall do no more work, but I will perhaps nevertheless be able to help or advise someone or other now and then.

GERLACH: Conditions in GERMANY make no difference whatsoever. To be there is the most important thing in the world.

17. GERLACH has, for the past week or two, been in a particularly nervy state, but in the last few days has been a little quieter.

VII. TECHNICAL

18. There were colloquia on Tuesday and Friday, both of which were continuations of that begun last week on 'Molecular structure of various chemical compounds.'

19. There was nothing else of technical interest during the week.

FARM HALL
GODMANCHESTER
20th November 1945

(Sgd) P.L.c Brodie
Captain,
for Major T.H. RITTNER

Copy No. 1.

Ref: App/FH18

To: Mr. M. PERRIN and Lt. Cdr. WELSH
From: Major T. H.RITTNER

OPERATION "EPSILON"

(Appendix to F.H.18.)

I. SPEECHES IN HONOUR OF HAHN

1. Attached are copies of various speeches, jokes, songs etc. composed by the guests to celebrate the news of the award to HAHN of the 1944 NOBEL prize for chemistry.†

2. These have not been translated as they would lose too much of their character in process.

3. They are not of operational value but are passed on for interest's sake only.

FARM HALL
GODMANCHESTER
26 Nov 45

(Sgd) P.L.C. Brodie
Captain,
for Major T.H. RITTNER

† One learns from Bagge's diary (loc. cit. pp 66-68, entry 18 November 1945) that Captain Brodie contributed a bottle of gin and two bottles of wine for this party; that fictitious newspaper extracts were invented and read by Heisenberg, von Weizsäcker and Wirtz; and that the story with aleatoric adjectives was Bagge's contribution. FCF

<u>Tischrede, gehalten in Farmhall, Godmanchester am 16.11.1945 zur Feier der Verleihung des Nobelpreises an Otto Hahn.</u>

VON M.v. LAUE.

Lieber Otto Hahn!

Je laenger sich unsere Detainung hinzieht, um so schlechter werden unsere Sitten; jetzt gibt es sogar schon Tischreden. Aber ich muss Dir heute ein kleines Geheimnis verraten.

Vor zehn oder mehr Jahren schenktest Du mir Dein Bild. Ich liess es rahmen und haenge es in mein Instituts-Arbeitszimmer; dort in Hechingen haengt es heute noch – hoffentlich. Auf seine Rueckseite schrieb ich: Otto Hahn, geb. 8. Maerz 1879, aufgenommen 1933 (was wenigstens einigermassen stimmen duerfte). Solche biographische Notizen schreibe ich naemlich auf alle Bilder von Bekannten, damit diese auch ueber mein Leben hinaus einen gewissen Wert behalten, und damit sich nicht die Goettinger Geschichte wiederholt, bei der Bilder von Gauss und Bessel verwechselt wurden. Aber hinter <u>Deinem</u> Bild steht ausserdem noch ein Distichon:

Gaben, wer haette sie nicht? Talente, Spielzeug fuer Kinder!
Erst der Ernst macht den Mann, erst der Fleiss das Genie.

Dies dichtete einst Theodor Fontane auf Adolf v. Menzel zu dessen 70-ten Geburtstag. Wenn ich die Gelehrten ueberblicke, deren Lebensweg ich einigermassen zu kennen glaube, so finde ich keinen, auf den es so gut passte, wie auf Dich. Denn – um nur einen Punkt zu erwaehnen – zu Schnellanalysen von rasch zerfallenden radioaktiven Stoffen, die Du so oft ausgefuehrt hast, reichen Gaben und Talente nicht aus; vielmehr muss noch eine Ausbildung der angeborenen Faehigkeiten hinzutreten, die tiefen Arbeitsernst und ungeheuren, lebenslaenglichen Fleiss erfordert. Aber weil Du Beides hast, so laesst sich die Reihe Deiner Entdeckung auch einer Kurve vergleichen, die auf hohem Niveau – mit der Entdeckung des Radiothor, die Du in dieser Gegend machtest beginnend ueber die Entdeckung des Mesothoriums (alias Megatheriums) dauernd ansteigt, um in der Entdeckung der Uran-Spaltung ein Maximum, aber kein Ende, zu finden. Und als aeussere Kroenung Deines Lebenswerkes hast Du nun heute die schoenste Ehrung erfahren, die dem naturwissenschaftlichen Forscher zu Teil werden kann, den Nobel-Preis. Ich glaube, man kann Dir in keiner besseren Form dazu graturlieren, als mit jenem Verslein.

Aber meine Rede waere arg unvollstaendig, wollte ich nicht noch einer anderen Persoenlichkeit gedenken: Deiner Frau. Auch Sie muss ja die Nachricht erhalten haben; wie widerstreitende Empfindungen moegen da heute Abend auf sie einstuermen! Aber ich hoffe doch, ueberwiegen wird bei ihr schliesslich die Freude, die stolze Freude, die Gattin eines solchen Mannes zu sein.

Meine Herren! Wir erheben unsere Glaeser und trinken auf das Wohl von Otto und Edith Hahn.

Sie leben hoch!

Auszug aus der 'Times' vom 17. November 1945.

Prof. Dr. Otto Hahn, German radiologist, has been awarded the 1944 Nobel prize for chemistry. It has been stated in official circles, that Otto Hahn has been detained since the end of the war. No further comment was available.

Uebersetzung aus 'Sunday Pictorial' vom 18. November 1945.

NOBEL ATOM ACE MYSTERY. WHERE IS OTTO HAHN?

Die neueste Atombombe ist die Nachricht aus Stockholm, dass Hitler's Atomexpert, Prof. Otto Hahn aus Berlin, den Nobelpreis fuer Chemie erhalten hat. Wie wir aus zuverlaessiger Quelle erfahren, hat die schwedische Akademie gleichzeitig mit der Preisverleihung einen Preis ausgeschrieben fuer diejenige Person, die zweckdienliche Angaben ueber den gegenwaertigen Aufenthaltsort Otto Hahns machen kann. Zahlreiche Loesungen sind eingegangen und harren der Bearbeitung. Die naechstliegende Annahme war, dass Otto Hahn vom U-Boot 530 gleichzeitig mit Hitler und Eva Braun in Patagonien abgesetzt worden ist und dort an der Herstellung neuer und besserer Atombomben arbeitet. Diese Vermutungen haben sich jedoch bisher nicht bestaetigen lassen; eine andere Quelle bringt jetzt die aufsehenerregende Mitteilung, dass Otto Hahn vor wenigen Tagen in Tel Aviv gesehen worden ist, wo er offenbar mit anderen dort lebenden Frankfurtern und Dahlemern ueber die Weiterfuehrung seiner Arbeiten beraten will. Der Bedarf an Fachleuten fuer Spaltung scheint dort aber nicht mehr gross zu sein, und eine sichere Bestaetigung seiner Anwesenheit in Tel Aviv liegt bisher nicht vor.

Um zuverlaessige Nachrichten ueber den Aufenthaltsort O. Hahns zu erhalten, haben wir unseren Sonderkorrespondenten nach Deutschland entsandt; dieser hat zunaechst das kleine wuerttembergische Staedtchen Tailfingen besucht, in dem Otto Hahn in den letzten Kriegsjahren seine Atomspaltungen durchgefuehrt hat. Da die Familie des Professors offenbar in der letzten Zeit mehrfach die Wohnung gewechselt hat, gelang es unserem Korrespondenten erst nach langer Muehe, bis zu einer kleinen, bescheidenen Wohnung vorzudringen, in der ihn zwar nicht Prof. Hahn, wohl aber seine reizende Schwiegertochter empfing und freundlich bewirtete. Leider konnte aber auch diese junge Dame keine weiteren Angaben machen, als dass der Professor im April von amerikanischen Truppen weggefuehrt worden sei und sich wahrscheinlich in England oder Amerika aufhalte. Nach laengeren erfolglosen Reisen, die unseren Korrespondenten nach Heidelberg, Versailles und ins Maastal bei Luettich fuehrten, stiess er auf ein Geruecht, dass Otto Hahn in England in einem kleinen Staedtchen Godmanchester zusammen mit anderen Warcriminals unter schaerfstem Arrest gehalten wuerde.

Ein Besuch in Godmanchester zeigte denn auch, dass dort ein roter Backsteinbau mit schwer vergitterten Fenstern durch Geheimpolizisten in Civil und Soldaten in Uniform so scharf bewacht war, dass jede Annaeherung waehrend des Tages voellig unmoeglich schien. Unser Korrespondent versuchte daher, waehrend der Nacht sich ueber die benachbarten Wiesen an den rueckwaerts gelegenen Gefaengnishof anzuschleichen, und es gelang ihm, in der ersten Morgendaemmerung einen mit Stacheldraht umwundenen eisernen Zaun zu erreichen, von dem er einen Rasenplatz hinter dem Gefaengnis ueberschauen konnte. Hier bot sich ihm ein merkwuerdiger Anblick: Eine nackte Gestalt mit muedem Gesichtsausdruck lief ununterbrochen an einem Rosenbett auf und ab, auf und ab, dem Eisbaer im zoologischen Garten vergleichbar. Auf einen leisen Anruf unseres Korrespondenten zuckte die Gestalt zusammen und rannte mit einem Schrei des Entsetzens in wilder Flucht in das Haus. Wer diese bemitleidenswerte Gestalt war, hat sich bisher nicht mit Sicherheit ermitteln lassen. Ein Schnappschuss, den unser Korrespondent noch eben nehmen konnte, fuehrte zu einer schlechten Aufnahme, auf der einige Wissenschaftler, um Rat gefragt, tatsaechlich die Zuege Otto Hahns erkennen wollten. Sic transit gloria mundi. Aber auch diese Feststellungen haben keine Sicherheit ueber den Aufenthalt Otto Hahns gebracht.

Wir sind daher noch anderen Geruechten nachgegangen, die behaupten, dass Otto Hahn in England gut untergebracht sei und dort an seinen Memoiren schreibe, die er unter dem Titel: 'Von Oxford Street nach Farm Hall' in Buchform herauszugeben gedenkt. Dieser Hinweis auf Oxford Street hat uns zu einer Umfrage in den Kaufhaeusern und Geschaeften dieses Viertels veranlasst. Dabei stiessen wir auf eine aeltere, wuerdig aussehende Matrone, Leiterin der Verkaufsabteilung eines Bekleidungshauses, die sich an Prof. Hahn gut erinnern konnte. Sie erzaehlte mit einem gluecklichen Laecheln (happy smile) ueber die Zeit ihrer Bekanntschaft mit Otto Hahn, aber sie war nicht in der Lage, naehere Einzelheiten aus dieser Zeit mitzuteilen (she was not in a position, to tell the more intimate details of this time). Irgendein Anhaltspunkt fuer den gegenwaertigen Aufenthaltsort Otto Hahns liess sich also auch hier nicht gewinnen. WHEN THE MYSTERY OF OTTO HAHN WILL BE SOLVED, NOBODY KNOWS.

FRANKFURTER ZEITUNG

VEREINIGT MIT DEM

FRANKFURTER GENERALANZEIGER

<u>Von GOETHE bis HAHN</u>

In unserer Reihe "Grosse Frankfurter" umspannen wir heute zwei Jahrhunderte Frankfurter Geschichte mit einem erlauchten Zwillingspaar: unsere Stadt darf einerseits Goethe, den Spalter des Herzen, andererseits zugleich, um der Sache, eine neue Wendung zu geben, Hahn, den Spalter der Atome, zu Ihren rechnen. Gehen wir ueber die Lebensschicksale des Aelteren der beiden, die den meisten unserer Leser bekannt sein duerften, rasch hinweg, und richten wir unsere Blicke auf die neue Saeule im deutschen Chemikerwalde, Otto Hahn.

1879 in unserer Stadt geboren, verehren wir in ihm den Entdecker des Mesozoikums, den Erfinder der Hahnenkammeinheit und des Cocktails, den Gruender mehrer Sportklubs und langjaehrigen unermuedlichen Leser unserer Zeitung. Er ist Ehrenvorsitzender des Vereins fuer Sparsamkeit und Gewerbefleiss und Inhaber verschiedener Patente fuer die Erhoehung der Lebensdauer von Rasierklingen, Zigaretten und Anekdoten. Dem Fass seiner Verdienste hat er aber erst unlaengst die Krone der Uranspaltung ins Gesicht geschlagen. Wie wir von der Sekretaerin der Liga fuer Koerperkultur erfahren, soll er dabei ein neues Verfahren zur Faellung strahlender Koerper angewandt haben. Er stand sozusagen gewissermassen mit dem einen Bein fest zwischen den Stuehlen der Tatsachen, mit dem anderen aber griff er nach der goldenen Leiter des Ruhms. Mit ihm sonnt sich seine Vaterstadt im bunten Abglanz der verdienten Ehrung, die ihn soeben ereilt hat, und moechten wir wie ein Vogel an mehreren Orten zugleich sein, um sowohl dem Jubilar unsere aufrichtig gemeinten Glueckwuensche zu Fuessen legen als auch unsern Lesern einen Augenzeugenbericht von dem juengsten und groessten Weltereignis geben zu koennen.[+]

[+] Wir verweisen darauf, dass im Feuilleton dieses Blattes von der berufenen Hand von Wilhelm Westphal die wissenschaftliche Verdionste Otto Hahns gewuerdigt werden, waehrend wir im Wirtschaftsteil einen Beitrag aus der geschaetzten Feder unseres Sonderkorrespondenten P.H. ueber die finanzielle Bedeutung des Nobelpreises bringen. (Red.)

WIRTSCHAFTSTEIL der FRANKFURTER ZEITUNG

Nobelpreis und Atomkernenenergie

Otto Hahn, der Entdecker der Uranspaltung, hat den Nobelpreis fuer Chemie erhalten. Die erste Frage, die sich dem unbefangenen Beobachter aufdraengt, lautet: Steht die Geldsumme, die er erhaelt, in einem angemessenen Verhaeltniss zu der Summe, die er der Menschheit durch seine Entdeckung geschenkt hat?

Der Nobelpreis betraegt heute etwa 6.000 Pfund Sterling. Das waeren unter der Vorraussetzung des Goldstandards RM 120.000. Die Unsicherheit aller Waehrungsverhaeltnisse legt es nahe, einen anderen Wertmasstab zu waehlen, der auf den elementaren menschlichen Beduerfnissen basiert. Setzen wir den Preis eines Pfundes Zucker zu 40 Pfennigen an, so erhielte Prof. Hahn den Gegenwert von 150.000 kg Zucker. Diese Zahl ist mit der Energieausbeute, die seine Entdeckung zur Verfuegung stellt, deshalb noch schwer zu vergleichen, weil man das Uran nicht essen kann. Wir fuehren deshalb den Zucker und das Uran auf das gemeinsame Mass der Kalorie zurueck. Ein kg Zucker enthaelt etwa 4000 Kcal; der Nobelpreis betraegt demnach rund 600 Millionen Kilokalorien.

Andererseits koennen aus einer Tonne Uran 235 nach Hahn etwa 10^{13}

Kcal. freigemacht werden. Die Gesamtmenge des Urans auf der Erde ist nur schaetzungsweise bekannt. Wir nehmen an, dass der hundert-millionste Teil der Materie der Erdoberflaeche aus Uran besteht. Setzen wir voraus, dass das Uran bis in eine Tiefe von lkm ausgebeutet wuerde, so erhalten wir, da die trockene Erdoberflaeche rund 100 Millionen Quadratkilometer betraegt, l00 Kubikkilometer ausbeutbare Materie, von der ein Kubikkilometer Uran ist. Das sind, da das spezifische Gewicht von Uran rund 20 ist, 20 Milliarden Tonnen Uran. Etwa der hundertste Teil des Urans ist Uran 235, so dass wir 200 Millionen Tonnen Uran 235 haetten. 2.10^{8} to Uran 235 zu je 10^{13} Kcal geben 2.10^{21} Kcal. Die 6.10^{8} Kcal des Nobelpreises sind davon der dreibillionste Teil oder der Bruchteil 3.10^{-13}. Diese Zahl hat man vermutlich gewaehlt, weil sie gerade der Radius eines Atomkerns ist.

THE FARMHALLER

Nobel-Prize Song

(Melodie: Studio of seiner Reis')

Detained since more than half a year
Sind Hahn und wir in Farm Hall hier.
Und fragt man, wer ist Schuld daran,
So ist die Antwort: Otto Hahn.

The real reason nebenbei
Ist weil we worked on nuclei.
Und fragt man, wer ist Schuld daran,
So ist die Antwort: Otto Hahn.

Die nuclei waren fuer den Krieg
Und fuer den allgemeinen Sieg.
Und fragt man, wer ist Schuld daran,
So ist die Antwort: Otto Hahn.

Wie ist das moeglich, fragt man sich,
The story about seems wunderlich.
Und fragt man, wer ist Schuld daran,
So ist die Antwort: Otto Hahn.

Die Feldherrn, Staatschefs, Zeitungsknaben,
Ihn everyday im Munde haben.
Und fragt man, wer ist Schuld daran,
So ist die Antwort: Otto Hahn.

Even the sweethearts in the world(s)
Sie nennen sich jetzt: Atom-girls.
Und fragt man, wer ist Schuld daran,
So ist die Antwort: Otto Hahn.

Verliert man jetzt so seine Wetten,
So heisst's, you didn't split the atom,
Und fragt man, wer ist Schuld daran,
So ist die Antwort: Otto Hahn.

Ein jeder weiss, das Unglueck kam
Infolge splitting von Uran.
Und fragt man, wer ist Schuld daran,
So ist die Antwort: Otto Hahn.

Die energy macht alles waermer,
Only die Schweden werden aermer.
Und fragt man, wer ist Schuld daran,
So ist die Antwort: Otto Hahn.

Auf akademisches Geheiss
Kriegt Deutschland einen Nobel-Preis.
Und fragt man, wer ist Schuld daran,
So ist die Antwort: Otto Hahn.

In Oxford Street, da lebt ein Wesen,
Die wird das heut' mit Thraenen lesen.
Und fragt man, wer ist Schuld daran,
So ist die Antwort: Otto Hahn.

Es fehlte damals nur ein atom,
Haett er gesagt: I marry you, madam.
Und fragt man, wer ist Schuld daran,
So ist die Antwort: Otto Hahn.

Dies ist nur unsre erste Feier,
Ich glaub die Sache wird noch teuer,
Und fragt man, wer ist Schuld daran,
So ist die Antwort: Otto Hahn.

Und kommen wir aus diesem Bau,
We hope, we'll be quite lucky now.
Und fragt man, wer ist Schuld daran,
Stets ist die Antwort: Otto Hahn.

MOOSEJAW – HERALD, 16.11.45.

Recent Nobelpreistraeger for chemistry ist famous Dr. Otto Hahn. His name ist nicht unbekannt in Moosejaw, wie wir erfahren. In the remarkable year 1905, aus offiziell nicht bekannten reasons, unterbrach er hier eine seiner importanten Welt-voyages fuer einige days, – (not exactly known how many) -, und, first of all, besuchte einen hiesigen Friseurladen. Der Friseur erinnert sich lebhaft, dass Hahn gemaess der Landessitte eine Kopfwaesche unter Verwendung von 14 fresh eggs verlangte. Er erinnert sich ferner very well, dass der Kopf von Professor Hahn was very geeignet fuer die Spaltung der Eier und er findet es sehr einleuchtend, dass er spaeter sich auch der Spaltung von Atomkernen zugewendet hat. One of the most important problems fuer die Mitbuerger von Moosejaw ist die question, why Dr. Hahn seine voyage hier interruptete. Der Friseur hat eine dunkele Erinnerung, dass nicht ein wissenschaftlicher, sondern mehr ein privater Anlass vorlag, und dass der Verbrauch von 14 Eiern mehr eine Vorbereitung war for that purpose. (Note of the editor: Already at that time Canada and especially Moosejaw have been famous for Pin-Ups and other beautiful girls).

Zusammenfassung und Schlussansprache auf der Nobelfeier:

In der Tat und so war es denn auch. Der alte Mann bekam seinen Hut wieder, die Feuerwehr holte das Grammophon vom Dach. Hoijoho! riefen die Matrosen, und der Kapitaen schwang sich jubelnd von Mast zu Mast. Im Hintergrund sah man einen Einjaehrigen muehsam sein Jahr abdienen. Der kleine Scheich von Nubien aber warf den Kaese zum Fenster hinaus, hob die Duzbruederschaft mit seiner Schwiegermutter auf und reiste ploetzlich ab, nicht wissend, war dies nun Zufall oder Absicht, kluge Ueberlegung oder Heuchelei.

In der hier mitgeteilten Geschichte wurden die unterstrichenen Adjektive in der von den Festteilnehmern zugerufenen Reihenfolge an fuer sie freigelassenen Plaetzen eingesetzt. Den Zurufern war dabei Inhalt und Anordnung des Stoffes unbekannt. Fuer den Inhalt verantwortlich: Seine Majestaet, der Zufall!

Die einwandfreie Geschichte der unbrauchbaren

Kernphysiker seit ihrer ueberfluessigen

Detainung.

Am 27. April starten die unwiderstehlichen Vertreter des durchwachsenen KWI.f.Chemie in Tailfingen und des masslos traurigen KWI.f.Physik in Hechingen zu einer suessen Autofahrt in die sauere Detainung. Durch krumme Staedte, vorbei an unendlichen Mengen von groben Panzern, zog die duestere Reisegesellschaft westwaerts, bis sie am Nachmittag im uebertriebenen Heidelberg eintraf. Sie waehnte sich schon am Ziele ihrer stummen Fahrt, aber sie sollte erst im Laufe der schnieken Zeit bemerken, dass dies nur die erste Haltestelle auf einer langen Reise war, die sie damit antrat. Das naechste Ziel war Reims, wo sie wenige Tage spaeter eintraf und zunaechst mal das Ende des Krieges abwartete. Aber dann gings smart weiter. In einem blauen Flug von Reims nach Versailles zog sie hinweg ueber Nordfrankreich, um nur fuer einige Tage in einem himmlischen Chateau zu hausen, das ganz darauf eingerichtet war, die idiotischen Atomphysiker ueberheizt zu empfangen. Hier geschah etwas Unerwartetes. Nachdem sie die erste Nacht und den ersten Tag auf eisig kalten Britschen und Feldbetten verbracht haben, kam Verstaerkung. Der jammervolle Herr Heisenberg und der ebens fabelhafte Herr Diebner trafen ein. Trotzdem jedoch missfiel es den Detainten, so lange in Chesnay zu bleiben und so beschlossen sie aufgeregt von da nach Vesinet zu uebersiedeln, zumal sich ihnen am neuer Aufenthaltsort gewisse kuemmerliche Chancen boten, mit ihren tobsuechtigen Familien in Verbindung zu treten. Inzwischen war es Ende Mai geworden und weil man ja doch die raetselhafte Abwechslung liebte und weil man in vielen 100-Runden Laeufen um die dumme Villa Argentina diese gefraessige auch schon gut kannte, packte man abermals die souveraenen Koffer und reiste nach Faqueval. Man kam damit den rucksichtsvollen Wuenschen unserer alkoholischen Detainer auch sehr entgegen, zumal man inzwischen auch den schamlosen Herrn freundlichst und angebrannt eingeladen hatte, sich an dieser scheinheiligen Fahrt zu beteiligen, was er bereitwilligst und mit geistvollem Gesicht auch tat. Im luesternen Faqueval schliesslich erfuhr der inzwischen scon auf neun uebelriechende Personen angewachsene wohlschmeckende Kreis der Atomisten seine letztes und endgueltige Erweiterung, als sich ihm auch noch der boshafte Herr Gerlach zugesellte. – Reisefreudig wie man war, hielt man es in Faqueval aber nicht lange aus. Man wollte weiter. Ja, und nach einigem schauerlichen Zoegern begab man sich – stets im besten und schrecklichen Einvernehmen mit den uns eigentuemlichen

Behoerden wie im trichinoesen Fluge in das nun historisch gewordene mieselsuechtige Godmanchester. – Hier verbringen nun alle in ehrenwerter Stimmung seit ihrem laecherlichen Eintreffen em 3. Juli 1945 ihre ohbeinigen Tage. Die zehn Detainten sind seitdem eifrig bemueht, sich diesen blumigen Aufenthalt so verwegen wie moeglich zu gestalten und weil die Charaktaere der Beteiligten nicht gerade in allen Punkten uebereinstimmen, erfolgt die Freizeitsgestaltung auf verschiedene Weise. Da ist zunaechst der kuehne Senior unseres Kreises, der bescheidene Herr Prof. Hahn. Er bearbeitet – wenn man hier aus der Schule plaudern darf – seine von ihm so getauften Memoroiden, sofern er nicht gerade in einem gleisnerischen 10 km Lauf die verschlafene Bewunderung seiner haarstraeubenden Genossen von Farm Hall erregt. Aber nicht nur das ist's, was alle Leute seiner Umgebung von ihm zu erzaehlen wissen. Wenn er nicht noch ganz andere Dinge auf dem Kerbholz haetter – von denen wir noch reden muessen – so muss doch berichtet werden von seinen gefraessigen Cocktails, die er in seiner zarten Guete zum besten gibt, von den sieben gespaltenen Eiern auf seinem lauen Haupt in Mossejaw, von der zutiefst empfundenen Lady in der Oxford Street, von den Berliner Fangkuchen und von den verlorenen 300 Talern bei einer Bahnfahrt und von der praesidialen Wendung, die Tischgespraeche nehmen koennen, aber nicht muessen, wenn er zur Unterhaltung beitraegt. Man koennte vieles von den anderen mittelpraechtigen Herren dieses Kreises erzaehlen, was fuer spaetere Generationen von Detainten als grundlose Anleitung zum wildromantischen Zeitvertreib beitragen koennte. Wir muessen uns aber kurz fassen. Immerhin sind die mollerten Verdienste des uebermuetigen Herrn v. Laue ganz unbestritten, der es verstanden hat, fuer das von ihm geleitete trockene Kolloquium seit einem halben Jahre woechentlich zwei eifoermige Vortragende zu bekommen und so zur schleimigen unseres gewuerzten Aufenthaltes ganz wesentlich beizutragen. Unser anmutiger Herr Gerlach hingegen vertreibt sich die Zeit mit der ganz heldenhaften Beschaeftigung mit den humoristischen Erscheinungen des Magnetismus, ausserdem traegt er zur dunstigen Verschoenerung unserer Zimmer bei, indem er sich zum verzuckerten Vorsatz gemacht hat, pro Tag etwa 20 Blumenvasen zu betreuen. Wieder anders sieht die sentimentale Freizeitgestaltung bei unserem froschaehnlichen Herrn Heisenberg aus. Von der loecherigen Quantentheorie und der Kernphysik kam er zur dickbaeuchingen Supraleitung. Die ist aber jetzt schon fertig. Er war im ersten Augenblick ganz bekuemmert, weil er nichts mehr zu tun hatte, aber seitdem er den gewuerzten Trollop liest†, leidet er nicht mehr supra, sondern nur noch ein ganz klein wenig, aber auch dies wird ueberwunden, weil der immer freundliche und zweideutige Herr Harteck, ihm da stets mit einem gemischten Wort in sportfreudige Laune versetzt. Und wenn das nichts hilft, versteht es Herr Harteck, die atemlose Stimmung dadurch zu heben, dass er uns vorrechnet

† Heisenberg found and read at Farm Hall the complete works of Anthony Trollope, as I learnt on the occasion of my lunch visit there. FCF

wie langhaxert es uns geht, indem er den Kaloriengehalt unserer vertrauenseligen Speisen bestimmt. So koennte man noch von vielen Leidensgenossen reden, von unseren feuchtfroehlichen Detainern selbst, an ihrer Spitze unser unrasierter Capt. Brodie, vom luesternen Weizsacker mit seinen schnippischen Schuettelreimen, vom erhabenen Wirtz und seinen erhebenden Eiweissen, vom launischen Diebner und seinen gutmuetigen Witzen und vom hochmuetigen Korsching mit seinem pathetischen Barte. Aber wir muessen damit schliessen. Denn wenn auch das Leben allein schon durch diese Dinge bei den hohlen Detainten sehr interessant ist, wirklich aufregend ist es geworden, seit Herr Heisenberg heute im 'Daily Telegraph' entdeckte, dass unser gruenschnaebliger Prof. Hahn zum gottwollten Nobelpreistraeger bestimmt wurde. Wir suchen katastrophal nach den richtigen Adjektiven, um diesen genialen Sachverhalt kolossal zu beschreiben. Unsere sonst so alluebertreffende Stimmung hat ploetzlich unerreicht strengeHoehen erklommen und wir koennen wider unglaublich in die unerschuetterliche Zukunft blicken und in diesem Sinne begluecwuenschen wir unseren nichtendenwollenden Meister zu dieser nebligen Ehre!

Farm Hall, Godmanchester, am 16. November 1945.

Die Detainten.

AMERICAN EMBASSY
OFFICE OF THE MILITARY ATTACHE
1, GROSVENOR SQUARE, W. 1,
LONDON, ENGLAND

HKC/rr

28 November 1945

SUBJECT: Transmittal of Document.

TO: Major A. E. Britt, Room 5004, New War Department Building, Washington, D. C.

Inclosed herewith is Copy No. 1 of Farm Hall Report No. 19, covering the period of 19th - 25th November 1945, also inclosed is an appendix to Farm Hall Report No. 18, which was forwarded to your office 26 November 1945.

For the Military Attache:

HKC Calvert
H. K. CALVERT,
Lt.Col., C.E.,
Assistant to the Military Attache.

2 Incls:
Incl 1 - App FH #18
2 - FH #19

11 December 1945

Subject: "Operation Epsilon" (19th to 25th November 1945).

MEMORANDUM to Major General L. R. Groves.

1. As requested by you, submitted herewith is a short report on Farm Hall Report 19 and appendix to Report 18.

2. The main activity of the week concerns HAHN's Nobel Prize. The guests are querulous and impatient, which is expected to culminate in a minor crisis in a week or so. Only conversations of HAHN, HEISENBERG, and WIRTZ were reported and are summarized below:

HAHN: Wishes confirmation of Nobel Prize Award and to reply to Swedes at least indirectly. Is generally politely restrained, but WIRTZ and HEISENBERG egged him to make an indignant request for information and permission to write Sir Lawrence BRAGG on the theory that denial would prove that they were being detained as war criminals. He was temporarily calmed although given little encouragement concerning correspondence.

HEISENBERG: Advised HAHN to write Sir John ANDERSON on Nobel Prize and proposed threatening to withdraw parole to force action. Feared WIRTZ's proposed letter demanding reunion with families would be held against them by "American militarists who are already at loggerheads with scientific opinion" as evidence of Nazi arrogance, but thought family angle would help their case with local authorities.

WIRTZ: Urged HAHN to take strong stand on Nobel Prize matter. Believed group should make written protest before Christmas voicing indignant resentment at being detained on theory that this would annoy the military and force them to take action. Wants more information in order to plan action to arouse sympathy. HEISENBERG agreed. Believes that in future scientists must be organized for political action, but has no specific proposals.

SHULER

Per Clarke

2nd Copy destroyed 9 Oct 47 R.W.G.

Copy No. 1.

Ref: F.H. 19

To: Mr. M. PERRIN and Lt. Cdr. WELSH
From: Major T.H. RITTNER

OPERATION "EPSILON"

(19th - 25th Nov. '45)

I. GENERAL

1. A rather more interesting week, the lack of confirmation of HAHN's NOBEL prize and general impatience with their present condition are making the guests querulous and, on past form, should lead to another minor crisis within the next week or so.

II. NOBEL PRIZE

2. On Wednesday morning, 21st November, HAHN asked me if it were not possible to get confirmation of the award to him of the 1944 NOBEL prize for chemistry and I told him, which was true, that I was doing what I could. HAHN felt that, at the least, the Swedes would think his silence discourteous.

3. In the evening he discussed the matter with HEISENBERG in the following terms:

4. HAHN: If a letter from Sweden has arrived for me, which they won't give me until the Commander's return, it must seem rude if I don't reply to it.

HEISENBERG: It is really nonsense that the Captain relies on the papers for confirmation of the award. It is not his business to contact the Swedish correspondent of the 'Times' or someone like that, but what he should do is to approach his superiors in order to get them to make enquiries with some higher authorities.

HAHN: Well, he did say he would make some further enquiries but I can't tell him very well where to go for them.

HEISENBERG: I think you could mention some of the higher-placed people whom we know, such as BLACKETT, who is bound to know the facts.

HAHN: Yes, I could do that. I suppose Sir John ANDERSON has not returned from CANADA yet?

HEISENBERG: Well, he is really on too high a level for the Captain to approach him. In any case, I would mention some of those names or he should ask that American Colonel. (Pause) After all, a whole week has already gone.

HAHN: There must eventually be a communication from SWEDEN as well. It would just not do for them to keep that from me.

HEISENBERG: Well, it might be that these people have said that, for the time being, nothing should be done until an official decision has been made on a high level. It could be that they have waited for the return of ATTLEE and Sir John ANDERSON.

HAHN: In any case, I could have another talk with the Captain before he goes to town again. On the other hand, I don't want to trouble him too much after I have talked to him only this morning. I know he is doing his best.

5. On Thursday HAHN discussed the matter further with HEISENBERG and WIRTZ who tried to egg him on to take a strong line with me. WIRTZ in particular likes to run with the hare and hunt with the hounds.

6. WIRTZ: The Captain is lying to you. He knows perfectly well that letters for us are in LONDON and everything else. Only nothing is being done about it as the Commander is still away.

HAHN: He told me that. He does not know whether the letters are already in LONDON. The day before yesterday, he told me that he thought the letters had not yet arrived but he said that, in any case, they would have to remain there until the Commander's return which will be some time in the middle of next week.

WIRTZ: I do think we should compile a statement about this whole case.

HAHN: I could tell him that I am of the opinion that, in a matter which concerns me so directly, they should give me permission to write to Sir Lawrence BRAGG if they have no means of finding out for themselves.

HEISENBERG: I would even ask whether you could not write to Sir John ANDERSON.

HAHN: That is out of the question.

WIRTZ: I think we should clearly show our resentment and a deterioration of morale. That depresses the Captain more than anything else.

HAHN: I shall tell him: "We are most surprised that in a matter such as this, we cannot get any definite news although anybody should easily be able to understand that we are most curious about it." I want to write that to BRAGG as well. It is possible that they won't allow me to send that letter, but then we would know for sure that we are considered as war criminals after all.

WIRTZ: I would be even more to the point. The only thing that has ever impressed the Commander and the Captain was that occasionally we have shown our resentment.

> 7. On Friday, 23rd November, HAHN came to see me in the afternoon, having obviously worked himself into a state of courageous fury. He was red in the face and was shaking all over even when he first came into the room. He said that he wanted to write to Sir Lawrence BRAGG for confirmation of the award, that it was monstrous to keep him in doubt and that, in fact, our guests here are treated worse than war-criminals in that they had no proper communication with their families and are detained without even a charge being brought against them. I tried to say that I had no objections to his writing such a letter, but that I did not think that the chances of it being delivered were particularly bright. HAHN, however, was hardly in a mood to consider my replies and left the room abruptly.
>
> 8. Allowing himself half an hour to cool off, he discussed the interview with HEISENBERG, who produced his usual theory about the withdrawal of parole. They also seemed to think that a firm line with me would best serve their purpose:

9. HEISENBERG: That would be quite useful. Do go for him for a change. After all, he could approach some other people in this matter. There must be someone to take the Commander's place when he is away, that is obvious.

HAHN: Well, I can't do a thing at the moment.

HEISENBERG: I don't agree, but, in any case, it is perfectly right that you have gone for him. He must be made to realize that things can't go on in this way.

HAHN: Let us not talk about this to the others. I shall just be frigidly polite to the Captain.

HEISENBERG: I feel we should not put up with just everything from these chaps. This constant reference to the Commander is perfectly ridiculous. There must be someone to take his place, that is obvious. When Captain BRODIE goes to LONDON, he goes to see somebody! He has also been able to arrange everything about the power of attorney for my wife. Probably the Commander, before leaving, has specially impressed on him not to do <u>a thing</u> in his absence.

HAHN: But I have said quite a lot.

HEISENBERG: That is good. And he hasn't said anything in particular at all, not even what is going to happen to you?

HAHN: No, nothing at all.

HEISENBERG: I always think the first thing to do would be for all of us to withdraw our parole. In that case he could not wait for the Commander either, he would have to act.

HAHN: I don't think that will serve any purpose, Mr. HEISENBERG, since these people are so very set on their plans. That would only make matters worse. In any case, he knows my opinion now and perhaps he will think it over and, when he is at his office next time, he might talk to somebody sort of semi-officially, or at least when the Commander does return, that rogue will tell us something.

HEISENBERG: (bitterly) By that time it will be Christmas and, besides, the celebrations in STOCKHOLM will be over. I think we must apply pressure.

HAHN: Well, let us see first whether he hits on an idea.

10. I, after a further half hour, sent for HAHN and discussed everything with him rather more calmly and was able to satisfy him for the time being that I really was trying to get him the information he required but that there were certain difficulties in my way and so on. After this he decided apparently to exercise a little more patience.

11. I think HAHN is prepared to wait for about a week before 'going for' me again.

III. GENERAL IMPATIENCE

12. On Sunday morning, 25th November, WIRTZ and HEISENBERG had conversations expressing their dissatisfaction. To begin with, WIRTZ said we were making enemies of them and suggested a written protest as follows:

13. WIRTZ: The result of our long detention may well be that we shall be filled with a desire for revenge. I mean, if they kept us here another two years, I might well say: "My sole object in life is to drop an atomic bomb on LONDON." I think that around about Christmas we should make some sort of demarche with the object of re-uniting us with our families. Otherwise we shan't get it done in the next ten years. We should have to say: "We admit, there are arguments which justify the detention of German nuclear physicists, firstly because they are nuclear physicists and secondly because GERMANY has to some extent justified such measures. But it must be said on the other hand that we were neither Nazis nor particularly for the war and that we are civilians - (int.)

HEISENBERG: The American militarists who are already at loggerheads with scientific opinion would get this letter. They would not accept any argument of ours but would merely say something like this: "There you are, the usual Nazi arrogance. They pretend they have some rights. They have none, we'll show them!"

WIRTZ: It could hardly be called arrogance, when we say: "We have been shut up for seven months without trial although we are not war-criminals or prisoners of war and it is particularly hard in that our families are left helpless.

HEISENBERG: I agree.

> 14. Discussing the best tactics to use in order to put an end to their detention, HEISENBERG and WIRTZ complained that they don't know enough about what is being done about them. WIRTZ has frequently said that they would all be very much happier if they knew what was going on. This conversation suggests that something more than happiness was what he was after:

15. HEISENBERG: We must find some dodge to alter our position.

WIRTZ: Yes, the most important thing for us would be to find out what is being done about us. If we only knew, we could settle everything to our own advantage in three weeks. We have had plenty of experience of that sort of thing in HITLER's GERMANY. All our contacts with our detainers so far have been failures. This Nobel Prize business is not our cue. Our cue is that your mother has died or your children are starving.

HEISENBERG: All my fights were on that score. HAHN's NOBEL prize is really also a personal problem which is certainly very important. If he gets it, his whole future life is changed, but I agree one's family is really more important. The whole trouble in this business between us and the militarists is that they are able to arrange that we know nothing of what is going on so that every move we make is a mistake. The worst of it is that we have nobody at hand, neither Captain BRODIE nor Major RITTNER nor Commander WELSH, to give us the tip when to do what.

> 16. Their appreciation of their situation seems somewhat misguided in that they think the big decisions lie with them and they feel they should stress the family problem as being the most likely to influence us:

17. WIRTZ: The only chink in the Commander's armour is the difficulty in which our families are placed, because these things, in the long run, will make him unpopular with the rest of the scientific world. So I should say: "Let us keep stabbing at it."

HEISENBERG: I quite agree.

WIRTZ: That is the sort of headline we want to see: "Poor little anti-Nazi toddlers" or something like that. That would finish them.

IV. THE FUTURE

18. The guests are inclined to see themselves as exerting great political influence in the future. They fairly frequently come back to the idea that in the future government will be by scientists and not by politicians. Apparently the programme is not yet defined:

19. WIRTZ: Scientists should be able to form a political organisation but there is much more to it than being an expert in one line. It is not only a matter of organisation, but there must be a definite programme. Even our own political ambitions are more than vague. We are always saying, just as HITLER used to: "First power and then the details", but clear concrete plans we just haven't got.

V. TECHNICAL

20. HEISENBERG gave a lecture on Thursday, 22nd November on 'The Structure of organic compounds, their symmetry etc.'

21. There was nothing of technical interest during the week except the above colloquium.

FARM HALL
GODMANCHESTER
26 Nov 45

(Sgd) PLc Brodie
Captain,
for Major T.H. RITTNER

AMERICAN EMBASSY
OFFICE OF THE MILITARY ATTACHÉ
1, GROSVENOR SQUARE, W. 1
LONDON, ENGLAND

WLW/rr

7 December 1945

SUBJECT: Transmittal of Document.

TO: Colonel W. R. Shuler, Room 5004, New War.Department Building, Washington, D. C.

1. Inclosed herewith is copy #1 of Farm Hall Report #20, covering the week from 26th November to 2nd December 1945.

2. On 4 December 1945, Colonel Calvert and myself visited the guests at Farm Hall and spent the evening with them. Their morale seemed considerably improved over the last time either of us saw them. Professor Hahn had requested previously permission to go to Stockholm to receive the Nobel Prize award. This permission was denied him, but in lieu thereof he was permitted to write a letter stating that he would be unable to receive this award, at least for the present time. This letter, of course, was in full compliance with all security measures.

3. The guests will probably be returned in a group by the first of next year to the British Army of the Rhine Headquarters and then be discharged individually.

For the Military Attache:

W. L. Warner

1 Incl:
FH Rpt #20
Copy #1

W. L. WARNER,
1st Lt, C.E.,
Assistant to the Military Attache.

Copy No. 1

Ref: F.H.20.

To: Mr. M. PERRIN and Lt.Cdr. WELSH
From: Major T.H. RITTNER

OPERATION "EPSILON"

(26th Nov. - 2nd Dec. 1945)

I. GENERAL

1. This week started badly with disappointment over the last lot of letters sent home and ended more cheerfully with a further mail and a successful visit from Lt.Cdr. WELSH.

II. LETTERS

2. On Wednesday, 28th November, I informed the guests that the American authorities in FRANKFURT had not been able personally to deliver the letters to their families, which had been sent from here on Tuesday, 23rd October, and that, therefore, no answers could be expected.

3. This caused general gloom and annoyance.

4. HAHN, in conversation with HEISENBERG and von LAUE expressed his disappointment as follows:

5. HAHN: I am so disgusted about this letter business. Perhaps it is not BRODIE's fault, it may be a dirty trick of the Commander. Now suddenly we are calmly informed that there is no courier. It is such a dirty trick to lie to us. If I went to the Captain now, I would - (very agitated) - no, I must not!

6. Colonel CALVERT, however, agreed to arrange a two-way mail in time for Christmas. This somewhat lightened the gloom, the guests reacting characteristically; for instance, the following conversation between HAHN, WIRTZ and BAGGE:

7. HAHN: BRODIE made efforts to have the letters delivered by the Americans via PARIS.

WIRTZ: BRODIE does not take any trouble; the American Colonel proposed that to him. They are a dirty lot of scoundrels.

HAHN: I would not say that it is the Captain's fault.

WIRTZ: All these fellows, including BRODIE, behave badly. That is the crux of the matter.

HAHN: Captain BRODIE cannot help that.

BAGGE: He reports regularly about us to the Commander.

HAHN: I am sure of that; therefore it is no use abusing him in the Captain's presence. That only makes the Commander more furious.

WIRTZ: Unfortunately, we don't abuse him at all.

8. Letters were duly written and taken to LONDON on Thursday 29th November.

III. HEISENBERG'S FAMILY

9. I was able to tell HEISENBERG on Tuesday 27th November that Colonel CALVERT had met his wife and to reassure him about her welfare, since when HEISENBERG has been much relieved and more cheerful.

IV. Lt. Cdr. WELSH's VISIT

10. Lt.Cdr. WELSH came down on Saturday, 1st December, for the night and cheered everybody considerably. The crisis forecast last week now seems much more remote.

11. Von LAUE was particularly pleased to get a letter from his son in the U.S.A. which he carried about proudly.

12. The results of the visit were thought satisfactory as the following conversations show:

13. HAHN: He did give us a lot of useful information after all.

HEISENBERG: The remarkable thing is that he really does know all these people like BOHR and so on.

14. HEISENBERG: The conversation at breakfast this morning with the Commander was extremely pleasant.

GERLACH: I thought, he was more friendly than usual last night.

HEISENBERG: I think, he was quite satisfied himself. The visit may not have been 100% valuable, but, nevertheless, I should say 50%. He really did say a lot that was useful.

15. There were no adverse comments.

V. TECHNICAL

16. On 27th and 30th November HEISENBERG gave lectures on the subject of 'Molecular structures from a mathematical point of view'.

17. There was nothing else of technical interest during the week.

FARM HALL
3rd December 1945

(Sgd) P.L.c. Brodie
Captain
for Major T.H. RITTNER

Copy No. 1.

Ref. F.H.21

To: Mr. M. PERRIN and Lt.Cdr. WELSH
From: Captain P.L.C. BRODIE†

OPERATION "EPSILON"

(3rd – 9th Dec. 1945)

I. GENERAL

1. In spite of the visit from Lt.Col. CALVERT and Lt.Cdr. WELSH on 4th December, this has been a very quiet week without difficulties or happenings of interest.

II. THE VISIT

2. Lt.Col. CALVERT and Lt.Cdr. WELSH came down on Tuesday, 4th December to persuade HAHN to write a letter to SWEDEN accepting the NOBEL Prize awarded to him but regretting that he did not think that it would be possible to go to STOCKHOLM himself. HAHN was not at all keen to write without saying that he was prevented from visiting STOCKHOLM by his detention at the hands of the Anglo-Americans. However, he was eventually persuaded to write as required.

III. THE FUTURE

3. In the early morning of 5th December WIRTZ, in conversation with VON WEIZSACKER, gave vent to his now usual anti-British spleen, and VON WEIZSACKER expressed a more definite opinion than usual:

4. WIRTZ: From what I have seen in ENGLAND during these months, I should not like to settle down here on my own free will or to volunteer for work here, no matter what conditions in GERMANY may be. It might be different, of course, if they were to offer me some terrific position in MANCHESTER or somewhere, but I would never actually apply for it.

WEIZSACKER: No, for that sort of thing one could only consider AMERICA or RUSSIA.

---

† From here on all the reports are from Capt Brodie and signed by him. Previous reports are from Major T H Rittner, but from FH6 (15 September 1945) onwards signed P L C Brodie, Captain, for Major T H Rittner. From Bagge's diary (loc. cit. pp 60–64, entries 12 September, 19 September, 29 October 1945) one learns that by 12 September Major Rittner was away, at home, ill and by 29 October was recovered, and expected to return on 23 December. But in fact FH23/24) it was Commander Welsh who relieved Captain Brodie at Farm Hall from 22–27 December and Rittner does not reappear. FCF

5. There is at present a fairly general tendency to regard the British as the authors of all their discomforts and the Americans as their one hope of salvation.

IV. TECHNICAL

6. DIEBNER, on 7th December, used the work of BALDWIN and KOCH of the university of ILLINOIS to hold a lecture on 'Kern-Photoeffekte'.

7. There was nothing else of technical interest during the week.

FARM HALL
GODMANCHESTER
10th December 1945

(Sgd) P.L.C. Brodie
Captain

Copy No. 1.

Ref: F.H.22

To: Mr. M. PERRIN and Lt.Cdr. WELSH
From: Captain P.L.C. BRODIE

OPERATION "EPSILON"

(10th - 16th Dec. 45)

I. GENERAL

1. Nothing of interest has happened this week, either in direct contact with the guests or in the monitoring service, although they are entering upon one of their periodic fits of gloom.

II. LETTERS

2. The promised courier from GERMANY is eagerly though rather sceptically awaited. The effect, if the letters from the guests' wives do not arrive before Christmas, will be extremely disagreeable.

III. DR. ING. HANSEN

3. At the request of Lt.Col. CALVERT, I asked HAHN and finally everybody else whether they knew anyone of this name. Nobody did.

IV. TECHNICAL

4. HEISENBERG, on 11th December, gave a lecture on the absorption of light by molecules.

FARM HALL
GODMANCHESTER
17th December 1945

(Sgd) P.L.C. Brodie
Captain

Copy No. 2.

Ref: FH 23/24

To: Mr. M. PERRIN and Lt.Cdr. WELSH
From: Captain P.L.C. BRODIE

OPERATION "EPSILON"

(17th - 30th Dec. 45)

I. GENERAL

1. This report covers two weeks, including the Christmas period 22nd - 27th December, when Lt.Cdr. WELSH nobly relieved me.

2. There was nothing to report before the arrival of Lt.Cdr. WELSH; the guests, however, were considerably cheered by the news of their impending return to GERMANY, which the Commander conveyed to them, and this, together with Christmas celebrations, produced a general feeling of good will.

II. CHRISTMAS

3. Lt.Cdr. WELSH brought Lt.Col. DEAN and Lt. WARNER down for the night on 22nd December. Our guests were gracious enough to speak well of the Colonel:

4. WIRTZ: His quiet thoughtfulness made a very deep impression on me. Compared to him we are a lot of hysterical children.

HEISENBERG: A fine type!

5. Lt. Cdr. WELSH brought the gratifying news of our guests' impending departure to GERMANY. WIRTZ, however, made a last and not very earnest attempt to keep his querulous colours flying:

6. WIRTZ: Who knows, he may have just made this up so that he will have a pleasant Christmas here.

The Company: Rubbish!

7. The news of the return had a good effect on the guests' opinion of us all:

8. WEIZSACKER: To be quite honest, I would not have minded being here another six months. They have looked after us marvellously.

HEISENBERG: If one had only known that we were to go back after a certain time, there would have been absolutely nothing to complain about.

(And a ragged broadside from:)

WIRTZ: There is one solid comfort, the Commander and BRODIE are going to fly with us, so at least they don't intend to have us crash.

9. About their eventual dispersal in GERMANY they have the following ideas:

10. WEIZSACKER: I am convinced that you (HEISENBERG) and HAHN will get away at once and most of us not too long thereafter.

HEISENBERG: I am not yet sure.

WIRTZ: I think, WEIZSACKER is right and I should add HARTECK. The problem is, of course, what is going to happen to people like DIEBNER. We know what sort of man he is, but I don't think, we should just drop him.

HEISENBERG: I agree. He has behaved himself very well recently.

WEIZSACKER: It ought to be possible to do something for him. He is not a physicist in the real sense of the word, but as it seems that we shall be able to work again, we shall need somebody to get hold of apparatus, to look after it and so forth. We could very well use him for something like that.

HEISENBERG: That is roughly my idea.

11. Werewolves still lurk in the background and the guests are anxious to avoid giving an impression in GERMANY that they have done anything for us.

12. WIRTZ: It would be a mistake, when we get back to GERMANY, to say how marvellous everything has been.

HEISENBERG: Yes, but, of course, there is a danger of that happening with people like LAUE. That should be avoided at all costs, but, on the other hand, we must do justice to the British who really have treated us extremely well.

WIRTZ: We should say that we were physically very well treated and mentally wretchedly.

III. LETTERS

13. Letters arrived in time for Christmas Day which gave very great pleasure to everybody.

IV. WIRTZ HAULS DOWN HIS COLOURS
OR
HE WHO FIGHTS AND RUNS AWAY, LIVES TO FIGHT ANOTHER DAY

14. WIRTZ: There is a lot to be said for the Commander after all, no matter how much we may have cursed him. In any case, it may be wise to be in his good books. We never know when we may have another use for him.

FARM HALL
GODMANCHESTER
31st December 1945

(Sgd) P.L.C. Brodie
Captain.

EPILOGUE I

Album given to Captain Brodie by the detainees for Christmas 1945†

† Translation by FCF. Titles, offices, names and positions have been left as they were written in the original handwritten German text.

The Farm Hall community sends you, dear Captain, very best wishes for Christmas and hopes you will have a few happy days of holiday in the circle of your family.

We ask you to accept this little 'Album' compiled with the restricted facilities of Farm Hall, as a mark of thanks for the readiness to help you have so often shown.

There was intended to be a picture on the front of each page, belonging with the text on the reverse. Perhaps it will be possible for each of us to provide a copy of photographs made by us for the purpose, for the missing ones.

Die Detainten

22.xii.45†

† Opening page: handwriting probably of Gerlach, in a more formal upright style than in his personal album entry.

Best wishes for Christmas. I would like, *Herr Hauptmann Brody* to express my thanks for having, as far as lay in your power eased my position.

H. Korsching

Note: Dr Korsching's contribution to the album, unlike the others, omitted biographical information. His entry in Kürschner's Gelehrter-Kalender translates as:

Korsching, Horst, *Dr rer. nat.*, scientific co-worker at the *Max-Planck-Institut für Physik und Astrophysik*; Germaniastrasse 31, W. 8000 München 40. Born Danzig, 11.8.12. *Promoviert* University of Berlin '38, Physical Electrochemistry. 5 articles in *Die Naturwiss.*, '39–'57; 1 in *Chem. Ber.*, '40; 12 in *Zs. Naturforsch.*, '46–'74, and others. (Editors).

I was born on 8th March, 1879 in Frankfurt a. M. After my schooling I studied Chemistry in Marburg and Munich and *promoviert* in July 1901 with an organic Doctor thesis.

After two years as an *Assistent* I went, principally to learn English, to London in the laboratory of Sir William Ramsay. He gave me a project in an area quite new to me, in radioactivity. By chance I found a radioactive substance which I named Radiothorium. Ramsay advised me to continue with radioactivity and so I went in winter 1905 to Montreal in Canada, to Professor E Rutherford. In summer 1906 I came to Berlin, *habilitierte* myself there and received the title Professor in 1910.

On the foundation of the *Kaiser Wilh. Gesellschaft* I became a member of the K.W.I. for Chemistry. Except for the war years 1914–1918 when I was on active service, I remained continuously in the *Kaiser Wilh. Institut*; since 1928 as Director. In 1933 I was for a few months a 'visiting Professor' in Cornell University (U.S.A.). At the beginning of 1944 my Institute in Dahlem was totally destroyed by bombs. I transferred activities to Tailfingen in Wurthemberg. From there I was taken away by American soldiers on 25th April, 1945.

Otto Hahn

Max Theodor Felix Laue

Born 9.10.1879 in Pfaffendorf near Koblenz became acquainted in his youth with a large realm by way of Brandenburg, Altona, Posen and Berlin finally to Strasburg i.E. There I attended the long-famous Protestant *Gymnasium* from 1893 to 1898. I studied in Strasburg, Göttingen, München and Berlin, *promovierte* in July 1903 in Berlin with a dissertation set by Planck on certain optical interference phenomena. In 1905 I became Planck's *Assistent*, 1906 *Privatdozent* at the University of Berlin, 1909 at that in München. There I wrote my book on relativity theory, and came upon X-ray interference. In 1912 I came as *a.o. Professor* to the University of Zürich, 1914 as *Ordinarius* to Frankfurt a.M., 1919 to Berlin. I became Emeritus in 1943.

Walther Gerlach

Born 1 August 1889 at Biebrich on the Rhine as son of a doctor, both parents came from old Frankfurt - Nassau families.

During my schooling at the *Humanische Gymnasium* in Wiesbaden, I interested myself equally in Natural Science, Classics and Religion. The pleasure learnt from my father in observing Nature has remained with me to this day. From 1908 I studied in Tübingen Mathematics, Philosophy, Physics and Chemistry, *promovierte* to *Dr. rer. nat.* on 29th February, 1912 with Professor F. Paschen and became *Assistent.* After service at the front in the war I was 2 years in Industry, then *Dozent*, later Professor, in Frankfurt a.M., from 1925 in Tübingen. Since 1929 I am Professor of Experimental Physics and Director of the Physical Institute of München University. Lectures and Conferences brought me a number of times to England, France, Holland, Italy, Poland, Switzerland, Spain. My scientific work is concerned with problems of radiation, atoms and magnetism. I take particular pleasure in developing the education of the young. For leisure I make music with my wife, or I wander, for choice in wooded mountains or the Bavarian mountain-foothills.

I, Werner Carl Heisenberg was born on 05.12.1901 in Würzburg where my father was a teacher at the *Gymnasium* and *Dozent* at the University. In 1909 my father was called to München; there I grew up, learning languages, mathematics and music, and studied from 1920 onwards — after a short interruption fighting as a volunteer — Physics with Sommerfeld. At the same time I familiarized myself in the youth-movement with wandering through the homeland and with many kinds of sport. In 1924 I became a *Dozent* in Göttingen, and worked out the quantum mechanics during a holiday stay on Heligoland. In 1926 and 1927 I was a *Lektor* in Copenhagen, as pupil and friend of the great physicist and philosopher Niels Bohr. From 1927 till 1941 I was a Professor in the University of Leipzig, where I instructed many young people, German and foreign, in Atomic Physics. In 1929 I gave lectures and courses in America, Japan and India. I have a family since 1937. In the war, 1941, I was called to the *Kaiser-Wilhelm-Institut für Physik*.

P. Harteck

Born 20.vii.02; Vienna

Learning and wander-years: Universities of Vienna and Berlin, *Technische Hochschule* Breslau.

1928–1933 at the Kaiser-Wilhelm Institut for Physical Chemistry with *Geh.* Haber.

1933–1934 Cambridge with Lord Rutherford

from 1934 Hamburg; Director of the Institute for Physical Chemistry of the University.

I, Kurt Diebner was born on the 13 May 1905. After an *Oberrealschule* I studied Natural Sciences in Halle and Innsbrück. I *promovierte* in November of the year 1931 as *Dr rer. nat.*. My teacher was Professor Hoffmann, known for his researches in Cosmic Rays.

From April 1931 I was an *Assistent* in the Physics Institute of the University of Halle, working particularly on matters of Nuclear Physics. In 1934 I became a collaborator of the *Physikalisch-Technische Reichsanstalt* and *Referent* in the research department of the H.W.A.† for questions of Atomic Physics.

Since 1942 I was in the *Reichsforschungsrat*, at first *Kommissarisch*, then *Hauptamtlich* as personal *Assistent* to Professor Gerlach and active as leader of a research department.

† = *Heereswaffenamt* = Army Ordnance Office. FCF

I was born on the 24 April 1910 in *Köln a. Rhein*. My father is a judge, my mother comes from South Germany. As a young boy I enjoyed physical–technical experiments. I studied in Bonn, Freiburg and Breslau, Physics, Mathematics and Chemistry. In the year 1935 I became an *Assistent* in the Physico-chemical Institute of the University of Leipzig, partly moved by the wish to acquaint myself with modern physics in the neighbourhood of Heisenberg, whose relationship to the intellectual development of our time interested me. Then I became an *Assistent* and finally a section leader in the *Kaiser Wilhelm Institut für Physik* in Berlin. In addition I became in 1941 a *Dozent* in the University of Berlin. My scientific work was chiefly on the molecular physics of liquids, and on problems at the border regions of physics and chemistry.

K Wirtz

I was born on 30 May 1912 in Neustadt near Coburg. My youth was spent in my home town, from which from 1922 to 1931 I attended the *Realgymnasium* in Sonneberg. After the *Abiturienten* examination I studied at the *Technische Hochschule*, München in the faculty of technical physics, where I (with one year in between in Berlin) carried out an experimental research on cosmic rays, to finish my studies, begun in 1935, with the *Diplomexamen.* Then I moved to the Institute for Theoretical Physics of Professors Heisenberg and Hund in Leipzig and *promovierte* 1938 with a research on nuclear vibrations. Until my *Habilitation* 1941 in the same University I was mainly occupied with carrying out theoretical investigations on problems of the cosmic radiation. Then followed my posting to the *institut de chimie nucleaire* of Professor Joliot in Paris. Following that I received a position as *Assistent* in the *K.W.I.f.Physik* in Berlin to carry out into practice a method of isotope-separation proposed by myself.

Erich Bagge

Carl Friedrich Freiherr von Weizsäcker, born 28.6.1912 in Kiel from a Swabian family. My childhood was spent, because my father was first a naval officer and then a diplomat, at varying places in Germany, Holland, Switzerland, and Denmark. In my school time I was interested in Astronomy and Philosophy. Early acquaintance with Heisenberg made it clear to me that in the present day Atomic Physics would have a decisive influence on both these branches of learning. *Studium* with Heisenberg and Bohr, *Dr. Phil.* Leipzig 1933, there 2 years *Assistent* with Heisenberg, 6 years *Assistent* at the *K.W.I.f.Physik*, Dahlem, since 1937 at the same time *Dozent* in the University of Berlin, since 1942 Professor of Theoretical Physics in Strassburg. Since 1937 married to a Swiss wife, 3 children. I have worked above all on the structure of the atomic nucleus and on the evolution of stars. Still more than the abstract science it is its meaning for the spirit of the time, and its relationship with Philosophy and Religion which engages me.

EPILOGUE II

What happened to the detainees after their release

What happened to the detainees after their release

The ten detainees were brought back to Germany on 3 January 1946, initially to the small village of Alswede. Harteck and Diebner left immediately for Hamburg, the former to resume his chair of Physical Chemistry at the University there and Diebner to set up a private Institute for Measuring Instruments. Gerlach departed to become a guest-Professor in Bonn, thence after a year or so to resume his chair of Experimental Physics in Munich, later to become Rector of the University. He died in 1979. On 12 March 1946 von Laue, von Weizsäcker, Wirtz, Korsching and Bagge left Alswede for Göttingen, Hahn and Heisenberg having gone on ahead of them to confirm the suitability of the space there, in Prandtl's former Aerodynamics Research Establishment (available because aerodynamics research was forbidden to the Germans under the control laws) which had been allocated by the British control authorities for the relocation of the Kaiser Wilhelm Institute for Physics and the Administration of the *Kaiser-Wilhelm-Gesellschaft.*

Bagge was appointed to a Professorship in Hamburg in 1948, later to become Professor and Head of the Department of Physics at the University of Kiel. In 1951 Harteck left Hamburg for the United States, becoming a Research Professor at the Rensselaer Polytechnic Institute in Troy, NY. He died in 1985.

Concurrently with their posts in the Kaiser Wilhelm Institut für Physik (presently to become the Max Planck Institut für Physik), Heisenberg, von Weizsäcker and Wirtz also held honorary Professorships at the University of Göttingen.

Already before his return to Germany Hahn had been invited to accept the presidency of the *Kaiser-Wilhelm-Gesellschaft* in succession to the aged and ailing Max Planck (who was to die on 4 October 1947), and had assented. The British control authorities gave permission for the Society to carry on with its work only on condition that the name of Kaiser Wilhelm should be dropped. Hahn secured Max Planck's agreement to lending his name to the Society, and foundation of the *Max-Planck-Gesellschaft* became official in the British Zone of Occupation on 11 September 1946. Hahn's presidency was no mere honorific sinecure: it took all his persistence and diplomacy to win over the American control authorities, who had resolved that the Kaiser Wilhelm Society should be dissolved, and re-establishment of its Institutes not permitted. But he got his way in the end. Soon after American objections were lifted in February 1948 the French also came round, and by July 1948 the *Max-Planck-Gesellschaft* had permission to function in all three Western zones of occupation, proceeding to re-establish most of the old Kaiser Wilhelm Institutes, and some new ones, at various points in Germany. The second general meeting of the Max Planck Society, at Munich in September 1951, celebrated both the fortieth anniversary of the foundation of the Kaiser Wilhelm Society and the centenary of its original founder, Adolf von Harnack, Professor of Theology and Chaplain to the Kaiser. With Heisenberg as Director of the German Research Council (*Deutsche Forschungsrat*) created early in 1946, the two of them played the leading parts in the post-war restoration of German scientific research. Hahn died in 1968.

In 1951 von Laue left Göttingen for Berlin-Dahlem to become Director of the Kaiser Wilhelm Institute for Physical Chemistry and Electrochemistry, which continued to call itself a Kaiser Wilhelm Institute until it too came into line in 1953 as the *Fritz Haber Institut* of the *Max-Planck-Gesellschaft*. Von Laue died in 1960.

Heisenberg moved his Max Planck Institute for Physics (Astrophysics having by now been added to its subject title) to Munich in 1958. He held an adjunct Professorship at the University there, and continued to find time for productive contributions to theoretical physics as well as for organizational matters till forced by illness to retire in 1970. He died in 1976.

Von Weizsäcker, pursuing his combined interests in cosmology and philosophy, was appointed Professor of Philosophy at the University of Hamburg in 1957. *Inter alia* he was to play a leading role in the German branch of the Pugwash movement.

In 1957 Wirtz was appointed Professor at the Karlsruhe *Technische Hochschule* and Director of the Institute for Neutron Physics and Reactor Technology at the Nuclear Research Centre (*Kernforschungszentrum*), Karlsruhe.

Korsching spent the rest of his career at the Max Planck Institute for Physics and Astrophysics.

Diebner, together with Bagge, was the driving force behind the foundation in 1956 of the *Gesellschaft für Kernenergieverwertung in Schiffbau und Schifffahrt* (Society for Exploitation of Nuclear Energy in Shipbuilding and Shipping) in Hamburg. This developed into the *GKSS-Forschungszentrum-Geesthacht GmbH*, which built the first German nuclear-powered ship, launched in 1968 and named the 'Otto Hahn.' But by then Diebner had died, in 1964.

Bagge, Korsching, von Weizsäcker and Wirtz are still living.

F C Frank

Index

THE POEM AS SACRAMENT

LOUVAIN THEOLOGICAL & PASTORAL MONOGRAPHS
26

THE POEM AS SACRAMENT

The Theological Aesthetic of Gerard Manley Hopkins

Philip A. Ballinger

PEETERS PRESS
LOUVAIN

ISBN 90-429-0807-6 (Peeters Leuven)
D. 2000/0602/12

ACKNOWLEDGMENTS

There are so many people whose love and support form the foundation of this work. First of all, I thank the Rev. Robert Turner for giving me the initial impetus to begin the doctoral research upon which this book is based. I offer further thanks to the Faculty of Theology at the Katholieke Universiteit Leuven for their willingness to allow an alumnus 'at large' to pursue his scholar's dream. To Professor Terrence Merrigan, who stands out brightly in my story, I owe a debt of gratitude. He continually gave me the confidence and optimism I needed to stay on task and move forward. Providence chose the director of my doctoral research well.

Even though I studied *for* Leuven, I studied *at* Gonzaga University, and there could be no better place to pursue research about Gerard Manley Hopkins. To the Special Collections staff at the Foley Center, Stephanie Edwards and Sharon Prendergast, go my warm thanks for their assistance, suggestions, and enthusiasm. Most of all, however, I thank a person of vast knowledge, erudition, and love of Hopkins — Mrs. Ruth Seelhammer. Ruth is the heart of the Hopkins Collection at Gonzaga and she shared that heart with me. Additionally, Ruth introduced me to some of the foremost Hopkins scholars in the world, Professors Michael Allsopp, Joseph Feeney, Francis Fennell, and David Downes. I thank them all for their kind support, suggestions, and 'grist for the mill'.

Finally, and most essentially, I wish to thank my wife Kathryn. Of all the support and encouragement I received, hers was the most constant, patient, and steadfast.

November 15, 1999

TABLE OF CONTENTS

GENERAL INTRODUCTION

A. General Comments

An old Jesuit lay brother, remembering Gerard Manley Hopkins (1844-1889) thirty years after his death, could only say, "Ay, a strange yoong man crouching down that gate to stare at some wet sand. A fair natural 'e seemed to us, that Mr. 'Opkins."[1] Many considered Hopkins odd. Though beloved by friends, particularly those from his Oxford days, he was deemed peculiar at best for having become a Jesuit, much less a Catholic. The Jesuits themselves, as hard as they tried to assimilate him, could not fully do so. The harsh reality was that his own religious brethren never understood him. He was an oddity. He attracted attention to himself through his innocent eccentricities — hovering over a frozen pond to absorb the pattern of trapped bubbles, putting his face down to a cup of hot chocolate to observe the grey and grained look of the film on its surface, sprinting out of a building after a shower to stoop down and study the glitter of crushed quartz before the water could evaporate.[2] Hopkins' oddness was eventually deemed uniqueness following the publishing of his poems by Robert Bridges three decades after his death.[3] It was only with the

[1] Humphry House, ed. *The Journals and Papers of Gerard Manley Hopkins*, 2nd ed. (London: Oxford University Press, 1959), p. 408.

[2] Robert Bernard Martin, *Gerard Manley Hopkins: A Very Private Life* (New York: G.P. Putnam's Sons, 1991), p. 202.

[3] In fact, Bridges was not enthusiastic about Hopkins' poetry. In his introduction to the poems, Bridges apologized for "the rude shocks of [Hopkins] purely artistic wantonness," while an early review in *Spectator* dismissed the poetry as "needlessly obscure, harsh, and perverse." It was not until 1930 that Hopkins'

benefit of looking at his life and work as a whole that the many, instead of the few, came to see him as *sui generis*, an original.

Hopkins' disregard of conventional behavior, traditional poetic rhythms, diction, syntax, and ways of perception lay at the heart of the originality of his poetry, and it is for his poetry that he is mainly renowned. Nonetheless, to consider Hopkins a poet only is to miss the man. No doubt he was a poet, perhaps one of the greatest Victorian poets, but before, through, and beyond his poetry he was a subtle philosopher and theologian whose poetic uniqueness arose in part because of his desire to speak as aptly as possible of the immanent God, of the Incarnate Christ. His tool for this task, his artistry, was not well perceived by others during his lifetime. To Robert Bridges, his life-long friend and post-humous editor of his poetry, he wrote: "If you do not like [my compositions] it is because there is something you have not seen and I see. That at least is my mind, and if the whole world agreed to condemn it or see nothing in it I should only tell them to take a generation and come to me again."[4] And so we do.

Hopkins' theological contribution is as timely as it is unique in its expression. His is a theological approach that emphasizes the perception of beauty in creation as revelation through an intensely Christological lens. Creation, in the Hopkinsian approach, is not to be taken generically. Creation most often has an utterly individuated sense. Truth is in the particular more than the universal. Each individual thing, each being in its self-expressiveness is a potential revelation of the Divine. This revelation occurs, for Hopkins, in the perception of a thing's or experience's beauty— a perception which he ultimately held to be a kind of 'knowing' of the incarnate Word in matter.

poems achieved wide regard. See Murray Roston, *Victorian Contexts: Literature and the Visual Arts* (New York: New York University Press, 1996), p. 133.

[4] Claude Colleer Abbott, ed. *The Letters of Gerard Manley Hopkins to Robert Bridges*, 2nd ed. (London: Oxford University Press, 1955), p. 214.

Part of Hopkins' uniqueness as a religious poet and subtle theologian flows from his concentration on the unique, on each being's or experience's individuality. The uncanny single-mindedness with which he stays with experiences of beauty's perception, seeking understanding in each unique moment of revelation, sets him apart. Yet, even more, his genius manifests itself in his use of words to convey this experience, this perception. Hopkins exploited the latent possibilities of language and radicalized the meanings of words he knew so well to break through to an expression and conveyance of a unitive perception of being, beauty, and divinity. His poetry, in many instances, is his theology. For Hopkins, poetry is the use of language best suited for an approach to the transcendent. Even more, it seems that Hopkins tended towards a view of poetic language as 'sacramental' in function.

A legacy was born when Hopkins' Ruskinian and signic approach to nature met the formative influence of Ignatian spirituality and the theology of Duns Scotus. In Hopkins, "aesthetic and religious experience became one in the sacramental apprehension of beauty."[5] Thus, the senses and the physical world became primary in Hopkins' attempt to bring beauty back to faith. This attempt is perhaps Hopkins' greatest contribution to our age.

Hans Urs von Balthasar greatly admired Hopkins for this reason. In his *The Glory of the Lord: A Theological Aesthetics*, von Balthasar asks how Christianity became such a universal power if it has always been as humourless, anguished, and grumpy as it is today. His answer is that today there is a missing element — the element of beauty:[6]

[5] Robert Bernard Martin, *Gerard Manley Hopkins: A Very Private Life* (New York: G.P. Putnam's Sons, 1991), p. 207.

[6] See John Coulson, "Hans Urs von Balthasar: Bringing Beauty Back to Faith," in *The Critical Spirit and the Will to Believe: Essays in Nineteenth-Century Literature and Religion*, ed. David Jasper and T.R. Wright (New York: St. Martin's Press, 1989), p. 218.

> Beauty must be restored to faith and to its traditional place with truth and goodness as one of the transcendental attributes of the Christian faith. What follows, theologically, is that God's self revelation in history and in the Incarnation must now become for us the very apex and archetype of beauty in the world, whether men see it or not. The supreme form of the beautiful is made manifest in Jesus Christ and particularly in his Resurrection. He is the visibleness of the Invisible One, *the* definitive and determinant form of God in the world. He is to be distinguished from all other forms of worldly beauty as their primal, archetypal source.[7]

This 'Balthasarian' appeal could have been made by Hopkins in the 1870s. He too, in his unique way, concerned himself with the reunion of beauty and faith. In this sense, finding a theological basis for discerning beauty and for carrying to others the experience of beauty through words became for Hopkins what von Balthasar refers to as a 'theological aesthetics'.

The 'unity' between the physical and spiritual is at the heart of Hopkins' aesthetic theory — a theory that had its beginnings in Hopkins' talent and practice of paying close attention to the uniqueness of everything he observed. A word and concept at the center of his theory is *inscape*. This Hopkinsian contribution to the English language is difficult to define for it never had a static meaning for Hopkins. It developed out of his experience that "what you look hard at seems to look hard at you."[8] As a starting point for our understanding of *inscape*, let us use R. B. Martin's effort at interpretation:

> ...[Hopkins] was expressing his belief that when one understands a person, an object, or even an idea, through close study, that which is studied radiates back a meaning, one that is necessarily unique because each manifestation of the world is somehow different from

[7] Ibid., pp. 218-19.

[8] Robert Bernard Martin, *Gerard Manley Hopkins: A Very Private Life* (New York: G.P. Putnam's Sons, 1991), p. 205.

> any other, so that no two meanings can be precisely the same. Inscape is that meaning, the inner coherence of the individual, distinguishing it from any other example. It is perceived only through close examination or empathy, but it is not dependent upon being recognized; rather, it is inherent in everything in the world, even when we fail to notice it.[9]

Norman MacKenzie, the famed Hopkins scholar, offers this succinct view:

> Inscape is not a superficial appearance; rather it is the expression of the inner core of individuality, perceived in moments of insight by an onlooker who is in full harmony with the being he is observing.[10]

A perception of beauty is for Hopkins a perception or grasping of inscape. One could say that beauty, in this context, is the self-expressiveness of a thing, or, as Hopkins might say, its 'selving'. For Hopkins, this is where the Divine may be met.

A corollary to inscape is Hopkins' concept of *instress*. Instress is an energy or "stress of being" which holds the inscape together. Hopkins also refers to instress as "the force which also, as an impulse from the inscape, carries it whole into the mind of the perceiver."[11] W.H. Gardner surmises that Hopkins, a classicist of high achievement at Oxford, found remote inspiration for his concept of instress in the "plastic stress" of Plato's One Spirit, "which sweeps through the whole world of dull matter to impose upon it the predestined forms of the Prime Good."[12] Whatever the origin, instress ultimately became for Hopkins the stress of God's Will in and through all things. Effectively, Hopkins' *inscape-instress* dynamic expressed a deep conviction not only in the Trin-

[9] Ibid.

[10] Norman H. MacKenzie, *A Reader's Guide to Gerard Manley Hopkins* (Ithaca: Cornell University Press, 1981), p. 233.

[11] W.H. Gardner and N.H. MacKenzie, eds. *The Poems of Gerard Manley Hopkins*, 4th ed. (London: Oxford University Press, 1967), p. XX.

[12] Ibid., p. XXI.

ity's ever present and expressive creative act, but also in human kind's ability to perceive this creative act and even to express this perception to others through language.

Hopkins hoped to convey inscape through his poetry. In his arguably better moments, he saw his poems as a kind of 'sacramental medium', a communication of the incarnate and creative divinity to others. Yet his views and aesthetic inclinations caused him confusion and suffering. In Hopkins, aestheticism and asceticism collided. His aesthetic valuing of the senses collided with his ascetic fear of them. His stress upon, empathy with, and at times, deep love for individual things and people could, so he feared, draw him away from the primary duty of love of God. So Hopkins struggled in a dilemma of conflict between the phenomenal world and the realm of the spirit. His artistry developed in the context of this moral and religious tension. Yet, as W.H. Gardner offers, the tension between the free creative personality of the artist and the acquired, dedicated character of the Jesuit priest in Hopkins was not unfortunate, rather "if there was, deep down, some conflict between aesthetic and ascetic ideals, there was also, at the conscious level, a remarkable reconciliation and fusion, which gave depth and spiritual power to everything Hopkins wrote."[13]

B. Gerard Manley Hopkins (1844-1889): A Life Rediscovered

Gerard Manley Hopkins was born in Stratford, Essex on July 28, 1844.[14] He was the eldest of nine children born to Manley and

[13] Ibid., p. XXXV.

[14] There are many works that are biographical in nature. The consensus of Hopkins scholars is that the definitive work has yet to be done. Nonetheless, the following works cover most aspects of Hopkins' life: Michael Allsopp, "Gerard Manley Hopkins: The Oxford Years (1863-1867)," *Gregorianum* 70, no. 4

Kate Smith Hopkins. The parents were refined and of artistic inclination. They shared with their children an active appreciation of the arts, particularly poetry, music and painting.

In 1852, the Hopkins family moved to Hampstead, a respectable London suburb. There, Gerard attended Highgate, a solid, though by no means elite, grammar school. He was noted as a smallish, somewhat fragile looking boy who was both witty and intelligent. Receiving an excellent preparation in classical languages and literature while at Highgate, Gerard attained an Exhibition (scholarship) to Balliol College at Oxford.

Oxford was a garden for Hopkins. He immersed himself, socially and academically, gaining notice as a budding scholar and popular companion. While at Oxford, the rather pliable Anglican-

(1989): 661-687; Bernard Bergonzi, *Gerard Manley Hopkins* (New York: Macmillan Publishing Co., Inc., 1977); David Anthony Downes, *The Ignatian Personality of Gerard Manley Hopkins*, 2nd ed. (Lanham: University Press of America, 1990); Wendell Stacy Johnson, *Gerard Manley Hopkins: The Poet as Victorian* (Ithaca: Cornell University Press, 1968); Robert Bernard Martin, *Gerard Manley Hopkins: A Very Private Life* (New York: G.P. Putnam's Sons, 1991); Jude V. Nixon, *Gerard Manley Hopkins and His Contemporaries: Liddon, Newman, Darwin and Pater*, ed. Todd K. Bender, 6 vols., *Origins of Modernism: Garland Studies in British Literature*, vol. 5 (New York: Garland Publishing, Inc., 1994); John Pick, *Gerard Manley Hopkins: Priest and Poet*, 2nd ed. (New York: Oxford University Press, 1966); Graham Storey, "Gerard Manley Hopkins," in *British Writers: Elizabeth Gaskell to Francis Thompson*, ed. Ian Scott-Kilvert (New York: Charles Scribner's Sons, 1982), 361-82; Alison G. Sulloway, *Gerard Manley Hopkins and the Victorian Temper* (New York: Columbia University Press, 1972); Alfred Thomas, S.J., *Hopkins the Jesuit: The Years of Training* (London: Oxford University Press, 1969); and Norman White, *Hopkins: A Literary Biography* (Oxford: Clarendon Press, 1992). Although there are no current comprehensive bibliographies concerning Hopkins, the following are very useful: Edward H. Cohen, *Works and Criticism of Gerard Manley Hopkins* (Washington, D.C.: The Catholic University of America Press, 1969); Tom Dunne, *Gerard Manley Hopkins: A Comprehensive Bibliography* (Oxford: Clarendon Press, 1976); and Ruth Seelhammer, *Hopkins Collected at Gonzaga* (Chicago: Loyola University Press, 1970).

ism adopted from his parents was put to the test and transformed by religious turmoil. The Oxford movement was still active under the shadow of E.B. Pusey and the ministrations of the popular preacher and mentor, Henry Parry Liddon. On the other side was Hopkins' first tutor, the brilliant and embattled Benjamin Jowett, the best known classicist at Oxford and the outstanding member of the Broad Church party.[15] Gerard struggled in their midst, but early on associated himself with the inclinations of the Tractarians.

In the midst of this religious foment, Hopkins gained the respect of his dons for his brilliance as a classical scholar. Additionally, he demonstrated a strong attraction to the arts, particularly painting, although his own gifts emerged through his verse. He created a number of poems while at Oxford, demonstrating a keen aesthetic grasp of the writings of Shakespeare, Milton and Keats. His philosophical essays of the time show that he absorbed well the thought of Plato and Aristotle, while continuing his pre-Oxford attachment to the work of the well known Victorian aesthete, John Ruskin. In fact, his second tutor at Oxford, Walter Pater, was himself strongly influenced by Ruskin.

The religious turmoil at Oxford mirrored the maelstrom of Hopkins' inner struggle. The year before he left Oxford, he came to the same conclusion that John Henry Newman had reached years before — he felt compelled to become a Roman Catholic. He wrote to Newman entreating him for a meeting and Newman agreed. After additional correspondence and two meetings with Newman, Hopkins was received into the Roman Catholic Church by the famous Oratorian.

The religious transition proved painful and difficult. His parents and dons were dismayed and exerted pressure on Hopkins to

[15] See Lesley J. Higgins, "Hopkins and "The Jowler"," *Texas Studies in Literature and Language* 31, no. 1 (1989): 143-167.

reconsider. Liddon pursued Hopkins continually and E.B. Pusey himself wrote a bitter letter to Hopkins. Once again the High Church party was losing one of its stars to Rome. Hopkins' last year at Oxford was stressful, yet with the encouragement of Newman he continued his preparation for Greats at Oxford. In the Trinity term, Hopkins took a distinguished double first in Greats, and on September 13, 1867, he accepted an offer from Newman to teach at the Oratory School as a Master.

Hopkins stayed at the Oratory for a short time. He struggled with deciding whether or not he was for the priesthood or religious life. On April 27, 1868, Hopkins made a retreat at the Jesuit Novitiate in Roehampton. The director of the retreat, Father Henry Coleridge, S.J., impressed him greatly. During the retreat, Hopkins decided to become a priest. The only question which remained was what religious order would be best for him. He had contemplated the Benedictines, but ultimately decided for the Jesuits. He received his acceptance from the Society of Jesus on May 30, 1868. In response to his decision, Newman wrote:

> I am both surprised and glad at your news....I think it is the very thing for you. You are quite out, in thinking that when I offered you a "home" here, I dreamed of your having a vocation for us. This I clearly saw you had <u>not</u>, from the moment you came to us. Don't call "the Jesuit Discipline" hard, it will bring you to heaven. The Benedictines would not have suited you. Ever yours affectionately, John H. Newman.[16]

Hopkins began his life in the Society of Jesus on September 7, 1868. Before he entered, he burned copies of the poems he had written. Luckily, friends and family had letters containing most of

[16] Claude Colleer Abbott, ed. *Further Letters of Gerard Manley Hopkins Including his Correspondence with Coventry Patmore*, 2nd ed. (London: Oxford University Press, 1956), p. 408.

them. Upon entering the Jesuits, Hopkins saw the arts as no longer open to him, in the main, and the burning was his symbolic act of sacrifice. He did not write another poem for seven years, and then only at the suggestion of his Rector.

Hopkins began the long process of Jesuit formation at Manresa and was there as a novice for two years. During this time he was absorbed in the daily schedule of worship, meditation, spiritual reading, receiving spiritual direction and doing a variety of menial tasks. Additionally, he made his first 'Long Retreat', a four-week period devoted to *The Spiritual Exercises of Ignatius of Loyola*. During these thirty days of intense examination of conscience, detailed meditations, prayers and direction, Hopkins tested his Jesuit vocation. This was Hopkins' time to create the foundation of his future life as a Jesuit and it culminated in his taking first vows in 1870.

After first vows, Hopkins began the second stage of his training — three years of philosophical studies at St. Mary's Hall in Stonyhurst. Upon completing his Philosophate, Hopkins was assigned to teaching Rhetoric at Manresa for a year. Then it was off to St. Bueno's College in North Wales for three years of theological studies. St. Bueno's was the place of greatest intellectual development and personal happiness for Hopkins during his Jesuit life. It was during this time that he discovered some writings of John Duns Scotus and began to create a kind of synthesis which may be called a theological aesthetics. It was also during this time that he once again wrote poetry.

In 1875, the superior at St. Bueno's suggested he write a poem about the tragic wreck of the *Deutschland* in which five German Franciscan nuns, expelled from Germany, were drowned. The work he wrote was the creative masterpiece, *The Wreck of the Deutschland.* After seven years of poetic silence and personal transformation, this effort was an explosion of not only a new poetic idiom, which he called *sprung rhythm*,[17] but also of a new religious vision tapping the influences of John Ruskin, Ignatius

[17] 'Sprung Rhythm', introduced by Hopkins in *The Wreck of the Deutschland*, "consists in scanning a poem by accents and stresses alone, without any count of the number of syllables, so that a foot may be one strong syllable or it may be many light and one strong..." See Claude Colleer Abbott, ed. *The Correspondence of Gerard Manley Hopkins and Richard Watson Dixon*, 2nd ed. (London: Oxford University Press, 1955), p. 14. Behind this technical definition offered by Hopkins, there is an effort to recapture a muted or lost function of language. M. Roston describes the context of 'sprung rhythm' as follows: "[Hopkins] practice...rested upon deeper foundations than mere wilfulness; for if, as he believed, words had originated onomatopoeically, then alliteration and assonance should by right be privileged above everyday speech-patterns which had, by his day, lost the pristine force of those sounds; and the rhythms should, where necessary, subjugate socially accepted forms in order to restore the vitality of earlier patterns." Murray Roston, *Victorian Contexts: Literature and the Visual Arts* (New York: New York University Press, 1996), p. 139. The "vitality" Hopkins wanted to recapture in poetry was the meaning of sound itself, not simply words. Assonance, internal rhyme, alliteration, and stress were all important to Hopkins in as much as they lent power to the "speaking out" of a poem. The sounds of words and not only the words, in Hopkins' approach, were a mediation of meaning. Roston continues: "His reintroduction of sprung rhythm he saw, therefore, as bringing verse genuinely closer to those unformulated pristine impulses of thought which — as stream-of-consciousness writings were soon to demonstrate — tend to thrust the emotionally charged word to the forefront, allowing the rest of the sentence to form...around it, regardless of normal order." Roston, *Victorian Contexts*, p. 139. For example, in his poem *Harry Ploughman* [Norman H. MacKenzie, ed. *The Poetical Works of Gerard Manley Hopkins* (Oxford: Clarendon Press, 1990), p. 193], Hopkins wants the sounds of the words in the poem to help create a vivid figure in the mind's eye:

> Hard as hurdle arms, with a broth of goldish flue
> Breathed round; the rack of ribs; the scooped flank; lank
> Rope-over thigh; knee-nave; and barrelled shank —
> Head and foot, shoulder and shank —
> By a grey eye's heed steered well, one crew, fall to;
> Stand at stress...

Roston perceptively notes that through 'sprung rhythm' and other unusual poetic techniques, Hopkins foreshadowed, in many ways, the priority which Wittgen-

and Duns Scotus. In *The Wreck*, Hopkins' developing Christocentric vision of created reality takes its first voice:

> I kiss my hand
> To the stars, lovely-asunder
> Starlight, wafting him out of it; and
> Glow, glory in thunder;
> Kiss my hand to the dappled-with-damson west:
> Since, tho' he is under the world's splendour and wonder,
> His mystery must be instressed, stressed;
> For I greet him the days I meet him, and bless when I understand.[18]

From this time on, Hopkins wrote poetry sporadically as his duties allowed and thus left the English-speaking world a legacy of rare descriptive beauty and unique theological depth. However, his poetry, as his life, remained essentially unknown until 1917, when his longtime friend, the poet laureate Robert Bridges, posthumously and somewhat reluctantly published a selection of his poetry.

Hopkins was ordained a priest on September 23, 1877. Although he was considered brilliant, his outspoken affinity for the thought of Duns Scotus apparently did not endear him to his Suarezian Jesuit theological examiners, therefore he was barred from a fourth year of theological studies.[19] Thus, Hopkins was

stein and subsequent linguists would accord to speech discourse. See Murray Roston, *Victorian Contexts: Literature and the Visual Arts* (New York: New York University Press, 1996), p. 149.

[18] Norman H. MacKenzie, ed. *The Poetical Works of Gerard Manley Hopkins* (Oxford: Clarendon Press, 1990), p. 120.

[19] See Joseph Feeney, S.J., "Hopkins' Failure in Theology: Some New Archival Data and a Reevaluation," *The Hopkins Quarterly* 13, no. 3 and 4 (1987): 99-114. Also see Norman White, "'He Played the Droll Jester': Hopkins the Unorthodox," in *Gerard Manley Hopkins: Tradition and Innovation*, ed. G. Marra, P. Bottalla, F. Marucci (Ravenna: Longo Editore, 1991), 149-59. White describes multiple examples of difficulty brought on by Hopkins' attachment to

never considered a theologian, *per se*, nor was he encouraged by his order to pursue a theological career. Nonetheless, as we shall see, his writings show a mind of deep theological insight, particularly in his joining of aesthetics and theology.

Deeply committed to his vocation as a Jesuit priest, Hopkins nonetheless struggled in the Society. Though well-treated and liked, he was considered somewhat eccentric by his colleagues and superiors.[20] His varied early assignments as a priest included parochial duties in London, Liverpool, Glascow and Oxford, as well as teaching responsibilities at Stonyhurst College. His health, always less than robust, affected his ability to meet his duties. Additionally, his success as a preacher was mixed because of a chronic inability to read congregations. One wonders what a working-class congregation in Liverpool made of his sermon on the Sacred Heart "with its intricate wordplay on *heart*, or the reaction of a high-society congregation to his use of a cow's udder and teats as an *exemplum* for the Church and the seven sacraments."[21] Nonetheless, his sermons were courageous (he never dodged difficult topics) and full of fresh theological perspectives.

Hopkins' final and longest assignment took him to the University College in Dublin in 1884. He was assigned the Chair of Classics at this remnant of Newman's Catholic University. His time in Dublin was the worst of his Jesuit life. Although more than adequately educated for the position, he had little ability as a teacher among Irish youth. His high-pitched voice of a clearly Oxonian

Scotus, e.g., certain sermons being rejected for espousing 'questionable' Scotist doctrine (p. 155).

[20] With regard to how Hopkins was seen by his fellow Jesuits, see the excellent study by Joseph J. Feeney, S.J., "Hopkins in Community: How His Jesuit Contemporaries Viewed Him," in *Saving Beauty: Further Studies in Hopkins*, ed. Michael E. Allsopp (New York: Garland Publishing, 1994), 253-94.

[21] David Anthony Downes, *The Ignatian Personality of Gerard Manley Hopkins*, 2nd ed. (Lanham: University Press of America, 1990), p. 10.

refinement, as well as his lack of sympathy for the Irish cause, did little to endear him to his students. His health and spirits declined in Dublin and he was often in a state of exhaustion and nervous fatigue. All of the many scholarly projects he initiated came to nothing. Even his poetic inspirations were subdued due to a lack of encouragement from his superiors (which he had hoped for as a legitimization of his desire to write poetry). Downes summarizes the Dublin period by writing:

> Outwardly, Hopkins' last years were filled with interests both artistic and intellectual, but inwardly he experienced recurring periods of desolation. His continuing bad health, his inability to bring any of his pursuits to any real completion, the arduousness of his teaching duties, to which must be added tepid periods in his spiritual life, made these years unbalanced with tensions, weaknesses, and frustrations. He recorded some of these moments in the sonnets dubbed "terrible" because they were born of grief and suffering, "unbidden and against my will."[22]

One of the "terrible sonnets", *Carrion Comfort*, expresses Hopkins' state as the Dublin years trudged on and the end of his life approached:

> NOT, I'll not, carrion comfort, Despair, not feast on thee;
> Not untwist — slack they may be — these last strands of man
> In me or, most weary, cry *I can no more*. I can;
> Can something, hope, wish day come, not choose not to be.
>
> But ah, but O thou terrible, why wouldst thou rude on me
> Thy wring-world right foot rock? lay a lionlimb against me? scan
> With darksome devouring eyes my bruised bones? and fan,
> O in turns of tempest, me heaped there; me frantic to avoid thee and
> flee?

[22] David Anthony Downes, *The Ignatian Personality of Gerard Manley Hopkins*, 2nd ed. (Lanham: University Press of America, 1990), p. 12

Why? That my chaff might fly; my grain lie, sheer and clear.
Nay in all that toil, that coil, since (seems) I kissed the rod,
Hand rather, my heart lo! lapped strength, stole joy, would laugh, cheer.
Cheer whom though? The hero whose heaven-handling flung me, foot trod
Me? or me that fought him? O which one? is it each one? That night, that year
Of now done darkness I wretch lay wrestling with (my God!) my God.[23]

In other sonnets of the time, he describes himself as a stranger among strangers, "a lonely began," gall and heartburn. His sense of desolation found expression in a poem:

I Wake and feel the fell of dark, not day.
What hours, O what black hours we have spent
This night! what sights you, heart, saw; ways you went!
And more must, in yet longer light's delay.

With witness I speak this. But where I say
Hours I mean years, mean life. And my lament
Is cries countless, cries like dead letters sent
To dearest him that lives alas! away.

I am gall, I am heartburn. God's most deep decree
Bitter would have me taste: my taste was me;
Bones built in me, flesh filled, blood brimmed the curse.

Selfyeast of spirit a dull dough sours. I see
The lost are like this, and their scourge to be
As I am mine, their sweating selves; but worse.[24]

[23] Norman H. MacKenzie, ed. *The Poetical Works of Gerard Manley Hopkins* (Oxford: Clarendon Press, 1990), p. 183.

[24] Ibid., pp. 181-2.

Still, his basic vision, a vision of a 'Christed' reality, even in the midst of extrinsic and intrinsic desolation, remained essentially untouched.[25] In his 1888 poem, *That Nature is a Heraclitean Fire and of the comfort of the Resurrection*, he wrote:

> CLOUD-PUFFBALL, torn tufts, tossed pillows flaunt forth then chevy on an air-
> built thoroughfare: heaven-roysterers, in gay-gangs they throng; they glitter in marches.
> Down roughcast, down dazzling whitewash, wherever an elm arches,
> Shivelights and shadowtackle in long lashes lace, lance, and pair.
>
> Delightfully the bright wind boisterous ropes, wrestles, beats earth bare
> Of yestertempest's creases; in pool and rutpeel parches
>
> Squandering ooze to squeezed dough, crust dust; stanches, starches
> Squadroned masks and manmarks treadmire toil there
> Footfretted in it. Million-fueled, nature's bonfire burns on.
> But quench her bonniest, dearest to her, her clearest-selved spark
> Man, how fast his firedint, his mark on mind, is gone!
> Both are in unfathomable, all is in an enormous dark
> Drowned. O pity and indig | nation! Manshape, that shone
> Sheer off, disseveral, a star, death blots black out; nor mark
> Is any of him at all so stark

[25] Cf. David Anthony Downes, "Gerard Manley Hopkins' Christed Vision of Ultimate Reality and Meaning," *Ultimate Reality and Meaning* 12, no. March (1989): 61-80. 'Christed' refers to Hopkins' radical Christocentric vision of creation. Downes contextualizes the term when he writes: "The ontological end of every self [for Hopkins] is to be self-expressive. Creation is multifarious selves 'selving'. If Creation is an expression of God, then Creation is a procession of divine selving, first Christ, then the Angels, and finally the universe. Since Christ was the first selving of God as his son, and everything else we call Creation followed from this first 'outstress' of God's power, Creation is a conglomerate of selves selving in Christ's selving of God" (p. 63).

But vastness blurs and time beats level. Enough! the Resurrection,
A heart's-clarion! Away grief's gasping, joyless days, dejection.
Across my foundering deck shone
A beacon, an eternal beam. Flesh fade, and mortal trash
Fall to the residuary worm; world's wildfire, leave but ash:
In a flash, at a trumpet crash,
I am all at once what Christ is, since he was what I am, and
This Jack, joke, poor potsherd, patch, matchwood, immortal diamond,
Is immortal diamond.[26]

Hopkins was a unique, singular poet. He was also a unique, singular Jesuit. His life and his work were hidden until long after his death, which came from typhoid in 1889. He died, seemingly in peace, and is said to have whispered to those surrounding him, "I am so happy, so happy."[27] His funeral was well-attended and he was buried in a common grave at Glasnevin on the outskirts of Dublin, being remembered as a genuinely good and kind man.

In 1918, Robert Bridges, a life-long friend of Hopkins, but never an admirer of his poetry, cautiously released his poems. Slowly they gained fame, and by 1930 Hopkins was being honored as one of the greatest Victorian poets — an honor he was loathe to seek during his lifetime. Because of the singular nature of his poetry, and because his works were essentially introduced anachronistically, Hopkins was claimed as a precursor of the Moderns in English poetry. In fact, for many in the 1930s, "it came as a shock to realize that [Hopkins'] poems dated from so long before — that the work of a poet who had been dead for nearly half a century was achieving recognition not in terms of an

[26] Norman H. MacKenzie, ed. *The Poetical Works of Gerard Manley Hopkins* (Oxford: Clarendon Press, 1990), pp. 197-8.

[27] David Anthony Downes, *The Ignatian Personality of Gerard Manley Hopkins*, 2nd ed. (Lanham: University Press of America, 1990), p. 12.

antiquarian discovery but as Modernist poetry of immediate relevance for their own time."[28] Certainly, Hopkins was emulated and he influenced the writings of Dylan Thomas, W.H. Auden, Seamus Heaney, John Berryman, Robert Lowell, David Jones and Sylvia Plath.[29]

Thus he was claimed as a poet by a later age which did not always understand the centrality of his vocation as a Jesuit priest. Nonetheless, at the heart of his art, there lay the beginnings of a theological vision that bound beauty to the Christian experience. This has slowly come to light, not only in theological circles but also in the disciplines of English studies, literature, and philosophy, particularly aesthetics. Because of his effort to bring beauty back to faith, Hopkins has become a catalyst for an uncommon interdisciplinary dialogue, and has helped these various disciplines overlap their usually discrete critical circles. Thus one finds theologians, whether Hans Urs von Balthasar, Walter Ong, or Michael Allsopp, joining the dialogue of the critical worlds of English literature and Victorian history, while English scholars and social historians such as David Downes, Hilary Fraser, and Alison Sulloway enter the theological realms of protology, soteriology, and the gnoseology of the doctrine of creation. Finally,

[28] Murray Roston, *Victorian Contexts: Literature and the Visual Arts* (New York: New York University Press, 1996), p. 133.

[29] See Pilar Abad, "Hopkins and the Modern Sonnet Tradition: Dylan Thomas, W.H. Auden and Seamus Heaney," in *Gerard Manley Hopkins: Tradition and Innovation*, ed. G. Marra, P. Bottalla, F. Marucci (Ravenna: Longo Editore, 1991), 223-34; Joseph Feeney, S.J., "Gerard Manley Hopkins, Poeta," *La Civilta Cattolica* 141, no. 11 (1990), p. 442; Tung-jung Chen, "Preoccupations of the Poet: A Reading of Gerard Manley Hopkins and Seamus Heaney," *Journal of Humanities East/West* 8 (1990): 155-76; Gerry Murray, "Gerard Manley Hopkins: His Influence on John Berryman," *Studies* 85, no. 338 (1996): 125-35; and Desmond Egan, "Hopkins' Influence on Poetry," in *Saving Beauty: Further Studies in Hopkins*, ed. Michael E. Allsopp (New York: Garland Publishing, 1994), 295-311.

these circles join with those of philosophers, such as Frank Burch Brown, Leonard Bowman, and Christopher Devlin, to discuss Hopkins' ontology, epistemology, and aesthetics. This interdisciplinary world created by the response to Hopkins' work tends toward what may be called a 'theological aesthetics'. Here, as we shall see, rests Hopkins' particular theological contribution, for I conclude that within his 'theological aesthetic' is a vision of poetry as 'sacrament'.

Hopkins opened his last complete poem with the stanza:

> The fine delight that fathers thought; the strong
> Spur, live and lancing like the blowpipe flame,
> Breathes once and, quenched faster than it came,
> Leaves yet the mind a mother of immortal song.[30]

So Hopkins has fathered thought through the inspiration of his vision of reality, expressed on the limits of language, pointing to and invoking the depth and ultimate beauty of Being as 'Worded'.

C. Structure and Method of this Study

Gerard Manley Hopkins was not a formal theologian. His extant writings come in the form of sporadic journal entries, letters and assorted correspondence, a few undergraduate essays, poems and fragments of poetry, sermons and devotional writings. Furthermore, he died prematurely and before he had an opportunity to flesh out some of his essential theological thought. Therefore, any effort to develop Hopkins' theological strands into something more systematic is somewhat akin to the paleontologist's exercise of reconstructing a skeleton from a few major bones. Still, the

[30] Norman H. MacKenzie, ed. *The Poetical Works of Gerard Manley Hopkins* (Oxford: Clarendon Press, 1990), p. 204.

remnants are sufficient to glean some content, sense directions, and follow the connections Hopkins made in his various writings. Thus, in this book I intend to flesh out what I believe is the main area of Hopkins' theological contribution, namely, his theological aesthetics and, ultimately, his unspoken vision of the poem as 'sacrament'.

In his aesthetics, Hopkins approached all created reality as potential revelation in a real sense. For Hopkins, reality was 'Christic' at its very core. This belief centered upon a somewhat unique view of the Incarnation which Hopkins probably derived from Duns Scotus. Additionally, Hopkins believed that human beings can know created things directly and intuitively. Thus, he placed great emphasis upon the importance of the senses in not only 'knowing' created reality, but also, in some sense, perceiving the divine ground of such reality. Essentially, in the perception of something's self-expressiveness, its beauty, its inscape, Hopkins believed one perceives the unique 'Christic stem or selving' given in an individual reality. Furthermore, Hopkins held that this unique revelation given in each individual thing or experience could be conveyed to others through language, or more precisely, through poetic language. Therefore, for Hopkins, poetry had a truly sacramental function. Thus, one might refer ultimately to Hopkins' 'sacramental aesthetics'.

The theological, or sacramental, aesthetics of Hopkins center around his concept of *inscape*. Although I have given a very initial description of this Hopkinsian term above, the fact remains that *inscape*, along with its corollary *instress*, is a very elusive term in Hopkins. Therefore, I believe that a 'circular' method is best used in seeking an understanding not only of these terms, but also of Hopkins' main theological themes. By 'circular' I mean a method in which the scholar approaches terms and themes initially in one context, developing aspects given and emphasized in that context, then moving onto another context and returning again to the terms and themes. The hope is that layers of meaning in various contexts will build a more complete sense of the terms and themes in Hopkins as opposed to offering Procrustean definitions which may be misleading.

In using this method, the various contexts will be the significant catalysts and influences upon Hopkins' thought. Although there is a vast amount of literature on a myriad of possible influences upon Hopkins, I believe three figures of influence are inescapable for understanding both Hopkins' writings and the critical literature. These figures are John Ruskin, Ignatius of Loyola, and John Duns Scotus. Therefore, in developing the main themes of Hopkins' theological aesthetics, I will trace and emphasize aspects stressed by each figure of influence and thereby hopefully add depth in each successive 'circle'. The final chapter will synthesize the results and develop Hopkins' contention that poetry indeed can be 'sacramental' and may be the language most appropriate for our approach to the divine ground of reality, or, as Hopkins writes, the "Ground of being, and granite of it...."[31]

[31] Norman H. MacKenzie, ed. *The Poetical Works of Gerard Manley Hopkins* (Oxford: Clarendon Press, 1990), p. 127.

BIOGRAPHICAL CHRONOLOGY

1844 Gerard Manley Hopkins born at Stratford, Essex on July 28.

1852 Hopkins family moves to Oak Hill, Hampstead.

1854 G.M. Hopkins attends Highgate school (9 years).

1862 Receives gold medal for prize-poem *A Vision of Mermaids.*

1863 Wins an Exhibition at Balliol College, Oxford and enters in April.

1864 First Class in "Moderations".

1866 August, writes to J.H. Newman concerning his conversion. Newman receives G.M. Hopkins into the Roman Catholic Church on October 21.

1867 June, graduates as BA, with a double-first in 'Greats'; teaches at Oratory School in Birmingham for two terms.

1868 May, decides to become a priest and a Jesuit. Enters the Jesuit Novitiate in Roehampton on September 7 (2 years).

1870 Jesuit Philosophate, St. Mary's Hall, Stonyhurst, Lancashire (3 years).

1873 Teaches Rhetoric at Roehampton (1 year).

1874 Theologate, St. Bueno's, North Wales (intended for four years, but 'fails' examination at end of third year).

1875 Writes *The Wreck of the Deutschland.*

1877 Ordained priest in September. Preacher at Farm Street Church, London. Teacher at Mt. St. Mary's College, near Sheffield for 7 months.

1878 Teaches at Stonyhurst College (3 months). Curate at Farm Street Church from July through November. Curate at St. Aloysius' Church, Oxford (10 months). Renews acquaintance with Canon Dixon.

1879 Curate at St. Joseph's Church, Bedford Leigh, Lancashire (3 months).

1880 Curate at St. Francis Xavier's Church, Liverpool (1 year and 7 months).

1881 Temporary curate at St. Joseph's Church, Glascow (2 months). Enters Tertianship at Roehampton in October (1 year).

1882 September, starts teaching Classics at Stonyhurst College (1 year and 4 months). Meets Coventry Patmore.

1884 February, appointed Professor of Greek at University College, Dublin, and Fellow in Classics of the Royal University of Ireland.

1889 June 8, dies of typhoid in Dublin.

CHAPTER I

RUSKIN: HOPKINS' "SILENT DON"

A. Introduction

Understanding may come to some seekers as a kind of revelation, but for many it is more akin to the slow growth of roots through soil. The soil in which the understanding of Hopkins may be sought is that of context and influence. What and whose ideas did he encounter, assimilate, reject or transcend? What, or perhaps who, provided him a foundation upon which his own genius could unfold?

Over decades of Hopkinsian studies, the question of context and influence has been raised consistently, whether by Pick concerning the effect of Ignatius upon the Jesuit Hopkins, or by Devlin illuminating the poet's self-identification with Duns Scotus, or by Downes claiming a Romantic influence through Coleridge and Wordsworth, or by Sulloway portraying Hopkins as a classic Victorian man, or, more recently, by Nixon, sifting the peripheral influences of Liddon, Darwin and Pater.[1] Nonetheless,

[1] See John Pick, *Gerard Manley Hopkins: Priest and Poet* (Oxford: Oxford University Press, 1942); Christopher Devlin, S.J., "Hopkins and Duns Scotus," *New Verse* 14 (1935): 12-5; David Anthony Downes, *Hopkins' Sanctifying Imagination* (Lanham: University Press of America, 1985); Alison G. Sulloway, *Gerard Manley Hopkins and the Victorian Temper* (New York: Columbia University Press, 1972); and Jude V. Nixon, *Gerard Manley Hopkins and His Contemporaries: Liddon, Newman, Darwin and Pater*, ed. Todd K. Bender, 6 vols., *Origins of Modernism: Garland Studies in British Literature*, vol. 5 (New York: Garland Publishing, Inc., 1994).

a relative gap remains in this study of influence. Foundationally speaking, who taught Hopkins, that artist of keenest vision, how to 'see'? The answer is immediately encountered upon entering the study of Hopkins — it was John Ruskin.[2]

B. Early and Enduring Attachments

Ruskin's influence upon Hopkins was as indelible as that of Oxford or even, perhaps, as that of Ignatius Loyola. Certainly it pre-dated both. Yet, relatively speaking, very little comprehensive study has been devoted to this essential and formational influence upon Hopkins' vision and artistry.[3]

[2] Cf. Jean-Georges Ritz, "Hopkins, Défenseur de la Singularité des Êtres," in *Du Verbe au Geste* (Nancy: Presses Universitaires de Nancy, 1986), 223-230. Ritz has done excellent work on Hopkins among French English literature scholars. He writes: "Tout commence chez Hopkins par son admiration de Ruskin, qui était un passionné du regard, prêchant l'observation la plus attentive de ce que nous voyons… Ce que nous venons de dire de Ruskin est pleinement valable pour Hopkins. Il affronté le réel comme le fait l'auteur des *Modern Painters*. Mais il va aller plus loin" (pp.223-24). The 'plus loin' for Ritz is Hopkins success in finding philosophical and theological ground for his aesthetic vision. Ritz emphasizes the influence of Duns Scotus in Hopkins' eventual joining of truth and beauty (p. 227), but this all starts with Ruskin and his 'seeing'.

[3] Although pioneering works on Hopkins, including John Pick's *Gerard Manley Hopkins: Priest and Poet* (1942) and W.H. Gardner's *Gerard Manley Hopkins (1844-1889): A Study of Poetic Idiosyncracy in Relation to Poetic Tradition* (1948), noted a connection between Hopkins and Ruskin, no parsing of the relationship was done. This 'mention without elaboration' has become characteristic of Hopkinsian scholarship and there are few works of note which develop Pick's and Gardner's early lead. The only dissertation of substance I have discovered is Francis Fike's *The Influence of John Ruskin upon the Aesthetic Theory and Practice of Gerard Manley Hopkins* [Ph.D. dissertation, Stanford, 1964]. As for published works, even though Ruskin is oft mentioned in some studies, only Patricia Ball in *The Science of Aspects: The Changing Role of Fact in the Work of Coleridge, Ruskin and Hopkins* [1971], Alison Sulloway in *Gerard*

Hopkins may well have used Ruskin's *The Elements of Drawing* when it was first published in 1857.[4] The book was very popular and ran through three editions between 1857 and 1860. Hopkins' family, an artistic clan, certainly knew of *The Elements.* Moreover, in 1857 Hopkins was thirteen, the age recommended by Ruskin for the use of the book.[5] In any case, Hopkins' journals and letters are full of Ruskinian comments, pronouncements and admonitions.[6] "As Ruskin says" is a kind of underlying mantra for the aesthetic Hopkins of Oxford.[7] Certainly his friend and fel-

Manley Hopkins and the Victorian Temper [1972], and Hilary Fraser in *Beauty and Belief: Aesthetics and Religion in Victorian Literature* [1986] offer quality academic grist. For some analysis of aspects of Ruskin's influence, including visual examples, cf. Jerome Bump, "Hopkins' Drawings," in *All My Eyes See: The Visual World of Gerard Manley Hopkins*, ed. R.K.R. Thornton (Tyne and Wear: Ceolfrith Press, 1975), 69-87. Also see Norman White's, "The Context of Hopkins' Drawings," in *All My Eyes See*, pp. 53-67.

[4] Though I find no reference to a Ruskin work in any remnant collection of the Hopkins' family library, G. M. Hopkins' early sketches already point to the influence of Ruskin. Additionally, his youthful poem, *The Escorial*, eludes to *The Stones of Venice, II* in stanzas 6 and 7 where Hopkins refers to the Doric, Corinthian and Gothic styles with a Ruskinian judgment in favor of the Gothic (cf. Sulloway, p.72).

[5] Bernard Bergonzi, *Gerard Manley Hopkins* (New York: Macmillan Publishing Co., Inc., 1977), pp. 20-21.

[6] For example, Ruskin had an aesthetic law which he referred to as 'the innocence of the eye', by which he meant a sort of 'childish perception of nature, as a blind man would see the world around him if suddenly gifted with sight' Cf. Alison G. Sulloway, *Gerard Manley Hopkins and the Victorian Temper* (New York: Columbia University Press, 1972), p. 67. Hopkins uses this same Ruskinian phrasing in several places in his papers and journals. The artist who lacked this first Ruskinian prerequisite was one whose "eye had not been trained to look severely at things, apart from their associations, *innocently* and *purely* as painters say." See Humphry House, ed. *The Journals and Papers of Gerard Manley Hopkins*, 2nd ed. (London: Oxford University Press, 1959), p. 77. For a further example see p. 80 of *Journals and Papers* where Hopkins uses the same Ruskinian phrase in discussing the innocent eye of the uneducated and children.

[7] See Humphry House, ed. *The Journals and Papers of Gerard Manley Hopkins*, 2nd ed. (London: Oxford University Press, 1959), p. 13.

low art enthusiast, Alexander Baillie, knew of Hopkins' intense interest in Ruskin, for Hopkins was earnest in his efforts to convert his colleague to a "Ruskinese point of view."[8]

By the late eighteen-fifties, Oxford was a "stronghold of Ruskin worship" and thus continued to nourish the young Hopkins' enthusiasm for him. Sulloway pronounces Ruskin "the silent don" of Hopkins' Oxonian years.[9] Certainly the influential Master of Hopkins' Balliol College, Benjamin Jowett, as well as another of Hopkins' tutors, Walter Pater, had a high regard for Ruskin.[10] Public and private readings of Ruskin's lectures and descriptive prose were common throughout Oxford. In Hopkins' undergraduate essays, *On the Signs of Health and Decay in the Arts* (1863) and *On the Origin of Beauty* (1864), it seems as if he is quoting Ruskin unconsciously, so imbued is he with the thought of the Victorian aesthete.[11] It is highly likely that some of Ruskin's ideas came to Hopkins through his Oxonian guides, particularly Pater.[12]

[8] See Claude Colleer Abbott, ed. *Further Letters of Gerard Manley Hopkins Including his Correspondence with Coventry Patmore*, 2nd ed. (London: Oxford University Press, 1956), p. 202.

[9] Alison G. Sulloway, *Gerard Manley Hopkins and the Victorian Temper* (New York: Columbia University Press, 1972), p.42.

[10] Jowett, normally immensely reserved, initiated a chorus of praise for Ruskin's work. He wrote: "I have read it [*Modern Painters*] all through with the greatest delight; the minute observation and power of description it shows are truly admirable… Since I have read it I fancy I have a keener perception of the symmetry of natural scenery. The book was written by Ruskin, a child of genius, certainly." This praise was joined by such figures as Tennyson, Wordsworth, Charlotte Bronte, George Eliot, etc. (cf. Sulloway, p.66).

[11] Cf. Wendell Stacy Johnson, "From Ruskin to Hopkins: Landscape and Inscape," *The Hopkins Quarterly* 8, no. 3 (1981), p. 98.

[12] Ruskin may have exerted a great influence upon Pater, particularly in his "impressionable years," including those in which he tutored Hopkins. See William S. Knickerbocker, *Creative Oxford: Its Influence in Victorian Literature* (Syracuse: Syracuse Bookstore, 1925), p.174. For a specific example of Ruskin's influence on Pater, see Paul Barolsky, *Walter Pater's Renaissance* (University

Ruskin's influence continued through Hopkins' time of religious distress at Oxford and his ensuing period of poetic silence. Although he left Oxford and his poetry temporarily behind, under the enduring and silent tutelage of Ruskin, Hopkins continued to sketch, verbally and pictorially, the world flooding his eyes. In fact, though he temporarily sacrificed his poetry upon entering the Society of Jesus, Hopkins' Ruskin-influenced sketching and descriptive writing continued until 1875, the year he started working upon *The Wreck of the Deutschland*. Sulloway notes that "these Ruskinian sketches and verbal descriptions form a bridge between Hopkins' last undergraduate poem and his first mature work as a Jesuit, and they helped him develop as a poet during the

Park: The Pennsylvania State University Press, 1987), p. 170. Barolsky speaks of Ruskin as being present in Pater's writing by implication. Also see Donald William Bruce, "Hopkins the Observer, 1844-1889," *Contemporary Review* 123, no. November (1989), pp. 247-48. Bruce hints at Pater's influence upon Hopkins and his terms *inscape* and *instress*; however, the analysis offered reads like a description of Ruskin's influence. Ruskin is mentioned nowhere in the article. Additionally, see Jude Nixon's chapter on Pater in his *Gerard Manley Hopkins and His Contemporaries: Liddon, Newman, Pater and Darwin*. (New York: Garland Publishing, 1994). I am not convinced by Nixon that Pater's unique ideas ultimately influenced Hopkins. I believe there is more significant material to harvest by concentrating on Pater's conveyance of Ruskinian ideas to Hopkins (Nixon mentions Ruskin only three times in his discussion of Pater and Hopkins). Though Hopkins was a student of Pater's, his mentor in the truest sense was Ruskin. This opinion is thoroughly confirmed by Lesley Higgins exhaustive work on the relationship between Pater and Hopkins, "Hidden Harmonies: Walter Pater and Gerard Manley Hopkins" (Ph.D. Dissertation, Queen's University, 1987). Higgins' definitive work contends that Pater influenced Hopkins as a foil. By holding an opposing position, Pater forced Hopkins to hone his metaphysical preferences and contentions that art is religiously and morally purposeful (cf. pp. 184-85). Additionally, Higgins, unlike most commentators, recognizes the common denominator shared by Hopkins and Pater, namely, Ruskin (cf. pp. 229-30). Higgins contributes greatly to Hopkins studies by including a critical edition of previously unpublished essays which Hopkins wrote for Pater as a major appendix to her dissertation.

nine years of poetic silence."[13] From beginning to end, Hopkins was expressive of his debt to Ruskin. Even on his death bed, Hopkins dictated a letter in which he referred to thoughts Ruskin had on the brightness of colors in nature.[14]

C. John Ruskin's 'Signic' Worldview

The descriptive prose, the sketches, and the early aesthetic mindset of Hopkins reflect the tutelege of John Ruskin. That this is so cannot be surprising. Ruskin was a Victorian giant whose

[13] Alison G. Sulloway, *Gerard Manley Hopkins and the Victorian Temper* (New York: Columbia University Press, 1972), p.62. Jerome Bump visually demonstrates Sulloway's point in his article, "Hopkins' Drawings," in *All My Eyes See: The Visual World of Gerard Manley Hopkins*, ed. R.K.R. Thornton (Tyne and Wear: Ceolfrith Press, 1975), 69-87. He writes: "Drawing made a great contribution to his literary career... his early exercises in truth to nature in drawing counterbalanced his preoccupation with ideal beauty and enabled him to achieve that fusion of ideal and real, universal and particular which distinguishes his finest poetry", p. 82. From early on, Hopkins' drawings were often accompanied by words, or vice-versa. For example, he adds the following commentary to sketches of waves he made on July 23, 1963: "The waves of the returning wave overlap, the angular space between is smooth but covered with a network of foam. The advancing wave already broken, and now only a mass of foam, upon the point of encountering the reflux of the former." Bump concludes that Hopkins is interested in drawing as a means of visual research to help him perceive, if not represent, a larger unity (p. 74). In drawing, along with his verbal sketches, Hopkins attempts to find "the distinctively *unifying* design, the 'returning' or recurrent pattern, the internal 'network' of structural relationships which clearly and unmistakably integrates or *scapes* an object or sets of objects, and thus reveals the presence of integrating laws throughout nature and a divine unifying force or 'stress' in this world"(pp. 74-75). This commonality of purpose in drawing and writing was certainly encouraged by Ruskin's own practice.

[14] See Claude Colleer Abbott, ed. *Further Letters of Gerard Manley Hopkins Including his Correspondence with Coventry Patmore*, 2nd ed. (London: Oxford University Press, 1956), p. 199.

influence engulfed his society and left few subjects untouched. Hilary Fraser describes well his place in Victorian society:

> It is not an unreasonable claim to say that Ruskin literally reshaped Victorian taste in art... He was responsible for establishing what we now regard as the characteristic tenor of Victorian art and aesthetics. He affirmed the significance of the aesthetic dimension of life in an age when, under the pressure of modern industrial development and burgeoning capitalism, that dimension was becoming marginal to the real business of life and perilously close to being lost altogether. He encouraged artists and laymen alike to appreciate the beauty of nature, art, and architecture, and to deprecate the ugliness of all that deformed nature and human creativity.[15]

Ruskin's works, including *Modern Painters, The Seven Lamps of Architecture, The Stones of Venice* and *The Elements of Drawing*, were eagerly devoured by Victorian England's rising middle class. As mentioned above, by 1863 Hopkins probably had read Ruskin's *The Elements of Drawing,* and he included *Modern Painters* on his February 1865 reading list.[16] *The Elements of Drawing*, according to Norman White, was the source of many Hopkinsian ideas about nature and how to look at nature:

> The whole book is crucial to Hopkins' ways of looking at nature and expressions of his reaction to it. Ruskin aimed at teaching to see, teaching to draw being secondary in importance; the aim of drawing was not learning a skill but learning to appreciate nature. From Ruskin's great principle, that art should be made, not by learning from general ideas or words, but by looking at natural objects, stem all Hopkins's debts to Ruskin.[17]

[15] Hilary Fraser, *Beauty and Belief: Aesthetics and Religion in Victorian Literature* (Cambridge: Cambridge University Press, 1986), p. 112.

[16] Humphry House, ed. *The Journals and Papers of Gerard Manley Hopkins*, 2nd ed. (London: Oxford University Press, 1959), p. 56.

[17] Norman White, *Hopkins: A Literary Biography* (Oxford: Clarendon Press, 1992), p. 75.

Ruskin's attention to detail was characteristic of the age. Patricia Ball notes that the science of the first half of the nineteenth century (which influenced Ruskin and Hopkins alike) centered upon the collection, classification, and close observation of phenomena. It was, in short, obsessed with detail.[18] However, Hopkins' unique obsession with detail stemmed not only from Ruskin's emphasis on recording detail, but also from his stress on *seizing* perceptions of beauty in the natural world to preserve fleeting reality. For example, Ruskin minutely described moments of pleasure derived from natural beauty:

> The sun is setting on [the lake], and I am sitting at my room window watching the opposite outline. The snow on the high point, fresh, is dazzingly bright… Brighter yet! Now it is running left, glowing on the pastures and pines. Oh, beautiful! The hills are all becoming misty fire, and all is grey beneath them and above. Yet redder… it is bursting into conflagration, over purple shades. Now the light has left the bases, but it is far along to the left on the broad field of snow, less and less but redder and redder. Oh, glorious! It is going fast; only the middle peak has it still — fading fast — fading — gone.[19]

Natural beauty, according to Ruskin, was "unique, momentary, and vulnerable to time and man's impiety."[20] Ruskin believed that nature's smallest part held grand significance and that one should engage the senses with detail:

[18] Patricia M. Ball, *The Science of Aspects: The Changing Role of Fact in the Work of Coleridge, Ruskin and Hopkins* (London: The Athlone Press, 1971), p. 53. See also Tom Zaniello, "Hopkins' Scientific Interests: "Face to Face with the Sphinx"," *Thought* 65, no. 259 (1990), pp. 511-14.

[19] John Ruskin, *The Diaries of John Ruskin*, ed. J. Evans and J.H. Waterhouse, 3 vols. (Oxford: Clarendon Press, 1956), vol. I, pp. 200-201 (June 9, 1841).

[20] Norman White, *Hopkins: A Literary Biography* (Oxford: Clarendon Press, 1992), p. 76.

> Men usually see little of what is before their eyes... I have to prove to them that there are more things in heaven and earth than are dreamed of..., and that the truth of nature is a part of the truth of God; to him who does not search it out, darkness, as it is to him who does, infinity.[21]

Ruskin uniquely emphasized, for his time, the hyperphysical in the physical. As White writes, "there was no hyperbolic leap from the visual perception of an object to a moral vision" for Ruskin.[22] The beauteous forms of nature, seen rightly, were signs of divinity. He wrote that nature had "a body and a soul like man; but her soul is the Deity."[23] Ruskin held that aesthetic experience was divinely inspired and that the impact of such experience was moral, engendering motivation to love and serve God. For Ruskin, beauty was theophany. He wrote:

> [Beauty] is either the record of conscience, written in things external, or it is a symbolizing of Divine attributes in matter, or it is the felicity of living things, or the perfect fulfilment of their duties and functions. In all cases it is something Divine; either the approving voice of God, the glorious symbol of Him, the evidence of His kind presence, or the obedience to His will by Him induced and supported.[24]

All beauty, then, some how relates to the nature of God for Ruskin. Because beauty was the reflection of God's nature in visible things, Ruskin believed that the most profound values are

[21] John Ruskin, *Modern Painters*, I, in *The Complete Works of John Ruskin*, 15 vols. (New York: T.Y. Crowell and Co., 19 —), vol. I, pp. 50-51.

[22] Norman White, *Hopkins: A Literary Biography* (Oxford: Clarendon Press, 1992), p. 78.

[23] See Hilary Fraser, *Beauty and Belief: Aesthetics and Religion in Victorian Literature* (Cambridge: Cambridge University Press, 1986), p. 113. Fraser is quoting from Ruskin's *Modern Painters* (vol. 1).

[24] John Ruskin, *Modern Painters*, II, in *The Complete Works of John Ruskin*, 15 vols. (New York: T.Y. Crowell and Co., 19 —), vol. I, p. 210.

learned through the eye. He went so far as to claim that a deficiency in the perception of nature affects the moral and emotional life as well as the whole scale of values in a society. This was, for him, a source of the ills of Victorian industrialized England. Urban dwellers, particularly the lower classes, were isolated from a proper perception of nature and, therefore, of God. Fundamental to Ruskin was a "real, discriminating and complete encounter with material phenomena" as a way to aesthetic judgments and moral insights.[25] He approached the theological in his belief that only an impartial, intense, and fearless seeing leads to "noble emotion" which in turn leads to a revelation of nature as God's work.[26] The following excerpt from a manuscript originally meant for inclusion in the second volume of *Modern Painters* captures this Ruskinian belief and emphasis. While gazing upon a storm in the Alps near Chamonix, he wrote:

> Suddenly, there came in the direction of Dome du Gouter a crash — of prolonged thunder; and when I looked up, I saw the cloud cloven, as it were by the avalanche itself, whose white stream came bounding down the eastern slope of the mountain, like slow lightening. The vapour parted before its fall, pierced by the whirlwind of its motion; the gap widened, the dark shade melted away on either side; and, like a risen spirit casting off its garment of corruption, and flushed with eternity of life, the Aiguilles of the south broke through the black foam of the storm clouds. One by one, pyramid above pyramid, the mighty range of its companions shot off their shrouds, and took to themselves their glory — all fire — no shade — no dimness. Spire of ice — dome of snow — wedge of rock — *all* fire in the light of the sunset, sank into the hollows of the crags — and pierced through the prisms of the glaciers, and dwelt within them — as it is in clouds. The ponderous storm writhed and moaned

[25] Patricia M. Ball, *The Science of Aspects: The Changing Role of Fact in the Work of Coleridge, Ruskin and Hopkins* (London: The Athlone Press, 1971), p. 64.

[26] Ibid., pp. 70-71.

> beneath them, the forests wailed and waved in the evening wind, the steep river flashed and leaped along the valley; but the mighty pyramids stood calmly — the very heart of the high heaven — a celestial city with walls of amethyst and gates of gold — filled with the light and clothed with the Peace of God. And then I learned — what till then I had not known — the real meaning of the word Beautiful. With all that I had ever seen before there had come mingled associations of humanity — the exertion of human power — the action of the human mind. The image of self had not been effaced in that of God... It was then that I understood that all which is the type of God's attributes...can turn the human soul from gazing upon itself...and fix the spirit...on the types of that which is to be food for eternity; — this and this only is in the pure and right sense of the word BEAUTIFUL.[27]

Through a kind of 'selfless' seeing, Ruskin claimed to experience an "absorption of soul and spirit — the prostration of all power — and the cessation of all will — before, and in the Presence of, the manifested Deity." He went on to write that it "was then only I understood that to become nothing might be to become more than Man."[28] Thus, Ruskin formulated a theory of the beautiful which, other than the act of 'seeing', downplayed the importance of human elements, stressed the role of the objectively given, and derived its meaning from the "eternal, the unchanging, and the infinite" revealed in nature.

Ruskin further held that great art conveys this revelation of nature. He said that "all great art is praise" and can be a revelation to humanity if fidelity to fact exists.[29] Ruskin primarily wrote such

[27] Quoted in George P. Landow, *The Aesthetic and Critical Theories of John Ruskin* (Princeton: Princeton University Press, 1971), pp. 111-12.

[28] Quoted in George P. Landow, *The Aesthetic and Critical Theories of John Ruskin*, p. 112.

[29] Patricia M. Ball, *The Science of Aspects: The Changing Role of Fact in the Work of Coleridge, Ruskin and Hopkins* (London: The Athlone Press, 1971), p. 71. The quote is from Ruskin's epilogue to *Modern Painters*. Ruskin claimed

things of painting and drawing, but Hopkins saw them to be true for poetry as well.[30] Ruskin wrote that "to find even in all that appears most trifling or contemptible, fresh evidence of the constant working of the Divine power for glory and for beauty, and to teach it and proclaim it to the unthinking and the unregarding" was the "peculiar province" of the artist.[31] Ruskin held that art is didactic, but not essentially so. Art is didactic by being beautiful, and because the beautiful, for Ruskin, is a symbol of God, "its perception becomes an essentially religious act which 'indirectly and occultly' informs the beholder of the nature of God."[32]

These Ruskinian tenets discussed above took poetic flesh throughout Hopkins' work. However, his poem, *Pied Beauty*, offers a good example of how much Hopkins, following Ruskin, looked for "the soul of the Deity" upon the flesh of particular

that the painter, J.M.W. Turner, was the superlative example of the artist who best accomplished this task. Ruskin literally beseeched artists "to see and feel" so as to relate the ever closer study of fact (nature) to the immediate, instinctive awareness of a divine creator (cf. *Modern Painters*, III, chapter 17).

[30] Ruskin seemed to hold the principle of *ut pictura poesis*, seeing a very close relationship between painting and poetry. Cf. George P. Landow, *The Aesthetic and Critical Theories of John Ruskin* (Princeton: Princeton University Press, 1971), p. 43.

[31] John Ruskin, *Modern Painters*, in *The Complete Works of John Ruskin*, 15 vols.,(New York: T.Y. Crowell and Co., 19 —), vol.I, p. 317. The prevalent opinion is that this kind of thinking, when found in the writings of the Jesuit Hopkins, is Ignatian. I do not dispute Ignatian influence, but I take exception to its commonly accepted degree of strength. Formation through the *Spiritual Exercises* affected Hopkins deeply, but his 'signic' vision of nature predates his entry into the Society of Jesus. It developed intellectually and artistically under Ruskin's touch and was later heightened and refined by Ignatius. Its final development came with Hopkins' enthrallment with Duns Scotus. Undoubtedly, too much has been made of the 'Jesuit worldview' in this matter.

[32] George P. Landow, *The Aesthetic and Critical Theories of John Ruskin* (Princeton: Princeton University Press, 1971), p. 68. Landow draws his commentary here from Ruskin's *The Queen of the Air*.

reality, even in those things that appeared to be "trifling or contemptible":

Glory be to God for dappled things —
For skies of couple-colour as a brindled cow;
For rose-moles all in stipple upon trout that swim;
Fresh-firecoal chestnut-falls; finches' wings;
Landscape plotted and pieced — fold, fallow, and plough;
And all trades, their gear and tackle and trim.
All things counter, original, spare, strange;
Whatever is fickle, freckled (who knows how?)
With swift, slow; sweet, sour; adazzle, dim;
He fathers-forth whose beauty is past change:
Praise him.[33]

Clearly, Hopkins followed Ruskin in viewing nature as a "holy book." In taking this approach to nature, Ruskin, and Hopkins following him, no doubt fell in the Romantic tradition.[34] At the heart of this tradition was the need to draw a kind of quintessence from things. Yet Ruskin distinguished himself sharply from classic Romanticism by shifting the emphasis from the subject observing nature to the object observed. For Ruskin, proper tribute to the object and everything the object offers is not possible if the observer is more concerned with self reaction, emotional response,

[33] Norman H. MacKenzie, ed. *The Poetical Works of Gerard Manley Hopkins* (Oxford: Clarendon Press, 1990), p. 144.

[34] I am equally tempted to say, however, that he reflects a 'medieval' tradition. There is a certain 'sacramental energy' to Ruskin's vision that hearkens back to what Marucci calls a "medievalist code." Cf. Franco Marucci, *The Fine Delight That Fathers Thought: Rhetoric and Medievalism in Gerard Manley Hopkins* (Washington, D.C.: The Catholic University Press, 1994). Ruskin's belief that God intended natural phenomena to bear moral meaning is more congenial to the Middle Ages than it is to the Victorian age. In Hopkins, this code originated with Ruskin, developed under Ignatius, and was finely honed by Scotus. Also see Philip Ballinger, "Review Article," *Studies* 85, no. 338 (1996): 195-97.

"or any other centre which seems to deprive the object of the full attention it merits in its independent existence."[35] Hopkins followed Ruskin in this. They both focused upon 'selfhood', but the selfhood of things seen. For Ruskin, and, it would seem, for Hopkins also, emotional, moral, and spiritual insight depended, to some degree, on the observer's total submission to the selfhood of the object observed "in the eloquence of its self-expression."[36] Seeing leads to feeling; fact leads to religious emotion. Ruskin believed that seeking 'pure fact', derived through careful and abiding observation, is the first indispensible artistic act as well as the first act of gleaning the 'Deity' from nature. This approach opposed Romanticism's tyranny of the ego. Ruskin held as conceit the kind of Romantic approach which claimed that it matters little what things are in themselves, but only what they are to the individual. This impoverishing egotism was the 'pathetic fallacy' of Romanticism for Ruskin. As George Eliot commented, Ruskin taught his contemporaries "a truth of infinite value," and that truth was "realism — the doctrine that all truth and beauty are to be attained by a humble and faithful study of nature, and not by substituting vague forms, bred by imagination on the mists of feeling, in place of definite, substantial reality."[37] For Ruskin, there was a certain supremacy and even 'sacredness' of fact. Perhaps the young Hopkins at Oxford, having read Ruskin's plea to artists to pay the physical universe at least as much honor as the scientist does, took the plea to heart. Certainly, Hopkins was loyal to Ruskin's contention that artists should find their mission in adver-

[35] Patricia M. Ball, *The Science of Aspects: The Changing Role of Fact in the Work of Coleridge, Ruskin and Hopkins* (London: The Athlone Press, 1971), p. 55.

[36] Ibid., p. 74.

[37] Alison G. Sulloway, *Gerard Manley Hopkins and the Victorian Temper* (New York: Columbia University Press, 1972), p. 65.

tising nature's *inherent* qualities and powers rather than altering nature to suit their anthropomorphic instinct.[38] Ruskin wrote:

> Every creature of God is in some way good, and has a duty and specific operation providentially accessory to the well-being of all; we are to look, in this faith, to that employment and nature of each, and to derive pleasure from their entire perfection and fitness for the duty they have to do, and in their entire fulfillment of it; and so we are to take pleasure and find beauty…in every creature for the doing of that which God has made it do.[39]

Of course, under the influence of Ignatius of Loyola and Duns Scotus, Hopkins' definition of *inherent* became purely Christological. This was not the case for Ruskin. While referring throughout his writings to a deliberate [divine] mystery in nature behind all her ascertainable laws, he was, in Hopkins' eyes, "unsatisfactorily vague about the characteristics of God."[40] Still, Ruskin did offer Hopkins a keenly 'signic' view of the world. Ruskin held that there was a bond between the human mind and all visible things in terms of a revelation of divine laws. He insisted that the discovery of these 'laws' is only arrived at by close study of natural detail. The divine 'laws', or the divinely given intelligibility of things, are not beyond the facts for Ruskin, rather, they are somehow within them. Effectively, he pursued what he held to be revelatory truth and beauty, "the divine soul upon the flesh of things," by using "the most objective kind of observation."[41] An example of

[38] Patricia M. Ball, *The Science of Aspects: The Changing Role of Fact in the Work of Coleridge, Ruskin and Hopkins* (London: The Athlone Press, 1971)., p. 79.

[39] John Ruskin, *Modern Painters* in *The Complete Works of John Ruskin*, 15 vols.,(New York: T.Y. Crowell and Co., 19 —), vol. II, pp. 101-102.

[40] See Alison G. Sulloway, *Gerard Manley Hopkins and the Victorian Temper* (New York: Columbia University Press, 1972), pp. 72 and 76.

[41] Patricia M. Ball, *The Science of Aspects: The Changing Role of Fact in the Work of Coleridge, Ruskin and Hopkins* (London: The Athlone Press, 1971), p. 57.

this view is evident in an 1842 encounter with a "small aspen tree against a blue sky":

> Languidly, but not idly, I began to draw it; and as I drew, the languor passed away: the beautiful lines insisted on being traced, — without weariness. More and more beautiful they became, as each rose out of the rest, and took its place in the air. With wonder increasing every instant, I saw that they 'composed' themselves, by finer laws than any known of men.[42]

Thus, Ruskin held that the Divine shows itself through the inherent qualities of individual things. Using the method and eye of an observational scientist, Ruskin awoke to the revelatory 'beauty of form' in things. The Deity, so to speak, was 'in the details' for Ruskin and was waiting to be 'seen' and thus 'felt' and communicated by the artist. Ruskin's aim in this method was to relate the "ever closer study of fact to the immediate, instinctive awareness of a divine creator" and to bring "such a sense of the presence and power of a Great Spirit as no mere reasoning can either induce or controvert."[43]

I believe this Ruskinian context was the root of Hopkins' vision of and unique approach to a 'signic' world. However, this contention is not universally affirmed in the critical literature. The beginnings of this 'signic' view of the world found in Hopkins' work, particularly in its more mature incarnational and Christological expressions, have been attributed by some to Tractarian influence. For example, C.M. Brittain claims that the Tractarian incarnationalism of Pusey and Liddon was the basis of Hopkins'

[42] Quoted from Ruskin's *Praeterita* by Patricia Ball in *The Science of Aspects: The Changing Role of Fact in the Work of Coleridge, Ruskin and Hopkins* (London: The Athlone Press, 1971), p. 56.

[43] Patricia M. Ball, *The Science of Aspects: The Changing Role of Fact in the Work of Coleridge, Ruskin and Hopkins* (London: The Athlone Press, 1971), p. 71.

aesthetic synthesis.[44] I strongly disagree. Such a view disregards three important facts. First of all, Hopkins himself attributes no such influence to Pusey, Liddon, or any other Tractarian. He does, however, attribute influence to Ruskin. Secondly, Brittain's thesis disregards pre-Oxonian roots for Hopkins' inclinations, whether they be his early exposure to Ruskin or simply the general ethos of High Anglicanism. Certainly Hopkins did not enter Oxford as a *tabula rasa*. Finally, Brittain overlooks the fact that when Hopkins does refer to the Tractarians, it is only to make little of their poetic efforts. For example, Hopkins thought that the poetic works of Keble and Newman, as aesthetic efforts, were feeble at best.[45] They were, in his opinion, a weak transference of the Romantic tradition of the Lake poets (Wordsworth, etc.) into the Victorian Church.[46] I agree that Liddon, Pusey, and the vestiges of Tractarianism at Oxford did influence the development of Hopkins' religious inclinations and his journey towards Roman Catholicism. Hopkins says as much. However, the Hopkins scholar cannot legitimately say more than this concerning Tractarian influence. As we shall see, the Incarnation became the central mystery for Hopkins and, as Brittain claims, "the Tractarians knew the Incar-

[44] Cf. C.M. Brittain, "God's Better Beauty: Hopkins, Pusey, and Tractarian Aesthetics," *Christianity and Literature* 40, no. 1 (1990):7-22.

[45] Cf. Claude Colleer Abbott, ed. *The Correspondence of Gerard Manley Hopkins and Richard Watson Dixon*, 2nd ed. (London:Oxford Univrsity Press,1955), p.99. For this reason, I find Alison Sulloway's description of Hopkins as "the Tractarian poet turned Jesuit" baffling. See *Gerard Manley Hopkins and the Victorian Temper* (New York: Columbia University Press, 1972), p. 23. For a fine contextualization of possible Tractarian influence on elements of Hopkins' aesthetics and poetry, see Maureen Moran, "Manl(e)y Mortal Beauty: Hopkins as Tractarian Aesthete," *The Hopkins Quarterly* 22, no. 1-2 (1995): 3-29.

[46] Cf. Stephen Prickett, *Romanticism and Religion: The Tradition of Coleridge and Wordsworth in the Victorian Church* (Cambridge: University Press, 1976). My comment is an expansion of Hopkins' own. He actually said that the Lake poets feebly expired in the poetic works of Keble and Newman.

nation to be the summit of God's economy of redemption for the world;"[47] nonetheless, this similarity is not an equation and cannot infer causality. The aesthetic inclination to see the world as signic, and eventually as sacramental, came, at least in part, from Hopkins' pre-Oxonian exposure to Ruskin, **then** was probably religiously echoed for Hopkins by Pusey and Liddon, knitted into him spiritually by Ignatius, and, finally, philosophically and theologically confirmed for him by Scotus. I shall further develop this latter theme below. Presently, let us continue with Ruskin's specific influence upon Hopkins.

D. 'Seeing' and the Laws of Beauty: Ruskin's Specific Influences upon Hopkins

Ruskin's influence upon Hopkins was not simply generic; rather, it was specific, directly touching his genius. I hold that the roots of Hopkins' creations, *inscape* and *instress*, tap Ruskinian aesthetics.[48] The terms were not originally theological or spiritual referents. Hopkins coined them before his entry into the Society of

[47] C.M. Brittain, "God's Better Beauty: Hopkins, Pusey, and Tractarian Aesthetics," *Christianity and Literature* 40, no. 1 (1990), p. 18.

[48] John Ruskin, *The Elements of Drawing*, in *The Complete Works of John Ruskin*, 15 vols.,(New York: T.Y. Crowell and Co., 19 —),vol. IV, p. 66. N. White hints that the point of origin for Hopkins' *inscape* may be in Ruskin's writings. Ruskin used the word *signalement* to refer to the "vital truth" of a thing. This is a French legal and scientific term for "a schedule of particulars serving to identify an individual or a genus." It is the one word in Ruskin closest to the meaning of *inscape*. Ruskin's true artist was one who could seize upon the vital truth of a thing. Though this is a potential catalyst for Hopkins' development of the concept of *inscape*, and thus of his aesthetic vision, the term itself must be related to Hopkins' study of Greek. See Norman White, *Hopkins: A Literary Biography* (Oxford: Clarendon Press, 1992), p. 77.

Jesus and his discovery of Scotus. He first revealed the terms in some notes he made on Parmenides dated February 9, 1868:

> [Parmenides'] great text... is that Being is and Not-being is not — which perhaps one can say, a little over-defining his meaning, means that all things are upheld by instress and are meaningless without it... His feeling for instress, for the flush and foredrawn, and for inscape/ is most striking... But indeed, I have often felt when I have been in this mood and felt the depth of instress or how fast the inscape holds a thing that nothing is so pregnant and straightforward to the truth as simple *yes* and *is*.[49]

That Hopkins later integrated these concepts into his theology and that they shaped his spirituality is certain. Nonetheless, we must look for their inspiration in Hopkins' earliest efforts to develop his own fundamental aesthetic principles. Again, the person met on such a search is Ruskin.

As we have discussed, **seeing** was Ruskin's first imperative. The aesthete must learn, above all, how to see:

> Nothing must come between Nature and the artist's sight; nothing between God and the artist's soul. Neither calculation nor hearsay, — be it the most subtle of calculations, or the wisest of sayings, — may be allowed to come between the universe, and the witness

[49] Humphry House, ed. *The Journals and Papers of Gerard Manley Hopkins*, 2nd ed. (London: Oxford University Press, 1959), p. 127. Hopkins did not begin reading Scotus until 1870. His use of the terms in these notes does not mean that Parmenides' thought inspired them. He uses the terms in a familiar way which suggests that they pre-date this particular usage. Bernadette Ward in "Newman's Grammar of Assent and the Poetry of Gerard Manley Hopkins," *Renascence* 43, no. 1-2 (1990): 105-120, claims that Newman's linguistic theories are "Scotist-like" and acted as a catalyst for Hopkins' creation of *inscape*. That Hopkins first recorded use of the term occurred during his sojourn at the Oratory with Newman is undeniable, but this does not mean causality. In fact, one could make as much of the differences between Newman and Scotus, or, for that matter, between Newman and Hopkins, as of the similarities. Rooting *inscape* in Newman's influence is a stretch and totally disregards the pre-Newmanian Ruskinian roots.

> which art bears to its visible nature. The whole value of that witness depends on its being eye-witness; the whole genuineness, acceptableness, and dominion of it depend on the personal assurance of the man who utters it. All its victory depends on the veracity of the one preceding word, "Vidi."[50]

This facility is not natural.[51] The senses must be trained. Uneducated senses cannot discern the truth of nature. Clearly, Hopkins took these admonishments to heart and followed Ruskin's advice. Through descriptive journaling and sketching, Ruskin held that a person's perceptive powers become refined — Hopkins was clearly Ruskin's disciple in this regard. Ruskin saw nature as a tutor for beauty. By "imitating" nature, at least at first, the aspiring artist could learn what is essential, namely, that "all beauty is founded on the laws of natural forms."[52] Ruskin held that if the artist can paint a leaf, he can paint the world.[53] He emphasized vegetation as being the best of tutors in this regard. It is not by accident that Hopkins spent so much time concentrating on plants and trees.

[50] John Ruskin, *The Stones of Venice*, in *The Complete Works of John Ruskin*, 15 vols.,(New York: T.Y. Crowell and Co., 19 —),vol. V, p. 37.

[51] Cf. Mary Ann Caws, "Cognitive Poetics and Passionate Reading in Ruskin and Hopkins," *Rivista di letterature moderne e comparate* 44, no. 3 (1991), pp. 244-5. Caws refers to Ruskin saying that "for every hundred who know, only one can see." Caws continues: "Ruskin's supreme value is always placed on seeing — the rarest gift fortune bestows, and at the same time, the major effort to be labored at." Hopkins voices the same thing differently: "I thought how sadly beauty of inscape was unknown and buried away from simple people and yet how near at hand it was if they had eyes to see it and it could be called out everywhere again." Humphry House, ed. *The Journals and Papers of Gerard Manley Hopkins*, 2nd ed. (London: Oxford University Press, 1959), p. 128.

[52] For example, see Ruskin's chapter on "The Lamp of Beauty" in *The Seven Lamps of Architecture*, in *The Complete Works of John Ruskin*, 15 vols.,(New York: T.Y. Crowell and Co., 19 —), vol. I, pp. 94-136.

[53] John Ruskin, *Modern Painters*, in *The Complete Works of John Ruskin*, 15 vols.,(New York: T.Y. Crowell and Co., 19 —), vol. III, p. 35.

Ruskin believed that beauty was a composition of symmetry and variety. This law of composition, so important to art, is embodied in natural forms. In observing natural forms, the artist must first master generic form. "The task of the painter, in his pursuit of ideal form, is to attain accurate knowledge... of the peculiar virtues, duties, and characters of every species of being; down even to the stone..."[54] Thus, Ruskin's first concern was with the "perfect *idea* of the form and condition in which all the properties of the species are fully developed."[55] Nonetheless, because of the variations of complexity among individuals within various species (oaks for example), the artist must ultimately focus on the form characteristic of individuals, and must "show the individual character and liberty of the separate leaves, clouds, or rocks." Therefore, both the law and individuality are essential to masterly work, but the individuality is more essential. To understand Ruskin "it is important to note that form is characteristic not only of species, but of individuals, and that the truths of specific form are thus the first and most important of all."[56] Fike demonstrates that Ruskin's 'law' refers to regularity while individuality implies irregularity.

Beauty is composed of symmetry and variety. Hopkins held this Ruskinian principle. In his 1865 essay, *The Origin of Beauty: A Platonic Dialogue*, written for Walter Pater at Oxford, Hopkins held that proportion is the source and seat of beauty, and that proportion is the proper combination of *regularity* ("consistency or agreement or likeness, either of a thing to itself or of several things to each other") and *irregularity* ("difference or disagree-

[54] John Ruskin, *Modern Painters*, in *The Complete Works of John Ruskin*, 15 vols.,(New York: T.Y. Crowell and Co., 19 —),vol. II, p. 109.

[55] Ibid., p. 102.

[56] Francis G. Fike, "The Influence of John Ruskin upon the Aesthetic Theory and Practice of Gerard Manley Hopkins" (Ph. D. Dissertation, Stanford University, 1964), p. 71.

ment or change or variety"). He summarized this idea of the nature of beauty by writing that "all beauty may…be called rhyme," for rhyme embodies the principle of regularity in relation to irregularity.[57] Here we see a specific and early influence upon Hopkins by Ruskin. Here we begin to approach the matrix in which Hopkins' *inscape* and *instress* developed.

Following Ruskin, Hopkins believed that form has a reality beyond its embodiment in individual objects; that is, form has ontological status beyond the physical world independent of a perceiver.[58] As Ruskin taught, the artist's first task is to discover the 'typical' or ideal forms in the natural objects observed. Hopkins certainly devoted immense effort in isolating these typical forms from the detail that surrounded them. These typical forms paralleled what Hopkins and Ruskin referred to as "regularity" in their aesthetic principles. Hopkins early on identified this aesthetic reg-

[57] Humphry House, ed. *The Journals and Papers of Gerard Manley Hopkins*, 2nd ed. (London: Oxford University Press, 1959), pp. 90 and 102. The expansion of this thought to "all things rhyme in Christ" is the poetic and theological center for Hopkins. See J. Hillis Miller, *The Disappearance of God: Five Nineteenth-Century Writers* (Cambridge: Harvard University Press, 1963), pp. 277ff.

[58] We can see here why some refer to Ruskin and Hopkins as Platonists. Cf. Alison G. Sulloway, *Gerard Manley Hopkins and the Victorian Temper* (New York: Columbia University Press, 1972), p. 65. They believed that there are forms and laws in Creation. Ruskin held nature to be "the art of God." The art of humanity is the expression of a rational and disciplined delight in the forms and laws of Creation, both generic and specific. The artist must learn the generic and specific forms of nature in order to perceive the wealth of beauty in Creation. Nonetheless, this does not mean that they refuted change. Ruskin glorified change and movement as *Vital Beauty* and a divine law. For Ruskin and Hopkins, nature is at her best when in motion. Their writings are full of descriptions of energetic nature, nature moving. Thus, Ruskin and Hopkins may qualify as Platonists in one sense, but a contrarian could claim them as Heraclitean. After all, Hopkins entitled a poem as *That Nature is a Heraclitean fire*. See Shelley King, " 'But Meaning Motion': Movement and Change in G.M. Hopkins"(Ph.D. Dissertation, University of Toronto, 1988), pp. 64-130.

ularity as a kind of obedience to law (law being the resemblance of several things or parts of a thing to each other).[59] Thus, in searching for the typical forms of things, their regularity, he was searching for their structural law. Throughout his Oxford journals, Hopkins referred to the 'form' or 'type' of particular things.[60] Trees particularly seemed to catch his attention (following Ruskin's preferences).[61]

[59] Humphry House, ed. *The Journals and Papers of Gerard Manley Hopkins*, 2nd ed. (London: Oxford University Press, 1959), pp. 90 and 102.

[60] E.g., "those tretted mossy clouds *have their own law* more in helices, wave-tongues, than in anything else and it is pretty perceptible" (*Journals and Papers*, p. 142); "... the short stroke of eyes, nose, mouth, repeated hundreds of times I believe it is which gives the visible law: looked at in any one instance it flies" (*Journals and Papers*, p. 139). Sulloway refers to this emphasis on individual reality and their particular forms in Ruskin and Hopkins as a manifestation of "the doctrine of specific creation." She writes: "The doctrine of specific creation, which Hopkins first came upon in Ruskin and later in Duns Scotus, teaches that God created each living thing with its own specific peculiarities that would never be exactly recapitulated in another created thing." She continues: "There is another paradox in these twin laws, for just as all creatures must live according to the laws of their type, they must also live according to the laws of their specific being. Equally, all artists must try to reproduce not only the laws of generic plenitude in picturing nature, but also the laws of each creature's specific being." See Alison G. Sulloway, *Gerard Manley Hopkins and the Victorian Temper* (New York: Columbia University Press, 1972), pp. 72-3. Sulloway thus notes an interesting parallelism between Ruskin and Scotus in their mutual emphasis on 'the individual'. While I agree with the observation on parallelism here, I would not call Ruskin's emphasis a 'doctrine'. Still, Sulloway adds to the argument that there was a logical flow for Hopkins from the general aesthetic context of Ruskin to the more specific philosophical and theological contexts of Scotus. Though Sulloway is on the mark here, I find it unfortunate that she did not add Ignatius in this observation. He too, within the context of the importance of the Incarnation, stressed the value of the particular (as we shall see below).

[61] Cf. Rossana Bonadei, "From Diaries to Poetry — Focusing on the Tree," in *Gerard Manley Hopkins: Tradition and Innovation*, ed. G. Marra, P. Bottalla, F. Marucci (Ravenna: Longo Editore, 1991), 101-119. Bonadei studies the place of 'the tree' in Romantic and Victorian writers as an entry into the metaphysics and aesthetics of Nature. Ruskin's preference was not *ex nihilo*.

An 1867 entry in his journal demonstrates Hopkins' appreciation of beauty in relation to form. During a trip to France he noted that "the trees were *irregular*, scarcely expressing *form*."[62] In the Bois de Boulogne, he noted "the thinness of French foliage, weakness in the *general type* of the tree, and naked shrub-like growth of the oaks."[63] He concluded that the trees lacked both fundamental elements of beauty. Not only did they insufficiently embody typical form, but they also lacked a sufficient amount of irregular detail. Thus, their effect on the beholder was weak. In any case, this search for regularity or law and its combination with irregularity in individual things, that is to say, beauty, was at the heart of Hopkins' minute observation of nature. We find this, for example, in the following journal entry:

> Oaks: the organisation of this tree is difficult. Speaking generally no doubt the determining planes are concentric, a system of brief contiguous and continuous tangents, whereas those of the cedar would roughly be called horizontals and those of the beech radiating but modified by droop and by screw-set towards jutting points. But beyond this since the normal growth of the boughs is radiating and the leaves grow some way there is of course a system of spoke-wise clubs of green — sleeve-pieces. And since the end shoots curl and carry young and scanty leaf-stars these clubs are tapered, and I have seen also the pieces in profile with chiseled outlines, the blocks thus made detached and lessening towards the end. However the star knot is the chief thing: it is whorled, worked round, a little and this is what keeps up the illusion of the tree: the leaves are rounded inwards and figure out ball-knots. Oaks differ much, and much turns on the broadness of the leaf, the narrower giving the crisped and starry and Catherine-wheel forms, the broader the flat-pieced mailed or shard-covered one, in which it is possible to see composition in

[62] Humphry House, ed. *The Journals and Papers of Gerard Manley Hopkins*, 2nd ed. (London: Oxford University Press, 1959), p. 147 (my italics).

[63] Ibid., p. 148 (my italics).

> dips etc. on wider bases than the single knot or cluster... I shall study them further... [64]

This passage was for Hopkins' use and is, therefore, difficult to understand. Nonetheless, it demonstrates his search for structure amidst complexity. He seeks and finds some order through careful observation and descriptive analysis. Later in the journal he claims, "I have now found the law of the oak leaves."[65] He thus mastered another form, another kind of beauty. He sought the "law" of a thing, e.g., an oak, the more to appreciate the variation of that law or form in other oaks. The typical form acts as a basis to perceive irregularities in individual members of a species. This is the basis for perceiving beauty in a thing. As Fike writes, for Hopkins, "the apprehension of natural beauty presupposes the combination of regularity and irregularity in the object, and that combination in turn presupposes in the beholder a sound knowledge of typical form or law as the norm from which individual variations may part."[66] This is clearly Ruskinian.

Hopkins was prepared for his vocation as poet by following the method of Ruskin and accepting the basic tenets of his aesthetics.

[64] Ibid., pp. 144-45.

[65] Ibid., p. 146. In a letter to his friend Baillie he wrote "I think I have told you that I have particular periods of admiration for particular things in Nature; for a certain time I am astonished at the beauty of a tree, shape, effect, etc., then when the passion, so to speak, has subsided, it is consigned to my treasury of explored beauty, and acknowledged with admiration and interest ever after, while something new takes its place in my enthusiasm. The present fury is the ash, and perhaps barley and two shapes of growth in leaves and one in tree boughs and also a conformation of fine-weather cloud" (July 10, 1863). See Claude Colleer Abbott, ed. *Further Letters of Gerard Manley Hopkins Including his Correspondence with Coventry Patmore*, 2nd ed. (London: Oxford University Press, 1956), p. 202.

[66] Francis George Fike, "The Influence of John Ruskin upon the Aesthetic Theory and Practice of Gerard Manley Hopkins" (Ph. D. Dissertation, Stanford University, 1964), p. 85.

They highly refined his powers of observation and description. These aesthetics also led him to a unique appreciation of individual realities, of each thing's pristine distinctiveness. For example, in a May 9, 1871 journal entry, Hopkins describes handling a bluebell:

> The bluebells in your hand baffle you with their inscape, made to every sense: if you draw your fingers through them they are lodged and struggle / with a shock of wet heads; the long stalks rub and click and flatten to a fan on one another, making a brittle rub and jostle like the noise of a hurdle strained by leaning against; then there is the faint honey smell and in the mouth the sweet gum when you bite them. But this is easy, it is the eye they baffle.[67]

In an unembarrassed seeing and caring, he goes on to describe detail upon detail to such a degree and with such heavy imagery that one feels the bluebell will disintegrate under his intense observation. The plant is described with images such as panpipes, trombones, sheephooks, waves riding through a whip being snapped, knights at chess, the heads of snakes. In the same paragraph these "crisped ruffled bells" are attached to such descriptive verbs as stiffen, strengthen, droop, gape, spike, strike, cluster, taper, flatten and rise. Hopkins declares that he has never seen anything more beautiful than the bluebell he is looking at, and that he knows the beauty of the Lord by it. In this kind of 'contemplative' mode, Hopkins not only affirmed Ruskin's aesthetic vision, he also created new terms to describe better his heightened experience of beauty in nature.

[67] Humphry House, ed. *The Journals and Papers of Gerard Manley Hopkins*, 2nd ed. (London: Oxford University Press, 1959), p. 209.

E. Ruskin and Hopkins' Ideas of *Inscape* and *Instress*

As we have noted, Hopkins first used his characteristic aesthetic terms, *inscape* and *instress*, in the 1868 notes on the philosopher Parmenides which he made while teaching at Newman's Oratory School in Birmingham. There is no indication that the terms were specifically derived from or inspired by Parmenides. In fact, they appear to be familiar terms to Hopkins when he first wrote of them. We know that Hopkins was still absorbed at the time by the Greek literature and philosophy he had studied for the Classical 'Greats' final he passed with honors at Oxford in June of 1867. In a February 12, 1868 letter to his friend, Baillie, he spoke of his preference for Plato and Aristotle over modern philosophers. "An interest in philosophy is almost the only one I can feel myself quite free to indulge in still."[68] Throughout his life, Hopkins was keenly interested in rooting his aesthetics in sound philosophical principles. It is, therefore, not unlikely that these terms, while exhibiting a Ruskinian core, were created by Hopkins in the context of his study of classical Greek philosophy.[69] As Fike suggests,

[68] Claude Colleer Abbott, ed. *Further Letters of Gerard Manley Hopkins Including his Correspondence with Coventry Patmore*, 2nd ed. (London: Oxford University Press, 1956), p. 231.

[69] There may be a prototype of Hopkins' theory of *inscape* in one of his unpublished undergraduate essays. Reacting to Pater's denigration of metaphysics, Hopkins wrote: "There are certain [designs *del.*] forms wh. have a great hold on the mind and are always reappearing and seem imperishable, such as the designs of Greek vases and lyres, the cone upon Indian shawls, the honeysuckle moulding, the fleur-de-lys, while every day we see designs both simple and elaborate wh. do not live and are at once forgotten; and some pictures we may long look at and never grasp or hold together, while the composition of others strikes the mind with a conception of unity wh. is never dislodged: and these things are inexplicable on the theory of pure chromatism or continuity — the forms have in some sense or other an absolute existence." See Lesley Jane Higgins, "Hidden Harmonies: Walter Pater and Gerard Manley Hopkins" (Ph.D. Dissertation, Queen's University, 1987), pp. 183-84.

perhaps the term *inscape* is a cognate of the Greek verb *skopein* (to look attentively) or the noun *skopos* (that upon which one fixes his or her look).[70] In any case, the essential point is that Hopkins' use of *inscape* implies many of the elements of Ruskinian aesthetics discussed above. It implies unity and, in that unity, the typical form by which one species of thing is distinguished from other species. It also implies individual form which distinguishes an individual object from other objects of the same kind. As Fike states:

> The term is thus designed to cover precisely the discovery that Hopkins had made at the climax of his attempt to train himself as an artist; namely, that there are individual characteristics in an oak, for example, which make it unlike any other oak in existence. There was no term in English to express this kind of reality, so Hopkins coined a term. Regularity and irregularity are thus implied in the term..."[71]

The idea of *inscape,* then, arose out of Hopkins' aesthetic concerns and, as we have seen, his aesthetics were shaped by Ruskin. *Inscape*, in its deepest root, refers to beauty. As Hopkins wrote in his sermon on *Creation and Redemption: The Great Sacrifice*, "... This song of Lucifer's was a dwelling on his own beauty, an instressing of his own inscape..."[72] Thus, Ruskin was the central initial influence on a key aspect of Hopkins' thought, an aspect which ultimately took on great significance in the formation of his theological thought and spirituality.[73]

[70] Francis George Fike, "The Influence of John Ruskin upon the Aesthetic Theory and Practice of Gerard Manley Hopkins" (Ph. D. Dissertation, Stanford University, 1964), p. 94. Fike suggests that **in-scape** is derived from a combination of **in**dividual and **scape** to mean "that total complex of unified characteristics in an object which reveals through impressions upon the eye (and other senses) of the beholder a knowledge of its individual being."

[71] Ibid., p. 97.

[72] Christopher Devlin, S.J., ed. *The Sermons and Devotional Writings of Gerard Manley Hopkins* (London: Oxford University Press, 1959), pp. 200-201.

[73] Cf. Lesley Jane Higgins, "Hidden Harmonies: Walter Pater and Gerard Manley Hopkins" (Ph.D. Dissertation, Queen's University, 1987), p. 229. As we

There is also continuity between Hopkins' unique term *instress* and his early aesthetic theory. Hopkins used *instress* variably, but it has two essential meanings. It refers to the force that holds the inscape of an object together as well as to the effect or feeling produced by inscape within the beholder of a particular object. As W. A. M. Peters wrote:

> The original meaning of instress... is that stress or energy of being by which 'all things are upheld,' and strive after continued existence. Placing 'instress' by the side of 'inscape' we note that the instress will strike the poet as the force that holds the inscape together; it is for him the power that ever actualizes the inscape. Further, we observe that in the act of perception the inscape is known first and in this grasp of the inscape is felt the stress of being behind it, is felt its instress... We can now understand why and how it is that 'instress' in Hopkins' writings stands for two distinct and separate things, related to each other as cause and effect; as a cause 'instress' refers for Hopkins to the core of being or inherent energy which is the actuality of the object; as effect 'instress' stands for the specifically individual impression the object makes on man.[74]

Fike connects the second sense of the term *instress* to Hopkins' early aesthetic theory derived from Ruskin, namely, that instress, as the effect of inscape, corresponds to Ruskin's views on the apprehension of beauty in an object.[75] Through careful observa-

shall see, Hopkins' concept of *inscape* developed into a theological concept. As Thornton writes, Hopkins "came to think of the inscape of natural things not just as the principle of their organization but also as an implication of an organizing power behind them and the expression of the nature of God in them." Cf. R.K.R. Thornton, *Gerard Manley Hopkins: The Poems* (London: Edward Arnold, 1973), p. 20.

[74] W.A.M. Peters, *Gerard Manley Hopkins: A Critical Essay towards the Understanding of his Poetry* (London: Oxford University Press, 1948), pp. 14-15.

[75] Francis George Fike, "The Influence of John Ruskin upon the Aesthetic Theory and Practice of Gerard Manley Hopkins" (Ph. D. Dissertation, Stanford University, 1964), pp. 106-107.

tion and the discovery of law and variety, or regularity and irregularity, in a thing, Ruskin held that the artist could experience a sense of beauty and the pleasure accompanying it.[76] This parallels Hopkins' schema whereby the inscape of a thing is discovered through close examination and the corresponding effect of that inscape (instress) is felt. Here we see the objective-subjective polarity that can be found in Ruskin continued in Hopkins' ideas about the perception of beauty. Inscape is the objective reality that exists independent of the beholder while instress is partly the response of the beholder and partly the force of being which links the object and the beholder. So Hopkins' comment that "what you look hard at seems to look hard at you" becomes intelligible.[77] Part of the work of the artist is not only to convey inscape, but also to convey the moral and aesthetic energy of a thing, its instress. Inscape, then, centers on *seeing* as Ruskin defines it, while instress is partially the internal alchemy or response of God's artist that occurs when a thing is truly seen. The artist 'feels' a thing's instress as God's plan behind its inscape and internally responds to the divine will expressed there, submits to

[76] The ultimate goal here is not pleasure, but in some sense God, or as Sulloway puts it, the "searching for the hint of God's soul upon the flesh of things." Cf. Alison G. Sulloway, *Gerard Manley Hopkins and the Victorian Temper* (New York: Columbia University Press, 1972), p. 70. The trained 'searcher' sees *inscape* then feels *instress*, which, according to Sulloway, is "God's plan behind nature's inscapes and man's submission to that plan" (p. 71).

[77] Humphry House, ed. *The Journals and Papers of Gerard Manley Hopkins*, 2nd ed. (London: Oxford University Press, 1959), p. 204. Although the theological import of *instress* for Hopkins is not discussed here, it is important to see that this principle has a 'divine' character. As Sulloway writes, "Hopkins used the term instress to mean many things, but its composite meaning encompasses God's plan for the world as it is revealed in the looks and conduct of natural things — as opposed merely to those things themselves; and instress also signifies man's response to the divine plan as he praises it and makes copies of it in his art." See Sulloway, *Gerard Manley Hopkins and the Victorian Temper*, p. 46.

it.[78] Thus, art, in its truest sense, is more than faithful reproduction of what the artist 'sees'. It is also an expression of the divine energy or will in a reality and the artist's response to it.[79]

It was here, with his idea of instress, that Hopkins began to move beyond Ruskin. Ruskin held that 'seeing' would help the great artist to separate purely personal emotion from appropriate responses to objects. Keeping the eyes fixed firmly on *pure fact* would assure that any feelings coming to the artist are true ones.[80] Ruskin dogmatically attributed the truth of visual perception "not to the organizing intelligence of the viewer but to the organizing force of nature."[81] Thus, for Ruskin, what the artist does is objectively study appearance in order to reveal the structure of an object and God's design underlying the object. However, without an organizing principle this 'dogmatic' objectivism is a kind of subjectivism still. In centering upon the study of a thing's appearance, its effect on us, there is still the matter of whether we can trust that the effect is not self-generated. For Hopkins, assurance that particular sense impressions have an identity with their generating object came from his concept of instress. As C. Christ holds, "Hopkins is the only Victorian poet who resolves the tension between the meaning objects have in themselves and our impressions of them."[82]

Instress ties the perceived and the perceiver together. Hopkins saw instress as the organizing principle by which a thing outward

[78] This may be a root for the 'bidding' function of Hopkins' poetry which will be developed in the final chapter of this study.

[79] Alison G. Sulloway, *Gerard Manley Hopkins and the Victorian Temper* (New York: Columbia University Press, 1972), p. 71.

[80] Carol T. Christ, *Victorian and Modern Poetics* (Chicago: University of Chicago Press, 1984), p. 66.

[81] Ibid., p. 66.

[82] Ibid., p. 70.

and objective becomes a thing inward and subjective. This was something absent in Ruskin. As we shall see, Hopkins eventually based the identity he created between object and apprehension upon his conceptions of God and being. Summarily put, for Hopkins, God charges the universe with a 'rhyming' capacity "which enables man's imagination, his capacity of instress, to realize the divinely ordained instress of the world."[83] In Hopkins' vision, "language's chime and rhyme ideally manifest the chime and rhyme of divine creation. By miming the shape of instress, the poet can recreate the moment when impression realizes its object."[84] Ultimately, as we shall see, this vision becomes ground for a kind of theological aesthetic. Nonetheless, it is important to realize that the basic structure of 'inscape' and 'instress' pre-dates any kind of theological systemization in Hopkins. The terms became theological and, more specifically, Christological only after Hopkins' internal dialogue with the mores and thought of Ignatius and Duns Scotus.

E. Conclusion

Certainly, Hopkins 'changed' after his entrance into the Society of Jesus and his enthrallment with the person and thought of Duns Scotus. This affected his interests and the subject matter of his poetry. He leaned more toward explicitly theological concerns. Thus, nature became more and more a revelation of the Creator. Nonetheless, the early aesthetic principles that he formed, refined, and applied through his study of Ruskin and, thus, of nature, remained. He refined this early influence in his two new metaphysical terms, *inscape* and *instress*. He grew to use them in a the-

[83] Ibid.
[84] Ibid., p. 71.

ological as well as the original aesthetic context. Thus, beauty remained an ultimate preoccupation of Hopkins, but in its association with religious experience. A revisiting of a stanza from his poem *The Wreck of the Deutschland* emphasizes this development:

> I kiss my hand
> To the stars, lovely-asunder
> Starlight, wafting him out of it; and
> Glow, glory in thunder;
> Kiss my hand to the dappled-with-damson west:
> Since, tho' he is under the world's splendour and wonder,
> His mastery must be instressed, stressed;
> For I greet him the days I meet him, and bless when I understand.[85]

Ruskin believed that the creation of beauty was a duty owed to society. Marcel Proust described Ruskin as an aesthete who dedicated his life to an unswerving cult of Beauty: "This Beauty, to which he elected to consecrate his life, was not conceived as a means of embellishing or sweetening existence, but as a reality infinitely more important than life itself..."[86]

[85] Norman H. MacKenzie, ed. *The Poetical Works of Gerard Manley Hopkins* (Oxford: Clarendon Press, 1990), p. 120.

[86] Peter Quennell, "John Ruskin," in *British Writers: Elizabeth Gaskell to Francis Thompson*, ed. Ian Scott-Kilvert (New York: Charles Scribner's Sons, 1982), p. 183. Quennell is quoting Proust's comment in *Gazette des Beaux-Arts*, no. 514 (April, 1900), p. 315. It is a lesser known fact that Ruskin markedly influenced Proust's aesthetic sensibility. In fact, Proust translated some of Ruskin's work into French in collaboration with his mother, Mme. Adrien Proust (see *The London Times Literary Supplement, October 7, 1955, p. 596)*. The prolific French biographer, André Maurois, emphasizes the affinity between Proust and Ruskin. See André Maurois, *The Quest for Proust*, trans. Gerard Hopkins (London: Jonathan Cape, 1950), pp. 112-21. Maurois writes: "Ruskin taught [Proust] how to look at things and, more especially, how to describe what he saw. A natural taste for the infinitely small variations of fine shades, a gift for registering the 'slow-motion' notation of the emotions, and what amounted to *greed* in savouring all colours and all forms — these were common to both men" (p. 116). Writ-

Clearly, Hopkins was strongly influenced by Ruskin and was, as Ball writes, ready for Ruskin's gifts. In Hopkins' work,

ing about Ruskin's influence, Proust stated that "I suddenly saw the universe as something of infinite value. My admiration for Ruskin gave such high importance to the objects he made me love, that they seemed as though charged with greater richness even than life itself." See Marcel Proust, "John Ruskin," in *Marcel Proust: A Selection from His Miscellaneous Writings*, ed. and trans. Gerard Hopkins (London: Allan Wingate, 1948). The following example from Proust's writing exemplifies this influence: "I would now gladly remain at the table while it was being cleared, and, if it was not a moment at which the girls of the little band might be passing, it was no longer solely towards the sea that I would turn my eyes. Since I had seen such things depicted in water-colours by Elster, I sought to find again in reality, I cherished, as though for their poetic beauty, the broken gesture of the knives still lying across one another, the swollen convexity of a discarded napkin upon which the sun would patch a scrap of yellow velvet, the half-empty glass which thus showed to greater advantage the noble sweep of its curved sides, and in the heart of its translucent crystal, clear as frozen daylight, a dreg of wine, dusky but sparkling with reflected lights, the displacement of solid objects, the transmutation of liquids by the effect of light and shade, the shifting colour of the plums which passed from green to blue and from blue to golden yellow in the half-plundered dish, the chairs, like a group of old ladies, that come twice daily to take their places round the white cloth spread on the table as on an altar at which were celebrated the rites of the palate where in the hollows of oyster-shells a few drops of lustral water had gathered as in tiny holy water stoops of stone; *I tried to find beauty there where I had never imagined before that it could exist, in the most ordinary things, in the profundities of 'still life'*(my italics). See Marcel Proust, *Remembrance of Things Past*, trans. C.K. Scott Moncrieff, 3 vols., vol. 1 (New York: Random House, 1934), p. 653. Clearly, Proust's 'seeing' has been formed by Ruskin's aesthetic emphasis. Maurois writes, "Read any description by Ruskin of a wave, of a precious stone, of a tree, of a flower-species. If it be well translated it might come straight out of Proust" (p. 118). There are definite parallels here with Ruskinian passages in Hopkins' descriptive writing. The influence Ruskin had upon Proust resembles greatly that which he had upon Hopkins. This is certainly a proof against discounting Ruskin's influence on Hopkins. N.B.: I use Moncrieff's translation for easier comparison of style with Hopkins and Ruskin. Proust collaborated with Moncrieff in the making of this translation of his defining work. Also notice that some of Proust's works here quoted, as well as Maurois' biography about him, were translated by Gerard Hopkins, incidentally the nephew of Gerard Manley Hopkins.

Ruskin's gifts and their poetic possibilities were eagerly taken up and developed.[87] For Hopkins, before Ignatius and Duns Scotus there was John Ruskin. The awakening crafted in Hopkins by Ruskin was the foundation for Hopkins' essential aesthetic. Using the formational, philosophical, and theological brick and mortar of Ignatius and Duns Scotus, Hopkins built upon this foundation to create the principles that underpinned his artistry and, ultimately, his theological thought.[88]

[87] Patricia M. Ball, *The Science of Aspects: The Changing Role of Fact in the Work of Coleridge, Ruskin and Hopkins* (London: The Athlone Press, 1971), p. 102.

[88] Quoting Ashton Nichols, De Angelis emphasizes the movement from epiphany to theophany in Hopkins: "His theology emerges not before but after his sensuous apprehension of the world, an apprehension so intense that it demanded new words and new metrical forms to embody its power in language." Cf. Palmira De Angelis, *L'Imagine Epifanica* (Rome: Bulzone Editore, 1989), p. 133, n. 95. Ruskin is at the heart of the 'epiphanic' movement of this progression, so to speak. Thus, I find it curious that De Angelis does not recognize the centrality of Ruskin's influence on Hopkins' ultimate theological aesthetics, an aesthetics which had its roots in Ruskin's 'seeing'.

CHAPTER II

CREATED TO PRAISE: GERARD MANLEY HOPKINS AND IGNATIUS

A. Introduction

In the realm of aesthetic expression, few explicitly Catholic artists have captured the consciousness of the modern age as has Gerard Manley Hopkins. As we have seen, he died in relative obscurity in 1889, and his poetry was not published until 1917. Even then, it did not gain wide recognition until after 1930. Thus, an oft misunderstood and ill-appreciated artist in his own time, Hopkins became, anachronistically and ironically, the celebrated poet of modernity in the twentieth century. Since 1930, hundreds of articles have been written about Hopkins' influence and meaning by scholars from diverse academic domains. Though his work is claimed and mined by critics and admirers from areas as varied as English literature, theology, ethics, aesthetics, and social history, one contentious question continually arises in the midst of the field — how and to what degree was Hopkins influenced in his thought and poetry by Ignatius of Loyola and the *Spiritual Exercises*? Clearly, there was change in Hopkins as he moved from being the Ruskinian aesthete to being the Jesuit priest. In this chapter, I intend to present and critique the main lines of thought on this central issue in Hopkins studies. I additionally hope to present an alternate and new point of view on this touchstone issue in Hopkinsian criticism. Finally, I hope to demonstrate the place of Ignatian influence in Hopkins' aesthetic synthesis.

B. Poet vs. Priest

Poet and priest — in the history of Hopkins literary criticism, these vocations have been seen both in union and in conflict. Certainly, the early disparagement and even negation of the Jesuit influence on Hopkins' poetry erupted in part because of the overt anti-Catholicism present in Victorian England.[1] Nonetheless, one still finds lament over Hopkins having been a Jesuit priest.[2] "Never was a squarer man in a rounder hole."[3] Imagine what

[1] For example, see E.E. Phare's *The Poetry of Gerard Manley Hopkins* (Cambridge: Cambridge University Press, 1933). This early classic of Hopkinsian criticism approaches the poet with an overtly anti-Catholic bias. Such bias helps to explain the hagiographic character of another early work in Hopkinsian criticism, Gerard F. Lahey, S.J.'s *Gerard Manley Hopkins* (London: Oxford University Press, 1930).

[2] Cf. Sophie Herzburg, "The Winter World of Gerard Manley Hopkins," *Faith and Freedom* 44, Spring/Summer (1991), p. 22. She writes that "in joining the Jesuit Order, he deliberately isolated himself from familiar places and his own family and, without realising it at first, began to destroy those artistic roots which were so necessary a part of his personality." A.R. Coulthard writes that "what limits his literary achievement...is not the brevity of his canon but the fact that he lived his curtailed life as a priest... Religion constricted Hopkins' poetry..." Coulthard contends that Hopkins' priesthood kept him "one rung short of greatness" and made him "a poet of excellent passages more than excellent poems." His vocation and theology were the spoilsports of his poetry. Furthermore, Coulthard claims that Hopkins' "Dionysian instincts" and "self-abnegating moralism" were in a constant conflict which muted his brilliance and led to "schizophrenic poetry". See A.R. Coulthard, "Gerard Manley Hopkins: Priest vs. Poet," *The Victorian Newsletter* 88, Fall (1995), pp. 35-36.

[3] Humphry House, "A Note on Hopkins' Religious Life," *New Verse* 14, April (1935), p. 3. Also see John Gould Fletcher, "Gerard Manley Hopkins: Priest or Poet?," *American Review* 6, no. 3 (1936): 331-346. Other comments from the same period are quoted in Martin C. Carroll, S.J., "Gerard Manley Hopkins and the Society of Jesus," in *Immortal Diamond: Studies in Gerard Manley Hopkins*, ed. Norman Weyand (New York: Octagon Books, 1969), 3-50: "Humanly speaking he made a grievous mistake in joining the Jesuits for on further acquaintance his whole soul must have revolted...;" "the rigid and pitiless Society of Jesus

could have been if he had exclusively followed his true calling, that of poet, goes the cry. Variations on this theme abound. The strict discipline and duties of Jesuit life sapped energies which would have otherwise fostered the realization of his poetic gift. The repression and even social isolation necessary to live the Jesuit vocation in Victorian England psychologically damaged Hopkins, deadening his creativity and art. Finally, the sheer, potentially sacrilegious freedom necessary for Hopkins to be a good poet was circumscribed by the priestly vocation, thus his Jesuit life cannot be seen as an influence, rather, it must be seen as an unfortunate detriment.[4] Hopkins himself gives cause for doubt as to the compatibility of these two vocations.[5] He clearly struggled with their

exhausted, discouraged, and perhaps destroyed him...;" finally, "the pathos and waste of such a situation have so affected the well-meaning world of literary criticism that the twenty-one years which Hopkins spent as a member of the Society of Jesus are generally considered to have left him a frustrated voluptuary, a genius blasted by asceticism, a soured and disappointed man" (p. 3).

[4] Robert Graves, *The White Goddess: A Historical Grammar of Poetic Myth*, American ed. (New York: Farrar, Straus and Giroux, 1966), pp. 425-26. Graves writes with reference to Hopkins and others: "It has become impossible to combine the once identical functions of priest and poet without doing violence to one calling or the other... The poet survived in easy vigour only where the priest has been shown the door..." Graves insists on the necessity of an almost 'pagan' freedom for the creation of true poetry. Graves' point of view seems unwittingly taken up by Canon H.J. Hammerton in "The Two Vocations of G.M. Hopkins," *Theology* 87, May (1984): 186-89. Hammerton also posits a stark opposition between the two "functions" in Hopkins' life and concludes that "we can be thankful that his first vocation to poetry saved him in the end" (p. 89). More recently, and in a more subtle fashion, B. Belitt follows this trend in *The Forged Feature: Toward a Poetics of Uncertainty* (New York: Fordham University Press, 1995), pp. 229-79. He implies that viewing Hopkins as a 'scientific' rather than priestly poet is the more useful paradigm.

[5] Hopkins tried to explain his view of the relationship between his vocation as a Jesuit and writing poetry in a letter to his friend and admirer, R.W. Dixon: "Our Society values...and has contributed to literature, to culture; but only as a means to an end. Its history and its experience show that literature proper, as poetry, has seldom been found to be to that end a very serviceable means. We

co-existence in his life and never achieved a peaceful synthesis of the two.[6]

Struggle and conflict — here is the rub of the alternative view. Artistic genius often consorts with struggle, and conflict can be creative as well as destructive. Hopkins was a true artist in this sense. One aspect of his personality led him to delight in the lavishness of nature's beauty while the other drew him to asceticism and the denial of the senses necessary to his art. In Hopkins, the aesthete wrestled with the ascetic. He struggled not only with seemingly contradictory vocations, but also with the conundrum forced by his aesthetic and theological center, namely, how to unify the ideal of sacrifice and selfless endeavor with the Romantic artist's intense

have had for three centuries often the flower of the youth of a country in numbers enter our body: among these how many poets, how many artists of all sorts, there must have been! But there have been very few Jesuit poets and, where they have been, I believe it would be found on examination that there was something exceptional in their circumstances or, so to say, counterbalancing in their career. For genius attracts fame and individual fame St. Ignatius looked on as the most dangerous and dazzling of all attractions." See Claude Colleer Abbott, ed. *The Correspondence of Gerard Manley Hopkins and Richard Watson Dixon*, 2nd ed. (London: Oxford University Press, 1955), pp. 93-94.

[6] R.K.R. Thornton writes: "At times you might think that Hopkins set no value on poetry or art at all, so strongly does he reject it or belittle it; at other times he gives the appearance of being fascinated by it but only as a meaningless if superb irrelevance." See R.K.R. Thornton, "Gerard Manley Hopkins: Aesthete or Moralist?," in *Saving Beauty: Further Studies in Hopkins*, ed. Michael E. Allsopp (New York: Garland Publishing, 1994), p. 40. Thornton recognizes, however, that Hopkins was not consistent in his disdain. He desired to create art but he chose to be a priest or, as Thornton concludes, Hopkins had an inability to reconcile the elective and affective will (p. 57). Canon Dixon's pleading letter to Hopkins dated November 4, 1881 demonstrates the reaction his non-Jesuit friends had to Hopkins' struggle. In response to Hopkins' statements marginalizing the place of writing poetry in his life as a Jesuit, Dixon hit the mark in his remonstration that "surely one vocation cannot destroy another." See Claude Colleer Abbott, ed. *The Correspondence of Gerard Manley Hopkins and Richard Watson Dixon*, 2nd ed. (London: Oxford University Press, 1955), p. 90.

self-consciousness and dwelling on true selfhood as beauty and ultimate perfection.[7]

Can a priest be a poet and remain faithful to his vocation? This was a valid question for Hopkins himself and he often treated his poetic inclinations and poetry with apparent indifference. Nonetheless, one answer to this question holds that the creative struggle brought on by his faithfulness to Jesuit life and the priestly vocation led to a unique poetic brilliance. The possibility that Jesuit spirituality, formed by the *Spiritual Exercises* and example of Ignatius of Loyola, acted as an influence on Hopkins' work thrives within this point of view.[8]

[7] For a reflection on conflict in Hopkins' life, see John Hawley, S.J., "Hopkins and the Christian Imagination," in *Gerard Manley Hopkins Annual 1993*, ed. Michael Sundermeier and Desmond Egan (Omaha: Creighton University Press, 1993), 45-55. Additionally, see R.A. Jayantha's, "Gerard Manley Hopkins: Devotional Poet of 'Unpropitious' Times," *The Literary Criterion* 24, no. 1-2 (1989): 133-46. Jayantha offers valuable thoughts on the difficulty of being either priest or religious poet in the Victorian age, a philosophically and theologically disjunctive time fraught with doubt and characterized by Matthew Arnold as "deeply unpoetical"(p. 136).

[8] For a summation of this viewpoint, see Martin C. Carroll, S.J., "Gerard Manley Hopkins and the Society of Jesus," in *Immortal Diamond: Studies in Gerard Manley Hopkins*, ed. Norman Weyand (New York: Octagon Books, 1969), 3-50. "The contribution which the Society of Jesus made to [Hopkins'] poetry is incalculable. It could not make Hopkins a poet in spite of himself, but in wisely guiding and tenderly caring for Hopkins the Jesuit and the priest, the Society of Jesus, more than can ever be known, formed Hopkins the poet… Hopkins' intense spiritual life, born of the *Spiritual Exercises* of St. Ignatius… finds its full expression in the passionate and personal force of his verse… Had Hopkins never become a Jesuit he would, without doubt, still have achieved distinction as a poet. But it would have been a different Hopkins, and much that his poems now possess to perfect measure, would strangely be missing" (pp. 49-50). For an early example of a defense of this point of view, see Joseph Keating, S.J., "Fr. Gerard Manley Hopkins and the Spiritual Exercises," *The Month* 166, no. September (1935): 268-270. The classic defense of this approach to Hopkins was made in John Pick's *Gerard Manley Hopkins: Priest and Poet*, 2nd ed. (New York: Oxford University Press, 1966). In the first edition of 1942, Pick argued the position that "the

I take exception to both extremes expressed above. Proving influence can be difficult, particularly if the claim is broad. The first point of view is suspect because it is subjective and based on certain biases that are anachronistic when applied to Hopkins. As the argument goes, Hopkins was a priest and, therefore, sexually, and otherwise, repressed. Such thought is based on assumptions that are clearly contestable and can lead to no certain conclusions. Additionally, the presumption that Hopkins would have produced a larger opus of poetry of even higher quality had he not been a Jesuit priest is an opinion beyond proof. Such opinions are for spirited dinner conversation, not scholarly discourse.

The simple fact is that Hopkins chose to become not only a Catholic, but also a Jesuit priest in Victorian England. He not only made this choice and was proud of it as the highest of vocations, but he also reaffirmed the choice continually, even in the darkest moments of his life.[9] Furthermore, being a Jesuit was not all darkness for Hopkins. Recently published letters by Hopkins, discovered in the Bischoff Collection at Gonzaga University, reflect an enjoyment of and keen interest in the various 'goings-on' in the English Society of Jesus.[10] It was in this life that he created most

secret of Hopkins's art lies in the unity in him of the poet, priest and Jesuit... There is a gulf between [the poetry of] his pre-Jesuit and Jesuit years. The main influence during the in-between-time was the *Spiritual Exercises* — they gripped and fashioned him" (pp. 439-40). Pick claimed Hopkins as the "poet of the *Spiritual Exercises*" (p. 444).

[9] In a letter to Dixon, Hopkins wrote, "My vocation puts before me a standard so high that a higher can be found nowhere else." Just before taking his final vows, Hopkins wrote, "I have never wavered in my vocation, but I have not lived up to it." See Claude Colleer Abbott, ed. *The Correspondence of Gerard Manley Hopkins and Richard Watson Dixon*, 2nd ed. (London: Oxford University Press, 1955), p. 88.

[10] Joseph Feeney, "My Dearest Father: Some unpublished letters of Gerard Manley Hopkins," *The Literary Supplement*, December 22, 1995, 13-14. The letters are found among the papers of the late Reverend Anthony D.

of his poetry as well as the beginnings of a valuable theological synthesis. It is hard to imagine that a person could live and reaffirm a religious vocation so consciously and not be deeply affected by the core thoughts and mores that give it meaning. Even presumption must go to the possibility that Jesuit spirituality and life were formative influences on the thought and work of Hopkins.

The difficulty with this view is not whether Hopkins' vocational choice and way of life had value for, and influence upon, the work he produced. Rather, the difficulty centers upon the issue of what is and what is not specifically Jesuit, or more precisely, Ignatian influence.[11] Care must be taken for a variety of reasons. Hopkins came to the Jesuits and Ignatius' *Spiritual Exercises* with a solid aesthetic foundation formed through both a strong Ruskinian inclination and a superior Oxonian education in classical studies. Some of his aesthetic inclinations were akin to emphases found in the *Exercises*. However, similarity does not infer causality. While not denying that there are specific and identifiable Ignatian influences that can shed light upon Hopkins' worldview and the meaning of his work, the foundational nature of the twenty-four years of life and experience he brought to the Society of Jesus and the *Exer-*

Bischoff, S.J., a Hopkins scholar and Professor at Gonzaga University in Spokane, Washington. He discovered them in England and Ireland, probably in 1947, and kept them for a never completed biography. Fr. Joseph Feeney is the appointed consultant for the Collection and recently uncovered these letters at Gonzaga.

[11] Here I agree with P. Endean who warns, following Karl Rahner, that Ignatian spirituality cannot be fully reduced to a list of particular principles. He argues for care in discussing Ignatian influence on Hopkins, especially when the issue is Hopkins' particular way of relating to God as a Jesuit. See Philip Endean, S.J., "How Should Hopkins Critics Use Ignatian Texts?," in *Gerard Manley Hopkins: Tradition and Innovation*, ed. G. Marra, P. Bottalla, F. Marucci (Ravenna: Longo Editore, 1991), p. 166.

cises must be emphasized.[12] Additionally, the fact that Hopkins was unique in the Society is of consequence. In artistic sensibility, background, and theological inclinations, Hopkins stood alone among English Jesuits of his time.[13] Because of this, some have likened him to Teilhard de Chardin.[14] Therefore, though I will identify instances of probable Ignatian influence, my main intent will be to

[12] This is the general failure of David Anthony Downes in his *The Ignatian Personality of Gerard Manley Hopkins*, 2nd ed. (Lanham: University Press of America, 1990). Although Downes has written prolifically and usefully on Hopkins, he has overstated the Ignatian influence. He writes: "Hopkins, the poet, was so deeply influenced by the ideals of Ignatius that one cannot fully read him without taking *The Spiritual Exercises* into account, without understanding Ignatian spirituality (p. 58)." Though I cannot disagree with this general thesis, it becomes inaccurate if removed from the wider context of Hopkins' life. Downes, at least in his early work, tends to approach Hopkins as if he came to the Jesuits a *tabula rasa*, thus mistakenly naming everything in Hopkins writings Ignatian. Additionally, Downes claims much to be 'Ignatian' or 'Jesuit' which is commonly Christian, e.g., "Ignatius believed in the Trinity and in the Trinity at work in the world." To identify such things as 'Ignatian' or Jesuit adds nothing to an attempt to trace the Ignatian 'turn' in Hopkins' thought and work. See Philip Endean, S.J., "The Spirituality of Gerard Manley Hopkins," *The Hopkins Quarterly* 8, no. 3 (1981): 107-129. Downes' approach is not without pedigree in the field of Hopkinsian criticism. John Pick already proceeded in this direction in his *Gerard Manley Hopkins: Priest and Poet*, (New York: Oxford University Press, 1942).

[13] In a letter to R. Bridges dated April 27, 1881, Hopkins wrote: "You give me a long jobation about eccentricities. Alas, I have heard so much about and suffered so much for and in fact have been so completely ruined for life by my alleged singularities that they are a sore subject." See Claude Colleer Abbott, ed. *The Letters of Gerard Manley Hopkins to Robert Bridges*, 2nd ed. (London: Oxford University Press, 1955), p. 126.

[14] Mario DiCicco, "Hopkins and the Mystery of Christ" (Ph.D., Case Western Reserve University, 1970), pp. 137-39. DiCicco finds strong affinities between de Chardin and Hopkins, not only in life but also in thought. For a well-crafted comparison of the thought of Teilhard de Chardin and Hopkins, see Roslyn Barnes, "Gerard Manley Hopkins and Pierre Teilhard de Chardin: A Formulation of Mysticism for a Scientific Age" (Master of Arts, State University of Iowa, 1962). See also Avery Dulles, S.J., "St. Ignatius and the Jesuit Theological Tradition," *Studies in the Spirituality of Jesuits* 14, no. 2 (1982), pp. 14-15.

show how the *Exercises* and Jesuit life heightened and transmuted pre-existing elements of Hopkins' thought.

As we have seen above, Hopkins came to the Society of Jesus in 1868 with strong "Ruskinian tendencies."[15] John Ruskin, the pre-eminent Victorian aesthete and social critic, held that nature, especially in the aspect of individuality, was God's holy book, a place of revelation. Beauty, as a reflection of the divine, was the artistic target and praise was the artist's purpose. For Ruskin, the senses, particularly sight, and a keen awareness of the effect of a thing on the perceiving self, that is, an awareness of the 'stress' of its beauty, were important tools for the artist. A signic or even sacramental view of reality was also part of Ruskin's gift to Hopkins, though this view was by no means Christological. I will claim that through the *Spiritual Exercises*, Hopkins ultimately recontextualized or 'christened' these early foundations.

Additionally, it must be noted that with his conversion to Catholicism and his decision to become a priest, Hopkins also brought a guilt-laden moralism and a sense that his artistic aspirations and their "dangerous" excitation of the passions must be put aside. In 1868, the year he burned his Oxford poems, and just prior to his entrance into the Jesuit novitiate, he wrote to his friend, Alexander Baillie:

> You know that I once wanted to be a painter. But even if I could I wd. not now, for the fact is that the higher and more attractive parts of the art put strain upon passions which I shd. think it unsafe to encounter. I want to write still, and as a priest I very likely can do that too, not so freely as I should have liked, e.g. nothing or little in the verse way, but no doubt what wd. best serve the cause of my religion.[16]

[15] See Philip Ballinger, "Ruskin: Hopkins' "Silent Don"," *Studies* 85, no. 338 (1996): 116-24.

[16] Claude Colleer Abbott, ed. *Further Letters of Gerard Manley Hopkins Including his Correspondence with Coventry Patmore*, Second ed. (London: Oxford University Press, 1956), p. 231.

These were among the inclinations Hopkins brought with him to the Jesuits. Clearly, at the outset of his life as a Jesuit, Hopkins saw only dichotomy between his earlier aesthetic passion and the religious life. This dichotomy was not to persist in an absolute form. Eventually, with the assistance of the *Spiritual Exercises*, Hopkins created the beginnings of a synthesis between delight in beauty and selfless dedication to God, between aesthetic experience and the experience of the Logos as the meaning-giving and 'beautiful' center of creation.

Although his words are couched in characteristically idealistic terms and could apply to non-Jesuit traditions as well, David Downes captures something essential about Hopkins in writing the following:

> The Ignatian man is a sensuous man. Unlike many other Christian disciples, he does not withdraw from the world, but rather plunges into it. He is overwhelmed by the beauty of things, not only because they are beautiful in themselves, but also because they are manifestations of God...[17] The Ignatian man trusts all created things because they lead him back to God... Because of his sacral view, he sees all creation as a sign of, a message from, a beckoning to, the Divine.[18]

[17] David Anthony Downes, *The Ignatian Spirit in Gerard Manley Hopkins* (Ph.D. Dissertation, University of Washington, 1955), p. 76. The union of aesthetic experience and the experience of God in Hopkins does not mean that Hopkins came to a peace about writing poetry. Throughout his life as a Jesuit, he regarded poetry as a somewhat frivolous activity in view of his other duties. In fact, he destroyed his poems upon entering the Jesuits and did not write verse again until creating *The Wreck of the Deutschland* seven years later.

[18] David Anthony Downes, *The Ignatian Personality of Gerard Manley Hopkins*, 2nd ed. (Lanham: University Press of America, 1990), p. 57.

C. The *Spiritual Exercises* as Context for Hopkins' Christocentrism

Gerard Manley Hopkins entered Manresa House, and the beginning of his training as a Jesuit novice, on September 7, 1868. Almost immediately, he began the "Long Retreat," the time of his intensive introduction to the *Spiritual Exercises* of Ignatius of Loyola. At the outset of the *Exercises*, the "Principle and Foundation" reads:

> Human beings are created to praise, reverence, and serve God our Lord, and by means of this to save their souls. The other things on the face of the earth are created for human beings, to help them in working toward the end for which they are created. From this follows that I should use these things to the extent that they help me toward my end, and rid myself of them to the extent that they hinder me. To do this, I must make myself indifferent to all created things, in regard to everything which is left to my freedom of will and is not forbidden. Consequently, on my own part I ought not to seek health rather than sickness, wealth rather than poverty, honor rather than dishonor, a long life rather than a short one, and so on in all other matters. I ought to desire and elect only the thing which is more conducive to the end for which I am created.[19]

This "Foundation" identifies the reason for the creation of humanity; it describes the proper relationship the exercitant should have toward created things; finally, it directs the will toward the fulfillment of the purpose for which humanity is created. Hopkins took each emphasis and made it uniquely his own.

Hopkins referred to a rough draft of a commentary on *The Exercises* which he hoped to present to his Provincial. Unfortunately,

[19] George E. Ganss, S.J., ed. *Ignatius of Loyola: The Spiritual Exercises and Selected Works* in *The Classics of Western Spirituality* (New York: Paulist Press, 1991), p. 130.

the draft is no longer extant.[20] However, roughly ninety-two pages of notes for the commentary remain in his interleaved copy of *The Exercises,* most of which were written during the Long Retreat of his Tertianship (November-December, 1881). Although the presumption might be that a commentary on the *Spiritual Exercises* would certainly show the mark of Ignatius, this appears not to be the case in Hopkins' commentary. On the contrary, as we shall see, the commentary plainly shows the influence of Hopkins' theological hero — John Duns Scotus.[21] The effect of Ignatius' teaching on the commentary seems to be catalytic in nature rather than theological in any formal sense. *The Exercises* guided, but did not form, Hopkins' theological reflections in the commentary. For example, Ignatius' statement that human beings are created to praise became the seed of a theological meditation on the gift and nature of existence for Hopkins in terms that can hardly be classified as 'Ignatian'. In his commentary notes, Hopkins reflected on what existence means to the self. He wondered at the uniqueness of his own "self-being":

> I find myself both as man and as myself something most determined and distinctive, at pitch, more distinctive and higher pitched than anything else I see; I find myself with my pleasure and pains, my powers and my experiences, my deserts and guilt, my shame and sense of beauty, my dangers, hopes and fears, and all my fate, more important to myself than anything I see. And this is much more true when we consider the mind; when I consider my selfbeing, my con-

[20] See Claude Colleer Abbott, ed. *The Letters of Gerard Manley Hopkins to Robert Bridges*, 2nd ed. (London: Oxford University Press, 1955), p. 150.

[21] Cf. Franco Marucci, *The Fine Delight That Fathers Thought: Rhetoric and Medievalism in Gerard Manley Hopkins* (Washington, D.C.: The Catholic University Press, 1994), pp. 188-98. Marucci argues that Hopkins' commentary on the *Spiritual Exercises* is his most Scotist text: "[The *Spiritual Exercises*] quite often function as a starting point, if not as a pretext, for theological reflections that are as foreign to the Ignatian letter as they are close to Scotus's theology" (p. 189).

> sciousness and feeling of myself, that taste of myself, of I and me above and in all things, which is more distinctive than the taste of ale or alum, more distinctive than the smell of walnut-leaf or camphor, and is incommunicable by any means to another man (as when I was a child I used to ask myself: What must it be to be someone else?). Nothing else in nature comes near this unspeakable stress of pitch, distinctiveness, and selving, this self-being of my own. Nothing explains it or resembles it, except so far as this, that other men to themselves have the same feeling. But this only multiplies the phenomena to be explained so far as the cases are like and do resemble. But to me there is no resemblance: searching nature I taste self but at one tankard, that of my own being. The development, refinement, condensation of nothing shows any sign of being able to match this to me or give me another taste of it, a taste even resembling it.[22]

This experience of uniqueness, viewed through the aspect of the first three words of the Foundation, "homo creatus est," led Hopkins to muse on its source:

> And when I ask where does all this throng and stack of being, so rich, so distinctive, so important, come from, nothing I see can answer me. And this whether I speak of human nature or of my individuality, my selfbeing. For human nature, being more highly pitched, selved and distinctive than anything in the world, can have been developed, evolved, condensed, from the vastness of the world not anyhow or by the working of common powers but only by one of finer or higher pitch and determination than itself and certainly than any that elsewhere we see, for this power had to force forward the starting or stubborn elements to the one pitch required.[23]

[22] Christopher Devlin, S.J., ed. *The Sermons and Devotional Writings of Gerard Manley Hopkins* (London: Oxford University Press, 1959), pp. 122-123.

[23] Ibid., p. 123. The term *pitch* needs explanation here. Hopkins always had an interest in music, and in sound as a major medium of perception. As there is a scale of notes, there was also a proportionate scale for beauty, or distinctiveness, in Hopkins' aesthetics. By *pitch*, he meant "that degree of distinct particularity in a scale, that refinement of organization and height of individual development which the various modifications of a fixed form or inscape has." See David Anthony Downes, *The Ignatian Personality of Gerard Manley Hopkins*,

He derived three possible conclusions:

> From what then do I with all my being and above all that taste of self, that selfbeing, come? Am I due (1) to chance? (2) to myself, as self-existent? (3) to some extrinsic power?[24]

After investigating each, he concluded that only "the third alternative then follows, that I am due to an extrinsic power."[25] In faith, he held this extrinsic power to be God.[26] Truth demands that humanity and all creation recognize their essential dependence:

2nd ed. (Lanham: University Press of America, 1990), p. 18. *Pitch* is a fluid term in Hopkins writings; however, his following 'definition' is somewhat useful: "So also *pitch* is ultimately simple positiveness, that by which being differs from and is more than nothing and not-being, and it is with precision expressed by the English *do* (the simple auxiliary), which when we employ or emphasize, as 'he said it, he did say it,' we do not mean that the fact is any more a fact but that we the more state it..." See Christopher Devlin, S.J., ed. *The Sermons and Devotional Writings of Gerard Manley Hopkins* (London: Oxford University Press, 1959), p. 151.

[24] Christopher Devlin, S.J., ed. *The Sermons and Devotional Writings of Gerard Manley Hopkins* (London: Oxford University Press, 1959), p. 123.

[25] Ibid., p. 128.

[26] For Hopkins, any 'differentiated form' can only "have been developed, evolved, condensed, from the vastness of the world not anyhow or by the working of common powers but only by one of finer or higher pitch and determination than itself." Ibid., pp.122-23. "If man were one of the lesser creatures of the world, a stone or a kingfisher, he could see in the creation many things of higher determination than himself, and could imagine them to be his source. But man looks in vain for anything in creation which could have made him... Therefore I must have been made by some being more highly pitched than I — some being outside the creation, who exists as an extrinsic power." See Joseph Hillis Miller, *The Disappearance of God: Five Nineteenth-Century Writers* (Cambridge: Harvard University Press, 1963), p. 272 and Christopher Devlin, S.J., ed. *The Sermons and Devotional Writings of Gerard Manley Hopkins* (London: Oxford University Press, 1959), p. 128. God is defined by Hopkins as the most "exquisite determining, selfmaking, power." Ibid., p. 125. As Miller writes, interpreting Hopkins: "Only such a power could have created man. God's infinity and his selfexistence consist not so much in his possession of some universal quality like 'being', 'power', or 'will' (though he has these too), as in his possession of the most highly

> God's utterance of himself in himself is God the Word, outside himself in this world. This world then is work, expression, news of God. Therefore its end, its purpose, its purport, its meaning, is God and its life or work to name and praise him.[27]

How do we name and praise God? Hopkins continued his reflection on this first principle by writing:

> The world, man, should after its own manner give God being in return for the being he has given it or should give him back that being he has given. This is done by the great sacrifice. To contribute then to that sacrifice is the end for which man is made.[28]

In idiosyncratically developing a central theme of the "Principle and Foundation," Hopkins formulated the root of his Christocentric vision of creation. Essentially, he held that we are created through God's self-utterance, the Word. Existence, each particular being, is patterned by the Christ.[29] This is uniquely true of human beings. Giving praise and reverence to God, then, means that each thing, after its own manner, lives out its truest nature, its 'Christed' nature. Hopkins expresses this in his sonnet, '*As kingfishers catch fire*' when he says that all things exist as particular selves and Cre-

patterned self of all. His pattern is infinitely complex, and therefore he contains in himself the matrices for all possible and actual creatures, including man. God vibrates [chimes] simultaneously at all possible pitches." See Miller, *The Disappearance of God: Five Nineteenth-Century Writers*, p. 272. This latter theme will be discussed in more detail below.

[27] Christopher Devlin, S.J., ed. *The Sermons and Devotional Writings of Gerard Manley Hopkins* (London: Oxford University Press, 1959), p. 129.

[28] Ibid.

[29] In the *Wreck of the Deutschland*, Hopkins described Christ, the Incarnate God, as 'the ground of being and granite of it'. The emphasis on ultimacy and even the language Hopkins used shows some similarity with later themes in Paul Tillich. I believe this to be coincidental. I am not aware of any knowledge Tillich had of Hopkins' poetry. Cf. R.A. Jayantha, "Gerard Manley Hopkins: Devotional Poet of 'Unpropitious' Times," *The Literary Criterion* 24, no. 1-2 (1989), p. 137.

ation consists of selves being themselves — "Deals out that being indoors each one dwells".[30]

The ontological end of every self, then, is to be self-expressive and, in this self-expression, each being is a self-expression of Christ as well. Christ is the first selving of God as his Son, and everything in Creation follows from this first 'outstress' of God. Creation is multifarious selves selving in Christ's selving of God.[31] For Hopkins, Christ is the prototypic pattern of creation. Downes expresses this differently:

> To Hopkins… God manifested His divine Selfness in the most distinctively selved presence in Creation — the union of divinity and humanity in Jesus Christ. Another way of saying this is that Christ possessed the most being (Hopkins' word is "pitch") of any creature. But the lesser being in mankind, and all other things, is the same being that is in Christ. Thus any human being enacting selfness, as do all other things at their level of being, perforce enacts Christ-being. While all existence is "selved" in Christ-being, only human beings are conscious of the "divine play" in their personalities. Hopkins put it in a sermon: "The sun and the stars shining glorify God…

[30] Norman H. MacKenzie, ed. *The Poetical Works of Gerard Manley Hopkins* (Oxford: Clarendon Press, 1990), p. 141.

[31] Humphry House, ed. *The Journals and Papers of Gerard Manley Hopkins*, 2nd ed. (London: Oxford University Press, 1959), p. 197. Also see David Anthony Downes, "Gerard Manley Hopkins' Christed Vision of Ultimate Reality and Meaning," *Ultimate Reality and Meaning* 12, March (1989): 61-80. This theme of selving and self-expression again finds echoes in Tillich. In various lectures on aesthetics, art and meaning, Tillich uses the term 'expressiveness' in place of the term 'beauty' (which he held had deteriorated and fallen into contempt). Certainly, this unintended connection is worthy of further investigation. See Tillich's lectures on *Art and Ultimate Reality* and *Religious Dimensions of Contemporary Art* in *On Art and Architecture* (New York: Crossroad Publishing Company, 1987). I find it interesting that Tillich settled on the term 'expressiveness' to convey what Hopkins referred to as beauty, and that beauty for Hopkins centered upon a thing's 'self-expressiveness'. Beauty thus becomes a Hopkinsian cipher for not only a thing's self-expressiveness, but also the Creator's self-expressiveness through the creation-mediating Logos.

> The birds sing to him, the thunder speaks his terror, the lion is like his strength, the sea is like his greatness, the honey like his sweetness, they are something like him, they make him known, they tell of him, they give him glory, but they do not know they know… But man can know God, can mean to give him glory. This then is why he was made man, to give God glory and to mean to give it."[32]

Is this theological vision, which claims a certain univocity of being, Ignatian? In spite of Downes' contention, this vision, *per se,* is more akin to what one finds in Scotus, not Ignatius. Nonetheless, the turning of Hopkins' 'generic' signic view of reality, adopted from Ruskin, toward the person of Christ, was certainly a result of the influence of *The Exercises.* As Downes writes, "to see life and see it whole became for Hopkins to see Christ in every particular of experience."[33] This was a formative goal of *The Exercises* which took root in Hopkins' life. The Christocentrism of *The Exercises* thus became the context for Hopkins' theological thought, not the content.

Saying this does not mean that Ignatian influence was peripheral in the formation of Hopkins' theological and aesthetic synthesis. The tenet that the experience of created beauty is a source of spiritual communion as well as a locus for theological truth was not self-evident. There were tensions for Hopkins as he forged the beginnings of a theological structure which allowed for the unity of the aesthetic and the theological. Hopkins struggled with his relation to what he termed 'mortal beauty'. Clearly, he followed Ignatius here — mortal beauty is meant to lead us back to God, and thus should be discerned and used accordingly. Immortal beauty is discerned in and through the natural. The danger in natural beauty for Hopkins was due to the lack of something Ruskin emphasized

[32] David Anthony Downes, *The Ignatian Personality of Gerard Manley Hopkins*, 2nd ed. (Lanham: University Press of America, 1990), p. 77.

[33] Ibid., p. 21.

— *seeing*. The danger of beauty rests in human spiritual shortsightedness; i.e., beauty might be seen as an end in itself rather than as a reflection of and means to something higher and more ultimate.[34] Ignatian spirituality helped Hopkins create a synthesis

[34] The beautiful and the good were not synonymous in Hopkins' mind, at least not in regard to humankind. In an 1883 letter to Coventry Patmore, Hopkins commented that "in nature outward beauty is a proof of inward beauty, outward good of inward good," but in human beings, the will can corrupt, making the human "wax" either "too cold and doughy...or too hot and boiling" to accept "the stamp of [God's] seal." See Claude Colleer Abbott, ed. *Further Letters of Gerard Manley Hopkins Including his Correspondence with Coventry Patmore*, 2nd ed. (London: Oxford University Press, 1956), pp. 306-307. Hopkins continued, "but why do we find beautiful evil? Not by any freak of nature, nature is incapable of producing beautiful evil. The explanation is to be sought outside nature; it is old, simple, and the undeniable fact. It comes from wicked will, freedom of choice, abusing beauty, the good of its nature." Beauty is not the danger for Hopkins, rather, human nature in its tendency toward sin is. In this context, Hopkins periodically worried about being overwhelmed by the alluring immediacy of 'mortal beauty' in its particular manifestations. He occasionally subjected himself to a penance of "not seeing." See Humphry House, ed. *The Journals and Papers of Gerard Manley Hopkins*, 2nd ed. (London: Oxford University Press, 1959), p. 190. Hopkins is in good company in struggling with this dilemma. Dostoevsky, in his *Brothers Karamazov*, has Dmitry Karamazov distinguish between the 'ideal of Sodom' and the 'ideal of Madonna' in speaking of the same tension as Hopkins: "Beauty is a fearful and terrible thing! Fearful because it is indefinable, and it cannot be defined because God sets us nothing but riddles. Here the shores meet, here all contradictions live side by side...It makes me mad to think that a man of great heart and high intelligence should begin with the ideal of Madonna and end with the ideal of Sodom. What is more terrible is that a man with the ideal of Sodom already in his soul does not renounce the ideal of Madonna, and it sets his heart ablaze, and it is truly,truly ablaze, as in the days of his youth and innocence...Is there beauty in Sodom? Believe me, for the great majority of people it *is* in Sodom and nowhere else. The awful thing is that beauty is not only a terrible, but also a mysterious, thing. There God and the devil are fighting, and the battlefield is the heart of man." See John Coulson, "Hans Urs von Balthasar: Bringing Beauty Back to Faith," in *The Critical Spirit and the Will to Believe: Essays in Nineteenth-Century Literature and Religion*, ed. David Jasper and T.R. Wright (New York: St. Martin's Press, 1989), p. 225. Coulson is quoting *The Brothers Karamazov*, trans. by David Magashack (Harmondsworth, 1958) vol. 1, pp. 123-4.

between the two movements, the two beauties, and did so through its unique centering on the Incarnation. Here is an instance where the *Exercises* did explicitly influence the theological thought and artistic purpose of Hopkins. Let us investigate this more closely.

D. Ignatian Incarnationalism and Hopkins' Theological Aesthetic

From early youth Hopkins was drawn to the beautiful in all its forms. Much of Hopkins' education at Oxford, under such prominent scholars as Walter Pater, emphasized beauty as a monistic, self-sufficient good. However, under Ruskin's indirect influence, Hopkins not only trained his eye for beauty and for the description of it, but also understood beauty to be of God and art to be a work of praise. Nonetheless, this attribution of beauty to God was generic and vague in Ruskin. In contrast, under the influence of the *Exercises*, Hopkins' vision of creation and beauty became intensely Christocentric. Let us now look at the evolution and specific character of Hopkins' Christocentric approach.

I hold that Hopkins developed a signic, 'sacramental' vision of reality, of beauty, under Ruskin.[35] Creation was 'God's holy book'.

[35] Unlike C.M. Brittain in his work "God's Better Beauty: Hopkins, Pusey, and Tractarian Aesthetics," *Christianity and Literature* 40, no. 1 (1990): 7-22, I do not agree that the Tractarians' Incarnationalism, whether incapsulated by Pusey or Liddon, was the theological basis for Hopkins' aesthetic synthesis. I agree that they influenced Hopkins' religious journey towards Catholicism and the development of his religious inclinations because Hopkins says as much in his journals. More than this, however, cannot be said regarding Tractarian influence on Hopkins. The Incarnation became the central mystery for Hopkins and, as Brittain claims, it was the central mystery for the Tractarians as well: "the Tractarians knew the Incarnation to be the summit of God's economy of redemption for the world"(p. 18). However, as I have demonstrated above, similarity should not infer an equation in this matter. Again, the aesthetic inclination to see the world

This characteristic seems to be classically Romantic to many Hopkins scholars, including David Downes and Robert Barth,[36] but such a designation is too generic. When Downes or Barth place Hopkins in the Wordsworthian or Coleridgean tradition because he had a "sacramental vision" of creation, they mislead by attempting to fit a very unique Hopkins into a kind of Romantic Procrustean bed. To say that Hopkins' vision of reality was Romantic, as was Wordsworth's or Coleridge's, is to say something of limited value because it does not say enough. Hopkins' vision was Romantic in the same sense as Ruskin's. Hopkins' "sacramental vision" of reality developed under Ruskin's tutelage, not that of Wordsworth or Coleridge. Hopkins followed Ruskin in his emphasis on the unique

as signic came from Hopkins' pre-Oxford exposure to Ruskin, **then** was probably religiously echoed for Hopkins by Pusey and Liddon, knitted into him spiritually by Ignatius, and, finally, philosophically/ theologically confirmed for him by Scotus.

[36] Cf. David Anthony Downes, *Hopkins' Sanctifying Imagination* (Lanham: University Press of America, 1985). Much of Downes' emphasis is on the importance of Wordsworth's and Coleridge's influence on Hopkins. He places Hopkins squarely in the Romantic tradition. Also see Robert J. Barth, "Hopkins as a Romantic: A Coleridgean's View," *The Wordsworth Circle* 25, no. 2 (Spring, 1994): 107-13. On p. 107, Barth writes: "Far more important... however, is what Hopkins derived from Wordsworth and Coleridge: insight into nature; a vision of nature which finds transcendent reality, the divine, revealed in and through the immanent world around us. It is this Romantic vision of nature, in the tradition of Wordsworth and Coleridge — a vision that looks through nature to the transcendent — that I wish to explore. It is what I... call in Wordsworth and Coleridge a 'sacramental vision', shaped and articulated by the 'symbolic imagination'." I do not argue that Hopkins was unaffected by Wordsworth. He clearly admired the man for his spiritual insight into nature and how that insight was communicated in Wordworth's poems. See Claude Colleer Abbott, ed. *The Correspondence of Gerard Manley Hopkins and Richard Watson Dixon*, 2nd ed. (London: Oxford University Press, 1955), p. 141. My argument here is that Hopkins' particular way of seeing nature and processing what he saw, was Ruskinian. It is this specific way of looking at reality that was the grist worked upon through Hopkins' assimilation of the *Exercises* and the thought of Duns Scotus.

self-hood of each being. This emphasis on reality as revelation *in its particularity* is what Hopkins derived from Ruskin. It is this kind of signic vision which he brought to his life as a Jesuit. How, then, was this vision transformed by the *Exercises*?

In the opening stanzas of his defining poem, *The Wreck of the Deutschland*, the first poem he wrote after becoming a Jesuit, we begin to appreciate Hopkins' "sacramental vision":

1

Thou mastering me
God! giver of breath and bread;
World's strand, sway of the sea;
Lord of living and dead;
Thou hast bound bones and veins in me, fastened me flesh,
And after it almost unmade, what with dread,
Thy doing: and dost thou touch me afresh?
Over again I feel thy finger and find thee.

2

I did say yes
O at lightning and lashed rod;
Thou heardst me truer than tongue confess
Thy terror, O Christ, O God;
Thou knowest the walls, altar and hour and night:
The swoon of a heart that the sweep and the hurl of thee trod
Hard down with a horror of height:
And the midriff astrain with leaning of, laced with fire of stress.

3

The frown of his face
Before me, the hurtle of hell
Behind, where, where was a, where was a place?
I whirled out wings that spell
And fled with a fling of the heart to the heart of the Host.
My heart, but you were dovewinged, I can tell,

Carrier-witted, I am bold to boast,
To flash from the flame to the flame then, tower from the grace
to the grace.

4
I am soft sift
In an hourglass — at the wall
Fast, but mined with a motion, a drift,
And it crowds and it combs to the fall;
I steady as a water in a well, to a poise, to a pane,
But roped with, always, all the way down from the tall
Fells or flanks of the voel, a vein
Of the gospel proffer, a pressure, a principle, Christ's gift.

5
I kiss my hand
To the stars, lovely-asunder
Starlight, wafting him out of it; and
Glow, glory in thunder;
Kiss my hand to the dappled-with-damson west;
Since, tho' he is under the world's splendour and
wonder,
His mystery must be instressed, stressed;
For I greet him the days I meet him, and bless when I
understand.[37]

[37] Norman H. MacKenzie, ed. *The Poetical Works of Gerard Manley Hopkins* (Oxford: Clarendon Press, 1990), pp. 119-20. This poem, comprised of 35 stanzas, was Hopkins' first effort after seven years of poetic silence. Concerning its origin, he wrote: "What I had written I burned before I became a Jesuit and resolved to write no more, as not belonging to my profession, unless it were by the wish of my superiors; so for seven years I wrote nothing but two or three presentation pieces which occasion called for. But when in the winter of '75 the Deutschland was wrecked in the mouth of the Thames and five Franciscan nuns, exiles from Germany by the Falck Laws, aboard her drowned I was affected by the account and happening to say so to my rector he said that he wished someone would write a poem on the subject. On this hint I set to work and, though my hand was out at first, produced one. I had long had haunting my ear the echo of a new

Here we find a sacramental vision that is clearly Christocentric. Along with the 'Host', the sacrificial meal so prominent in Hopkins' sermons and devotional writings, there is also the sacrament of the human person and of nature. The Eucharist, the poet, and the world are 'stressed' with the life of God, of Christ. Hopkins wrote that "stress" means "the making a thing more, or making it markedly, what it already is; it is the bringing out of its nature."[38] As Barth writes, "Humankind is by its very nature not only human but Godlike; that 'stress' from God makes mankind 'markedly' what he or she already is — both human and a sharer in the divine nature. The poet, like every sharer in human nature, is also a sacrament of God's presence to the world."[39]

Creation became a sacrament for Hopkins. Because of the Incarnation, God can be wafted out of thunder and starlight and the human heart. God is "under the world's splendour and wonder." The artistic mission for Hopkins is that "His mystery must be instressed, stressed;" that is, it must be impressed upon humanity that God's "stress," his will, life, and energy, is in all things. This stress of the Godhead is mediated through the Word, the Christ. For Hopkins, this incarnational mediation is in some sense 'prior' to creation.[40] In the writings of Hopkins, it is clear that the devel-

rhythm which now I realized on paper." Cf. Claude Colleer Abbott, ed. *The Correspondence of Gerard Manley Hopkins and Richard Watson Dixon*, 2nd ed. (London: Oxford University Press, 1955), p. 14.

[38] Claude Colleer Abbott, ed. *The Correspondence of Gerard Manley Hopkins and Richard Watson Dixon*, 2nd ed. (London: Oxford University Press, 1955), p. 141.

[39] Robert J. Barth, "Hopkins as a Romantic: A Coleridgean's View," *The Wordsworth Circle* 25, no. 2, pp. 108-9.

[40] This distinctive characteristic of Hopkins' interpretation of the doctrine of the Incarnation will be parsed below. At this time, suffice it to say that it is Scotian in origin. Hopkins' stress, when speaking of the Incarnation, was not so much on the union of two natures, human and divine. Rather, it was on the Scotian tenet that there are 'two' Incarnations: an Incarnation into 'aeonian' time and Christ's

opment of this sacramental vision is due to his fascination with the consequences of the Incarnation of Christ. Towards the end of his life he wrote in his spiritual journal that "my life is determined by the Incarnation down to most of the details of the day."[41] Spiritually speaking, he shared this fascination with Ignatius and derived it from the *Exercises*.

The incarnational principle in Hopkins is best expressed in his 'kingfisher' poem:

> As kingfishers catch fire, dragonflies draw flame;
> As tumbled over rim in roundy wells
> Stones ring; like each tucked string tells, each hung bell's
> Bow swung finds tongue to fling out broad its name;
> Each mortal thing does one thing and the same:
> Deals out that being indoors each one dwells;
> Selves — goes itself; *myself* it speaks and spells,
> Crying *What I do is me: for that I came.*
>
> I say more: the just man justices;
> Keeps grace: that keeps all his goings graces;

historical Incarnation into ordinary time. As Lichtmann says, "Christ's historical Incarnation is superseded by an Incarnation not into human existence but into simple matter." See Maria R. Lichtman, "The Incarnational Aesthetic of Gerard Manley Hopkins," *Religion and Literature* 23, no. 1 (1991), p. 38. In a sense, this was the reason the Eucharist held particular importance for Hopkins. He wrote that "the first intention of God outside himself or, as they say, *ad extra*, outwards, the first outstress of God's power, was Christ;...Why did the Son of God go thus forth from the Father not only in the external and intrinsic procession of the Trinity but also by an extrinsic and less than eternal, let us say aeonian one? — To give God glory and that by sacrifice, sacrifice offered in the barren wilderness outside of God... The sacrifice would be the Eucharist, and that the victim might be truly victim like, motionless, helpless, or lifeless, it must be in matter." See Christopher Devlin, S.J., ed. *The Sermons and Devotional Writings of Gerard Manley Hopkins* (London: Oxford University Press, 1959), p. 197.

[41] Christopher Devlin, S.J., ed. *The Sermons and Devotional Writings of Gerard Manley Hopkins* (London: Oxford University Press, 1959), p. 263.

Acts in God's eye what in God's eye he is —
Christ. For Christ plays in ten thousand places,
Lovely in limbs, and lovely in eyes not his
To the Father through the features of men's faces.[42]

In this poem, one discerns not only the stress on particularity found in the writings of Ruskin, but also an underlying incarnational theme revealing itself through particularity. For Hopkins, "the enfleshment of the divine took place not only in history but was taking place, literally, throughout all places of nature within all beings created to give God praise."[43] Through the Incarnation continually occurring in the world of nature and in human beings, Christ becomes, in some sense, the *inscape* or individuating 'pattern' of each particular being. This rather radical statement will be developed below. For now, suffice it to say that this incarnational principle in Hopkins was focused through the prism of the *Spiritual Exercises*.[44]

E. Ignatius and Hopkins' Christocentric Particularism

In the *Spiritual Exercises*, one finds the religious equivalent of Ruskin's stress on the particular. As Walter Ong writes, "the Exercises are full of particularities, and their particularism is relevant to Hopkins' particularist mindset."[45] Certainly, the particularities

[42] Norman H. MacKenzie, ed. *The Poetical Works of Gerard Manley Hopkins* (Oxford: Clarendon Press, 1990), p. 141.

[43] Maria R. Lichtman, "The Incarnational Aesthetic of Gerard Manley Hopkins," *Religion and Literature* 23, no. 1 (1991), p. 37.

[44] Robert J. Barth, "Hopkins as a Romantic: A Coleridgean's View," *The Wordsworth Circle* 25, no. 2 (Spring, 1994), p. 109.

[45] Walter J. Ong, S.J., *Hopkins, the Self, and God* (Toronto: University of Toronto Press, 1986), p. 67. Also see Virginia Ridley Ellis, *Gerard Manley Hopkins and the Language of Mystery* (Columbia: University of Missouri Press, 1991), pp. 56-7.

in the *Spiritual Exercises* are methodological and procedural. Ignatius is concerned about the details of the exercitant's life, whether with the position of the body in prayer, the forms of address one uses in prayer, what one eats or drinks, or how one works and plays. However, Ignatius' particularist mindset extended beyond organization and procedure. It extended even to how one uses the imagination and envisions the life of Christ. One instance, among many, is Ignatius' insistence on detailed representation during the 'composition of place'. For example, in the contemplation on the Nativity, the exercitant is directed to construct a mental representation stressing particulars:

> *The Second Prelude*. The composition, by imagining the place. Here it will be to see in imagination the road from Nazareth to Bethlehem. Consider its length and breadth, whether it is level or winds through valleys and hills. Similarly, look at the place or cave of the nativity: How big is it, or small? How low or high? And how is it furnished?[46]

Ignatius' call for the intensive application of the five senses, whether imagining the persons involved, hearing what they might say, or even touching or smelling something, shows his emphasis upon what Ong refers to as paying "imaginative attention to physical detail."[47]

There are many examples of Hopkins following Ignatius' instructions to particularize. In his retreat notes of 1888, he contemplates Jesus' baptism by John as suggested in the *Spiritual Exercises*:

> The penitents then went down into the water, but this was their own act and for the symbol this was far from enough. John was the Baptist and must baptise them. For this probably he used *affusion*, throw-

[46] George E. Ganss, S.J., ed. *Ignatius of Loyola: The Spiritual Exercises and Selected Works*, (New York: Paulist Press, 1991), p. 150.

[47] Walter J. Ong, S.J., *Hopkins, the Self, and God* (Toronto: University of Toronto Press, 1986), p. 71.

> ing water on them, and for this some shell or scoop, as he is represented. And he seems to allude to this in contrasting himself with Christ; *ego quidem aqua baptizo...cuius ventilabrum in manu eius* Luke iii. 16,17. — *he* baptises with breath and fire, as wheat is winnowed in the wind and sun, and uses no shell like this which only washes once but a fan that thoroughly and forever parts the wheat from the chaff. For the fan is a sort of scoop, a shallow basket with a low back, sides sloping down from the back forwards, and no rim in front, like our dustpans, it is said. The grain is either scooped into this or thrown in by another, then tossed out against the wind, and this vehement action St. John compares to his own repeated 'dousing' or affusion. The separation it makes is very visible too: the grain lies heaped on the side, the chaff blows away the other, between them the winnower stands; after that nothing is more combustible than the chaff, and yet the fire he calls unquenchable. It will do its work at once and yet last, as this river runs forever, but has to do its work over again...[48]

Ong rightly observes that Hopkins intensifies Ignatius' attention to physical and psychological detail.[49] He builds on it, especially when it comes to the explication of personhood and selfhood. For Ignatius, and more explicitly, for Hopkins, the most particular of particulars is the self. Each person is unique in the world and before God. Hopkins made the following salient notes while composing place for the contemplation of the Last Supper. He elaborates upon Ignatius' instructions to see the persons at the Supper and to listen to what they are saying:[50]

> As all places are at some point of the compass and we may face towards them: so every real person living or dead or to come has his

[48] Christopher Devlin, S.J., ed. *The Sermons and Devotional Writings of Gerard Manley Hopkins* (London: Oxford University Press, 1959), pp. 267-8. For further examples, see Hopkins' notes of 1881-2 on pp. 122-209.

[49] Walter J. Ong, S.J., *Hopkins, the Self, and God* (Toronto: University of Toronto Press, 1986), p. 74.

[50] George E. Ganss, S.J., ed. *Ignatius of Loyola: The Spiritual Exercises and Selected Works*, (New York: Paulist Press, 1991), p. 167.

> quarter in the round of being, is lodged onewhere and not anywhere, and the mind has a real direction towards him. We are to realise this here of 'the persons of the Supper': as we have got the orientation of the room, its true measurements and specifications, properly furnished it and so on, so now we are properly to people it and give it its true personallings. It is in this way that Scotus says God revealed the mystery of the Trinity that His servants might direct their thoughts in worship toward, determine them, pit them, upon real terms, which are Persons, of His being the object of that worship.[51]

As noted above, there is a broad incarnational theme underlying the particularism of Hopkins and this theme is focused through the prism of the *Spiritual Exercises*. The *Exercises* center upon the Incarnation of the Word. Ignatius' particularism, his attention to physical and psychological detail, is partially explained by his close attention to the belief that the Word of God became an historically real human being. As Ong writes:

> "Since Jesus' life consisted of real events in real places and time, to grasp its full meaning it helps to make his life as realistically concrete in detail as possible," for Ignatius believed that "each and every one of the specific actions in Jesus' life were, individually, infinitely salvific... and every one of his actions both spoke truth and was truth, being a concrete manifestation of the Father's Word."[52]

This Ignatian belief is brought home by a passage in his *Autobiography*:

> When this was over, returning to where he had been before, he felt a strong desire to visit Mount Olivet again before leaving... On Mount Olivet there is a stone from which Our Lord rose up to heaven, and his footprints are still seen there; this was what [Ignatius] wanted to see again.

[51] Christopher Devlin, S.J., ed. *The Sermons and Devotional Writings of Gerard Manley Hopkins* (London: Oxford University Press, 1959), p. 186.

[52] Walter J. Ong, S.J., *Hopkins, the Self, and God* (Toronto: University of Toronto Press, 1986), p. 74-5.

> So without saying anything or taking a guide (for those who go without a Turk as guide run a great risk), he slipped away from the others and went alone to Mount Olivet. But the guards would not let him enter. He gave them a penknife that he carried, and after praying with great consolation, he felt the desire to go to Bethphage. While there he remembered that he had not noted on Mount Olivet on what side the right foot was, or on what side the left. Returning there, I think he gave his scissors to the guards so they would let him enter.[53]

Even the question of which direction the feet of Jesus were pointing at the moment he ascended into heaven was of importance to Ignatius!

Like Ruskin, Ignatius wanted to get at the way things really are. However, the intentions of the two were different. Ruskin felt that attention to detail was necessary for the artist to see and thus feel the beauty of a thing, God's revelation in a thing. Ignatius' attention to detail centered on the person of Christ. As Ong comments:

> Ignatius zeroes in on detail as a means of implementing or recalling or extending a personal relationship… Ignatius, like the Victorians, valued an accurate report of reality, but whereas their concerns were most often essentially aesthetic, his were devotional and existential, profoundly personal, concerning the self of Jesus in relation to one's own self.[54]

Ong's comment is not totally accurate with respect to John Ruskin. Afterall, Ruskin's aim was that accurate seeing would lead to art that was truly revelatory and, thus, truly praiseful. Nonetheless, Hopkins' Ruskinian tendencies here were met and transformed by similar tendencies in Ignatius. Based on the Incarnation, the par-

[53] George E. Ganss, S.J., ed. *Ignatius of Loyola: The Spiritual Exercises and Selected Works*, (New York: Paulist Press, 1991), p. 88.

[54] Walter J. Ong, S.J., *Hopkins, the Self, and God* (Toronto: University of Toronto Press, 1986), p. 75.

ticularism of Ignatius turned that of Hopkins toward the person of Christ. All things glorify God by acting out their 'creative destiny', their *inscape*. The Incarnation of Christ, the kenotic nature of the Word become flesh, touches all creation, especially human beings. The vision of created reality becomes at the same time a vision of the Creator. When the 'kingfisher poem' claims that "Each mortal thing does one thing and the same: /Deals out that being indoors each one dwells" and "Crying *What I do is me: for that I came*", one should hear echoes of Jesus' words to Pilate, "for this was I born, for this I came into the world" (John 18:37).[55] Barth concludes:

> Clearly, Hopkins learned well the lesson of Ignatius, which is the lesson of the Incarnation itself, that it is in the specificities of creation and of our own experience that we most deeply encounter both the world and God. The Incarnation of Christ has touched the whole creation to new life.[56]

In stanza 29 of *The Wreck of the Deutschland*, Hopkins attributes the following to the tall nun, one of the five Franciscan nuns who went to their death on a stormy sea:

> Ah! there was a heart right!
> There was single eye!
> Read the unshapeable shock night
> And knew the who and the why;
> Wording it how but by him that present and past,
> Heaven and earth are word of, worded by?—
> The Simon Peter of a soul! to the blast
> Tarpeian-fast, but a blown beacon of light.[57]

[55] Robert J. Barth, "Hopkins as a Romantic: A Coleridgean's View," *The Wordsworth Circle* 25, no. 2 (Spring) (1994), p. 110.

[56] Ibid.

[57] Norman H. MacKenzie, ed. *The Poetical Works of Gerard Manley Hopkins* (Oxford: Clarendon Press, 1990), p. 126. Some commentators hold that "the structure of Ignatian meditation underlies the poem, governing the sequence of thought

Here is the new poet's vision under Ignatius' influence: "Wording it how but by him that present and past, Heaven and earth are word of, worded by." Through the Incarnation, creation is 'word' of Christ and 'worded' by Christ. Creation is 'sacramental' in this sense — its very meaning, each thing's very meaning, is rooted in Christ.

First, created reality, especially in its particularity, is 'word' of Christ; i.e., it is revelatory — it conveys the Christ. Second, it is 'worded' by Christ; i.e., its very intelligibility is 'Christically' rooted. The 'common denominator' or 'rhyming foundation' of created reality is the Christ, the meaning-giving and creation-mediating Logos, 'selving', and first 'outstress' of God.

and imagery at its deepest level." Cf. Jay L. Parini, "Ignatius and 'The Wreck of the Deutschland'," *Forum for Modern Language Studies* 11, no. April (1975): 98; David Anthony Downes, *The Ignatian Personality of Gerard Manley Hopkins*, 2nd ed. (Lanham: University Press of America, 1990); and Peter Milward, "'The Wreck' and the Exercises," *English Literature and Language*, no. 12 (1975): 1-19. The thesis is as follows: The structure of the poem displays "a direct correspondence to the *Spiritual Exercises* (whose four weeks of meditation actually constitute three phases: the purgative, the analytical, and the affective). *The Deutschland* begins with a meditation on the poet's own salvation, proceeds to the body of the poem — a central myth, then concludes with a colloquy of praise and a further meditation on salvation as it relates to the whole British nation. But as in the *Spiritual Exercises*, each section is a self-contained meditation," so, too, in *The Deutschland* (Parini, p. 101). I do not disagree with this contention, but it is clearly derivative. Louis Martz, in his *The Poetry of Meditation*, 2nd ed. (New Haven: Yale University Press, 1962), held that there were strong affinities between important early seventeenth-century writers, such as Herbert and Donne, and techniques of meditation then reaching England from the Continent. Most of these techniques were versions of Ignatian meditation. Although Hopkins is compared with Donne, he had no familiarity with him. Parini claims then that the commonality between the two is due to the shared influence of Ignatius (pp. 100, 102). Though this thesis demonstrates the extent to which Hopkins' rigorous formation under the *Exercises* influenced the structure and purposes of his spirituality and poetry, it does little to shed significant light on the nature and depth of his theological thought. These authors tend to make too much of the Ignatian influence. Hopkins was by no means an Ignatian mimic.

Furthermore, in light of this vision, the poet's purpose, life's purpose, is to 'word' reality, to word life, "by him" [i.e., Christ]. Thus, Hopkins' incarnational theology resides not only in his vision and experience of nature, but also in the making of his art. As Lichtmann writes:

> [The] incarnation, the embodiment of spirit in matter, also becomes a principle not only of Hopkins's life but also of Hopkins's poetry. Not merely in their explicit content and subject matter, but in their very form and material substance, his poems embody the meaning of incarnation.[58]

For Hopkins, poetry — wrought and wrenched language — has the capacity to 'capture' the spiritual 'stress' of the Incarnation in the physical world as well as in human experience. Downes observes that, in Hopkins, "poetic language can become a sacramental language in a unique sense."[59] This was in some sense possible for Hopkins because through the Incarnation everything 'rhymes in Christ'. Again, Hopkins found the philosophical grounding for such an esoteric view, as we shall see, in the writings of Duns Scotus.

The poet is to be as the tall nun and have the 'single eye', the eye by which a "symbolic, translucent vision of a sacramental world is perceived and articulated."[60] In a brief commentary on the *Contemplation for Obtaining Love*, Hopkins stated this in another way: "All things therefore are charged with love, are charged with God and if we know how to touch them give off sparks and take fire, yield drops and flow, ring and tell of him."[61] We see this

[58] Maria R. Lichtman, "The Incarnational Aesthetic of Gerard Manley Hopkins," *Religion and Literature* 23, no. 1 (1991), p. 37.

[59] David Anthony Downes, *The Ignatian Personality of Gerard Manley Hopkins*, 2nd ed. (Lanham: University Press of America, 1990), p. 86.

[60] Robert J. Barth, "Hopkins as a Romantic: A Coleridgean's View," *The Wordsworth Circle* 25, no. 2 (Spring, 1994), p. 112.

[61] Christopher Devlin, S.J., ed. *The Sermons and Devotional Writings of Gerard Manley Hopkins* (London: Oxford University Press, 1959), p. 195.

poetic purpose in two of Hopkins' masterful sonnets, *Hurrahing in Harvest* and *The Windhover*.

In *Hurrahing in Harvest*, the earth and the harvest are sacraments, transcendent symbols. The earth, the sky, the harvest and the poet all partake of God's being in some sense — they partake in the reality which they express and render intelligible. They are instressed with the life and energy of the savior:[62]

[62] Robert J. Barth, "Hopkins as a Romantic: A Coleridgean's View," *The Wordsworth Circle* 25, no. 2 (Spring 1994), p. 112. Barth defines the use of sacrament here in a Coleridgean sense. Coleridge speaks of a symbol for God, such as light, as "abiding itself as a living part of the Unity, of which it is representative." Symbols are 'translucent'; God passing through them is as light passing through a stained-glass window—the Eternal reveals itself in and through the temporal. "The colored window and the light of the sun are quite distinct, but in an act of vision they are not separate; the sun and the stained-glass are 'translucent' to one another. We perceive them, not separately, but in a single act of vision. This is 'symbolic' vision: not merely metaphor, in which one reality points to another that remains separate from it; but symbol, in which two realities — distinct but not separate — have become so intimately united that we cannot perceive one without the other. The same light at once reveals both; they are translucent to one another." Cf. p. 111. I find Barth's Coleridgean interpretation of Hopkins' theological and poetic vision useful, but not entirely accurate. Hopkins distilled his sacramental vision from a variety of sources, Ignatius and Duns Scotus being the most important. Sparsely put, Scotus held the theory that God and creatures can be included, in some sense, in the same metaphysical genus; i.e., they both have a concept of being that is univocal, the distinction being the difference of being *per se* and being *per participationem*. This fosters a sacramental view of the world, but the relationship between creature and Creator in this context is somehow much closer than Coleridge's stained-glass window and light. Hopkins put it this way: "Neither do I deny that God is so deeply present to everything...that it would be impossible for him but for his infinity not to be identified with them or, from the other side, impossible but for his infinity so to be present to them. This is oddly expressed, I see; I mean, a being so intimately present as God is to other things would be identified with them were it not for God's infinity or were it not for God's infinity he could not be so intimately present to things." Cf. Christopher Devlin, S.J., ed. *The Sermons and Devotional Writings of Gerard Manley Hopkins* (London: Oxford University Press, 1959), p. 128.

Summer ends now; now, barbarous in beauty, the stooks rise
Around; up above, what wind-walks! what lovely
behaviour
Of silk-sack clouds! has wilder, wilful-wavier
Meal-drift moulded ever and melted across skies?

I walk, I lift up, I lift up heart, eyes,
Down all that glory in the heavens to glean our Saviour;
And, eyes, heart, what looks, what lips yet gave you a
Rapturous love's greeting of realer, of rounder replies?

And the azurous hung hills are his world-wielding shoulder
Majestic — as a stallion stalwart, very-violet-sweet!—
These things, these things were here and but the beholder
Wanting; which two when they once meet,
The heart rears wings bold and bolder
And hurls for him, O half hurls earth for him off under his
feet.[63]

The poetic 'single eye', the poetic purpose, is not as overt in *The Windhover*. The poet and the Savior are 'hidden', but the communication is even more powerful. Dedicated 'To Christ our Lord' the poem reads:

I Caught this morning morning's minion, king-
dom of daylight's dauphin, dapple-dawn-drawn Falcon, in
his riding
Of the rolling level underneath him steady air, and striding
High there, how he rung upon the rein of a wimpling wing
In his ecstasy! then off, off forth on swing,
As a skate's heel sweeps smooth on a bow-bend: the hurl and
gliding
Rebuffed the big wind. My heart in hiding
Stirred for a bird, — the achieve of, the mastery of the thing!

[63] Norman H. MacKenzie, ed. *The Poetical Works of Gerard Manley Hopkins* (Oxford: Clarendon Press, 1990), pp. 148-9.

Brute beauty and valour and act, oh, air, pride, plume, here
  Buckle! AND the fire that breaks from thee then, a billion
Times lovelier, more dangerous, O my chevalier!

  No wonder of it: sheer plod makes plough down sillion
Shine, and blue-beak embers, ah my dear,
  Fall, gall themselves, and gash gold-vermillion.[64]

The poet is awestruck by a power almost beyond his imagining, but as he watches the union or 'buckling' of "brute beauty and valour and act, oh, air, pride, plume", the beauty "buckles" (the verb takes a different sense): the bird tucks its wings and arcs with great speed out of its hovering position. In this moment, the moment of catching beauty, the predator reveals its true self; it reveals its inscape, its 'destiny', its 'Christed' nature. As Corrington words it, the "falcon falcons" and in doing so reveals its inscape, its "Christic-stem."[65] So too does the plough ploddingly making a furrow — it shines in the muck, while the seemingly consumed, blue-black (like a crow's beak) embers in a fire tumble and show their inner radiance, they gash themselves and show gold vermillion. These things do not only point to Christ for Hopkins, they mediate Christ. More than Coleridge's stained-glass window and the light it mediates, the windhover and Christ, all creation and Christ "have so interpenetrated that they must be seen together;" though distinct, "they cannot be separated."[66] In the *Contemplatio ad Amorem*, Ignatius writes:

[64] Norman H. MacKenzie, ed. *The Poetical Works of Gerard Manley Hopkins* (Oxford: Clarendon Press, 1990), p. 144.

[65] Cf. Robert S. Corrington, "The Christhood of Things," *Drew Gateway* 52, no. 1 (1981): 41-47.

[66] Robert J. Barth, "Hopkins as a Romantic: A Coleridgean's View," *The Wordsworth Circle* 25, no. 2 (Spring, 1994), p. 113. Rene Gallet refers to this vision as a form of 'panchristism' in his ""The Windhover" and "God's First Intention Ad Extra"," in *Gerard Manley Hopkins: Tradition and Innovation*, ed. G. Marra, P. Bottalla, F. Marucci (Ravenna: Longo Editore, 1991), p. 58.

> I will consider how God dwells in creatures; in the elements, giving them existence; in the plants, giving them life; in the animals, giving them sensation; in human beings, giving them intelligence; and finally, how in this way he dwells also in myself, giving me existence, life, sensation, and intelligence; and even further, making me his temple, since I am created as a likeness and image of his Divine Majesty…
>
> I will consider how God labors and works for me in all creatures on the face of the earth; that is, he acts in the manner of one who is laboring. For example, he is working in the heavens, elements, plants, fruits, cattle, and all the rest…[67]

All of this is for a purpose, to draw the loved one to the Lover, the creature to its God, and to elicit love in return, and this love will express itself in the offering of self, in a sense, of Christ, to the Father. Through Hopkins' poetry, smithed by the intentions and influence of Ignatius and the *Exercises,* one draws close not only to creation, but to the Creator, not only to beauty, but to "beauty's self and beauty's giver." It was in the *Exercises*, therefore, that Hopkins found spiritual justification for his "remarkable sensitivity to the beauty of the world about him."[68] Sulloway warmly describes this justification:

> [Hopkins] believed that recalcitrant humans could sometimes be led to their creator if they were first enticed with a cornucopia of the Creator's gifts in nature and in the best of human nature, and if artists were then to recreate these gifts in their own works. His theories of inscape and instress form a lovely sanctified circle… The circle moves from the Creator to the inscapes of natural and human objects and on to the artist's selfless reproduction of them. Artists working reverently and ordinary human beings performing their assigned tasks

[67] George E. Ganss, S.J., ed. *Ignatius of Loyola: The Spiritual Exercises and Selected Works* (New York: Paulist Press, 1991), p. 177.

[68] Maurice B. McNamee, "Hopkins: Poet of Nature and of the Supernatural," in *Immortal Diamond: Studies in Gerard Manley Hopkins*, ed. Norman Weyand (New York: Octagon Books, 1969), p. 226.

with diligence and good will, then complete the loop or circle by dedicating their efforts to their creator. The internal selves of the inscapes of things and people cause morally sensitive artists to instress them in poetry, music, or painting. The results of this instress in turn become fresh inscapes, ready to be instressed by others, who respond to them, at least ideally, by rededicating themselves to the greater glory of God."[69]

F. Conclusion

Intuition and commonsense accept a deep relationship between the religious and the aesthetic. However, giving this association any kind of philosophical and theological basis is another matter, particularly in the wake of the ambivalence with which aspects of Christian history have held things sensory or bodily. Beauty has not always been in the Christian fold, so to speak, and Hopkins was keenly aware of this. He struggled with the relationship between the two in the context of his life as a Jesuit priest. The tenor of the spirituality and religious thought he brought with him to the Society of Jesus, as well as some of the philosophical emphases he encountered through Pater,[70] were also disjunctive in nature; i.e.,

[69] Alison G. Sulloway, ed. *Critical Essays on Gerard Manley Hopkins*, ed. Zack Bowen, in *Critical Essays on British Literature* (Boston: G.K. Hall & Co., 1990), p. 6.

[70] Pater emphasized the autonomous value of each aesthetic experience. In this sense, he was a proponent of aestheticism as religion. Thus, while sharing Ruskin's stress on particularism, he did not make aesthetic experience and its communication to others religiously purposeful; i.e., it was not something, per se, of God. Palmira De Angelis in his *L'Imagine Epifanica* (Rome: Bulzone Editore, 1989), p. 17, quotes from Pater's 1868 *Conclusion*: "Every moment some form grows perfect in hand or face; some tone on the hills or the sea is choicer than the rest; some mood of passion or insight or intellectual excitement is irresistibly real and attractive to us, — for that moment only. Not the fruit of experience, but experi-

they did not encourage an intimate joining of theology and aesthetics, of religious and aesthetic experience. Thus, Hopkins came

ence itself, is the end. A counted number of pulses only is given to us of a variegated, dramatic life. How may we see in them all that is to be seen in them by the finest senses? How shall we pass most swiftly from point to point, and be always at the focus where the greatest number of vital forces unite in their purest energy? To burn always with this hard, gem-like flame, to maintain this ecstasy, is success in life." De Angelis stresses Pater's influence on Hopkins (p. 90) and seems to limit Ruskin's influence on Hopkins to their common manner of rigorously and minutely describing what they see (p. 123). De Angelis fails to assess adequately the depth to which Ruskin affected the young Hopkins as well as how Ruskin influenced Pater's aesthetics. I do not accept the thesis, as put forward by Jude V. Nixon in *Gerard Manley Hopkins and His Contemporaries: Liddon, Newman, Darwin and Pater*, ed. Todd K. Bender, 6 vols., *Origins of Modernism: Garland Studies in British Literature*, vol. 5 (New York: Garland Publishing, Inc., 1994), that the influence of Pater was substantial. Hopkins clearly liked Pater and maintained contact with him during his pastoral assignment in Oxford; however, I see no evidence that Pater was a central influence. Rather, I see that from the earliest contact, Hopkins was looking for the intellectual path to overcome Pater's disjunctive approach to aesthetics and theology. For a specific example of the nature of the early relationship of Hopkins and Pater, see Gerald Monsman, "Pater, Hopkins, and Fichte's Ideal Student," *The South Atlantic Quarterly* 70, no. 3 (1971): 365-376. Also see Lesley Jane Higgins, "Hidden Harmonies: Walter Pater and Gerard Manley Hopkins" (Ph.D. Dissertation, Queen's University, 1987). Higgins contextualizes Pater's influence on Hopkins in relation to their common connection to Ruskin. Pater heightened certain Ruskinian elements for Hopkins (p. 233), but Hopkins always contested Pater's rejection of an aesthetics based in metaphysics. For Hopkins, art was religiously and morally purposeful while for Pater it was not. For Hopkins, beauty was an ontological referent and was of God while for Pater it was not. Art was for art's sake, in Pater's eye, and beauty was its own end and reward. Pater's influence then was that of a catalytic 'foil' — his persuasive, at times extreme, views forced Hopkins to think more clearly, scrutinize more penetratingly, and thus helped him harmonize his academic, religious, and aesthetic priorities. M. Billi arrives at a similar conclusion. In Hopkins' 1867 essay, *The Probable Future of Metaphysics*, Billi finds statements "which oppose Pater's philosophy, particularly his conception that what is important is not the permanence of fixed patterns but the transience of experience." Billi continues: "Hopkins challenges what he calls 'the prevalent philosophy of continuity or flux' and resists the Paterian idea of the individual ordering of reality. On the contrary,

by his struggle honestly, for he had a foot deeply planted in each 'world', worlds which in the Victorian age were rapidly dissociating. The towering exception of the age was Ruskin. As we have seen, Ignatius' influence, building on the 'purposeful' aesthetic Hopkins derived from Ruskin, offered the poetic priest the beginnings of a unifying foundation — a foundation built upon the centrality of the Christ and the Incarnation.

I must emphasize the term *beginnings* here. As I have argued, it is tempting to claim too much with regard to the Jesuit influence on Hopkins. I find Joseph Feeney's approach too strong in this regard. Part of what he calls the "Jesuit worldview" in Hopkins is the poet's incarnationalism.[71] This is valid within the context I have

he invokes the principle of a central unity, affirming fixity versus arbitrariness, the organic versus the fragmentary, the absolute versus the personal. If Pater saw a world full of beautiful things, with no significance outside of themselves, for Hopkins these things were part of God's creation, and visible beauty was for him inseparably connected with moral truth." Cf. Mirella Billi, "Hopkins and the Figurative Arts," in *Gerard Manley Hopkins: Tradition and Innovation*, ed. G. Marra, P. Bottalla, F. Marucci (Ravenna: Longo Editore, 1991), p. 75.

[71] Cf. Joseph Feeney, S.J., "The Collapse of Hopkins' Jesuit Worldview: A Conflict between Moralism and Incarnationalism," in *Gerard Manley Hopkins Annual 1992*, ed. Michael Sundermeier (Omaha: Creighton University Press, 1992), 105-126. Feeney takes one of Hopkins recent biographers, R.B. Martin, to task for claiming that Hopkins first realized that matter and God — "the phenomenal world and the realm of the spirit"—were not mutually exclusive when in July or early August of 1872 he discovered "a new framework" in the writings of Duns Scotus. Cf. Robert Bernard Martin, *Gerard Manley Hopkins: A Very Private Life* (New York: G.P. Putnam's Sons, 1991), p. 260. Feeney holds that this unitive "worldview" was established earlier through Hopkins formation via the *Spiritual Exercises* (p. 113, footnote 6). They are both correct. The religious orientation and essential incarnationalism was formed and fostered under Ignatius, but the true intellectual, philosophical grounding for this unitive view did not develop fully until Hopkins discovered Scotus. In this respect, Hopkins was not at all Jesuit, at least not in the eyes of his Jesuit contemporaries. Hopkins attachment to the philosophical and theological leanings of Scotus did not serve him well when it came to his superiors' judgment as to his fitness for further theo-

described above. However, if one must insist on a particular worldview in any theologically developed sense, and if one must further insist on associating this worldview with the traditions of a particular religious order, then I claim Hopkins' was more 'Franciscan' than 'Jesuit'.[72] The true philosophical and developed theological groundings for his incarnationalism were not essentially Jesuit, rather they were, as we shall see, more akin to what is found in the Franciscan traditions of Duns Scotus and even Bonaventure.[73] The

logical studies. Feeney himself argues this point convincingly in his article, "Hopkins' Failure in Theology: Some New Archival Data and a Reevaluation," *The Hopkins Quarterly* 13, no. 3 and 4 (1987): 99-114. This being said, I do think that Martin generally fails in his assessment of the Jesuit influence on Hopkins. His biography shows a general dismissal of the dynamics of the 'spiritual life' in his attempt to psychologically parse the personality and work of Hopkins.

[72] Cf. Renita Tadych, "The Franciscan Perspective in the Nature Poetry of Gerard Manley Hopkins Augmented by the Writings of John Duns Scotus" (Ph.D. Dissertation, Indiana University of Pennsylvania, 1992). Tadych claims that Hopkins' 'worldview', particularly as expressed in his nature poetry, is thoroughly Franciscan and serves to revitalize Scotus' Franciscan theories. Additionally, she claims that Ignatius took the essential content (not method) of the *Exercises* from the "Franciscan stream" of spirituality prevalent in the late Middle Ages (pp. 8-9, 30). Thus, for Tadych, even Ignatian Incarnationalism is, in its roots, Franciscan. Also see Dal-Yong Kim, "Gerard Manley Hopkins' 'Inscape'," *The Journal of English Language and Literature* 36 (1990), p. 630; Christine Goeme, *Jeans Duns Scot où la révolution subtile* (Paris: Fac Editions, 1982), p. 45; and W.H. Gardner, *Gerard Manley Hopkins: A Study of Poetic Idiosyncracy in Relation to Poetic Tradition*, vol. 2 (London: Oxford University Press), p. 229.

[73] With regard to parallels in Bonaventure's thought, see Leonard Bowman, "Bonaventure's Symbolic Theology and Gerard Manley Hopkins' 'Inscapes'," in *Atti del Congresso internazionale per il settimo centenario di san Bonaventura da Bagnoregio* (Rome: Pontificia facoltia teologica san Bonaventura, 1974), 611-18. Also see Hans Urs von Balthasar, *Studies in Theological Style: Lay Styles*, ed. John Riches, trans. Andrew Louth, John Saward, Martin Simon, Rowan Williams, 7 vols., *The Glory of the Lord: A Theological Aesthetics*, vol. 3 (San Francisco: Ignatius Press, 1986), p. 380. I am not aware of Hopkins' ever having read Bonaventure directly; nonetheless, Bowman makes a good case for essential similarities between what he calls Bonaventure's 'symbolic theology' and Hopkins'

Ignatian influence on Hopkins was not primarily theological; rather, it was spiritual and formational in nature.[74]

This conclusion seems utterly inevitable when scholars remember the purpose of the *Spiritual Exercises*. They are formational; they give method and context. They are not a kind of theological or aesthetic *summa*. This simple fact tends to be lost in much of the critical literature on this topic. Too much is laid upon the *Spiritual Exercises* as a major influence with respect to Hopkins' theology and artistry. Thus, we find critical work which summarily dismisses Ignatian influence in reaction to this overly wrought element in Hopkins studies.[75] The *Spiritual Exercises*, as an influence

incarnationalism centered around concepts like inscape and instress. Again, against Feeney, this diminishes the value of using the notion of a 'Jesuit worldview' as a primary cipher for understanding Hopkins theologically.

[74] Cf. Alfred Thomas, S.J., *Hopkins the Jesuit: The Years of Training* (London: Oxford University Press, 1969). This classic work in Hopkins studies offers a detailed description and analysis of how Hopkins was formed, practically speaking, through his years of training in the Society of Jesus. Suarezian Thomism was most likely the template for Hopkins' studies in the Philosophate and Theologate (pp. 96-99 and 157). He spoke little of it in contrast to his overwhelming enthusiasm for what he found in Scotus. His sole comment on the study of scholastic logic was that it "takes all the fair part of the day and leaves one fagged at the end for what remains. This makes the life painful to nature." Cf. Claude Colleer Abbott, ed. *Further Letters of Gerard Manley Hopkins Including his Correspondence with Coventry Patmore*, 2nd ed. (London: Oxford University Press, 1956), p. 234. His opinion of Suarez was offered in a letter to Dixon: "Suarez is our most famous theologian: he is a man of vast volume of mind, but without originality or brilliancy; he treats everything satisfactorily, but you never remember a phrase of his, the manner is nothing." Cf. Claude Colleer Abbott, ed. *The Correspondence of Gerard Manley Hopkins and Richard Watson Dixon*, 2nd ed. (London: Oxford University Press, 1955), p. 95.

[75] Cf. A.R. Coulthard, "Gerard Manley Hopkins: Priest vs. Poet," *The Victorian Newsletter* 88, no. Fall (1995): 35-40. In making this statement, I have the work of David Downes primarily in mind. In fairness to Downes, however, he himself recognizes the controversial extent to which he promotes Ignatian influence in Hopkins as well as the possible reaction that the "agnostic critic or reader" may have. Cf. David Anthony Downes, *The Ignatian Personality of Gerard Manley Hopkins*, 2nd ed. (Lanham: University Press of America, 1990), p. 140.

on Hopkins' theological thought, should be approached within the broader context of both Hopkins' pre-Jesuit formation and post-novitiate theological 'eccentricities'.[76] Succinctly put, Hopkins used other sources to supply the primary theological and philosophical underpinnings for his theologico-aesthetical synthesis. The two figures found among these underpinnings are primarily John Ruskin and John Duns Scotus, not Ignatius.

The writings of Hopkins resonate with the spiritual force and emphasis of Ignatius. He is Hopkins' mentor of the inner life. He transformed Hopkins' generic 'sacramental' vision of reality, derived from Ruskin, into a more specific Christocentric vision. Still, although Hopkins found encouragement in Ignatius to use the Incarnation as a kind of interpretive screen for his experience of reality and 'beauty', he did not find either the aesthetic or philosophical reasons for doing so in his spiritual father.

Hans Urs von Balthasar wrote astutely that for Hopkins "the *Exercises* were the breath of life, an ever fresh occasion for self-examination and conversion, an occasion too for the most daring speculations, which, however, betrayed more interest in Scotus than in the father of his order. Hopkins and Ignatius do not look one another full in the face."[77]

[76] Cf. Norman White, "Poet and Priest: Gerard Manley Hopkins, Myth and Reality," *Studies* 79, no. 314 (1990), pp. 144-46. White states boldly, yet correctly, that Hopkins was not orthodox in many elements of his life, including his philosophical and theological proclivities.

[77] Hans Urs von Balthasar, *Studies in Theological Style: Lay Styles*, ed. John Riches, trans. Andrew Louth, John Saward, Martin Simon, Rowan Williams, 7 vols., *The Glory of the Lord: A Theological Aesthetics*, vol. 3 (San Francisco: Ignatius Press, 1986), p. 374.

CHAPTER III

HOPKINS AND DUNS SCOTUS: "THE RAREST-VEINED UNRAVELLER"

A. Introduction

Towery city and branchy between towers;
Cuckoo-echoing, bell-swarmed, lark-charmed, rook-racked,
river-rounded;
The dapple-eared lily below thee; that country and town did
Once encounter in, here coped and poised powers;

Thou hast a base and brickish skirt there, sours
That neighbour-nature thy grey beauty is grounded
Best in; graceless growth, thou hast confounded
Rural rural keeping — folk, flocks, and flowers.

Yet ah! this air I gather and I release
He lived on; these weeds and waters, these walls are what
He haunted who of all men most sways my spirits to peace;

Of realty the rarest-veined unraveller; a not
Rivalled insight, be rival Italy or Greece;
Who fired France for Mary without spot. [1]

Hopkins thus paid poetic tribute to John Duns Scotus. Scotus was "the rarest-veined unraveller" because he gave philosophical

[1] Norman H. MacKenzie, ed. *The Poetical Works of Gerard Manley Hopkins* (Oxford: Clarendon Press, 1990), p. 156. This poem celebrates Hopkins' admiration for Scotus. It is entitled *Duns Scotus's Oxford*.

ground for Hopkins' essential empiricism and realism.[2] Hopkins held that the world had existence beyond the constructs of the human mind yet was directly and intuitively accessible to the mind. Scotus gave him reason, better than any other philosopher, to affirm these tenets.[3] Additionally, Scotus undertook the philosophical validation of the reality and revelatory uniqueness of individual things — a theme already dear to Hopkins the artist. Thus, Hopkins was exhilarated and grateful when, in August of 1872, during his Philosophate, he happened upon Scotus' two commentaries on the *Sentences* of Lombard. In his journal entry following this discovery he wrote:

> At this time I had first begun to get hold of the copy of Scotus on the Sentences in the Baddely library and was flush with a new stroke of enthusiasm. It may come to nothing or it may be a mercy from God. But just then when I took in any inscape of the sky or sea I thought of Scotus.[4]

Later, in a letter responding to his friend Bridges' enthusiasm for Hegel, he wrote:

> I have no time to read even the English books about Hegel, much less the original… I do not afflict myself much about my ignorance

[2] Cf. Hans Urs von Balthasar, *The Glory of the Lord: A Theological Aesthetics*, vol. 3, *Studies in Theological Style: Lay Styles*, ed. John Riches, (San Francisco: Ignatius Press, 1986), p. 354.

[3] Cf. Bernadette Ward, "Philosophy and Inscape: Hopkins and the Formalitas of Duns Scotus," *Texas Studies in Literature and Language* 32, no. 2 (1990), p. 214.

[4] Humphry House, ed. *The Journals and Papers of Gerard Manley Hopkins*, 2nd ed. (London: Oxford University Press, 1959), p. 221. Again, the parallel with Teilhard de Chardin is startling. In conversations with Gabriel Allegra, Teilhard discovered more about Duns Scotus' thought on the primacy of Christ and was excited to find the theology of Scotus converging with his own concept of the Cosmic Christ. Additionally, Allegra introduced Teilhard to Hopkins' poem about Duns Scotus. Teilhard was delighted. Cf. Mary Zoghby, O.S.M., "The Cosmic Christ in Hopkins, Teilhard, and Scotus," *Renascence* 24 (Autumn 1971), pp. 33-34.

> here, for I could remove it as far as I should much care to do, whenever it became advisable, hereafter, but it was with sorrow I put back Aristotle's Metaphysics in the library some time ago feeling that I could not read them now and so probably should never. After all I can, at all events a little, read Duns Scotus and I care for him more even than Aristotle and more *pace tua* than a dozen Hegels.[5]

In the then obscure and suspect theological forefather, Hopkins found a kindred intellect and spirit who helped him unite "the priest and the poet aspects of his being."[6] Shortly after his discovery, Hopkins began solidifying a bond between the aesthetic and the theological within the context of his interpretation of the Subtle Doctor.

In this chapter, I will demonstrate that Scotus is a central, if not determining, influence upon Hopkins' theological aesthetics. Nonetheless, long before Hopkins read Scotus he already embodied in his poems and journals a sense of the 'thisness' or individual distinctiveness of things, and the conviction that this characteristic was somehow revelatory of their divine origin.[7] The terms he coined in his pre-Scotian years — *inscape, instress, selving* — express his "feeling for the ultimate presence waiting to be discovered at the core of every created thing."[8] We have noted this

[5] Claude Colleer Abbott, ed. *The Letters of Gerard Manley Hopkins to Robert Bridges*, 2nd ed. (London: Oxford University Press, 1955), pp. 30-31.

[6] Rachel Salmon, "Frozen Fire: The Paradoxical Equation of "That Nature is a Heraclitean Fire and of the Comfort of the Resurrection"," in *Critical Essays on Gerard Manley Hopkins*, ed. Alison Sulloway (Boston: G.K. Hall & Co., 1990), p. 23. Also see K. R. Srinivasa Iynegar, *Gerard Manley Hopkins: The Man and the Past* (New York: Haskell House Ltd., 1971), p. 82.

[7] Cf. Alfred Thomas, S.J., "Was Hopkins A Scotist Before He Read Scotus?," in *Scotismus Decursu Saeculorum, De doctrina Ioannis Duns Scoti: Acta Congressus Scotistici Internationalis* (Rome: Cura Commissionis Scotisticae, 1968), pp. 617-29.

[8] Rachel Salmon, "Frozen Fire: The Paradoxical Equation of "That Nature is a Heraclitean Fire and of the Comfort of the Resurrection"," in *Critical Essays on Gerard Manley Hopkins*, ed. Alison Sulloway (Boston: G.K. Hall & Co., 1990), p. 23.

Hopkinsian emphasis in the previous study of Ruskin's influence on Hopkins. Therefore, once again, the influence of Hopkins' pre-Jesuit years must not be ignored.

While recognizing earlier influences, the Hopkins scholar must, nonetheless, honor the primacy of Scotus in Hopkins' mind and life. Hopkins says as much in his writings. Even more, I conclude that Hopkins sacrificed a potential future as a theologian in the Society of Jesus because of his Scotian convictions.[9] There can be

[9] Cf. Joseph Feeney, S.J., "Hopkins' Failure in Theology: Some New Archival Data and a Reevaluation," *The Hopkins Quarterly* 13, no. 3 and 4 (1987): 99-114. Feeney does an excellent job explaining the context for Hopkins' partial 'failure' in his final oral examinations in theology and his assignment to pastoral work soon after. Early biographers were inclined to gloss over this episode. For example, in 1930 G.F. Lahey wrote that Hopkins "completed a successful course of theology at St. Bueno's and left there with the reputation of being one of the best moral theologians among his contemporaries." Cf. Gerard F. Lahey, S.J., *Gerard Manley Hopkins* (London: Oxford University Press, 1930), p. 132. John Pick wrote that "there is no reason why Hopkins' admiration for Scotus should have brought him into any trouble with his fellow Jesuits." Cf. John Pick, *Gerard Manley Hopkins: Priest and Poet* (Oxford: Oxford University Press, 1942), p. 157. Even Alfred Thomas hesitated to lay Hopkins' failure totally at the feet of his penchant for Scotus. He raised the issue of Hopkins' "dubious state of health" as a contributor. Cf. Alfred Thomas, S.J., *Hopkins the Jesuit: The Years of Training* (London: Oxford University Press, 1969), pp. 181-82. Although the claim that there were pressing pastoral needs and that Hopkins had sufficient theological training to meet them may be true, it seems an insufficient reason to explain why he was not permitted a fourth year of theological studies. In various letters to friends, Hopkins certainly presumed that he would study theology for four years. Cf. Claude Colleer Abbott, ed. *Further Letters of Gerard Manley Hopkins Including his Correspondence with Coventry Patmore,* 2nd ed. (London: Oxford University Press, 1956), pp. 82, 122, 242; and *The Letters of Gerard Manley Hopkins to Robert Bridges,* 2nd ed. (London: Oxford University Press, 1955), p. 30. I believe Feeney's original contention that Hopkins' passion for Scotus is the reason for his theological 'failure' best fits the facts. Seemingly, Hopkins insisted on defending his Scotian leanings during his final examinations. His Suarezian professors were not impressed, gave him low marks, and thereby essentially barred Hopkins from pursuing advanced theological studies for the Society. Addition-

little doubt that the path to an analysis of Hopkins' theological aesthetics runs through his interpretation and possible assimilation of Scotist doctrine.

Still, even with regard to Scotus, Hopkins did not imitate. As with Ignatius and the *Spiritual Exercises*, Hopkins applied his interpretation of Scotus to an already existent, if nascent, framework of his own original thought.[10] The main question, then, does not cen-

ally, this corresponds with the fact that Hopkins was already seen as a singularly intelligent, if eccentric, individual by his superiors, and that those superiors ultimately assessed that Hopkins was best suited for the life of an academic. Why would an academic be held back from further studies unless there was some impediment? Finally, of his fourteen classmates, ten were allowed to enter the fourth year of theological studies when Hopkins was not. If pastoral needs were so great, it is difficult to imagine that this would have been the case. Therefore, I do not see the need to mistrust the assessment of those who knew him at University College in Dublin when they noted: "As a theologian [Hopkins'] undoubted brilliance was dimmed by a somewhat obstinate love of Scotist doctrine, in which he traced the influence of Platonist philosophy. His idiosyncrasy got him into difficulties with his Jesuit preceptors who followed Aquinas and Aristotle." Cf. Thomas, *Hopkins the Jesuit*, p. 183. Here Thomas is quoting *A Page of Irish History: Story of University College, Dublin, 1883-1909*, compiled by the Fathers of the Society of Jesus (Dublin and Cork, 1930), p. 105.

[10] Cf. Alfred Thomas, S.J., "Was Hopkins A Scotist Before He Read Scotus?," in *Scotismus Decursu Saeculorum, De doctrina Ioannis Duns Scoti: Acta Congressus Scotistici Internationalis* (Rome: Cura Commissionis Scotisticae, 1968), 617-629. Thomas contends that there was already a great deal of similarity between Hopkins' approach to reality and some of Scotus' key tenets. W.H. Gardner recognized this in his preface to Hopkins' poems: "In the subtle Schoolman's 'principle of individuation' and 'theory of knowledge' [Hopkins] found what seemed to be corroboration of his own theory of inscape and instress." Cf. W.H. Gardner and N.H. MacKenzie, eds. *The Poems of Gerard Manley Hopkins*, 4th ed. (London: Oxford University Press, 1967), p. xxi. Although Hopkins' enthusiasm was due to his discovery of a kindred intellect, I believe there was another reason. In the life of Scotus, Hopkins saw himself. Hopkins was aware that Scotus too had been marginalized and rejected over time. This opinion is substantiated by Hopkins' last reference to Scotus in his writings, which I take to be autobiographical: "And so I used to feel of Duns Scotus when I used to read him

ter on what Duns Scotus himself meant or did not mean; rather, the issue is how Hopkins understood and used Scotus' teachings relative to his own thought. This issue shapes the main content of this chapter.

B. Hopkins' 'Reading' of Scotus

Before delving into the question of how Hopkins understood and used Scotus' teachings, there is the corollary question of how Hopkins 'read' Scotus. Was Hopkins a critical reader of Scotus? In an unpublished letter contained in the Bischoff Collection and dated July 23, 1948, Anthony Bischoff, S.J. writes to Christopher Devlin, S.J. about this issue:

> The impression grows on me that GMH was, with few exceptions, a "flighty" reader, a skimmer, one who read nervously and with frequent interruptions. He seldom read thoroughly: he read enough to be stimulated, then often went off on speculations of his own. Hence, I am inclined to conjecture that he "browsed" through Scotus, noting mainly, if not only, those headings that were of particular interest to him, not reading the *Commentarius* in any particularly methodical, thorough manner. GMH's mind, whatever else it was, does not

with delight: he saw too far, he knew too much; subtlety overshot his interests; a kind of feud arose between genius and talent, and the ruck of talent in the Schools finding itself, as his age passed by, less and less able to understand him, voted that there was nothing important to understand…" Cf. Claude Colleer Abbott, ed. *Further Letters of Gerard Manley Hopkins Including his Correspondence with Coventry Patmore*, 2nd ed. (London: Oxford University Press, 1956), p. 349. Some scholars do not accept Hopkins' apparent self-confirmation of Scotus' influence. Bender, for example, writes that "the ideas of Hopkins are fortuitously congruent to those of Scotus rather than derived from him." Todd K. Bender, *Gerard Manley Hopkins: The Classical Background and the Critical Reception of His Work* (Baltimore: The Johns Hopkins University Press, 1966), p. 38.

> seem to have been methodical. As you will, I think, agree, he had come upon his notions of inscape, stress, and instress long before he "discovered" Scotus. His early diaries, including the parts I discovered last year, note the inscapes, the essences, the individuality, the "thisness" of things. Scotus came as a confirmation of much that he had already thought out. The basic attraction to Scotus was that in him Hopkins found a mind akin to his own, a man who delighted, as he did, in bypaths, originality, daring, etc. [11]

Bischoff's impression may be accurate. Although Hopkins praised Scotus consistently, he also continually referred to how little time he had for reading and for completion of the various scholarly works he hoped to produce.[12] This seems to have been the case throughout his training, pastoral work, and teaching career as a Jesuit. Additionally, although Scotus is given highest praise by Hopkins, for what specifically we are never told. This seems out

[11] Anthony Bischoff, S.J. was a pre-eminent and frustrating Hopkins scholar. Widely recognized as a foremost Hopkins expert in his time, he seldom published although there was always the promise and expectation of a definitive work on Hopkins by him. Nonetheless, he was a prominent collector of primary materials concerning Hopkins and a mentor to many budding Hopkins scholars. Christopher Devlin, S.J. is still considered to be the foremost scholar on Hopkins' Scotism. The Bischoff Collection is maintained at Gonzaga University in Spokane, Washington.

[12] E.g., Claude Colleer Abbott, ed. *Further Letters of Gerard Manley Hopkins Including his Correspondence with Coventry Patmore*, 2nd ed. (London: Oxford University Press, 1956), pp. 251-2; and *The Letters of Gerard Manley Hopkins to Robert Bridges*, 2nd ed. (London: Oxford University Press, 1955), p. 31. Numerous letters written during his years of training, pastoral ministry, and teaching may be cited. Between 1864 and 1889, the reader continually comes upon expressions such as "time, of which I have so little," "but the time is crowded," "I am so fallen into a sense of employments that I have given up doing everything whatsoever but what is immediately before me to do," "I cannot find time...," "I have no time hardly for anything but school reading," "I have not much time and almost no energy to do anything on my own account," "my time is short," "I reserve my most serious difficulties for want of time," "I have no abundance of time," etc.

of character. Normally, Hopkins did not hesitate to offer critical or praiseful comment on works, whether they were in paint or writing, that he analyzed carefully. This trait manifests itself throughout his letters to Bridges, Dixon, Patmore, and even his own brother. We find Hopkins offering no such specific criticism of Scotus on any point. This may be due to the kind of respect he had for Scotus as a kindred soul, but it may also be due to a lack of critical analysis. Furthermore, Hopkins cited Scotus infrequently and adopted Scotian ideas selectively. Thus, one cannot claim Hopkins as a complete Scotist based on the limited evidence.[13]

In any case, Hopkins seemed to find his life as a Jesuit constraining and exhausting relative to his pursuit of poetry and scholarly endeavors beyond necessity.[14] Certainly, in this context, detailed study of Scotus qualified as 'beyond necessity'. Therefore, I am inclined to give weight to Bischoff's contention that Hopkins may not have been a thorough scholar of Scotus. This does not necessarily lessen Scotus' influence on Hopkins, rather, it contextualizes the influence.

First, Bischoff's contention supports the premise that Hopkins was in no way derivative in his thinking. Hopkins, in short, found Scotus to be an appealing intellectual hero and a compatible substrate for his own thought.[15] This conclusion is not commonly shared in

[13] Cf. Clark M. Brittain, "Logos, Creation, and Epiphany in the Poetics of Gerard Manley Hopkins" (Ph.D. Dissertation, University of Virginia, 1988), p. 4.

[14] Cf. Donald Stanford, "The Harried Life of Gerard Manley Hopkins," *Review* 16 (1994), p. 211.

[15] There is a potential parallel here between Hopkins' 'relation' to Newman and to Scotus. Perhaps part of the reason Hopkins was so devoted to Newman was that their biographies were similar — they had passed through the same fires in their conversions to Catholicism. Similarly, what Hopkins came to feel for Duns Scotus may also have come from a perceived correspondence of biographies. As R.B. Martin surmises, Duns Scotus was an outsider in the Church, just as Hopkins felt himself to be: "If Hopkins in part recognized Newman as his clerical father, he surely felt that in Duns Scotus he had found his theological progenitor,

Hopkins criticism. Rather, Hopkins' clearly indicated intellectual passion for Scotus is more generally transformed into Hopkins having been an assiduous student of Scotus.[16] Although I do not deny this possibility, I also do not believe that one necessarily leads to the other. Influence does not require great assiduousness or specificity.

This leads to a second contextual conclusion inferred by Bischoff's contention. Namely, it is only with great caution that the Hopkins scholar should seek to identify specific terms and ideas used by Hopkins with specific terms and ideas used by Scotus. The fact may be that the influence was general or simply confirming in nature. How, then, was Scotus an influence upon Hopkins? Let us begin by tracing the answers to this question offered in the history of Hopkins criticism.

C. A Critical History of Hopkins' Scotism

Hopkins' Scotism is a widely accepted fact. Additionally, it is commonly accepted that Scotist philosophy became bound up in some way with Hopkins' poetic inspiration.[17] In spite of this, con-

and in an odd way he regarded him as a personal friend filling the gap left by his friends at Oxford." Cf. Robert Bernard Martin, *Gerard Manley Hopkins: A Very Private Life* (New York: G.P. Putnam's Sons, 1991), p. 207. The fact that Newman was dear to Hopkins certainly does not mean that Newman essentially influenced Hopkins thought — he did not. This acts as a cautionary note in the approach to Scotus' influence on Hopkins as well. 'Endearment' does not necessarily engender influence.

[16] For example, see Stephen Edward Wear, "John Duns Scotus and Gerard Manley Hopkins: The Doctrine and Experience of Intuitive Cognition" (Ph.D. Dissertation, The University of Texas at Austin, 1979), p. 223. Here Wear states that "we have good reason to believe that Hopkins was quite widely read in Scotus." Wear's question is whether or not Hopkins had success in reading Scotus; i.e., did Hopkins understand him.

[17] Christopher Devlin, S.J., "Time's Eunuch," *The Month* 1, no. 5 (1949), p. 306.

sensus on how Scotus *specifically* influenced Hopkins has not been achieved. This discordance can be explained by two facts. First, Hopkins wrote little about the direct connection between his thought and particular elements of Scotus' philosophy.[18] Second, the thought of Scotus himself is obscure much less subtle, and the sketchy nature of much of his work makes inquiry difficult. Therefore, attempts to make inarguable connections between specific Scotian concepts and some of Hopkins' unique terms have been few in number, depth, or quality. In any case, the early work in this area of inquiry determined much of what followed. In the main, Hopkinsian criticism on Hopkins' Scotism is derivative and based upon the work of one scholar, Christopher Devlin, S.J.

In 1935, Devlin wrote a short but seminal article on Hopkins and Scotus.[19] This article, written while Devlin was still a student, was his first work on the topic, and he obviously attempted to place as much content as he could into a constraining four page parameter. Unfortunately, the outcome is a dense, nearly indecipherable introduction to what Devlin called "some of the points where the minds of Scotus and Hopkins found themselves in unison."[20]

Devlin took as his starting point the Scotist formal distinction between the Nature in a thing and its individuality. Scotus' posi-

[18] In fact, "documentarily speaking, there are only four significant places in GMH's *spiritual* writings where he certainly refers to Scotus. From these, as from four small bones of a prehistoric monster, must be reconstructed the skeletal outline of the undoubted relations between the two men." Cf. Christopher Devlin, S.J., ed. *The Sermons and Devotional Writings of Gerard Manley Hopkins* (London: Oxford University Press, 1959), p. 338. Also see Sean Flinn, "Scotus and Hopkins: Christian Metaphysics and Poetic Creativity," in *Annual Report of the Duns Scotus Philosophical Association*, ed. Lynn Behl (Cleveland: Our Lady of Angels Franciscan Seminary, 1962), p. 51.

[19] Cf. Christopher Devlin, S.J., "Hopkins and Duns Scotus," *New Verse* 14 (1935): 12-15.

[20] Ibid., p. 13.

tion regarding the nature of things took the form of a kind of substance metaphysics. All substances are integrally comprised of a 'common nature' and a positive individuating principle called the 'haecceitas'. These two 'elements' are inseparable but distinguishable "realitates" of all substances.[21] They are inseparable in that they cannot exist apart, but they can be distinctly posited *in re*; i.e., they are not merely logical distinctions. The common nature is the formal reality of things, and is common in all members of the same species. The 'haecceitas' is the positive, though non-formal, principle of individuation. Being non-formal, its individuation of substance does not destroy the commonness of the nature.

Interpreting the Scotian tenet — "Singulare si esset natum movere intellectum nostrum, esset ad visionem universalem (If we could have vision of a self, we should have vision of all things)" — Devlin wrote:

> Each man's nature is the Nature of all the world, elemental, vegetative, sensitive, human. But one man differs utterly from another because by his Individuality he possesses the common nature in an especial degree. The individual degree is the degree in which he lacks the Infinite; it knits together in one man all his natural activities, animal, rational etc., and gives them direction God-wards.[22]

Additionally, in this first effort, Devlin emphasized the Scotian stress that it is only through humanity that earth can go back to its Creator, and that this "summation" and "mediation" reaches its height in Christ. Devlin wrote that for Scotus, "Christ as Man pos-

[21] Cf. Stephen Edward Wear, "John Duns Scotus and Gerard Manley Hopkins: The Doctrine and Experience of Intuitive Cognition" (Ph.D. Dissertation, The University of Texas at Austin, 1979), p. 228.

[22] Christopher Devlin, S.J., "Hopkins and Duns Scotus," *New Verse* 14 (1935), p. 13. Scotus' approach to "the all" through "the one" is, in method, similar to Ruskin's aesthetic approach. For example, as noted above, Ruskin contended that if an artist could truly see and draw a leaf, he or she could see and draw all.

sesses His created Nature in the highest possible degree summing up all other degrees."[23] Devlin noted a kind of sympathy with this Scotian stress in *The Wreck of the Deutschland* as well as in Hopkins' *Kingfisher* poem:

> Christ — For Christ plays in ten thousand places
> Lovely in limbs, and lovely in eyes not His,
> To the Father through the features of men's faces.[24]

In the article, Devlin then made a short leap from this apparent Hopkinsian sympathy to Scotus' epistemology. Again interpreting Scotus, he wrote:

> The source and the object of all knowledge in man is the common nature which he possesses: it is that which gives colour, warmth, meaning etc., in response to external excitations of the nervous system. Every distinct act of knowing which takes in the adapted world of habit and practical necessity, has been preceded by a first act wherein sense and intellect are one, a confused intuition of Nature as a living whole, though the effect of the senses is to contract this intuition to a particular 'glimpse,' which is called the 'species specialissima.'[25]

Here is where Devlin made the first identification between terms in Scotus and Hopkins. Devlin noted that ordinarily, conation enters in and by abstraction adapts this first 'glimpse', or '*species spe-*

[23] Christopher Devlin, S.J., "Hopkins and Duns Scotus," *New Verse* 14 (1935), p. 13.

[24] Norman H. MacKenzie, ed. *The Poetical Works of Gerard Manley Hopkins* (Oxford: Clarendon Press, 1990), p. 141.

[25] Christopher Devlin, S.J., "Hopkins and Duns Scotus," *New Verse* 14 (1935), p. 13. John Abraham, describing the same thing, writes: "This *primum cognitum* which more efficaciously and forcibly moves the senses and by means of which the singular is known intuitively as existing, Scotus calls the *species specialissima*, a term which the Scotist scholars understand as the "nature as nature" or as the common nature as it is first apprehended in a singular." See John Abraham, "The Hopkins Aesthetic II," *Continuum* 1, no. 2 (1963), p. 359.

cialissima', into knowledge to suit our needs. However, he continued by noting that "if this first act is dwelt on to the exclusion of succeeding abstractions, then you can feel, see, hear or somehow experience the Nature which is yours and all creation's as "pattern, air, melody, — what I call *inscape*."[26] Thus, Devlin made a first equation between Scotus' *species specialissima* and Hopkins' *inscape*.

Devlin concluded that Hopkins was able to hold this first act of knowing in place, as it were, and return to it with abstractive reasoning in order to express it via poetry. "That seems to be why Hopkins's images so tumble over each other intertwining so as to keep pace with and capture a single 'species' which has broken from his consciousness."[27] According to Devlin, in the 'glimpse' one touches Nature, and in Christ is the fullness of Nature. Thus in the glimpse, in the first act of knowing, in the '*species specialissima*', in the inscape, one touches Christ. Devlin concluded that here Hopkins found his justification when he knew the beauty of the Lord by a bluebell or in the flight of a windhover, and mediated their self-expressiveness, their beauty, their Christic-stem, their inscapes, in poetry.

One may now begin to see how aspects of Scotus' theory of knowledge could be adapted for aesthetic purposes by Hopkins. Devlin proposed that Hopkins connected Scotus' first act of knowing, which is a kind of intuitive cognition, with poetic inspiration. Furthermore, in Scotus' thought, this first act as a spontaneous expression of Nature is good of itself, but neither right nor wrong. Its moral 'rightness' comes when it is directed by the individual back to God by an act of love. Hopkins wrote: "Give

[26] Christopher Devlin, S.J., "Hopkins and Duns Scotus," *New Verse* 14 (1935), pp. 13-14.

[27] Ibid., p. 14.

beauty back, beauty, beauty, beauty, back to God, beauty's self and beauty's giver."[28] Only in this way can natural beauty become something more, something transformed, something of immortal value.

Devlin concluded his first article on the topic with brief statements about Scotus' striving to explain how the Son 'personifies' nature, and how, based on Scotus' teaching, beauty and truth became the same goal for Hopkins. Because of Scotus, Devlin claimed, Hopkins' poetry is much more than "affectation in metaphor."[29]

The initial response to Devlin's first foray into the developing world of Hopkinsian criticism came from none other than W.H. Gardner, an early and pre-eminent Hopkins scholar.[30] He agreed with Devlin that because of his theories of individuation and knowledge, Scotus offered Hopkins the poet an aesthetic sanction and Hopkins the priest a moral justification for his inordinate attachment to poetry.[31] Yet Gardner took issue with Devlin's article as all too short and general. Gardner went on to claim that much more must be said of Scotus' teaching on the univocity of being and how this influenced Hopkins' belief and art. In 1942, John Pick, in his excellent, widely read, and still useful biography of Hopkins, affirmed Devlin's general contentions.[32] However, following Gard-

[28] Norman H. MacKenzie, ed. *The Poetical Works of Gerard Manley Hopkins* (Oxford: Clarendon Press, 1990), p. 170. This is from Hopkins' poem *The Leaden Echo and the Golden Echo*.

[29] Christopher Devlin, S.J., "Hopkins and Duns Scotus," *New Verse* 14 (1935), p. 15.

[30] W.H. Gardner, "A Note on Hopkins and Duns Scotus," *Scrutiny* 5, no. 1 (1936): 61-6.

[31] Ibid., pp. 64-5.

[32] John Pick, *Gerard Manley Hopkins: Priest and Poet* (Oxford: Oxford University Press, 1942).

ner, he too wrote that Scotus gave Hopkins justification to "find the One ablaze in the many."[33]

Neither Gardner nor Pick attempted to develop their emphasis on the common thread of the univocity of being in Scotus and Hopkins. Both simply saw the Scotian principle as an apparent background for Hopkins' thought and poetry. Interestingly enough, even though Gardner and Pick held that Scotus' univocity of being was central to Hopkins' thought and work, Devlin never developed this proposed element of commonality between Hopkins and Scotus.

By 1942 then, general connections were being made, if not proven, between Hopkins' writings and the Scotian stresses on the particular, on 'being as being' as the 'object' of the intellect, on the summation of all Nature in Christ, and on the univocity of being. Additionally, Devlin linked Hopkins' '*inscape*' with Scotus' '*species specialissima*', thereby inferring connections in Hopkins between the discernment of beauty, and the discernment of truth, of Christ, and of the revelatory character of poetic expression. Nonetheless, the connections were neither grounded nor well developed.

Accordingly, not all writers were convinced by Devlin's first claims. For example, in 1945 Arthur Little, S.J., insisted that the depth of the relation between nature and God in Hopkins' poetry was due to a kind of mystical experience, and not to any philosophical justification he found in Scotus. He wrote:

> Scotus' authority persuaded [Hopkins] to continue his effort to find such a portrait [of God] in created things, encouraging him to another trial by professing to overthrow the proofs of its hopelessness. Though Scotus's case was unconvincing in itself it distracted a mind desirous to believe in it from the convincingness of St. Thomas's.[34]

[33] Ibid., p. 36.

[34] Arthur Little, S.J., "Hopkins and Scotus," *Irish Monthly* 71, no. 836 (1945), p. 53.

With regard to any claims about Hopkins' poetry, Little dismissed Devlin and Pick by saying that art [poetry] is simply a prereligious discipline, a propaedeutic to worship. "It gives experimental acquaintance with the spiritual, not God, but the human soul in its acts."[35] Hopkins' intention with regard to the purpose of his poetry, if otherwise, was mistaken. Certainly, according to Little, Pick's and Devlin's opinions of the nature of Hopkins' poetry were mistaken. It was unthinkable that Hopkins could have intended anything more than poetry giving an "experimental acquaintance" with God.

In a series of articles written between 1946 and 1950, Devlin finally developed in earnest the connection between the work of Scotus and the thought and poetry of Hopkins which he had hinted at in his 1935 article.[36] His corpus on this topic culminated in the commentaries and appendixes of his 1959 edition of Hopkins' sermons and devotional writings.[37] It is primarily upon these articles, commentaries, and appendixes that most further work on and conclusions about Hopkins' Scotism have been based. According to Devlin's mature work, then, what influence did Scotus have upon Hopkins' aesthetic vision?

[35] Ibid., p. 59. Perhaps Little's attitudes expressed in his article of 1945 do not fall far from the attitudinal tree of Hopkins' Jesuit contemporaries vis-à-vis Scotus and poetry.

[36] In order, the articles were "An Essay on Scotus," *The Month* 182, no. 954(1946): 456-66; "Time's Eunuch," *The Month* (ns)1, no. 5 (1949): 303-12; "The Image and the Word — I," *The Month* (ns)3, no. 2 (1950): 114-27; "The Psychology of Duns Scotus: A Paper Read to the London Aquinas Society on 15 March, 1950," *The Aquinas Papers* 15 (1950), and "The Image and the Word — II," *The Month* (ns)3, no. 3 (1950): 191-202.

[37] See Christopher Devlin, S.J., ed. *The Sermons and Devotional Writings of Gerard Manley Hopkins* (London: Oxford University Press, 1959). The long periods of time between Devlin's various works on Hopkins are explained by his services as a military chaplain during World War II and as a missionary in Africa. Cf. Madeleine Devlin, *Christopher Devlin* (London: Macmillan Co., 1970).

D. Two Centers of Scotian Influence

We have already seen that Devlin attached Hopkins' early emphasis on the individuality of things to Scotian epistemology and its stress on individuality, or 'thisness', as intrinsic to being. Earlier in this work, I attempted to show that Hopkins' general attachment to the particular stemmed from Ruskin's influence and was bolstered by Ignatius' own variety of particularism. In Scotus, Hopkins found some metaphysical ground for his own inclination. Being, beauty, truth, Christ — these are met in knowing the individual and particular, not the general and universal.

A second strain of influence developed by Devlin concerns the incarnationalism of Scotus and Hopkins.[38] Hopkins did not see God the Son's descent into creation primarily as a reparation for sin. He saw it as an 'aeonian' act of love which would have taken place whether or not there was sin.[39] This is the heart of a primary the-

[38] Christopher Devlin, S.J., ed. *The Sermons and Devotional Writings of Gerard Manley Hopkins* (London: Oxford University Press, 1959), pp. 6, 109-15, 338.

[39] Conceptualizing this temporally or spatially is trying. The incarnation is in what Hopkins refers to as 'aeonian' or angelic time (the scholastic aevum). The angelic aevum can be described as standing to our time as a three-dimensional space would stand to a two-dimensional surface. Every movement in the space can touch the whole or any part of the surface, while possessing a depth outside the receptive power of the surface. Thus a single event in angelic duration could be coincident with the whole of our time, and a series of events (in depth) in angelic duration could be coincident with a single instant in our time; therefore, one might argue that Christ's aeonian Incarnation and his earthly incarnation were parallel executions of the same divine intention. Hopkins refers to the aeonian Incarnation as *ensarkosis,* the 'taking of flesh', while he refers to the earthly Incarnation as *enanthropesis*, 'the becoming man'. Additionally, by this aeonian Incarnation, Hopkins thinks that Christ was 'materially' present in the same way as he is present in the Eucharist — as a material substance without actual extension. The reference here could be Scotus' *Oxoniense*, iv, dist. 10, qu. 4: "I say then, but without insisting on it, that before the Incarnation and 'before Abraham was',

ological theme for Hopkins, namely, 'the Great Sacrifice'. As God, the Word could not perform the act of an inferior nature; so from all eternity the Word willed to become a creature so as to express that aspect of love which was impossible as God alone. This sacrifice of love incidentally became a sacrifice of sorrow and reparation.[40] In a sense, for Hopkins, and apparently for Scotus, creation was dependent upon the decree of the Incarnation. This helps explain Hopkins' early interest in the presence of God's design or inscape (that is Christ in some sense) in inanimate nature as well as his later interest in the working out of this design, by stress and instress, in the minds and wills of people.[41] In any case, again one finds continuity between earlier emphases in Hopkins' thought and his connections with Scotian incarnationalism. The particular variety of Ruskin's romantic strain and the incarnationalism of Ignatius begged, in Hopkins' mind, for some metaphysical and theological base. This he found in Scotus.

in the beginning of the world, Christ could have had a true temporal existence in a sacramental manner. And if this is true, it follows that before the conception and formation of the Body of Christ from the most pure blood of the Glorious Virgin there could have been the Eucharist." Devlin further comments: "GMH appears to mean that Christ was present to the angels in the same sort of way as He is present in the Eucharist. Such substantial presence is impossible to ordinary human or angelic knowledge. But to God in His eternity all things are present in their actual substance. According to the thought of GMH God communicated this kind of presence to the angels." Cf. Christopher Devlin, S.J., ed. *The Sermons and Devotional Writings of Gerard Manley Hopkins* (London: Oxford University Press, 1959), p. 112.

[40] Christopher Devlin, S.J., ed. *The Sermons and Devotional Writings of Gerard Manley Hopkins* (London: Oxford University Press, 1959), p. 109.

[41] Ibid. Although it is not a particular theme of this dissertation, the interplay between Scotus and St. Paul is of interest here. Hopkins apparently leans heavily upon Paul, in associating Christ's created nature with an original pattern for creation. Creation is through the Word, the Christ — we and, in some sense, wider creation are patterned after Christ. Devlin offers Ephesians 2:10 as a summation of this Hopkinsian thesis (p. 341).

Effectively, Devlin ultimately centered his studies upon two central areas of commonality between Hopkins and Scotus — their incarnationalism and their epistemology.[42] In view of Hopkins' pre-Scotus inclinations, these 'centers of influence' are at least sensible points of entry into the question of how Hopkins may have assimilated Scotian principles.

D.1. Hopkins' Scotian Incarnationalism

On January 5th, 1888, Hopkins wrote: "My life is determined by the Incarnation down to most of the details of the day." Additionally he stated that he was "only too willing... to help on the knowledge of the Incarnation."[43] He said these things in a reflection on the pervasive influence of the Incarnation on the outward history of the world and the inward history of individuals. The Incarnation was foundational for Hopkins, not only as a religious belief, but ultimately as a metaphysical, aesthetic principle. All that happens in the world is marked "as a great seal" by the Incarnation.[44] For Hopkins, the permeating themes and the major questions of life and art are defined by the idea that the Word was made flesh. What he took from Scotus here is a kind of metaphysical expan-

[42] Devlin ultimately claims that Hopkins' central identity with Scotus is in Scotus' theology of the Incarnation. Cf. Christopher Devlin, S.J., ed. *The Sermons and Devotional Writings of Gerard Manley Hopkins* (London: Oxford University Press, 1959), p. 351. Curiously enough, however, Devlin spends the majority of his time attempting to parse Scotus' epistemology and its influence on Hopkins. Cf. Christopher Devlin, S.J., "The Image and the Word–I," *The Month (ns)* 3, no. 2 (1950): 114-27 and Devlin's "The Image and the Word–II," *The Month (ns)* 3, no. 3 (1950): 191-202. Giving him the benefit of the doubt, this could be due to what he terms as the "almost hopeless task to try and give a coherent account in modern terminology of Scotus' theory of knowing."

[43] Christopher Devlin, S.J., ed. *The Sermons and Devotional Writings of Gerard Manley Hopkins* (London: Oxford University Press, 1959), p. 263.

[44] Ibid.

siveness with regard to the Incarnation. For Hopkins, as for Scotus, the "enfleshment of the divine took place not only once in history but was taking place, literally, throughout all places of nature and within all the beings created to give God praise."[45]

The Incarnation is a continual reality because it is a root or primordial reality vis-à-vis creation. Following Scotus, as we have seen above, Hopkins held that the Incarnation was an 'aeonian' rather than 'historical' event. The first Incarnation was that of the Word into matter (Hopkins' *ensarkosis* or "word becoming flesh"). The Word 'seals' matter in a sacramental sense. The first 'Great Sacrifice' was not the *enanthropesis* ('the word becoming man'), rather it was the Word emptying itself into matter.[46] Hopkins

[45] Maria R. Lichtman, "The Incarnational Aesthetic of Gerard Manley Hopkins," *Religion and Literature* 23, no. 1 (1991), p. 37.

[46] Hopkins wrote: "The first intention then of God outside himself, or, as they say, *ad extra*, outwards, the first outstress of God's power, was Christ; ...Why did the Son of God go thus forth from the Father not only in the external and intrinsic procession of the Trinity but also by an extrinsic and less than eternal, let us say aeonian one? — To give God glory and that by sacrifice, sacrifice offered in the barren wilderness outside of God... This sacrifice and this outward procession is a consequence and shadow of the procession of the Trinity, from which mystery sacrifice takes its rise; ...It is as if the blissful agony or stress of selving in God had forced out drops of sweat or blood, which drops were the world, or as if the lights lit at the festival of the 'peaceful Trinity' through some little cranny striking out lit up into being one 'cleave' out of the world of possible creatures. The sacrifice would be the Eucharist, and that the victim might be truly victim like, motionless, helpless, or lifeless, it must be in matter." Christopher Devlin, S.J., ed. *The Sermons and Devotional Writings of Gerard Manley Hopkins* (London: Oxford University Press, 1959), p. 197. Hopkins' 'incarnational distinctions' *may* have been taken from Athanasius' *de Incarnatione Verbi*, thus apparently giving some support to Brittain's claim that the major influence in this theological area was not so much Scotus as the patristic theology Hopkins' absorbed from Tractarian sources at Oxford (Pusey, Liddon, etc.). Cf. Clark M. Brittain, "Logos, Creation, and Epiphany in the Poetics of Gerard Manley Hopkins" (Ph.D. Dissertation, University of Virginia, 1988). Again, I must disagree with Brittain. Even though some of Athanasius' writings contain images and a

inferred that this union of 'flesh and word' in the *ensarkosis* is a kind of eucharistic Incarnation. The Word, the Christ, incarnates continually in the world of nature and in people as their individuating design or 'inscape'. As Christ is the first outstress or 'selving' of God, so creatures are, in some sense, the outstress or selving of Christ in his created nature.[47]

This stress on the radical *kenosis* of the 'eucharistic' or aeonian Incarnation allowed Hopkins to put unusual emphasis on the human, the earthly, and the natural.[48] He expressed this in the phrase "God's infinity/ Dwindled to infancy" in his poem *The Blessed Virgin Compared to the Air We Breathe*.[49] God the Word's

unified vision of creation, incarnation, and redemption closely resembling elements found in Hopkins' writing, Hopkins makes no mention of Athanasius, or of particular Oxford influences, in this matter. If anyone from the patristic era influenced Hopkins directly, it may have been Origen. We know that Hopkins admired Origen for his fervor (as he did Savonarola). Additionally, Origen's writings contain images of Christ as 'the First Born', of a world (kosmos) containing principles (logoi), and calls the Son 'the image of the invisible God according to which all others have been made'. Cf. Michael E. Allsopp, "G.M. Hopkins, Narrative, and the Heart of Morality: Exposition and Critique," *The Irish Theological Quarterly* 60 (1994), p. 306, n. 25. Brittain's thesis offers an interesting comparative analysis, but remains purely conjectural in identifying a direct source of theological influence in Hopkins.

[47] Here I touch upon an area of Hopkins' work which is of great importance, but beyond the specific scope of this monograph. In human beings, created in and through the Word, and singular in their summative nature and in their freedom of will, how do will, grace, and Christic inscape mesh? Ultimate beauty for Hopkins was, in human beings, moral beauty. People becoming what they are — Christ — was ultimate self-expressiveness and beauty. People are created for Christ just as Christ's created nature is for God (Devlin, *Sermons*, p. 196). People take part in their own selving or creation by corresponding with grace and seconding God's Christic design in themselves (p. 197).

[48] Cf. Maria R. Lichtman, "The Incarnational Aesthetic of Gerard Manley Hopkins," *Religion and Literature* 23, no. 1 (1991), p. 40.

[49] Norman H. MacKenzie, ed. *The Poetical Works of Gerard Manley Hopkins* (Oxford: Clarendon Press, 1990), p. 173.

self-limitation is, as Lichtmann puts it, "nearly infinite" in Hopkins. She notes that Hopkins' version of the Incarnation "denies God not only divinity but humanity as well, tumbling God from the height of infinity through intelligence and life down to mere Being alone."[50]

Lichtmann's exaggerated expression is useful. By emphasizing the aeonian Incarnation, Hopkins gave himself a kind of philosophical/theological touchstone for both his priestly and poetic vocations. The *kenosis* involved here gave him not only a moral paradigm for his priesthood, but also an artistic license for delving into and dwelling upon earthly reality, for as a foundation of creation, the 'eucharistic' Incarnation consecrates created nature. For Hopkins, the world indeed is "charged with the grandeur of God."[51]

Hopkins' incarnational metaphysic and theology, adapted from Scotus, thus became a kind of aesthetic principle.[52] The Incarnation, the pervasive Word in matter, became a defining principle for Hopkins' poetry.[53] For example, in *Hurrahing in Harvest*, we read:

[50] Maria R. Lichtman, "The Incarnational Aesthetic of Gerard Manley Hopkins," *Religion and Literature* 23, no. 1 (1991), p. 40.

[51] Norman H. MacKenzie, ed. *The Poetical Works of Gerard Manley Hopkins* (Oxford: Clarendon Press, 1990), p. 139.

[52] Teilhard de Chardin shared this centrality of a 'primordial Incarnation'. He expressed a need in dogmatic theology for articulating a Christology which would be in keeping with the dimensions of the universe as we know it. This would mean "a recognition that, along with those strictly human and divine attributes chiefly considered by theologians up to now, Christ possesses, by virtue of the mechanism of the Incarnation, attributes which are universal and cosmic, and it is these which constitute him the personal Centre hypothetically invoked by the physics and metaphysics of evolution." Cf. Mary Zoghby, O.S.M., "The Cosmic Christ in Hopkins, Teilhard, and Scotus," *Renascence* 24, no. Autumn (1971), p. 38.

[53] Lichtmann puts this differently: "Hopkins's notion of a preexistent eucharistic Incarnation affects his way of seeing matter and his way of incarnating Christ in his poems." Cf. Lichtmann, "The Incarnational Aesthetic of Gerard Manley Hopkins," *Religion and Literature* 23, no. 1 (1991), p. 40. I find Prof. Lichtmann's

I walk, I lift up, I lift up heart, eyes,
Down all that glory in the heavens to glean our Saviour;
And eyes, heart, what looks, what lips yet gave you a
Rapturous love's greeting of realer, of rounder replies?

And the azurous hung hills are his world-wielding shoulder
Majestic — as a stallion stalwart, very violet-sweet —
These things, these things were here and but the beholder
Wanting; which two when they once meet,
The heart rears wings bold and bolder
And hurls for him, O half hurls earth for him off under his feet.[54]

Thus, if the purpose of poetry is to convey inscape, as Hopkins held, and if the discernment of inscape is, in part, the experience of Christ's Incarnation in matter, then poetry becomes more than a theological or religious medium. It becomes, as an *effective mediation* of the Word, sacramental.

We have yet to investigate how Hopkins held that words can be the 'matter' of this 'sacrament'. At this juncture, suffice it to say that Hopkins held that poetry, not only in its subject matter and

work thought-provoking but extreme. She holds that Hopkins' stress on the radical kenosis of the eucharistic (aeonian) incarnation denies God both divinity and humanity, thus creating a kind of absence of God which leaves only the merest trace, or, as she writes, "less than Bonaventure's footprints." Though she expresses a trend of thought found elsewhere in Hopkinsian criticism [see, for example, Joseph Hillis Miller's section on Hopkins in his *The Disappearance of God: Five Nineteenth-Century Writers* (Cambridge: Harvard University Press, 1963)], I believe that one would be more faithful to Hopkins' stress, particularly during the time of his life when he was enamored of Scotus, if one emphasized the fullness of the presence of the Logos in creation brought about by this aeonian Incarnation as opposed to stressing the absence implied by this *kenosis*. For an example of this emphasis, see Viola Papetti, "The Figure of Mary in the Poetry of Hopkins," in *Gerard Manley Hopkins: Tradition and Innovation*, ed. G. Marra P. Bottalla, F. Marucci (Ravenna: Longo Editore, 1991), p. 180.

[54] Norman H. MacKenzie, ed. *The Poetical Works of Gerard Manley Hopkins* (Oxford: Clarendon Press, 1990), p. 149.

content, but in its form and rhetoric, embodies inscape. He wrote that "poetry is in fact speech only employed to carry the inscape of speech for the inscape's sake — and therefore the inscape must be dwelt on… even over and above its interest of meaning."[55] He further wrote, with respect to poetry, that "each word is one way of acknowledging Being and each sentence by its copula *is*…the utterance and assertion of it."[56] Thus, Hopkins contended implicitly that poetic language can be sacramental, as is created being, and that the "merely formal, material character [of language] carries its most precious meaning — the logos of Christ."[57] Lichtmann perceptively contends:

> The poem, for Hopkins, is the Body of Christ. It is the Eucharist in the sense of bearing the motionless, lifeless Real Presence of Christ, of acting with sacramental, transforming instress on the reader as Hopkins has himself been instressed by nature…[58]

The Incarnation is the key to Hopkins' thought and art. Yet, to understand this is not to understand completely the foundations of Hopkins' theological aesthetics. Because of his Scotian Incarnationalism, Hopkins was able to justify his Ruskinian 'seeing'— his aesthetic focus upon nature (including humanity) — as being of spiritual worth and of value for others. However, how did he justify the validity of 'what' he saw? That is, epistemologically, how did Hopkins link his concept of 'inscape' and knowing? The answer to this question constitutes the second major potential influence of Scotus upon Hopkins.

[55] Humphry House, ed. *The Journals and Papers of Gerard Manley Hopkins*, 2nd ed. (London: Oxford University Press, 1959), p. 289.

[56] Ibid., p. 42.

[57] Maria R. Lichtman, "The Incarnational Aesthetic of Gerard Manley Hopkins," *Religion and Literature* 23, no. 1 (1991), p. 47.

[58] Ibid., p. 48.

D.2. Hopkins' Scotian Epistemology

We have seen above how Devlin, in his 1935 article, first made some general connections between aspects of Scotus' epistemology and Hopkins' concept of inscape. However, Devlin did not develop these connections until his articles of 1950.[59]

Devlin noted Hopkins' reaction after he had first begun to get hold of Scotus' *Sentences*. Hopkins stated that whenever he took in any inscape of the sky or sea, he thought of Scotus. Why would this be so? If Hopkins paged through *Book I* of the *Opus Oxoniense* available to him in 1872, then he probably started at the beginning, which deals with the origin of knowing. Even if Hopkins was cursorily skimming the early distinctions in *Book I*, he would most likely have been drawn to the topic, for he had just completed his first year of philosophy, which dealt in part with epistemology. Therefore, it is here that Devlin logically looked for Hopkins' first connections between inscape and Scotus' philosophy.[60]

As we have seen, Devlin focused early upon one specific connection between Hopkins and Scotus. That connection was Scotus' concept of the *species specialissima*. Sparsely put, Scotus implied that every distinct act of knowing is preceded by a first act wherein sense and intellect are one, a kind of confused intuition of Nature as a living whole, though the effect of the senses is to contract this

[59] See Christopher Devlin, S.J., "The Image and the Word–I," *The Month (ns)* 3, no. 2 (1950): 114-27, and "The Image and the Word–II," *The Month (ns)* 3, no. 3 (1950): 191-202.

[60] Interestingly enough, Scotus' epistemology, not his incarnationalism, was the first common thread Devlin gleaned from his study. Only later did Devlin claim that the essential core of Hopkins' and Scotus' commonality was their Incarnationalism. Additionally, even though Gardner and Pick, as we have seen, focused on 'univocity of being' as a common element, Devlin did not significantly take this up in his later studies. The development of this element was left to others.

intuition to a particular 'glimpse'. This glimpse is called the *species specialissima*, and for Scotus it is immediately given in the sensation of a thing.[61]

This specific connection between Scotus and Hopkins cannot be proven in my opinion. However, it does correspond with the **general** epistemological context one finds in Hopkins. Hopkins' poetry and writings express his acceptance of the 'intuitive cognition' implied by the *species specialissima*. Hopkins seemed to hold that through sensation there can be a kind of intuitive knowing or insight that puts one in immediate touch with what is real and actually existing. Granted, this sensation is vague and confused, but it is direct and non-abstractive.[62] Generally put, in this 'insight' there is no distinction between sense and intellect and no opposition between spirit and matter.[63] Perhaps, in part, this first act of knowing, which he found in Scotus, encouraged Hopkins to accept metaphysically that spiritual reality could be imbibed directly through the senses.[64]

I focus on this general aspect of Scotus' epistemology because it seems foundational for what Hopkins had in mind when he postulated *inscape*. Devlin chose to be more specific about Scotus' influence by claiming that "if there is any Scotist equivalent for

[61] Christopher Devlin, S.J., "The Image and the Word–I," *The Month (ns)* 3, no. 2 (1950), p. 117.

[62] Hopkins, like Scotus, certainly does not deny that there is abstractive or distinct knowing. He only infers that such knowing follows a more primitive 'intuitive' insight. Art generally, and poetry specifically, in their best form, can carry this insight. We can see how this epistemological contention could support Hopkins's early acceptance of Ruskin's aesthetical methodology that focused on intense 'seeing'.

[63] Christopher Devlin, S.J., "The Image and the Word–I," *The Month (ns)* 3, no. 2 (1950), p. 119.

[64] Christopher Devlin, S.J., "The Image and the Word–II," *The Month (ns)* 3, no. 3 (1950), p. 191.

Hopkins's inscape, it is the *species specialissima*."[65] Unlike Devlin, I cannot use the word 'equivalent' here.[66] Again, Hopkins coined the term 'inscape' before he discovered Scotus. We must remember that Hopkins was a self-taught Scotist who very well may have read into Scotus a vision of the world received, in part, from other sources. Attempting to interpret Hopkins as a 'strict' Scotist may lead to misinterpretations — particularly if the scholar attempts too much specificity. The best one can say is that Scotus gave some metaphysical affirmation for Hopkins' experience and developing thought. Nonetheless, Scotus spoke philosophically and formally

[65] Ibid., p. 197. Most Hopkins scholars reflecting on this topic summarily equate 'inscape' with another concept in Scotus, namely, *haecceitas* (or 'thisness'). See, for example, W. A. M. Peters, *Gerard Manley Hopkins: A Critical Essay towards the Understanding of his Poetry* (London: Oxford University Press, 1948), p. 23. Peters is a much quoted Hopkins scholar and many have adopted his identification of inscape with 'haecceitas'. The difficulty is that 'haecceitas' is not strictly associated with the first act of knowing as such, and it seems that Hopkins' aesthetic connection between sensation, being, Christ, inscape, and beauty must reside at this primary level. This is not to deny the importance of 'haecceitas' and its potential influence on Hopkins. Scotus' stress on the individual must have caused echoes in Hopkins who already tended in this direction (as did Ruskin). The point of fact missed by Peters, simply put, is that '*haecceitas*' is not something we can 'know' as such in Scotian thought. Therefore, it would be unlikely that Hopkins had this in mind as a Scotian correlation to inscape.

[66] One reason for my hesitance is that the *species specialissima* as a possible equivalent for Hopkins' *inscape* does not seem to honor Hopkins' emphasis on the importance of concrete detail in order to capture *inscape*. If, as Devlin ultimately claims in his "The Psychology of Duns Scotus: A Paper Read to the London Aquinas Society on 15 March, 1950," *The Aquinas Papers* 15 (1950), the glimpse of universal nature that is involved in the *species specialissima* is an "innate image of the Ideal [i.e., Christ]" toward which human intelligence strives, then, as Bowman chides, a "vision of Christ would be as readily obtained from the *species specialissima* of elm trees, leopard spots, garbage cans, or bits of morning toast floating on the Thames." The concreteness of a thing becomes somewhat dispensable in this case. Cf. Leonard Bowman, "Bonaventure and the Poetry of Gerard Manley Hopkins," in *Philosophica*, ed. Jacques Guy Bougerol (Rome: Collegio S. Bonaventura Grottoferrata, 1973), p. 556.

about intuitive cognition while Hopkins informally emphasized intuitive knowing as an *actual experience*.

In short, what Scotus does best for Hopkinsian criticism is to give the scholar an entrance into the heart and goal of Hopkins' aesthetics. Claiming the *species specialissima* as a specific gateway, Devlin writes:

> The *species specialissima* does not represent any particular individual, a woman, a horse, or a mulberry tree, as it is afterwards known by abstraction and secondary images; no more does inscape. It represents the Ideal Person to whom universal nature tends.[67]

While I cannot confirm Devlin's specific claim here, I can claim with him that ultimately inscape is 'Christic'. Hopkins certainly held, developing Scotus, that one can intuitively know and experience, via sensation, the Christic-stem of things. This is the heart of his aesthetic stress. As Devlin continues:

> In "inscape"... there is a momentary contact between the Creative Agent who *causes* habitual knowing in me, and the created individual who *terminates* it in my actual insight. And the medium of this contact, if I am correct, is the *species specialissima*, the dynamic image of nature being created. In easier language: the poet, if the original motion of his mind is unimpeded, does perhaps see things for a moment as God sees them...[68]

[67] Christopher Devlin, S.J., "The Image and the Word–II," *The Month (ns)* 3, no. 3 (1950), p. 197. Devlin may have misread Scotus here. Bowman points out that Scotus did not teach that there is a kind of "one absolutely first *species specialissima* that virtually includes all others" (i.e., a vision of totality and Christ). The *species specialissima* is simply "the knower's grasp of the aspect of the nature of a thing that most efficaciously and strongly moves the senses." See Leonard Bowman, "Bonaventure and the Poetry of Gerard Manley Hopkins," in *Philosophica*, ed. Jacques Guy Bougerol (Rome: Collegio S. Bonaventura Grottoferrata, 1973), p. 556.

[68] Christopher Devlin, S.J., "The Image and the Word—II," *The Month (ns)* 3, no. 3 (1950), p. 197. One could argue whether such an unimpeded state exists. Both Scotus and Hopkins gave examples which infer that it does (p. 198). A more

While honoring his specific claim about the importance of Scotus' *species specialissima*, placing the focus simply on the possibility of intuitive cognition is, in my opinion, Devlin's true contribution to understanding how Scotus may have influenced Hopkins and his theological aesthetics. Hopkins clearly claimed to have intuitive experience of things. It would be easy to accept his claim as a form of idiosyncratic mysticism and, therefore, dismiss it as singular and inapplicable.[69] However, Hopkins did not see that his experience was idiosyncratic, and he attempted to root his claim and experience in metaphysics and theology — thus the centrality of Scotus. Still, with his particular emphasis on the experience of intuitive knowing and how it relates to poetry, Hopkins moved in his own orbit beyond Scotus. Let us, therefore, take a closer look at how he understood the intuitive act.

E. Hopkins and Intuitive Cognition

Ruskin, as I have claimed, is the starting point in the study of Hopkins' theological aesthetics. Specifically, Ruskin is the fountainhead of the search for the meaning of intuitive cognition in

interesting question is whether or not this initial insight can be 'held' in such a way as to be the object of verbalization in poetry. Interestingly enough, here we find a definite parting of ways between Hopkins and Scotus. Scotus formally held that there is the possibility of intuitive cognition, but he seems to have downplayed the reality of such cognition due to original sin. Hopkins, on the other hand, believed and emphasized that such cognition is possible, though not easily.

[69] Gabriel Allegra, commenting on both Hopkins and Teilhard de Chardin, attributed their approach to 'phenomena' to their "mystical cast of mind." As mystical expressions, Allegra seemingly held their writings to be of great value, but as theological expressions, they were suspect. See Mary Zoghby, O.S.M., "The Cosmic Christ in Hopkins, Teilhard, and Scotus," *Renascence* 24, no. Autumn (1971), p. 38.

Hopkins.[70] We saw above how, like Ruskin, Hopkins, through intense 'sensation' or 'seeing', was after the 'visible laws' of things. He sought to master individual forms of specific realities, and in doing so, he sought beauty. For Hopkins, as for Ruskin, the beauty, or 'self-expressiveness', of individual forms was revelatory. Thus he took joy in finally discovering 'the law' of an oak tree or a bluebell. In this sense, for Ruskin and Hopkins, nature was a 'holy book' for the artist and 'theophany' for the seer.

At this early stage, Hopkins had not yet coined the term 'inscape'. Furthermore, following Ruskin, there was a certain emphasis on the accidental properties of things when speaking of their 'laws'. In fact, when we first find the term 'inscape' in his writings, Hopkins seemed to use it in part to describe the 'order' or 'law' that he found anywhere.[71] He did not always use the term in an 'essential' manner.

However, later in his journal, the use of 'inscape' shifted. Early on, he used the term to describe some basic form which gave him insight into a particular aspect of something.[72] The usage of the

[70] I find myself in agreement with V.R. Ellis when she states that we find a "a Scotist confirmation of Ruskin's ideas" in Hopkins' elucidation of inscape and intuitive cognition. See Virginia Ridley Ellis, *Gerard Manley Hopkins and the Language of Mystery* (Columbia: University of Missouri Press, 1991), p. 31.

[71] E.g., Hopkins speaks of inscape in the order of rows of trees on a river flat, or of the "flowing and careless inscape" of an evening sky, or of the "flowing and well marked inscape" of cut grass. For these examples, see Humphry House, ed. *The Journals and Papers of Gerard Manley Hopkins*, 2nd ed. (London: Oxford University Press, 1959), pp. 189, 218, and 227 respectively. These uses of inscape all focus on accidental attributes as opposed to a deeper sense of some 'internal' inscape he intuits.

[72] He used it synonomously with the term 'law'. Thus, an appropriate example is once again his study of oak trees. When he claimed that "I have now found the law of the oak leaves," he has grasped some kind of 'accidental commonality' of the trees. The 'law' of the oak leaves "is of platter-shaped stars altogether." See Humphry House, ed. *The Journals and Papers of Gerard Manley Hopkins*, 2nd ed. (London: Oxford University Press, 1959), p. 148.

term, however, became even more profound. Hopkins began seeing these particular 'laws' or aspects of things as aids in grasping deeper, more basic inscapes of whole entities.[73] Finally, Hopkins explicitly emphasized "the existence of one basic inscape over and above whatever particular aspects may be seen."[74] In the course of this development, there was a constant. Hopkins claimed a kind of intuitive experience of a thing accomplished by some sort of "holding of the intellect on the level of sensation, and not allowing it a further act on its own."[75]

How did Hopkins describe and justify this? There is only one place in his writings where Hopkins went into any detail or offered any context. It is in a section of his commentary on the *Spiritual Exercises* where he describes what contemplation of the persons, words, and actions surrounding the nativity means.[76] Here Hopkins couches his epistemology in the context of three classic intellec-

[73] He speaks, for example, of noticing the "flow and slow splaying of the stream of spots down the backbone" of a leopard, and that seeing this "flow" allows him "to inscape the whole animal." See Humphry House, ed. *The Journals and Papers of Gerard Manley Hopkins*, 2nd ed. (London: Oxford University Press, 1959), p. 244.

[74] Stephen Edward Wear, "John Duns Scotus and Gerard Manley Hopkins: The Doctrine and Experience of Intuitive Cognition" (Ph.D. Dissertation, The University of Texas at Austin, 1979), p. 105. The example that Wear gives here is a journal entry from 1875: "Rembrandt/...a master of scaping rather than of inscape. For vigorous rhetorical but realistic and unaffected scaping holds everything but no arch-inscape is thought of.' See Humphry House, ed. *The Journals and Papers of Gerard Manley Hopkins*, 2nd ed. (London: Oxford University Press, 1959), p. 245.

[75] Stephen Edward Wear, "John Duns Scotus and Gerard Manley Hopkins: The Doctrine and Experience of Intuitive Cognition" (Ph.D. Dissertation, The University of Texas at Austin, 1979), p. 132. Cf. Humphry House, ed. *The Journals and Papers of Gerard Manley Hopkins*, 2nd ed. (London: Oxford University Press, 1959), p. 75.

[76] Cf. Christopher Devlin, S.J., ed. *The Sermons and Devotional Writings of Gerard Manley Hopkins* (London: Oxford University Press, 1959), pp. 173-75.

tual powers: memory, understanding, and will. Hopkins, following Scotus, conflates the essence of the intellect and its powers in stating that "all three faculties are mind, intellect, *nous*."[77] He describes each intellectual faculty as follows:

> Memory is the name for that faculty which toward present things is Simple Apprehension and, when it is a question of the concrete only, the faculty of Identification; towards past things is Memory proper; and towards things future or things unknown or imaginary is Imagination.
>
> Understanding… applies to words; it is the faculty for grasping not the fact but the meaning of the thing… This faculty not identifies but verifies; takes the measure of things, brings word of them; is called *logos* or reason.
>
> By the will here is meant not so much the practical will as the faculty of fruition, by which we enjoy or dislike, etc., to which all intellectual affections belong.[78]

Intuitive cognition, or the potential experience of inscape, seems to involve primarily the aspects of memory and understanding. Hopkins expands on these two a bit. He writes that memory, in the sense of simple apprehension or identification, may "just do its office and falls back, barely naming what it apprehends," and then "gives birth to the second," or understanding.[79] Hopkins then goes on to state that the intellect can also maintain itself on the level of apprehension. Regarding the "grasp" by memory of present things, he writes that "when continued or kept on the strain the act of this faculty is attention, advertence, heed, the being *ware*…"[80] By doing this, Hopkins states, the intellect "cannot but continually beget" understanding.

[77] Ibid., p. 174.
[78] Ibid.
[79] Ibid.
[80] Ibid.

Hopkins' 'keeping on the strain' by the faculty of 'memory' rests at the heart of his intuitive cognition of inscape. Wear writes:

> It is the holding of the mind on the level of concrete apprehension, rather than allowing it to take its usual course where understanding reflects on the meaning of the thing only vaguely, momentarily identified in the first act of memory. What Scotus would have called the abstractive process is thus thwarted. The movement from sensation to general concept is replaced by continuing attention to the immediate presentation itself.[81]

Hopkins writes that such intuition occurs when the intellect is "employed upon the object of sense alone and not referring back or performing some wider act within itself."[82] This holding the intellect to the level of sensation Hopkins describes as "attendere, advertere, et contemplari / *to be on the watch for, take notice of, and dwell on.*"[83]

In another context, while describing forms of apprehension by the faculty of memory, Hopkins writes:

> The mind has two kinds of energy, a transitional kind, where one thought or sensation follows another, which is to reason, whether actively as in deliberation, criticism or passively, so to call it, as in reading etc.; (ii) an abiding kind for which I remember no name, in which the mind is absorbed (as far as that may be), taken up by, dwells upon, enjoys, a single thought.[84]

[81] Stephen Edward Wear, "John Duns Scotus and Gerard Manley Hopkins: The Doctrine and Experience of Intuitive Cognition" (Ph.D. Dissertation, The University of Texas at Austin, 1979), p. 135.

[82] Humphry House, ed. *The Journals and Papers of Gerard Manley Hopkins*, 2nd ed. (London: Oxford University Press, 1959), p. 75. This quote is from one of Hopkins' undergraduate essays. Clearly, at that time, he already informally accepted that intuitive cognition is possible and that it is the means by which we perceive beauty.

[83] Christopher Devlin, S.J., ed. *The Sermons and Devotional Writings of Gerard Manley Hopkins* (London: Oxford University Press, 1959), p. 175.

[84] Humphry House, ed. *The Journals and Papers of Gerard Manley Hopkins*, 2nd ed. (London: Oxford University Press, 1959), pp. 125-6.

This 'abiding kind' of mental energy refers to the 'keeping on the strain' by the faculty of the memory. Hopkins goes on to call this mental 'abiding' simply "contemplation."[85] It results in a heightened form of intellectual awareness: "It is a mood of great, abnormal in fact, mental acuteness, either energetic or receptive, according as the thoughts which arise in it seem generated by a stress and action of the brain, or to strike it unasked."[86] Here Hopkins refers to the way of knowing necessary for poetic inspiration. Additionally, he speaks of this contemplative knowing as somehow relying, in part, upon the 'action' of the object contemplated. The object 'strikes into it unasked'; i.e., no matter what pre-conditions must exist in the subject to experience inscape, there is some type of 'action' by the object — a sort of spontaneous and powerful striking of the mind. Although Hopkins does not speak of 'instress' in this context, it seems that this 'action by the object' is part of what he means by that term. His humorous observation that 'what you look hard at seems to look hard at you' should be seen in this light. It would seem, therefore, that even given the universality of inscape for Hopkins, and the personal attempt to remain at the level of sensation, contemplation is not the result only of intent observation. The spontaneous 'strike' by the object is also required. Therefore, even while accepting the possibility of intuitive cognition, Hopkins does not see it as simply a mechanical process.[87] In the subject-object dynamic that can lead to intuitive cognition, it appears that, for Hopkins, the experience of inscape involves a

[85] Humphry House, ed. *The Journals and Papers of Gerard Manley Hopkins*, 2nd ed. (London: Oxford University Press, 1959), p. 126.

[86] Claude Colleer Abbott, ed. *Further Letters of Gerard Manley Hopkins Including his Correspondence with Coventry Patmore*, 2nd ed. (London: Oxford University Press, 1956), p. 216.

[87] Stephen Edward Wear, "John Duns Scotus and Gerard Manley Hopkins: The Doctrine and Experience of Intuitive Cognition" (Ph.D. Dissertation, The University of Texas at Austin, 1979), p. 138.

kind of apprehension by the mind which is counter to its usual habit of abstraction. In short, intuitive cognition requires certain pre-conditions on the part of the subject, a kind of training, in order to overcome the habit of immediate abstraction. This hearkens back to Hopkins' acceptance of Ruskin's admonition that the artist must learn to see, and that true 'seeing' is not simply a natural talent but rather something that must be learned. Unfortunately, Hopkins does not offer much explanation on how this 'habit' of immediate abstraction is overcome. He simply affirms that the sensory world is too significant to be relegated to the subservient task of being only a source of ideas and examples — only 'food for thought'. Rather, for Hopkins, the sensory world is a locus of a more perfect form of knowing.

F. Hopkins' Experience of Inscape

Hopkins found philosophical and theological support in Scotus for his experience. Intuitive cognition, or 'contemplation', was indeed possible in Hopkins' mind. It was in this kind of knowing that inscape, and everything inscape carried, could be known and thus potentially carried over into art. Still, according to Hopkins, the experience of inscape was not common. The reader finds Hopkins inferring how difficult the perception of inscape can be for most people: "I thought how sadly beauty of inscape was unknown and buried away from simple people and yet how near at hand it was if they had eyes to see it..."[88] Additionally, the experience of inscape is not a constant, even for Hopkins. He writes that "unless you refresh the mind from time to time you cannot always remem-

[88] Humphry House, ed. *The Journals and Papers of Gerard Manley Hopkins*, 2nd ed. (London: Oxford University Press, 1959), p. 221.

ber or believe how deep the inscape in things is."[89] Finally, he even held that youth has advantage in gaining insight. Here we again find echoes of Scotus. A main reason Scotus believed that intuitive cognition was extremely difficult, if not impossible, in this life was because of the affect of sin. Thus, Hopkins wrote that insight "is more sensitive, in fact is more perfect, earlier in life than later."[90] The openness of youthful innocence is an aid in intuition. Hopkins expressed this in his poem *Spring*:

> What is all this juice and all this joy?
> A strain of the earth's sweet being in the beginning
> In eden garden.—Have, get, before it cloy,
>
> Before it cloud, Christ, lord, and sour with sinning,
> Innocent mind and Mayday in girl and boy...[91]

Still, with all the qualifications, Hopkins claimed that intuitive cognition, 'contemplation', intuition, is possible for anyone in this life. Further, he believed that such knowing could be fostered. First, Hopkins held that an "intense and prolonged activity of observation and meditation on objects prior to the intuiton of their inscape" is necessary.[92] Additionally, he held that experiencing inscape was not a matter of perceiving only one individual of a particular species. Rather it was a matter of experiencing many separate individuals of a species often and repetitively over time. Hopkins wrote

[89] Ibid., p. 205.

[90] Claude Colleer Abbott, ed. *The Correspondence of Gerard Manley Hopkins and Richard Watson Dixon*, 2nd ed. (London: Oxford University Press, 1955), p. 38.

[91] Norman H. MacKenzie, ed. *The Poetical Works of Gerard Manley Hopkins* (Oxford: Clarendon Press, 1990), p. 142.

[92] Stephen Edward Wear, "John Duns Scotus and Gerard Manley Hopkins: The Doctrine and Experience of Intuitive Cognition" (Ph.D. Dissertation, The University of Texas at Austin, 1979), p. 142.

that for the "production" of inscape, "repetition/oftening, over-and-overing, aftering of the inscape must take place in order to detach it to the mind."[93] He further wrote that in 'meditating', one needs to "stay long enough to let the inscape...grow on one."[94] Thus, by attentively and repeatedly seeing things, their activities, and even their various stages of growth or life, the observer allows inscape 'to grow on one'.

This seems to infer that the experience of inscape may come 'piecemeal'. Seemingly, the grasp of inscape for Hopkins is not an all-or-nothing experience. The experience can gain depth over time. Additionally, Hopkins seemed to hold that certain representatives of a species may be more revelatory of inscape than their peers. Thus, in his poems *Felix Randall*, *Harry Ploughman*, and *Henry Purcell*, we find him using examples of individuals he found more expressive of inscape than others. Inscape may also be more intensely conveyed through particular activities of individual species. Thus, Hopkins finds inscape in the swooping dive of the falcon in his poem *The Windhover*.

Hopkins also held something quite radical and quite beyond what he found in Scotus. Not only can inscape be 'caught' in sensible reality, it can also be conveyed by and gleaned from art — particularly poetry. Hopkins wrote that inscape is "the very soul of art."[95] How Hopkins meant this is particularly radical. We will investigate this further below; however, some preliminary comments may be useful now.

[93] Humphry House, ed. *The Journals and Papers of Gerard Manley Hopkins*, 2nd ed. (London: Oxford University Press, 1959), p. 289.

[94] Ibid., pp. 213-14.

[95] Claude Colleer Abbott, ed. *The Correspondence of Gerard Manley Hopkins and Richard Watson Dixon*, 2nd ed. (London: Oxford University Press, 1955), p. 135.

Hopkins claimed that inscape, with all the depth and theological significance that term had for him in his mature thought, could be caught and communicated through "speech framed to carry the inscape,"[96] namely poetry. He wrote that "as air, melody, is what strikes me most of all in music, and design in painting, so design, pattern or what I am in the habit of calling 'inscape' is what I above all aim at in poetry."[97] Hopkins had an unqualified belief in the realist possibilities of art and poetry. Poetry was in part a matter of "being in earnest with your subject: reality."[98] How, then, did Hopkins intend us to understand the nature and experience of inscape in poetry? Is an artistic inscape simply a representation of an 'objective' inscape in a descriptive sense? The answer must be that the relation between 'natural' and 'artistic' inscapes is much more intimate for Hopkins.[99]

Poetry, as conveyance of inscape, is much more than description. In what I believe is a type of sacramental structure, Hopkins actually held that poetry can 're-present' the inscapes he had gleaned **as such**. In his poem, *Henry Purcell*, Hopkins writes the following introduction:

> The poet wishes well to the divine genius of Purcell and praises him that, whereas other musicians have given utterance to the moods of man's mind, he has, beyond that, uttered in notes the very make and species of man as created both in him and in all men generally.[100]

[96] Humphry House, ed. *The Journals and Papers of Gerard Manley Hopkins*, 2nd ed. (London: Oxford University Press, 1959), p. 289.

[97] Claude Colleer Abbott, ed. *The Letters of Gerard Manley Hopkins to Robert Bridges*, 2nd ed. (London: Oxford University Press, 1955), p. 66.

[98] Ibid., p. 225.

[99] Cf. Stephen Edward Wear, "John Duns Scotus and Gerard Manley Hopkins: The Doctrine and Experience of Intuitive Cognition" (Ph.D. Dissertation, The University of Texas at Austin, 1979), p. 153.

[100] Norman H. MacKenzie, ed. *The Poetical Works of Gerard Manley Hopkins* (Oxford: Clarendon Press, 1990), p. 157.

In the poem itself, he expresses this by writing:

> Not mood in him nor meaning, proud fire or sacred fear,
> Or love or pity or all that sweet notes not his might nursle:
> It is the forged feature finds me; it is the rehearsal
> Of own, of abrupt self there so thrusts on, so throngs the ear.[101]

Hopkins claims that Purcell's music does not simply describe the moods of the human mind, rather it utters in notes a rehearsal "of own, of abrupt self"; that is, it 're-presents' human inscape itself via music.[102] This same possibility Hopkins holds true of poetry as well. Poetry, as art in general, can convey the 'abrupt self' of reality, whether that reality be human or any other thing in nature or experience.

In fairness to Hopkins, we must understand the holistic sense he had of poetry. Poetry, for Hopkins, was not simply wordplay. In a letter to his fellow poet and friend, Robert Bridges, Hopkins defends a poem he wrote. He educates Bridges on how to do a poem justice:

> "You must not slovenly read it with eyes but with your ears, as if the paper were declaiming it to you. For instance the line 'she had come from a cruise training seamen' read without stress and declaim is mere Lloyd's Shipping Intelligence; properly read it is quite a different thing. Stress is the life of it."[103]

[101] Ibid.

[102] Hopkins gives an example of another artistic medium, in this case sculpture, capturing inscape when he writes: "[I] caught that inscape in the horse that you see in the pediment especially and other bas reliefs of the Parthenon..." See Humphry House, ed. *The Journals and Papers of Gerard Manley Hopkins*, 2nd ed. (London: Oxford University Press, 1959), pp. 241-2.

[103] Claude Colleer Abbott, ed. *The Letters of Gerard Manley Hopkins to Robert Bridges*, 2nd ed. (London: Oxford University Press, 1955), pp. 51-2.

By stress, Hopkins means in part the full quality of dynamic language used to carry inscape.[104] This includes wordcraft, rhetoric, 'sprung rhythm', and declamation. Hopkins' poetry is meant to be a kind of word-music whose aim is to catch the flow and life of things in the flow and life of spoken sound.[105] In his poem, *The Windhover*, for example, Hopkins intends that the flight of the bird is "not only descriptively suggested in the poem, but obviously and intentionally *'re-presented'* in the rhythms and movements of the sound of the spoken poetry."[106]

The dynamics of how artistic inscapes correspond to natural inscapes need further development. Suffice it to say here that Hopkins held a radical correlation between the two, and that he intended to achieve this correlation in his poetry. Although the univocal approach to inscape here is purely Hopkins', he was able to proceed in this direction because of the metaphysical base he found in Scotus. In the context of Scotus' epistemology, Hopkins found ground for claiming that 'inscape' as such was not abstractively,

[104] Stephen Edward Wear, "John Duns Scotus and Gerard Manley Hopkins: The Doctrine and Experience of Intuitive Cognition" (Ph.D. Dissertation, The University of Texas at Austin, 1979), p. 155. Also see Franco Marucci, *The Fine Delight That Fathers Thought: Rhetoric and Medievalism in Gerard Manley Hopkins* (Washington, D.C.: The Catholic University Press, 1994). Marucci has done a service by parsing the rhetoric of Hopkins as part of his sacramental approach to poetry. Marucci shows that "Hopkins's rhetoric is also about the signic and unitive quality of his thought and expression. Hopkins' wordcraft cannot be separated from his skillful rhetoric. His rhetoric is as finely worked as his language and is of equal importance in realizing a signic function. Marucci holds and demonstrates that both aspects form the matter of Hopkins' sacrament — his poetry." See Philip Ballinger, "Review Article," *Studies* 85, no. 338 (1996), p. 197.

[105] Cf. Joseph Hillis Miller, *The Disappearance of God: Five Nineteenth-Century Writers* (Cambridge: Harvard University Press, 1963), p. 281.

[106] Stephen Edward Wear, "John Duns Scotus and Gerard Manley Hopkins: The Doctrine and Experience of Intuitive Cognition" (Ph.D. Dissertation, The University of Texas at Austin, 1979), p. 155.

but intuitively 'caught'. Hopkins believed that this intuitive experience could be re-produced and revived in art for the benefit of others. As Marucci writes:

> The poetic message...is first of all a translation into words of a mysterious and miraculous experience, of a form of intuitive knowledge that for an instant, and with an even painful pungency, lets one recognize things in all their reality and uniqueness, as if one saw them for the first time.[107]

G. An Undeveloped Theme: Hopkins' Univocity of Being

We have seen how Devlin identified two loci of Scotus' influence on Hopkins — his incarnationalism and epistemology. We have also noted that Devlin did not choose to develop another possible facet of commonality between Hopkins and Scotus, namely, the theme of the univocity of being. Although Hopkinsian scholars, such as Gardner, Pick, and J.H. Miller, presume this 'Parmenidean' idea as background for Hopkins, they do not develop the claim. I believe a study of the Scotian influences in Hopkins requires some addressing of this theme.

Scotus held that the primary natural object of human intellect is 'being as being'.[108] "Being is the one concept present in all knowledge."[109] 'Being', whether used for contingent being or necessary being, has a univocal meaning for Scotus in some sense. Desiring

[107] Franco Marucci, *The Fine Delight That Fathers Thought: Rhetoric and Medievalism in Gerard Manley Hopkins* (Washington, D.C.: The Catholic University Press, 1994), p. 199.

[108] Frederick Copleston, S.J., *Mediaeval Philosophy: Albert the Great to Duns Scotus, A History of Philosophy*, vol. 2 (Garden City: Image Books, 1962), p. 210.

[109] Bernadette Ward, "Means and Meaning: Gerard Manley Hopkins' Scotist Poetic of Revelation and Matthew Arnold's Poetic of Social Control" (Doctoral Dissertation, Stanford University, 1990), p. 164.

to safeguard the possibility of a natural theology, Scotus posited that a univocal concept of being is necessary for any kind of reasoning and discourse about God. Scotus contended that, based on human experience, we can predicate 'being' and its attributes, such as goodness and beauty, of God. Put in another more applicable way vis-à-vis Hopkins, Scotus seemed to hold that not only knowledge of God's likeness but even knowledge of God *per se* can come through the senses. Scotus seemed to apply this univocal sense of being primarily in its aspect of 'positiveness'. Being, as a univocal concept, is an opposition to nothingness. Hopkins reflected a similar theme in a note-book passage (regarding Parmenides) written prior to his discovery of Scotus:

> His [Parmenides] great text... is that Being is and Not-being is not — which perhaps one can say, a little over-defining his meaning, means that all things are upheld by instress and are meaningless without it... But indeed I have often felt when I have been in this mood and felt the depth of an instress or how fast the inscape holds a thing that nothing is so pregnant and straightforward to the truth as simple *yes* and *is*... There would be no bridge, no stem of stress between us and things to bear us out and carry the mind over: without stress we might not and could not say / Blood is red / but only / This blood is red / or / The last blood I saw was red / nor even that, for in later language not only universals would not be true but the copula would break down even in particular judgements.[110]

Hopkins' stated the same thing later in holding that all beings 'rhyme'. He wrote that "any two things however unlike are in something like."[111] Following this approach to Being made both

[110] Humphry House, ed. *The Journals and Papers of Gerard Manley Hopkins*, 2nd ed. (London: Oxford University Press, 1959), p. 127.

[111] Christopher Devlin, S.J., ed. *The Sermons and Devotional Writings of Gerard Manley Hopkins* (London: Oxford University Press, 1959), p. 123.

Hopkins and Scotus "linguistic realists."[112] They believed that language has a foundation in a world of intelligibility; i.e., language refers to something real and true that can be experienced. Language, in short, is attached to the sensible world, and not merely to the level of isolated concepts.[113]

This wispy 'Scotian' univocity was accentuated by Hopkins through another aspect of Scotist doctrine. Scotus seemed to hold that Christ is the common nature upon whom all creation is modeled. In Christ is the summation of all nature. This logocentric vision allowed Hopkins to "see the universe as a vast interlocking harmony, full of…echoes and resonances."[114] All things 'rhyme' because they are patterned by the Word. Hopkins, in this context, can affirm that God's being in Christ is "selving," "speaking," and "playing" in and through every individual thing.[115]

This aspect of commonality between Scotus and Hopkins is vague.[116] Certainly, Hopkins held for the univocity of being in some

[112] Bernadette Ward, "Means and Meaning: Gerard Manley Hopkins' Scotist Poetic of Revelation and Matthew Arnold's Poetic of Social Control" (Doctoral Dissertation, Stanford University, 1990), p. 164.

[113] This "linguistic realism" one finds in Scotus was formed in arguments against Averroes and the nominalists. As Ward puts it, Scotus fought the approach characteristic of nominalism; namely, that "it is easy to treat all concepts as equally valid if one does not admit any reality to the abstractions which function as the referents of language." Ibid., pp. 165-6. In short, language cannot and should not be detached from the sensible world.

[114] Joseph Hillis Miller, "The Theme of the Disappearance of God in Victorian Literature," in *Victorian Subjects* (Durham: Duke University Press, 1991), p. 65.

[115] David Anthony Downes, *Hopkins' Sanctifying Imagination* (Lanham: University Press of America, 1985), p. 37.

[116] Cf. Donald Walhout, "Scotism in the Poetry of Hopkins," in *Saving Beauty: Further Studies in Hopkins*, ed. Michael Allsopp (New York: Garland Publishing, 1994), p.122. Studying direct references to various Scotist doctrines in Hopkins' poetry, Walhout concludes, regarding the doctrine of the univocity of being, that "univocity is occasionally intended, but not uniformly and not abundantly."

way **prior** to discovering Scotus. Perhaps this is why Devlin chose not to develop the theme. It may not be of Scotian origin in Hopkins. Nonetheless, as a theme it may have profound implications for Hopkins' poetics and poetry of nature.

In their acceptance of the univocity of being, Scotus and Hopkins diverged significantly from Aquinas. Aquinas centered upon analogy. Things, for Aquinas, are analogous to the nature of God, and each thing in nature stands not for the whole nature of God, but for a particular attribute of the deity. As J. Hillis Miller puts it, "the book of nature is a set of hieroglyphs or symbols, each one of which tells us something specific about God."[117] However, for Scotus, and certainly for Hopkins, it seems that particular objects *in themselves* present Christ. Kingfishers, dragonflies, or stones are somehow more than symbols. They are presence-laden. As Hopkins could say, we can know the beauty of the Lord by them. Thus, Scotus' univocity of being is more than a ground for analogy. There is a realism about it which allows Hopkins to approach sensation, intuitive cognition, and the poetic expression and conveyance of inscape as a theological, even sacramental, enterprise. As Miller puts it, the notion of univocity implies a kind of metaphysical egalitarianism:

> Each created thing, in its own special way, is the total image of its creator. It expresses not some aspect of God, but his beauty as a whole. Such a view of nature leads to a poetry in which things are not specific symbols, but all mean one thing and the same: the beauty of Christ, in whom they are created.[118]

Scotus theory that God and creatures can be included, in some sense, in the same metaphysical genus, the distinction being the dif-

[117] Joseph Hillis Miller, *The Disappearance of God: Five Nineteenth-Century Writers* (Cambridge: Harvard University Press, 1963), p. 314.

[118] Ibid., p. 315.

ference of being *per se* and being *per participationem*, correlated with similar thoughts in Hopkins. This metaphysical fountainhead certainly offers rich intellectual ground for a sacramental view of the world. However, the relationship between creature and Creator in this 'univocal' context is somehow even closer, perhaps, than that afforded by the relation between a sacramental sign and a thing sacramentally signified. Certainly, this point of view exposed Scotus to 'charges' of pantheism, and in part explain why his writings were considered heterodox for so many centuries. Although Hopkins himself was never explicitly associated with pantheism[119], I believe he was aware of the possibility of a pantheistic interpretation of his writings. Attempting to clarify his thought for others, Hopkins wrote the following:

> "Neither do I deny that God is so deeply present to everything... that it would be impossible for him but for his infinity not to be identified with them or, from the other side, impossible but for his infinity so to be present to them. This is oddly expressed, I see; I mean, a being so intimately present as God is to other things would be identified with them were it not for God's infinity or were it not for God's infinity he could not be so intimately present to things."[120]

Clearly, Scotian univocity of being is potential 'background' for Hopkins. Yet again, we do not find Hopkins presenting this theme in a clearly Scotian manner. Hopkins' 'univocal stress' presents

[119] Still, it must be said that some of his sermons were 'censored' by his supervising pastors at times. The apparent reasons had to do with Hopkins' Scotesian leanings. These leanings, I hold, barred Hopkins from continued theological studies as well. Clearly, Hopkins' Scotism caused him difficulty, but in the main, was seen simply as another one of his 'idiosyncracies'. I find no indication, however, that Hopkins was ever labelled a 'pantheist'.

[120] Christopher Devlin, S.J., ed. *The Sermons and Devotional Writings of Gerard Manley Hopkins* (London: Oxford University Press, 1959), p. 128. For Hopkins' expanded 'proofs' for the distinction between what he terms 'self' and the 'universal being' or 'universal mind being', see pp. 125-29.

itself in his mature vision of inscape as a Christic theme. Although Hopkins would assert that there is meaning in saying that 'being is being' in some sense, whether speaking of contingent or necessary being, this is not the manner of expression we find in his later writing on inscape. Although the Scotian theme is present as another support of Hopkins' theological, even sacramental, aesthetic, I believe Devlin had reason for choosing not to develop it as a primary influence upon Hopkins' work.[121] The primary influential Scotian loci for Hopkins' thought, with which the theme of the univocity of being dovetails, are the unique approaches to the Incarnation and to an epistemology which allows for the possibility of intuitive cognition.

H. Conclusion

We have reviewed areas where Scotus may have influenced Hopkins. Additionally, we have seen how Hopkins already tended toward some Scotian principles prior to discovering Scotus, as well as how he developed these principles once he discovered them. In contending that Scotus influenced Hopkins primarily via his epis-

[121] Perhaps Devlin understood the difficulties of making too close a link between Hopkins' poetry and Scotus' arguments for the univocity of being. The cautionary point of view here is given by R.V. Young, "Hopkins, Scotus, and the Predication of Being," *Renascence* 42, no. 1-2 (1989), p. 48: "The poetry of Gerard Manley Hopkins poses in a particularly acute fashion the problematic nature of the relationship between philosophical propositions and literary vision. The poet's obvious fascination with Duns Scotus does not necessarily mean that the poetry is informed by or illustrative of Scotist metaphysics... Univocity, I submit, is less poetic [than the Thomist analogical alternative] because it is, finally, a less accurate account of reality. Hopkins' discovery of Scotus seems to have been spiritually and imaginatively salutary for him, but his poetry is compatible with Thomist analogical predication..."

temology and specific incarnationalism, I have claimed that Hopkins found in Scotus a philosophical and theological confirmation for his unique aesthetic vision. As Marucci perceptively claims:

> In 1872 Hopkins found in Scotus a philosopher who met his strong aesthetic demands, a philosopher who did not condemn, and on the contrary justified and in a certain sense even exalted... the possibility of a cohabitation — even of a symbiosis — of art and faith. Scotus, better than Aquinas and St. Ignatius, offered Hopkins a theological system in which it might seem easier to be both a priest and a poet.[122]

In light of Scotus, Hopkins was able to justify a vision of poetry as a conveyance in words of a form of intuitive knowledge that allows one to recognize things in all their reality and uniqueness. Developing Scotus, Hopkins held that one could come to a revelation of God through apprehending an individual reality's 'thisness'. Furthermore, Scotus justified for Hopkins the value and essential importance of sensation in this enterprise. In Scotus, Hopkins discovered that sensation could be a kind of "spiritual sense."[123] The moment of sensation, the moment of 'naming' or 'wording' a thing as opposed to 'defining' a thing, becomes an exalted moment, a moment of revelation. In short, Hopkins "glimpsed in Scotus the possibility of making poetry a theology."[124] Even more, in view of

[122] Franco Marucci, *The Fine Delight That Fathers Thought: Rhetoric and Medievalism in Gerard Manley Hopkins* (Washington, D.C.: The Catholic University Press, 1994), p. 198. Hopkins says little of Thomas, but clearly the Thomism he was taught appeared to be a philosophy which paid little attention to the aesthetic problem. Thomas claimed the superiority of philosophy over poetry and even condemned the latter as *infima scientia* (p. 199).

[123] Renita Tadych, "The Franciscan Perspective in the Nature Poetry of Gerard Manley Hopkins Augmented by the Writings of John Duns Scotus" (Ph.D. Dissertation, Indiana University of Pennsylvania, 1992), p. 156.

[124] Franco Marucci, *The Fine Delight That Fathers Thought: Rhetoric and Medievalism in Gerard Manley Hopkins* (Washington, D.C.: The Catholic University Press, 1994), p. 205.

the consequences of Scotus' incarnation of the Logos into matter (Hopkins' *ensarkosis*), Hopkins seemed to be heading towards a vision of poetry as 'sacrament'. Scotus, though not the origin of Hopkins' approach, allowed Hopkins to synthesize the beginnings of a theological, or even sacramental, aesthetic.[125] It is to this nascent synthesis that we now turn.

[125] Some Hopkins scholars chafe at the use of the terms 'sacrament' and 'sacramental' to identify Hopkins' approach to reality and poetry. Nathan Cervo offers the most direct criticism of those who claim that Hopkins had a "sacramental" view of things. He writes: "A grave ambiguity attends the use of this word, as one suspects that *sacrament* and not *sacramental* is intended. By definition, a *sacrament* is an outward sign instituted by Christ to give grace. A *sacramental* is an action or object of ecclesiastical origin that serves as an indirect means of grace by producing devotion. In neither case... is "sacramental" appropriate to Hopkins criticism — "things" are not of ecclesiastical origin, nor is a poet a priest endowed with the power to transubstantiate "things" into the equivalent of the Blessed Sacrament. *Piety* is the humbler, more accurate word — respect for all things. Hopkins is a pious poet." See Nathan Cervo, "Scotistic Elements in the Poetry of Hopkins," *Hopkins Quarterly* 10 (1983), p. 64. This is an excellent example of having one's mind made up before and in spite of the evidence. Simply to conclude that Hopkins was a "pious poet" discounts the philosophical and theological emphases of Hopkins' poetry. Furthermore, it may well be that scholars have in mind 'sacrament' when they refer to Hopkins' sacramentalism, but to conclude that they use the term in the same sense as the Baltimore Catechism is inaccurate. Cervo unfortunately limits the boundaries of possible discourse on this subject by insisting on a rather anachronistic definition. His point of view disallows the discussion of "sacrament" in any context other than that of the ordained seven. Christ as 'the sacrament of God' or the 'Church as sacrament', for example, would not be meaningful phrasings in Cervo's context. Cervo's position seems to forbid the possibility of created reality and language having 'sacramental' dimensions for Hopkins. In my opinion, filling the terms "sacrament" and "sacramental" with such a limited meaning offers little of value to Hopkinsian criticism.

CHAPTER IV

THE THEOLOGICAL AESTHETIC OF GERARD MANLEY HOPKINS

A. Introduction

"Beauty's self and beauty's giver"[1] — this is more than a simple poetic description of God for Hopkins. As we have seen, 'beauty', as 'self-expressiveness', is a kind of nexus for Hopkins' Incarnationalism, epistemology, poetic theory, and emphasis upon the importance of the senses. It is a theological cipher throughout his writings. Beauty as 'self-expressiveness' is ultimately divine 'Self-expressiveness' in Hopkins' work. Thus, the natural world, the world imbibed through the senses, can become revelatory. Hopkins deeply honors the sensory world because in its self-expressive depth and intelligibility it is ultimately Christic. The created world for Hopkins is much more than simply food for thought — it is divinely, because it is Christically, self-expressive.

These themes have been echoed above as we investigated Hopkins in the contexts of various influences. Now I will focus upon the synthesis of these themes in the beginnings of what I hold to be a 'theological aesthetic' in Hopkins. As mentioned above, Hopkins offers us only the beginnings of such an aesthetic. In this chapter, I hope to bring together various strands of thought investigated earlier, and with some extrapolation, offer a more developed synthesis. I will further explain why Hopkins' theological aesthetic is

[1] Norman H. MacKenzie, ed. *The Poetical Works of Gerard Manley Hopkins* (Oxford: Clarendon Press, 1990), p. 170.

more specifically a 'sacramental aesthetic', and that this aesthetic comes home in a particular form of language and communication in Hopkins, namely, poetry. Ultimately, I will claim that Hopkins' aesthetic leads to a view of 'the poem as sacrament'.

Very few authors have approached Hopkins with the specific aim of developing his theological aesthetic. The only theologian of note having done so to some degree is Hans Urs von Balthasar.[2] Most others who have circled this issue have been scholars of English literature.[3] In the course of this chapter, I will use von Balthasar and others as sources and 'dialogue partners' to assist with the parsing of Hopkins' aesthetic. First of all, however, the concept of a 'theological aesthetic' itself needs some expansion.

B. 'Theological Aesthetic' as Applied to Hopkins

I use the term 'theological aesthetic' with regard to Hopkins' thought. Admittedly, it is not a completely harmonious phrasing. There are two reasons for this. First of all, the marriage of Christian and aesthetic experience has not always been blessed in the annals of Church history. The 'true' and the 'good' have been, at

[2] See Hans Urs von Balthasar, *Studies in Theological Style: Lay Styles*, ed. John Riches, trans. Andrew Louth, John Saward, Martin Simon, and Rowan Williams, 7 vols., *The Glory of the Lord: A Theological Aesthetics*, vol. 3 (San Francisco: Ignatius Press, 1986), pp. 353-99.

[3] For example, see Virginia Ridley Ellis, *Gerard Manley Hopkins and the Language of Mystery* (Columbia: University of Missouri Press, 1991); Margaret R. Ellsberg, *Created to Praise: The Language of Gerard Manley Hopkins* (New York: Oxford University Press, 1987); Hilary Fraser, *Beauty and Belief: Aesthetics and Religion in Victorian Literature* (Cambridge: Cambridge University Press, 1986); Maria R. Lichtman, "The Incarnational Aesthetic of Gerard Manley Hopkins," *Religion and Literature* 23, no. 1 (1991): 37-50; and Franco Marucci, *The Fine Delight That Fathers Thought: Rhetoric and Medievalism in Gerard Manley Hopkins* (Washington, D.C.: The Catholic University Press, 1994).

times, much more acceptable referents to God than the 'beautiful'. For Augustine and others, aesthetic experience was suspect because of its association with and stress upon the senses and bodily delights.[4] This association was a taint of sorts, and we have seen how Hopkins himself struggled with it.[5]

The phrasing is difficult and controversial for a second reason. There is a strong philosophical tradition which emphasizes an utter autonomy of aesthetics from religion; i.e., the realms of the religious and the aesthetic do not correspond. This view is summarized as follows:

> For something to be aesthetic it must be simply autonomous and hence *not* religious or moral or anything of the kind. The more philosophically inclined among us may put it this way: What is recognized as genuinely aesthetic in a religious context is precisely what cannot be converted even in part to religion proper without losing its identity as aesthetic. Anything aesthetic placed at the service of religion becomes something else, because it no longer has a proper aes-

[4] Cf. Frank Burch Brown, *Religious Aesthetics: A Theological Study of Making and Meaning* (Princeton: Princeton University Press, 1989), p. 160. Brown gives two examples. Augustine approves of the singing of hymns, but primarily so that by indulging the ears, weaker spirits may be inspired with feelings of devotion. The spiritually mature will be more moved by the meaning and truth of the words sung than by the singing. Bernard of Clairvaux allows that 'material beauty' might be needed in churches so that the devotion of carnal people can be aroused. In the history of the Christian Church, there is a strand of 'Gnostic scorn' for the material world and for something so sensory as art (in any form).

[5] Recall that just prior to entering the Society of Jesus, Hopkins wrote: "You know that I once wanted to be a painter. But even if I could I wd. not now, for the fact is that the higher and more attractive parts of the art put strain upon passions which I shd. think it unsafe to encounter. I want to write still, and as a priest I very likely can do that too, not as freely as I should have liked, e.g. nothing or little in the verse way, but no doubt what wd. best serve the cause of my religion." Claude Colleer Abbott, ed. *Further Letters of Gerard Manley Hopkins Including his Correspondence with Coventry Patmore*, 2nd ed. (London: Oxford University Press, 1956), p. 231.

thetic function; to the extent that the function of religious art remains aesthetic, it is to that very extent not religious.[6]

The difficulty, thus controversy, surrounding the term 'theological aesthetics' starts with the aim and nature of aesthetics as a discipline. Derrida, Gadamer, and others call into question whether or not there can even be a meaningful aesthetics, *in se*, at all.[7] The difficulty is heightened, as seen in Brown's quote above, by attempting a connection between the philosophical discipline of aesthetics and the aims and nature of theological discourse.

The approaches to explaining the relationship between or attempting a unification of aesthetics and theology are varied, but for simplicity's sake they may be categorized into three groupings. First, there is what may be termed the 'purist approach'. This approach, as seen above, posits a chasm between aesthetics and theology which cannot be bridged. In short, it disallows communion

[6] Frank Burch Brown, *Religious Aesthetics: A Theological Study of Making and Meaning* (Princeton: Princeton University Press, 1989), pp. 7-8. Robert Graves, as we have seen above, offers an example of this dichotomous point of view 'in action'. His claim with reference to Hopkins in *The White Goddess: A Historical Grammar of Poetic Myth*, American ed. (New York: Farrar, Straus and Giroux, 1966), pp. 425-26, is that "it has become impossible to combine the... functions of priest and poet without doing violence to one calling or the other... The poet survived in easy vigor only where the priest has been shown the door." Graves insists that an absolute, or even pagan, freedom from religious constraint is necessary for true poetry. John Coulson observes that this desired disassociation is common place in modern literary criticism: "Has not religious belief always been incompatible with the free play of imagination? The aim of literature is to please, that of religion to press questions of truth. And even if a case could be made out in theory, directly we turn to religious poetry we shall see how religious belief inhibits imaginative power. Literary critics assume that it is a weakness for poetry to be explicitly religious, and religious poetry is accordingly regarded as a species of minor or attenuated poetry." John Coulson, *Religion and Imagination* (Oxford: Clarendon Press, 1981), pp. 115-6.

[7] Frank Burch Brown, *Religious Aesthetics: A Theological Study of Making and Meaning* (Princeton: Princeton University Press, 1989), pp. 9-11.

between experiences commonly termed 'religious' and 'aesthetic'. Second, there is an approach best voiced by Hans Urs von Balthasar who holds that only an aesthetic derived primarily from revelation can be useful to theology. For von Balthasar, 'theological aesthetics' means an aesthetics "which does not primarily work with extra-theological categories of worldly philosophical aesthetics…but which develops its theory of beauty from the data of revelation itself with genuinely theological methods."[8] Finally, there is a 'common sense' approach as put forward by Frank Burch Brown. He holds that Christian theology can be informed by aesthetics, and that there is much promise in analyzing the links between aesthetic and religious experience.[9]

I dare not dive into this area as a *philosophical* controversy. In Hopkins' life, however, it is clear that he moved from a place of immersion in aesthetic experience with vague religious connotations to a place of revelatory, religious experience with specific aesthetic roots. Effectively, for Hopkins there could be an overlap or correlation between aesthetic and religious experience. This progression towards a 'theological aesthetic' in Hopkins seems to validate Brown's 'common sense' approach to the issue. Clearly, Hopkins moved from an aesthetic that was not at heart theological and arrived at a theological aesthetic that incorporated the former in a

[8] Hans Urs von Balthasar, *Seeing the Form*, ed. Joseph Fessio and John Riches, trans. Erasmo Leiva-Merikakis, 7 vols., *The Glory of the Lord: A Theological Aesthetics*, vol. 1 (Edinburgh: T. & T. Clark, 1982), p. 117. More specifically and more practically vis-à-vis Hopkins, theological aesthetics is also defined by Balthasar as "a study of the perceptible forms in which God is manifested and of the subjective conditions that must be fulfilled if the individual is to apprehend the forms for what they are." Cf. Louis Roberts, *The Theological Aesthetics of Hans Urs von Balthasar* (Washington, D.C.: The Catholic University of America Press, 1987), p. 192.

[9] Cf. Frank Burch Brown, *Religious Aesthetics: A Theological Study of Making and Meaning* (Princeton: Princeton University Press, 1989).

transformed fashion. In effect, Hopkins seems to be an historical example of Brown's philosophical contention.

In addition to the philosophical and/or theological contention over the concept of 'theological aesthetics', there is also the fact that until relatively recently, there has not been much interest in the field. Brown, writing in the mid-eighties, explains:

> Yet even in our own time, when the ascetic strain within Christianity is by no means prominent, such prolific Christian theologians as Hans Küng, Karl Rahner, and Karl Barth — all of whom have at moments acknowledged and reflected on the religious import of the arts — have devoted but a minute proportion of their enormous output to considerations that are in any overt way aesthetic as well as religious. Theologians such as Tillich and Berdyaev or, earlier, Schleiermacher and Newman, have been exceptions; and even these thinkers, for all their aesthetic sensitivities, have not been nearly so read in aesthetics *per se* as in other areas of philosophy and theology.[10]

The exception to Brown's contention is Hans Urs von Balthasar (whose work is somewhat marginalized by Brown as being idiosyncratic). Although Brown finds much of value in von Balthasar, he holds the theologian's basic argument to be specious. The position that revelation shows us whatever is true about art and beauty, he claims, is false. Brown writes:

> Christian theology may indeed want to say that no truth can be inconsistent with fundamental understandings of self, others, world, and ultimate reality — truth having its source in the very God that is made known in Christ. It should be obvious, however, that Christian revelation is not what has discovered to us the nature of the truths of logic or quantum mechanics, of language or the interpretation of dreams. Nor is it, consequently, the sufficient means by which we can discover all other basic truths, such as those pertaining to art and

[10] Frank Burch Brown, *Religious Aesthetics: A Theological Study of Making and Meaning* (Princeton: Princeton University Press, 1989), p. 3.

> aesthetics. Thus there is little reason to believe that whatever insight and theological understanding the aesthetician requires is immediately available in the form of revealed theology. Nor does it stand to reason that, just because theology can indeed make its own contribution to aesthetics, only aesthetics that begin with revelation can be pertinent to theology itself.[11]

Brown's claim vis-à-vis von Balthasar is somewhat shortsighted. Certainly, with regard to Hopkins, von Balthasar is not as 'adversarial' to the contribution of aesthetics proper as Brown would have us believe. After all, von Balthasar was taken by Hopkins' work and approach, an approach subsuming clearly aesthetic, not primarily theological or 'revealed', roots. Additionally, while he clearly stresses the theological center of Hopkins' mature aesthetics, von Balthasar recognizes that there would be great value in further analysis of Hopkins' 'aesthetic' technique and practices in order to better understand his 'theological aesthetic'. He writes:

> For a complete picture [of Hopkins' aesthetics] it would be necessary to separate out the tangled and yet always lucid threads of his formal and critical reflections — on verse form; on rhythm; on the nature of poetry and prose; on the role of number, proportion, chromatics, counterpoint, fugue, canon in verse and music — it would be necessary to trace his word creations, his powerful conciseness, his poetry's assonance and inner rhymes, his subtle, critical remarks on his own poems and those of others; it would be necessary to build out of these countless fragmentary aspects a complete edifice, whose coherence with Hopkins' philosophy and theology would then have to be shown. We must content ourselves with a theological approach.[12]

[11] Ibid., p. 20.

[12] Hans Urs von Balthasar, *Studies in Theological Style: Lay Styles*, ed. John Riches, trans. Andrew Louth, John Saward, Martin Simon, and Rowan Williams, 7 vols., *The Glory of the Lord: A Theological Aesthetics*, vol. 3 (San Francisco: Ignatius Press, 1986), p. 385.

Granted, von Balthasar is primarily concerned with the theological core of Hopkins' aesthetics — a core which holds that all truth is grounded in Christ and that all beauty is related to him.[13] Furthermore, Hopkins' approach to beauty claims the material world as a potential 'theological font', if you will. Thus, Brown's essential claim about von Balthasar is not essentially errant, just a bit extreme. As the quote above demonstrates, von Balthasar's theological approach does not mean he lacks interest in many 'practical' elements associated with aesthetic study.

Ultimately, I accept Brown's 'common sense' approach because it best corresponds with the Christian experience in general and Hopkins' experience in particular. Holding either an absolute or methodological dichotomy between aesthetic and religious experience, and thus between aesthetics and theology, does not coincide with the constant historical attachment between religion and art, for example. The religious and the aesthetic have been deeply and universally related in space and in time no matter what the faith tradition. Hopkins essentially accepted this relationship and attempted to ground it philosophically and theologically. Thus, unlike von Balthasar, Hopkins did not begin with a theological center; rather, he arrived at and developed such a center. In the process, he unified his aesthetic approach with his central metaphysics, epistemology, and theology, thus creating what may be usefully described as a 'theological aesthetic'.

As a further aid to describing the meaning of 'theological aesthetics' as applied to Hopkins, I turn to Louvain's Herman-Emiel Mertens. In a thoughtful article discussing the relationship between religious, aesthetic, and faith experience, he quotes F.D. Martin:

> The aesthetic experience always disengages us from the ontical treadmill, liberates us from the frenzy of functions. Thus the ontical is seen

[13] Ibid., p. 386.

> from a detached angle that may bring *Being* into awareness. The aesthetic experience in its pure mode is rapt or intransitive attention to a presented thing, the "given," whether that thing be a color, a cup of coffee, wallpaper, a pretty girl, or a work of art. When the "given" attracts and holds our attention because of its intrinsic value, not for its utility in serving some end, it becomes an aesthetic object. Thus the red in the evening sky is enjoyed for its own sake, for its qualities such as brilliancy and texture, not as a sign of good weather tomorrow.[14]

One could read this quote and think of certain stresses found in Hopkins — his emphasis on the "giveness" of the object experienced, his descriptions of capturing 'inscape' as requiring a kind of contemplative awareness, etc. However, his aesthetic approach, very similar to that described here by Martin, is transformed by two factors (as we have seen). First, the "given" for Hopkins is a potential theological center. In its metaphysical roots and in its self-expressive inscape, the "given" is Christic. Additionally, Hopkins accepted that in the 'aesthetic experience' there can be a non-abstractive, intuitive cognition which, in some sense, connects the 'experiencer' to the Divine self-expressiveness, or beauty, present in the "given." Thus, the aesthetic experience can be not only a religious experience, but also a possible source of theological reflection. 'Beauty', in this context, becomes a primary 'name' of God and, as a cipher for the Divine, a possible theological referent. The aesthetic pursuit and experience, in Hopkins, can become a theological source. Finally and more specifically, the poem, as a carrier of inscape, can be a theological locus and even 'sacramental expression' for Hopkins.[15] Thus, I refer to Hopkins' 'theological aesthetic'.

[14] Herman-Emiel Mertens, "His Very Name Is Beauty: Aesthetic Experience and Christian Faith," *Louvain Studies* 20, no. 2-3 (1995), pp. 319-20. Prof. Mertens is quoting F.D. Martin's *Art and the Religious Experience: The "Language" of the Sacred* (Lewisburg: 1972), p. 63.

[15] I have referred to Hopkins' 'sacramental poetry' and even to 'the poem as sacrament' in various places above. In this light, Prof. Mertens identifies a felic-

C. Hans Urs von Balthasar and Hopkins' Theological Aesthetic

C.1 Von Balthasar's General Approach

In Catholic theological circles, a somewhat radical subsuming of aesthetics by theology was attempted by Hans Urs von Balthasar in his *The Glory of the Lord: A Theological Aesthetics*.[16] In this monumental work, von Balthasar's goal was to bring beauty back to faith. His aim was practical; namely, "that beauty must be restored to faith and to its traditional place with truth and goodness as one of the transcendental attributes of the Christian faith."[17] His

itous passage from Joyce's *A Portrait of the Artist as a Young Man* in which Joyce describes the poet as "a priest of the imagination, transmuting the daily bread of experience into the radiant body of everlasting life." Cf. Herman-Emiel Mertens, "His Very Name Is Beauty: Aesthetic Experience and Christian Faith," p. 329. Without much extrapolation, the poem, in this context, becomes the Body of Christ. This is certainly reminiscent of Maria Lichtmann's blunt claim that for Hopkins, the poem is the Body of Christ. Cf. Maria R. Lichtman, "The Incarnational Aesthetic of Gerard Manley Hopkins," *Religion and Literature* 23, no. 1 (1991), p. 47.

[16] This seven volume work was originally published under the title *Herrlichkeit: Eine theologische Ästhetik* by Johannes Verlag in Einsiedeln, 1962.

[17] See John Coulson, "Hans Urs von Balthasar: Bringing Beauty Back to Faith," in *The Critical Spirit and the Will to Believe: Essays in Nineteenth-Century Literature and Religion*, ed. David Jasper and T.R. Wright (New York: St. Martin's Press, 1989), p. 218. Von Balthasar reminds his readers that patristic writers did not hesitate to use the *beautiful* as the core of their theological methodology. Cf. Louis Roberts, *The Theological Aesthetics of Hans Urs von Balthasar* (Washington, D.C.: The Catholic University of America Press, 1987), pp. 123 and 195. Von Balthasar writes: "Beauty is the word that shall be our first. Beauty is the last thing the thinking intellect dares to approach, since only it dances as an uncontained splendor around the double constellation of the true and the good in their inseparable relation to one another." Hans Urs von Balthasar, *Seeing the Form*, ed. Joseph Fessio and John Riches, trans. Leiva-Merikakis, Erasmo, 7 vols., *The Glory of the Lord: A Theological Aesthetics*, vol. 1 (Edinburgh: T. & T. Clark, 1982), p. 18. For von Balthasar, 'the true' and 'the good' are crippled without 'the

aim was also specifically pastoral. Von Balthasar believed that the muting of beauty in Christianity has produced a humorless, anguished, and grumpy religion which is becoming incomprehensible to people:

> When it is no longer fostered by religion, beauty is lifted from its face as a mask, and its absence exposes features on that face which threaten to become incomprehensible to man.[18]

Von Balthasar stresses that the truth becomes objectively evident, even luminous, when it is beautiful.[19] Beauty engages the human imagination, and the engagement of imagination, for von Balthasar as for Newman, is part of the holistic experience which allows for conviction in faith.[20] Additionally, if beauty is a tran-

beautiful'. "Everything that exists is true because it can be grasped by the intellect; everything is good because it can become an object of love; everything is beautiful because it can provide pleasure in the act of apprehension;" however, "without beauty the true would be just what is logically correct and the good would be the same as what is useful. What asserts itself beyond the correct and useful is a delight in the apprehension of the object of beauty." See Louis Roberts, *The Theological Aesthetics of Hans Urs von Balthasar* (Washington, D.C.: The Catholic University of America Press, 1987), p. 233.

[18] Hans Urs von Balthasar, *Seeing the Form*, ed. Joseph Fessio and John Riches, trans. Erasmo Leiva-Merikakis, 7 vols., *The Glory of the Lord: A Theological Aesthetics*, vol. 1 (Edinburgh: T. & T. Clark, 1982), p. 18.

[19] Louis Roberts, *The Theological Aesthetics of Hans Urs von Balthasar* (Washington, D.C.: The Catholic University of America Press, 1987), p. 4.

[20] This point is of particular importance to von Balthasar. He speaks of "two wrong answers to the question about Christian faith": "One is that it rests on argument and the other that it rests on feeling. The theory that it rests on feelings is usually held by non-believers. Having found no cogency in the arguments themselves, and wondering what convinces others if not argument, they can think of no alternative but feelings... Those who do see that feeling cannot be a legitimate basis for belief usually conclude that for themselves it must be argument or nothing. Then they make the mistake of thinking that Christianity is a jumble of creeds, each of which must be considered and judged separately... Both groups fail to see the possibility of a global vision and global acceptance that involves an intuitive grasp of the faith which is both vivid and convincing and which compels assent

scendental attribute of God, then God's beauty must be the apex and archetype of beauty in the world, according to von Balthasar. Beauty, as God's self-revelation in the world, finds intense expression through the Incarnation of the Word. This, however, is not only an act of divine self-expression in history; it is even more a foundation of on-going creation. For von Balthasar, the Logos is "the first, innermost and ultimate ontological principle of creation."[21] More specifically, the Incarnate Word is the definitive and determinant form of God in the world.[22] The Incarnate Word thus becomes the primal and archetypal source of beauty in the world. As such, beauty is a theological referent and an aid to faith.

While stressing the importance of beauty in this context, von Balthasar urges great care. His emphasis on restoring a 'theological aesthetics' is not the same as developing an 'aesthetic theol-

while leaving the details to be explicated by subsequent reflection and inquiry." The beautiful is key in this intuitive grasp of the faith. See Louis Roberts, *The Theological Aesthetics of Hans Urs von Balthasar* (Washington, D.C.: The Catholic University of America Press, 1987), p. 192. John Coulson shares von Balthasar's viewpoint. He writes: "When the vital connection between religion and imagination is either overlooked or denied, it is not merely theology or the theologian that suffers. The very life of religion ebbs and becomes infertile." See John Coulson, *Religion and Imagination* (Oxford: Clarendon Press, 1981), p. 3.

[21] Louis Roberts, *The Theological Aesthetics of Hans Urs von Balthasar* (Washington, D.C.: The Catholic University of America Press, 1987), p. 35. Von Balthasar writes: "God's Incarnation perfects the whole ontology and aesthetics of created Being. The Incarnation uses created Being at a new depth as a language and a means of expression for the divine Being and essence." Hans Urs von Balthasar, *Seeing the Form*, ed. Joseph Fessio and John Riches, trans. Erasmo Leiva-Merikakis, 7 vols., *The Glory of the Lord: A Theological Aesthetics*, vol. 1 (Edinburgh: T. & T. Clark, 1982), p. 29.

[22] John Coulson, "Hans Urs von Balthasar: Bringing Beauty Back to Faith," in *The Critical Spirit and the Will to Believe: Essays in Nineteenth -Century Literature and Religion*, ed. David Jasper and T.R. Wright (New York: St. Martin's Press, 1989), p. 219.

ogy'. In this context, he refers to Bonaventure's lament that so much of the water of philosophy has become mixed with the wine of Scripture that the wine has turned to water, thus making for a "miserable miracle."[23] For von Balthasar, a 'theological aesthetics', as opposed to an 'aesthetic theology', "derives from the primal beauty inherent in 'the data of revelation' — that is, from Scripture and the self-revelation of God in Christ."[24] For von Balthasar, earthly beauty unrelated to this primal beauty is a perversion.

Von Balthasar's concern and plea in calling for a 'theological aesthetics', as opposed to an 'aesthetic theology', is based upon what he sees as philosophy's penchant to "make a meal of all things, the sacred science of theology included."[25] The powerful symbols, images, and dogmas of Christian tradition are irreducible, for von Balthasar, and should be left to speak for themselves instead of being subordinated by and insulated within particular philosophies. Confirming von Balthasar's contention, L. Roberts remarks:

> Modern aesthetics prefers to talk about art rather than beauty and the functioning of the mind rather than the transcendentals of being. Modern aesthetics claims the source of form is in the constructive powers of the mind. We receive shapeless and manifold impressions. By selection and synthesis we make for ourselves a world-view that has order and form... Further, we project symbolic meanings into things

[23] See Hans Urs von Balthasar, *Studies in Theological Style: Clerical Styles*, ed. Joseph Fessio and John Riches, trans. Andrew Louth, Francis McDonagh, and Brian McNeil, 7 vols., *The Glory of the Lord: A Theological Aesthetics*, vol. 2 (San Francisco: Ignatius Press, 1984), p. 347.

[24] Ibid., p. 38.

[25] John Coulson, "Hans Urs von Balthasar: Bringing Beauty Back to Faith," in *The Critical Spirit and the Will to Believe: Essays in Nineteenth-Century Literature and Religion*, ed. David Jasper and T.R. Wright (New York: St. Martin's Press, 1989), p. 224.

> and processes and people, and in terms of these symbols we construct mythologies that reflect our deepest loves, hates, and fears.[26]

In a 'theological aesthetics', von Balthasar means that what is 'given' to the senses is taken seriously, for the divine reveals itself in the sensible world.[27] In the awareness of the divine, the senses in relation to the memory and imagination are "seized and kindled" by the perceptible forms of God (ultimately Christ). "Theological aesthetics studies these perceptible forms and the conditions that must be present to apprehend them for what they are."[28] In this approach, faith precedes reason instead of reason preceding faith. Reflecting on von Balthasar's approach here, Coulson comments in agreement that "if the eyes of faith are lacking, what remains are arbitrary constructions."[29]

Seeing von Balthasar's uncompromising approach, which seems to bar philosophical aesthetics from the realm of a theology of

[26] Louis Roberts, *The Theological Aesthetics of Hans Urs von Balthasar* (Washington, D.C.: The Catholic University of America Press, 1987), p. 241.

[27] This is related to von Balthasar's *analogia entis*. Reality is analogously and simultaneously like and unlike divine being. It is like insofar as it 'is', but it is unlike insofar as it is not identical with divine being or necessary in its being. This is somewhat similar to the 'univocity of being' implied by Duns Scotus (see above). Again, this common denominator helps one understand why Hopkins' work is so attractive to von Balthasar. Von Balthasar and Hopkins take created reality seriously because of this *analogia entis*. The world is a reflection of the divine, and in such a context, one can understand why its beauty becomes a "springboard to transcendence." See Roberts, p. 37.

[28] Louis Roberts, *The Theological Aesthetics of Hans Urs von Balthasar* (Washington, D.C.: The Catholic University of America Press, 1987), p. 241. This is precisely what Hopkins does methodologically. There is no wonder then in von Balthasar's acceptance of Hopkins' method as a kind of theological aesthetics.

[29] John Coulson, "Hans Urs von Balthasar: Bringing Beauty Back to Faith," in *The Critical Spirit and the Will to Believe: Essays in Nineteenth-Century Literature and Religion*, ed. David Jasper and T.R. Wright (New York: St. Martin's Press, 1989), p. 228.

beauty, one can understand Brown's criticism. It seems as if von Balthasar begins with a 'closed system' which disallows any contribution coming from a philosophical aesthetic which is faith-neutral. While accepting Brown's wariness of this apparently disjunctive approach, I do not believe he appreciates fully that von Balthasar's 'closed system' encompasses the whole of creation in some sense. This is exemplified in von Balthasar's work when he refers to a practice attributed to St. Francis. Of Francis it was said: "He looked at him who is most beautiful of all in the things that are beautiful, and pursued the beloved in the traces he had imprinted everywhere."[30] Von Balthasar is starting from a standpoint of faith in order to understand. Creation is sacred and communicative. Reality is missed or circumvented if one does not start, in some sense, from this ground as a given. Perhaps it is the a priori acceptance of this given that makes von Balthasar's approach ultimately unacceptable to Brown.

Von Balthasar is not unaware of the objection lodged in Brown's complaint; namely, that he does not honor the valuable outcomes of purely aesthetic pursuits and fields of inquiry. Recognizing this, von Balthasar asks the question as to whether or not it is possible to provide an explanation of beauty which fulfills the "aspiration of worldly and pagan beauty while all glory is at the same time given to God in Jesus Christ?"[31] This is where von Balthasar presents Gerard Manley Hopkins' 'theological aesthetics' as a marvelous synthesis and 'style' of seeing God.

[30] Hans Urs von Balthasar, *Studies in Theological Style: Clerical Styles*, ed. Joseph Fessio and John Riches, trans. Andrew Louth, Francis McDonagh, and Brian McNeil, 7 vols., *The Glory of the Lord: A Theological Aesthetics*, vol. 2 (San Francisco: Ignatius Press, 1984), p. 347.

[31] Hans Urs von Balthasar, *Seeing the Form*, ed. Joseph Fessio and John Riches, trans. Erasmo Leiva-Merikakis, 7 vols., *The Glory of the Lord: A Theological Aesthetics*, vol. 1 (Edinburgh: T. & T. Clark, 1982), p. 180.

C.2 Von Balthasar's Interpretation of Hopkins

Von Balthasar proposes Hopkins' work as a 'style' of seeing God and as an example of a nascent theological aesthetics. In doing so, von Balthasar does not offer us a systematic study of Hopkins. Rather, he disjunctively comments on aspects of Hopkins' life and efforts in the context of his own agenda for *The Glory of the Lord: A Theological Aesthetics*. Nonetheless, von Balthasar stresses aspects of Hopkins' work and thought which are useful for this study. Let us begin by reviewing von Balthasar's general comments on Hopkins' 'style of seeing God'.

C.2.1 Hopkins' Aesthetic 'Style of Seeing God'

Von Balthasar recognizes that in Hopkins, the image and not the concept must say everything. Additionally, as Coulson highlights, von Balthasar sees in Hopkins' poetry a dynamic which is essentially theological in method. To ensure the most adequate expression of the truth (i.e., beauty) which he experiences, Hopkins requires of his poetry exactly what theological explanation requires; namely, "a constantly renewed language which is richly innovative in rhythm and syntax…and grounded in symbols which, far from being fantasies, are irreducible and uninvertible."[32] Von

[32] John Coulson, "Hans Urs von Balthasar: Bringing Beauty Back to Faith," in *The Critical Spirit and the Will to Believe: Essays in Nineteenth-Century Literature and Religion*, ed. David Jasper and T.R. Wright (New York: St. Martin's Press, 1989), p. 223. Coulson turns this around and says, in another work, that theology must do what poetry does: "The theologian, too, has to reclaim theological language from its inherent tendency to remain a 'projection' or to degenerate into clichés… Like the poet, the theologian has to say and unsay to a positive result. In each case the integrity sought, if it is achieved, is signalized by technical or linguistic accomplishment, since…it is within the successful poetic evocation of the experience that its meaning lies hidden." "It is the poet who

Balthasar attributes this character of Hopkins' work to what he terms "the English spiritual tradition and obstinate empiricism". He writes:

> All that in continental symbolism — from Flaubert and Baudelaire to Maeterlinck and d'Annunzio — was set aside in a very programmatic and polemical manner, was in England completely bathed in the ancient and medieval tradition. And where on the continent ecclesiastical modernism sought to take over the ideas of the imaginary character of concepts and also of dogmatic *symbola*, thereby earning the sharpest of censures — so that the two worlds were violently split in two — just there English theology, reared in an hereditary empiricism, sensed no danger and preserved the native rights of imagery in religious thought, and therefore in Christian theology, right up to the present day. Newman's *Grammar of Assent* cannot be understood apart from this tradition of the place of the imagination in thought...[33]

In Hopkins, von Balthasar both notes and echoes a Scotian "mistrust of the value of universal concepts" and an empirical emphasis on individual realities as the "basis for any consideration of universal constructions of things; only here do they touch the ground of reality."[34] In a sense, von Balthasar praises Hopkins for remaining 'primeval' in his approach to reality and God. Another way of putting this is that von Balthasar sees Hopkins removing layers of conceptualization and returning to a level of perception and expression which is as close to an 'unadulterated', intuitive encounter with

shows the theologian how to prevent his beliefs from ossifying into 'big-words', by showing him the method by which we avoid cliché and secure meaning. As for the poet, so for the theologian... " John Coulson, *Religion and Imagination* (Oxford: Clarendon Press, 1981), pp. 139 and 140.

[33] Hans Urs von Balthasar, *Studies in Theological Style: Lay Styles*, ed. John Riches, trans. Andrew Louth, John Saward, Martin Simon, and Rowan Williams, 7 vols., *The Glory of the Lord: A Theological Aesthetics*, vol. 3 (San Francisco: Ignatius Press, 1986), p. 354.

[34] Ibid., pp. 354 and 356.

created reality as possible.[35] He finds in Hopkins a basic trust of the senses, an emphasis on the individual as opposed to the universal, and a use of language which accepts the power of imagery and the role of the imagination in thought and in faith. Thus, von Balthasar, citing Wolfgang Clemen, is able to conclude that Hopkins is "a singular case in modern times in that with him the theoretical concept and 'poetic conceptus' are completely congruent."[36]

In addition to his perceptions about Hopkins' theological method and honoring of imagery, von Balthasar correctly realizes that Hopkins' 'landscape' is neither mythological nor romantically ideal. It is not a cultural landscape, rather it is unprocessed, wild and primeval.[37] It is "brute beauty" that Hopkins wants to

[35] In other words, art, or specifically poetry, is not a kind of conceptual abstractive knowledge; rather, it is rooted in sensible intuition. Von Balthasar's approach here bears remarkable resemblance to a passage from J. Maritain's *Art and Scholasticism*: "The splendour or radiance of the form glittering in the beautiful object is not presented to the mind by a concept or idea, but rather by the sensible object intuitively grasped — in which there is transmitted as through an instrumental cause, this radiance of a form...in the perception of the beautiful the intellect is, through means of the sensible intuition itself, placed in the presence of a radiant intelligibility...which insofar as it produces the joy of the beautiful cannot be disengaged or separated from its sense matrix and consequently does not procure an intellectual knowledge expressible in a concept." Quoted by Gerard Casey, "Hopkins — Poetry and Philosophy," *Studies* 84, no. 334 (1995), p. 165 from Jacques Maritain, *Art and Scholasticism*, trans. J.W. Evans (Notre Dame University Press, 1974), pp. 163-4.

[36] Hans Urs von Balthasar, *Studies in Theological Style: Lay Styles*, ed. John Riches, trans. Andrew Louth, John Saward, Martin Simon, and Rowan Williams, 7 vols., *The Glory of the Lord: A Theological Aesthetics*, vol. 3 (San Francisco: Ignatius Press, 1986), p. 358. The work cited by von Balthasar is Wolfgang Clemen's afterword in *Gerard Manley Hopkins' Gedichte, Schriften, Briefe* (Kösel, 1954), p. 732.

[37] Hans Urs von Balthasar, *Studies in Theological Style: Lay Styles*, ed. John Riches, trans. Andrew Louth, John Saward, Martin Simon, and Rowan Williams, 7 vols., *The Glory of the Lord: A Theological Aesthetics*, vol. 3 (San Francisco: Ignatius Press, 1986), p. 359.

sing.[38] Hopkins intends to convey an encounter with the 'given' in its self-expressive power. In one place, Hopkins even claims surprise on rereading a verse in a poem he had written: "It struck me aghast with a kind of raw nakedness and unmitigated violence I was unprepared for."[39] Thus, Hopkins exercises a "tense, utterly objective contemplation of the primal power of nature" and expresses it in a primitive form of rhythm free of any hindrance imposed from without (sprung rhythm). He desires an objectivity in his poetry, as von Balthasar puts it, which "lays claim to the whole man through and through."[40]

Ultimately, von Balthasar sees in Hopkins' poetry the possibility of people encountering 'in whole' what the poetic texts bring alive, and responding, in whole, with the real assent of faith. Thus, Hopkins' poems make present and felt that which, if 'captured' in argument and systematic philosophy, might become incomprehensible and inaccessible. In other words, the poetic stress of Hopkins is to put forward the self-expressive Word in such a way that the whole person (meaning the person as also imaginative) confronts "the visibleness of the Invisible one, *the* definitive and determinant form of God in the world and the primal archetypal source of all worldly beauty,"[41] and in doing so, may assent to the presence with faith, praise, and even delight.

[38] Norman H. MacKenzie, ed. *The Poetical Works of Gerard Manley Hopkins* (Oxford: Clarendon Press, 1990), p. 144.

[39] Claude Colleer Abbott, ed. *The Letters of Gerard Manley Hopkins to Robert Bridges*, 2nd ed. (London: Oxford University Press, 1955), p. 79. Hopkins is referring to *The Eurydice*.

[40] Hans Urs von Balthasar, *Studies in Theological Style: Lay Styles*, ed. John Riches, trans. Andrew Louth, John Saward, Martin Simon, and Rowan Williams, 7 vols., *The Glory of the Lord: A Theological Aesthetics*, vol. 3 (San Francisco: Ignatius Press, 1986), pp. 360-61.

[41] John Coulson, "Hans Urs von Balthasar: Bringing Beauty Back to Faith," in *The Critical Spirit and the Will to Believe: Essays in Nineteenth-Century Literature and Religion*, ed. David Jasper and T.R. Wright (New York: St. Martin's Press, 1989), p. 230.

This specific 'religious experiential' present in Hopkins' poetry typifies a main characteristic of what von Balthasar holds to be great art. For von Balthasar, all great art is 'religious' in the same sense as Hopkins' poetry; it is both an act of "adoration of the *doxa* of being"[42] as well as an effective mediation of that same *doxa*.[43] When this characteristic fades from art, it degenerates into "a superficial fascination with the sensuous."[44] Von Balthasar's sen-

[42] Louis Roberts, *The Theological Aesthetics of Hans Urs von Balthasar* (Washington, D.C.: The Catholic University of America Press, 1987), p. 28.

[43] Von Balthasar centers much around the idea of *Herrlichkeit,* i.e., the *doxa*, the *khabod Jahwe,* the *Glory of God*. For von Balthasar, 'beauty', as a theological referent for the divine, is bound up with the concept of *doxa*. God radiating gloriously through the veiled forms of created reality, that is, God self-expressing or self-revealing through the sensible world, is beauty. This approach to beauty as a form of revelation comes close to Hopkins' own emphasis on beauty as divine self-expressiveness. For von Balthasar, facilitating the encounter of the individual and *doxa* is the religious function of good art and literature. Certainly, Hopkins' own definition of poetry as words used to carry and communicate *inscape* (in all of this word's mature Christic meaning for Hopkins) corresponds with von Balthasar's idea here of "the radiance from within." One could say that for von Balthasar, a "theological aesthetics," as a study of the perception and experience of the self-revealing glory of God, is actually a theology of glory. Von Balthasar demonstrates his patristic roots here. Seeing God's glory is an experience of the beautiful as an irruption of the eternal into the material, whether the 'material' is person, thing, or poem. This 'vision' overcomes or 'enthuses' one, confronts one with a sense of otherness, draws one up into a higher realm. Beauty, in this sense, is God's divine kenosis or divine 'ec-stasy', the procession out of the Godhead into the world which manifests itself as glory. The 'seer' is seized by this glory and taken up into the Godhead in some sense. This 'dramatic' narrative is realized in the Christ who is ultimately the object of a theological aesthetics. Thus, one can conclude that for von Balthasar, much of the theological endeavor in general is actually subsumed under a theological aesthetics. As Roberts concludes, "Balthasar's aesthetics embraces the whole of dogma." Louis Roberts, *The Theological Aesthetics of Hans Urs von Balthasar* (Washington, D.C.: The Catholic University of America Press, 1987), p. 236.

[44] Louis Roberts, *The Theological Aesthetics of Hans Urs von Balthasar* (Washington, D.C.: The Catholic University of America Press, 1987), p. 28.

timents here mirror Hopkins' own. Certainly for Hopkins, following John Ruskin's emphasis, all great art is praise as well as communicative of that which elicits praise.

Von Balthasar finds the key to Hopkins' theological aesthetics, and the root for the 'religious experiential' present in great art, in the uniquely Hopkinsian terms 'inscape' and 'instress'. They are, as von Balthasar states, Hopkins' "worldview in cipher."[45] Here von Balthasar finds necessary 'bridges' in Hopkins — the bridge between "us and things," as well as the bridge between "being and word." The key text in Hopkins for von Balthasar's reflection on inscape and instress comes from some 'notes' about the thought of Parmenides dated 1868:

> His great text, which he repeats with religious conviction, is that Being is and Not-being is not — which perhaps one can say, a little over-defining his meaning, means that all things are upheld by instress and are meaningless without it. An undetermined Pantheist idealism runs through the fragments, which makes it hard to translate them satisfactorily... His feeling for instress, for the flush and foredrawn, and for inscape is most striking... But indeed I have often felt when I have been in this mood and felt the depth of an instress or how fast an inscape holds a thing that nothing is so pregnant and straightforward to the truth as simple *yes* and *is*. "Thou couldst never either know or say / what was not, there would be no coming at it." There would be no bridge, no stem of stress between us and things to bear us out and carry the mind over... Being and thought are the same. The truth in thought is Being, stress, and each word is one way of acknowledging Being and each sentence by its copula *is* (or its equivalent) the utterance and assertion of it.[46]

[45] Hans Urs von Balthasar, *Studies in Theological Style: Lay Styles*, ed. John Riches, trans. Andrew Louth, John Saward, Martin Simon, and Rowan Williams, 7 vols., *The Glory of the Lord: A Theological Aesthetics*, vol. 3 (San Francisco: Ignatius Press, 1986), p. 365.

[46] Humphry House, ed. *The Journals and Papers of Gerard Manley Hopkins*, 2nd ed. (London: Oxford University Press, 1959), pp. 127-29.

Von Balthasar infers that 'instress' and 'inscape', in this context, are rooted in being. Specifically, they are metaphysical terms which ultimately connect being and language. Developing this Hopkinsian 'bridging' further, von Balthasar points out that at roughly the same time as he produced these notes on Parmenides, Hopkins was also jotting down thoughts on the nature of words and language. He wrote: "A word then has three terms belonging to it... — its prepossession of feeling; its definition, abstraction, vocal expression or other utterance; and its application, 'extension', the concrete things coming under it."[47] Von Balthasar interprets this text in light of the Parmidean notes and concludes that "the first moment of prepossession has clearly to do with instress."[48] He validly claims that *instress*, as in-stress, im-pression, in-tention, refers to both the object and subject in Hopkins. It is the connecting force between the two, if you will. Objects express their instress, their deep, unique act, which establishes them and holds them together.[49] Furthermore, there is required in the subject, as von Balthasar puts it, "an answering stress, so that it can hold communion with the stress of things and experience them from within."[50]

[47] Von Balthasar is here conflating a difficult and somewhat rambling text by Hopkins found in Humphry House, ed. *The Journals and Papers of Gerard Manley Hopkins*, 2nd ed. (London: Oxford University Press, 1959), p. 125. See Hans Urs von Balthasar, *Studies in Theological Style: Lay Styles*, ed. John Riches, trans. Andrew Louth, John Saward, Martin Simon, and Rowan Williams, 7 vols., *The Glory of the Lord: A Theological Aesthetics*, vol. 3 (San Francisco: Ignatius Press, 1986), p. 365.

[48] Hans Urs von Balthasar, *Studies in Theological Style: Lay Styles*, ed. John Riches, trans. Andrew Louth, John Saward, Martin Simon, and Rowan Williams, 7 vols., *The Glory of the Lord: A Theological Aesthetics*, vol. 3 (San Francisco: Ignatius Press, 1986), p. 365.

[49] Humphry House, ed. *The Journals and Papers of Gerard Manley Hopkins*, 2nd ed. (London: Oxford University Press, 1959), p. 199.

[50] Hans Urs von Balthasar, *Studies in Theological Style: Lay Styles*, ed. John Riches, trans. Andrew Louth, John Saward, Martin Simon, and Rowan

This in-selfment of being, or experiencing of a being from within, via instress, produces a "prepossession of feeling" for the nature of a thing which allows the subject 'instressed' to find the word that exactly expresses the object 'instressed'. "The objective instress is taken up by the subject that is open to it, that is moved in its depths by the depth of its power of being."[51]

The 'wording' referred to here by von Balthasar is, however, incomplete without the corollary of instress, namely, inscape. For von Balthasar, *instress* refers to, in part, the communicative power of a thing, while *inscape* refers to its form.[52] Instress carries the form, or inscape, over to the 'seer'.

Von Balthasar's interpretation of *form* here is interesting. He writes that "what is intended is not a separate form, resting in itself, but a form released from its creative source and at the same time shaped and held by it…"[53] This form is also "species or individually distinctive beauty…"[54] Although von Balthasar does not go so far, it seems that he is hinting at a clear extension of Hopkins' thought which we have touched upon previously; namely, that the ultimate inscape of all things is Christic, and that the possibility of wording things is also Christic. We will return to this theme below.

Von Balthasar, in his attempt at capturing the sense of Hopkins' term *inscape*, is particularly taken by Hopkins' poem, and the com-

Williams, 7 vols., *The Glory of the Lord: A Theological Aesthetics*, vol. 3 (San Francisco: Ignatius Press, 1986), p. 365. Another way that von Balthasar puts this in his own work is that we experience being as it is "en-selved as person." Ibid.

[51] Ibid., p. 365.

[52] Ibid., p. 366.

[53] Ibid.

[54] Claude Colleer Abbott, ed. *Further Letters of Gerard Manley Hopkins Including his Correspondence with Coventry Patmore*, 2nd ed. (London: Oxford University Press, 1956), p. 373.

ments surrounding it, concerning the musician Purcell. He finds the meaning of inscape perfectly expressed there: "for genius unfolds its wings, and 'also unaware gives you a whiff of knowledge about his plumage, the marking of which stamps his species'."[55] Unique inscapes are everywhere — the world is full of inscape hinting, flashing, and 'whiffing'.[56] "All one has to do is find the point where it [inscape] discloses itself... The form thus grasped is the key to the word, is itself already an objective word — *formed* can also mean *worded*."[57] Hopkins himself can finish von Balthasar's thought here with his own question and answer: Worded by whom? "By him that present and past, Heaven and earth are word of, worded by..."[58]

Von Balthasar's description of Hopkins' aesthetic dynamic, whereby inscape is gleaned through the senses, in-selved via instress, and discovered as word, outlines a style or framework for a theological *aesthetic* — particularly if this dynamic is contextualized within the mature theological vision of Hopkins. Thus, once again we frame Hopkins' theological tenet:

> "God's utterance of himself in himself is God the Word, outside himself is this world. This world then is word, expression, news of God.

[55] Von Balthasar here quotes Hopkins from Claude Colleer Abbott, ed. *The Letters of Gerard Manley Hopkins to Robert Bridges*, 2nd ed. (London: Oxford University Press, 1955), p. 83.

[56] Humphry House, ed. *The Journals and Papers of Gerard Manley Hopkins*, 2nd ed. (London: Oxford University Press, 1959), p. 230.

[57] Hans Urs von Balthasar, *Studies in Theological Style: Lay Styles*, ed. John Riches, trans. Andrew Louth, John Saward, Martin Simon, and Rowan Williams, 7 vols., *The Glory of the Lord: A Theological Aesthetics*, vol. 3 (San Francisco: Ignatius Press, 1986), p. 366.

[58] Norman H. MacKenzie, ed. *The Poetical Works of Gerard Manley Hopkins* (Oxford: Clarendon Press, 1990), p. 126.

Therefore its end, its purpose, its purport, its meaning, is God and its life or work to name and praise him."[59]

This naming of God in the world and giving of praise is accomplished through human selves. As von Balthasar observes, only "the free word of a free self can...express this."[60] Here, von Balthasar begins a transition in his reflection upon inscape and instress by moving from a metaphysical context to a theological context. The ultimate core of Hopkins' theological aesthetics, and of the mature meaning of inscape and instress, is the Scotistic-Hopkinsian 'dramatic narrative' of 'the Great Sacrifice'.

C.2.2 Von Balthasar's Contextualization of Hopkins' 'Great Sacrifice'

The dynamic of God's self-revelation, creation, and even grace flows from the Scotus-inspired assertion of 'the Great Sacrifice'. Von Balthasar asserts:

> All this is founded on the 'great sacrifice' of the Son, to which Idea of sacrifice all the rest of the creatures are related at the level of idea. The 'great sacrifice' on which everything is founded is... the primary way in which the self-emptying, the pure being for another of God's personal, trinitarian being can be manifest externally.[61]

The external manifestation of God, God's sensible self-expressiveness, God's 'beauty' in this sense, is centered in the Great Sac-

[59] Christopher Devlin, S.J., ed. *The Sermons and Devotional Writings of Gerard Manley Hopkins* (London: Oxford University Press, 1959), p. 129.

[60] Hans Urs von Balthasar, *Studies in Theological Style: Lay Styles*, ed. John Riches, trans. Andrew Louth, John Saward, Martin Simon, and Rowan Williams, 7 vols., *The Glory of the Lord: A Theological Aesthetics*, vol. 3 (San Francisco: Ignatius Press, 1986), p. 376.

[61] Ibid., pp. 380-81.

rifice, which is itself a consequence of the Trinitarian procession. We recall once again Hopkins' description of this:

> The first intention then of God outside himself... the first outstress of God's power, was Christ; and we must believe that the next was the Blessed Virgin. Why did the Son of God go thus forth from the Father not only in the external and intrinsic procession of the Trinity but also by an extrinsic and less than eternal, let us say aeonian one? — To give God glory and that by sacrifice. This sacrifice and this outward procession is a consequence and shadow of the procession of the Trinity, from which mystery sacrifice takes its rise; but of this I do not mean to write here. It is as if the blissful agony or stress of selving in God had forced out drops of sweat or blood, which drops were the world, or as if the lights lit at the festival of the 'peaceful Trinity' through some little cranny striking out lit up into being one 'cleave' out of the world of possible creatures.[62]

Whatever Hopkins ultimately intends by the 'Great Sacrifice' in his aesthetic vision, it is clear that he sees this dynamic as a consequence of the nature of the Divine being itself — it is a consequence of "the stress of selving in God." The notion of 'sacrifice', in this context, is a "consequence of... the procession of the Trinity." From this center, Hopkins begins, as von Balthasar states, "his breathtaking speculations."[63]

Von Balthasar here recognizes in Hopkins an acceptance of a Scotian theological possibility. In view of the eucharistic presence of Christ, Scotus 'revised' Aristotle and asserted that 'quantity', "though secondarily signifying the being of parts outside them-

[62] Christopher Devlin, S.J., ed. *The Sermons and Devotional Writings of Gerard Manley Hopkins* (London: Oxford University Press, 1959), p. 197.

[63] Hans Urs von Balthasar, *Studies in Theological Style: Lay Styles*, ed. John Riches, trans. Andrew Louth, John Saward, Martin Simon, and Rowan Williams, 7 vols., *The Glory of the Lord: A Theological Aesthetics*, vol. 3 (San Francisco: Ignatius Press, 1986), p. 381.

selves, primarily signified only the relatedness of parts one to another within a being; that is, its harmony or corporeal form."[64] Therefore, Scotus held that God has the power to cause the eucharistic body of Christ to be present universally in any position in time and space throughout the cosmos:

> I say then…that before the Incarnation and 'before Abraham was', in the beginning of the world, Christ could have had a true temporal existence in a sacramental manner. And if this is true, it follows that before the conception and formation of the Body of Christ from the most pure blood of the Glorious Virgin there could have been the Eucharist.[65]

Apparently reflecting on this Scotian theological opinion, Hopkins extended the thought: "The sacrifice would be the Eucharist, and that the victim might be truly victim like, like motionless, helpless, or lifeless, it must be in matter."[66] Von Balthasar sees this extension on Scotus' thought as key in Hopkins' theological aesthetic. "This eternally universal temporality and spatiality of God sacrificing himself in matter is primary."[67]

This kenosis of the Son, even into matter, is the heart of all beauty, whether present generically in creation or specifically in the inner life of grace, according to von Balthasar. This seems to be a valid interpretation of Hopkins. In a letter to Robert Bridges,

[64] Ibid. Also see Eleanor McNees, "Beyond 'The Half-way House': Hopkins and Real Presence," *Texas Studies in Literature and Language* 31, no. 1 (1989), p. 86.

[65] Christopher Devlin, S.J., ed. *The Sermons and Devotional Writings of Gerard Manley Hopkins* (London: Oxford University Press, 1959), p. 113. This is a translation of Scotus' *Op. Oxoniense IV, dist. 10, qu. 4.*

[66] Ibid., p. 197.

[67] Hans Urs von Balthasar, *Studies in Theological Style: Lay Styles*, ed. John Riches, trans. Andrew Louth, John Saward, Martin Simon, and Rowan Williams, 7 vols., *The Glory of the Lord: A Theological Aesthetics*, vol. 3 (San Francisco: Ignatius Press, 1986), p. 382.

Hopkins comments on the kenosis of the Son who "could not but see what he was, God, but he would see it as if he did not see it, and be it as if he were not… [there] seems to me the root of all holiness and the imitation of this root of all moral good in other men."[68] The encounter with the 'Eucharistic Christ' in this Scotian sense, thus the encounter with the self-expressive, kenotic Word, leads those encountered to decide whether or not to "contribute towards the Incarnation," as Hopkins puts it.[69] That is, an encounter with the beauty of Christ, even in matter, is also an invitation to and an aid in affirming the same kenotic dynamic in self. Beauty, in this sense, is rooted in Christ and leads to Christ becoming more intensely incarnated in the believer. The 'just man', as Hopkins states, "plays at Christ, and Christ plays at him." Thus, Hopkins is able to write the following verse:

> Acts in God's eye what in God's eye he is —
> Christ. For Christ plays in ten thousand places,
> Lovely in limbs, and lovely in eyes not his
> To the Father through the features of men's faces.[70]

I will not dwell on von Balthasar's expansion of this theme into an abbreviated theology of grace. What is central to this study is von Balthasar's recognition that in Hopkins, "creation in its entirety is traced from the pinnacle of Christ," and that this 'tracing' is essentially kenotic in nature.[71] This leads von Balthasar to conclude

[68] Claude Colleer Abbott, ed. *The Letters of Gerard Manley Hopkins to Robert Bridges*, 2nd ed. (London: Oxford University Press, 1955), p. 175.

[69] Christopher Devlin, S.J., ed. *The Sermons and Devotional Writings of Gerard Manley Hopkins* (London: Oxford University Press, 1959), p. 200.

[70] Norman H. MacKenzie, ed. *The Poetical Works of Gerard Manley Hopkins* (Oxford: Clarendon Press, 1990), p. 141.

[71] Hans Urs von Balthasar, *Studies in Theological Style: Lay Styles*, ed. John Riches, trans. Andrew Louth, John Saward, Martin Simon, and Rowan Williams, 7 vols., *The Glory of the Lord: A Theological Aesthetics*, vol. 3 (San Francisco: Ignatius Press, 1986), p. 383.

that "if the creation of the world is seen in this way, as an implication of the decree of the Incarnation, then it follows that the cosmos as a whole possesses, either manifestly or secretly, a christological form. And it further follows that "through all the raging elements, all the wilderness of matter, all shipwrecks and ruins, Christ can be coming and truly is."[72] This vision is what led Hopkins, according to von Balthasar, to once again take up poetry after a seven year hiatus, but now not only as an aesthetic expression of 'feeling', but also as a means of communicating and eliciting in others this vision of materially present and soliciting beauty. Thus Hopkins is able to write in *The Wreck of the Deutschland*:

I admire thee, master of the tides,
Of the Yore-flood, of the year's fall;
The recurb and the recovery of the gulf's sides,
The girth of it and the wharf of it and the wall;
Stanching, quenching ocean of a motionable mind;
Ground of being, and granite of it: past all
Grasp God, throned behind
Death with a sovereignty that heeds but hides, bodes but abides
With a mercy that outrides
The all of water, an ark
For the listener; for the lingerer with a love glides
Lower than death and the dark;
A vein for the visiting of the past-prayer, pent in prison,
The-last-breath penitent spirits — the uttermost mark
Our passion-plunged giant risen,
The Christ of the Father compassionate, fetched in the storm
of his strides.[73]

[72] Hans Urs von Balthasar, *Studies in Theological Style: Lay Styles*, ed. John Riches, trans. Andrew Louth, John Saward, Martin Simon, and Rowan Williams, 7 vols., *The Glory of the Lord: A Theological Aesthetics*, vol. 3 (San Francisco: Ignatius Press, 1986), p. 383.

[73] Norman H. MacKenzie, ed. *The Poetical Works of Gerard Manley Hopkins* (Oxford: Clarendon Press, 1990), p. 127.

Von Balthasar concludes by stating that "this then is the theological centre from which we can develop the laws of Hopkins' aesthetics… The principle lies in the fact that all truth is grounded in Christ and that all beauty belongs to him, is related to him, is yielded to him in the 'great sacrifice' and must rest with him."[74] In this light, the "natural doctrine of instress and inscape" is changed, "for the true inscape of all things is Christ."[75] Thus, von Balthasar interprets Hopkins as holding that "all natures and selves are fashioned and determined for Christ, who is both their ultimate inscape and instress," and that this means "there is no possibility of reading them objectively and understanding them than in relation to this centre in which they are integrated."[76]

Here we find von Balthasar solidifying Hopkins' work as a true 'theological aesthetics' in the sense he intends. Although this contextualization is useful in understanding the main lines of Hopkins' theological aesthetic, I find two difficulties in von Balthasar's specific approach here. First, echoing F.B. Brown's concern once again,[77] I find that von Balthasar ignores and underplays certain early influences upon Hopkins' aesthetics. He stresses Scotus greatly, and integrates, to some degree, the Ignatian strain in Hopkins. However, he merely mentions Ruskin and Ruskin-related strains of influence, some of which are not 'theological' in nature at all (e.g., Walter Pater's indirect influence). Hopkins did not come to his 'style of seeing God' in a whole piece. He evolved the style from early roots that were, at best, generically religious in context,

[74] Hans Urs von Balthasar, *Studies in Theological Style: Lay Styles*, ed. John Riches, trans. Andrew Louth, John Saward, Martin Simon, and Rowan Williams, 7 vols., *The Glory of the Lord: A Theological Aesthetics*, vol. 3 (San Francisco: Ignatius Press, 1986), pp. 385-86.

[75] Ibid., p. 387.

[76] Ibid., p. 390.

[77] Cf. Frank Burch Brown, *Religious Aesthetics: A Theological Study of Making and Meaning* (Princeton: Princeton University Press, 1989).

but surely not 'theological' in the sense von Balthasar intends. This is important because I do not believe that Hopkins would conclude, as does von Balthasar, that any sensible reality or an aesthetic work, such as a painting or a poem, can only be understood in an 'objective' sense through a 'Christic lens'. Hopkins critiqued aesthetic works using non-theological criteria and technique throughout his life. His aesthetic did not 'begin with Christ' if you will, at least not for the entirety of his life. Von Balthasar perhaps goes to far in attempting to squeeze Hopkins into his parameters for a theological aesthetic.

Second, as seen above, von Balthasar equates Christ with both 'inscape' and 'instress'. This is not a valid interpretation and is no where found in Hopkins. Inscape is certainly Christic in nature, at least in the mature Hopkinsian vision. Instress, however, is not. Instress may well be related, as inscape, to the dynamic of the Trinitarian procession. However, unlike inscape, instress is not the 'structured', 'intelligible', or 'Worded' core of a thing. It is more of a founding and foundational creative energy holding inscape in being and carrying inscape over to the observer. If anything, within the Trinitarian context used by von Balthasar, instress is 'Pneumatologic' instead of 'Christic'.[78] As Alan Heuser accurately

[78] Ellis speaks of instress as *energeia* or even 'will' of God in the creation of individual beings. The following quote could certainly be interpreted as the action of the Holy Spirit in creation and in knowledge of a thing: Hopkins held that "the power of any created thing to give off instress and to instress derives specifically and absolutely from an original and divine source, God's stress of energy, which charges the world with life, sustains it, keeps inscapes tautly at tension, prevents disintegration of being. Instress is thus both the life-giving energy in any object, derived from God, and the energy given off by that object, news of its selfhood and of its maker, when it is perfectly fulfilling its God-given function. It is the outgoing energy of inscape, the *form* speaking." Virginia Ridley Ellis, *Gerard Manley Hopkins and the Language of Mystery* (Columbia: University of Missouri Press, 1991), p. 33. Also see Leonard Cochran, O.P., "Instress and Its Place in the Poetics of Gerard Manley Hopkins," *The Hopkins Quarterly* 6, no. 4 (1980): 143-82.

observes, inscape and instress are "the two words of Being...the two paracletes, Son and Spirit" in Hopkins.[79] In any case, von Balthasar carries his identification of everything with Christ too far here.

The above critiques aside, von Balthasar's interpretation of Hopkins reviewed thus far is not too extreme in the main. Indeed, for Hopkins, as von Balthasar writes, "out of the glory of the Incarnate God there breaks forth the truest and most inward glory of forms both of natures and persons."[80] In a kind of Platonic context, von Balthasar holds that within the context of faith, Hopkins' 'inscape-instress' dynamic presents Christ as the "eternal idea, which shines through the phenomena" of persons and things. This explains why 'beauty', in this context, is so key in the life of faith. Von Balthasar holds that "the Christian, who is able to read this picture of the manifestation of the glory of God knows that here all truth and therefore all beauty lies, that he owes it to himself to surrender in love to this archetype [Christ], because he owes him his being and existence and can therefore only glorify him in an ascent to him in his life and work."[81] This ascent, this 'deepening of the Incarnation', this 'becoming who we are' is the beauty and duty of faith. Von Balthasar insists that "it is precisely the duty of one who ascends to Christ in faith, hope and love to interpret all the forms of God's revelation in Christ throughout the universe," and that

[79] Alan Heuser, *The Shaping Vision of Gerard Manley Hopkins* (London: Oxford University Press, 1958), p. 49. Also see Clark M. Brittain, "Logos, Creation, and Epiphany in the Poetics of Gerard Manley Hopkins" (Ph.D. Dissertation, University of Virginia, 1988), pp. 101-4.

[80] Hans Urs von Balthasar, *Studies in Theological Style: Lay Styles*, ed. John Riches, trans. Andrew Louth, John Saward, Martin Simon, and Rowan Williams, 7 vols., *The Glory of the Lord: A Theological Aesthetics*, vol. 3 (San Francisco: Ignatius Press, 1986), p. 390.

[81] Ibid., p. 391.

these 'forms' to be interpreted "are not concepts (of 'universal', abstract truths), but images (of the unique, personal, divine-human truth)."[82]

This interpretive task, envisioned by von Balthasar, is achieved by Hopkins the poet. Indeed, for this endeavor of aesthetic interpretation of Christic forms, poetry, von Balthasar claims, is "the absolutely appropriate theological language."[83] Here, von Balthasar offers us some initial reflection on Hopkins' 'sacramental poetry', and how the aesthetic experience of beauty, and therefore of Christ, can be 'captured' in poetic language.

C.2.3 Von Balthasar's Vision of Hopkins' 'Sacramental Poetry'

Von Balthasar offers helpful reflections on the 'theological character' of Hopkins' poetic language and the nature of his sacramental poetry. However, he places these reflections in an initial context which I believe does not honor Hopkins' intent. Therefore, before investigating von Balthasar's more specific comments on these topics, I must separate them from what I term his 'imperative context'.

Von Balthasar stresses Hopkins' 'interpretation of' Christic forms via poetry. Thus Hopkins 'fulfills his Christian duty' by "learning to read" the forms of nature under a "higher Christian law."[84] Indeed, as von Balthasar contends, Hopkins speaks of 'reading inscapes' throughout his writings.[85] This 'reading' of things

[82] Ibid.

[83] Ibid.

[84] Ibid., p. 391.

[85] Speaking of a certain type of cloud, Hopkins writes that its "make is easily read." See Humphry House, ed. *The Journals and Papers of Gerard Manley Hopkins*, 2nd ed. (London: Oxford University Press, 1959), p. 210. Speaking of a moonlit 'brindled heaven', he writes that "I read a broad careless inscape flow-

suggests a unity between the poet and what the Romantics referred to as "the spirit of nature." Von Balthasar insists that "for the Christian, the final creative unity [between poet and object] lies higher":

> [The Christian] must raise himself to this in his faith, in the great sacrifice; his enthusiasm may have no other source than does his faith. It is of this exultation as the whole man's engagement and effort that the poetic form must speak...[86]

Von Balthasar insists that "the unprecedented character of Hopkins' language is a theological phenomenon and can be understood only in this way."[87] He claims that this language was "unintelligible to a non-Christian like Bridges."[88] Though not disagreeing at all with his stress on the "unprecedented character" of Hopkins' language, I find von Balthasar's limitation of the possibility of truly understanding Hopkins' language and meaning to believers quite unacceptable. It is one thing to contextualize and unify Hopkins' theology, aesthetics, and language with an overarching, even radically, Christological vision. This is valid and true to Hopkins. However, the 'usefulness' of Hopkins' work should not therefore be limited to those with the vision of faith. 'Beauty', as we have spoken of it, is universally present. Its power and possibilities, its ability to 'enthuse', are not limited to those with faith and with a particular cipher for understanding Hopkins' theological language. It

ing throughout." See *Journals and Papers*, p. 218. Inscapes, as von Balthasar further points out, are discovered, unfolded, and caught. Cf. respectively *Journals and Papers*, pp. 199, 200, and 227.

[86] Hans Urs von Balthasar, *Studies in Theological Style: Lay Styles*, ed. John Riches, trans. Andrew Louth, John Saward, Martin Simon, and Rowan Williams, 7 vols., *The Glory of the Lord: A Theological Aesthetics*, vol. 3 (San Francisco: Ignatius Press, 1986), p. 392.

[87] Ibid.

[88] Ibid.

is one thing to hold that someone cannot fully understand Hopkins outside of a Christian context (although this is already debatable); it is quite another to infer that the 'power' of Hopkins' poetry, and specifically his language, as conveyors of inscape and 'Christic beauty', is somehow voided by a lack of faith, or, dare I say, by a lack of special knowledge. Does understanding make for the content and force of beauty in the Hopkinsian sense? To answer yes, as von Balthasar appears to do here, is comparable to a kind of gnosticism.[89] Beauty, in the Hopkinsian context, is not 'in the eye of the beholder'. It is there to be seen, to be open to — by anyone.

I believe that von Balthasar takes the sacramental character of Hopkins' poetry too far in this instance. This approach by von Balthasar reminds me of aspects of theological debates surrounding the nature of the Eucharist during, among other times, the Reformation years, when apologists polemically argued over whether Christ was present in the Eucharist via faith or present

[89] The dynamic here is reminiscent of that which Duns Scotus' argued against when he claimed that theology is a science that may be pursued by 'wayfarers', i.e., people not yet vouchsafed of the Beatific Vision. 'Wayfarers', according to Duns Scotus, "could pursue theology scientifically through experience which did not necessarily include faith." B.W. Ward explains: "Henry of Ghent, [Scotus'] predecessor, had argued that knowledge of God could only be pursued scientifically by those who had received the divine illumination of faith… Scotus conceded that faith was the normal route to theology, but he recognized that Henry left a dangerous opening for [a] radical fideism… [which] proclaimed a God who is an inconceivable and unsayable mystery, utterly dissevered from…human knowledge." Through his univocal term for talk about God — 'being' — Duns Scotus allowed for some knowledge of God which could be derived from a wayfarer's sensible experience of a limited and contingent world. A primitive knowledge of being is immediately available to everyone through the senses. Cf. Bernadette Ward, "Philosophy and Inscape: Hopkins and the Formalitas of Duns Scotus," *Texas Studies in Literature and Language* 32, no. 2 (1990), pp. 217-19. Hopkins, in my opinion, follows this univocal path more than the 'fideistic' way of Von Balthasar here.

independent of faith. It strikes me that von Balthasar is being quite 'Barthian' in this peculiar emphasis at the outset of his reflections on Hopkins' 'sacramental poetry'. Beauty's essential power is not determined in either content or effect by one's mode of understanding.[90] This prequel to von Balthasar's more helpful reflections on Hopkins' use of language is an example of why F.B. Brown finds von Balthasar's general 'theological aesthetics' inaccessible to philosophic aesthetics and therefore 'idiosyncratic'. Let us now begin analyzing von Balthasar's more helpful criticism of Hopkins' 'sacramental poetry'.

Clearly, Hopkins was greatly occupied by the use of language in his poetry. His development of an utterly unique style of syntax and rhetoric caused him great concern at times. Upon sharing his work with friends and colleagues, he was often criticized or met with incomprehension. For example, Robert Bridges referred to his

[90] V.R. Ellis comments on this from the non-sectarian view of a scholar of literature: "[Hopkins'] poems, at their best, can at least partially transform the world for us, whether we are believers or non-believers, as Christian faith wholly transformed it for Hopkins." See Virginia Ridley Ellis, *Gerard Manley Hopkins and the Language of Mystery* (Columbia: University of Missouri Press, 1991), p. 7. Additionally, von Balthasar was either unaware of or ignored certain Hopkinsian texts in framing his 'Gnostic' approach. For example, Hopkins writes that in creating poetic language, or "the language of inspiration," there must be a "mood of great, abnormal in fact, mental acuteness, either energetic or receptive, according as the thoughts which arise in it seem generated by a stress and action of the brain, or to strike it unasked." There should correspond in the hearer of this language the stimulation of a similar acute mental and even physical reaction: "In a fine piece of inspiration [poetry] every beauty takes you as it were by surprise... every fresh beauty could not in any way be predicted or accounted for by what one has already read." This 'unbidden' or 'surprising' quality of poetic beauty does not correlate well with von Balthasar's 'context of belief' emphasis. Cf. Claude Colleer Abbott, ed. *Further Letters of Gerard Manley Hopkins Including his Correspondence with Coventry Patmore*, 2nd ed. (London: Oxford University Press, 1956), pp. 216 and 217.

work as obscure and affectatious.[91] The poet and colleague of Hopkins, Coventry Patmore, was only slightly more kind in his criticism:

> My dear Hopkins, I have read your poems — most of them several times — and find that my first impression is confirmed with each reading. It seems to me that the thought and feeling of these poems, if expressed without any obscuring novelty of mode, are such as often to require the whole attention to apprehend and digest them; and are therefore of a kind to appeal only to a few. But to the already sufficiently arduous character of such poetry you seem to me to have added the difficulty of following *several* entirely novel and simultaneous experiments in versification and construction, together with an altogether unprecedented system of alliteration and compound words; any one of which novelties would be startling and productive of distraction from the poetic matter to be expressed... I often find it...hard to follow you.[92]

Hopkins attempted not to be disappointed by this lack of comprehension and appreciation his poems found among his poet friends. In one place he states: "I laughed outright and often, but very sardonically, to think that you [Bridges] and the Canon [Richard Watson] could not construe my last sonnet...It is plain

[91] Claude Colleer Abbott, ed. *The Letters of Gerard Manley Hopkins to Robert Bridges*, 2nd ed. (London: Oxford University Press, 1955), p. 54. In this same letter, dated May 30, 1878, Hopkins does return the criticism to Bridges when he writes: "Obscurity I do and will try to avoid so far as is consistent with excellences higher than clearness at a first reading." This position is a echo of Ruskin's position when he writes: "Excellence of the highest kind, without obscurity cannot exist." Sulloway quotes this from *Modern Painters*. See Alison G. Sulloway, *Gerard Manley Hopkins and the Victorian Temper* (New York: Columbia University Press, 1972), p. 100.

[92] Claude Colleer Abbott, ed. *Further Letters of Gerard Manley Hopkins Including his Correspondence with Coventry Patmore*, 2nd ed. (London: Oxford University Press, 1956), pp. 352-53.

I must go no further down this road: if you and he cannot understand me who will?"[93] Nonetheless, 'down this road' he continued.

As von Balthasar highlights, because Hopkins framed poetry as language meant to carry inscape, he believed that language with this task must be revalued: "it must be equipped anew so as to be able to express" what it carries.[94] Along with the poet, poetic language must be able to reach out "beyond its immanence because the mystery of God does not hold sway as something incomprehensible *behind* the forms of the world," for the Word was made flesh.[95] For Hopkins, the world is 'Worded', and this 'Wording' is the source of the world's intelligibility. Hopkins moves from the divine 'Wording' of the world which brings it into being, to knowing the world via each thing's or person's unique 'Wording' (inscape via instress), to humanly 'wording' the world via poetic language, thereby 'capturing' and conveying inscape, and its potential transforming power, to others.

In this progressive dynamic, the concept of the sacramental comes into play for Hopkins. This is how the mystery of God takes form in the world. Von Balthasar offers a helpful reflection on how Hopkins sees his poetic language in this light. Poetic language, as 'sacramental', "certainly contains within itself the power of 'symbol', while it goes far beyond it;" for "the form of the image is a likeness to the primordial form in that it has the 'stress' of the latter in itself: *sacramenta continent quae signifi-*

[93] Claude Colleer Abbott, ed. *The Letters of Gerard Manley Hopkins to Robert Bridges*, 2nd ed. (London: Oxford University Press, 1955), p. 272.

[94] Hans Urs von Balthasar, *Studies in Theological Style: Lay Styles*, ed. John Riches, trans. Andrew Louth, John Saward, Martin Simon, and Rowan Williams, 7 vols., The Glory of the Lord: A Theological Aesthetics, vol. 3 (San Francisco: Ignatius Press, 1986), p. 392.

[95] Ibid., p. 393.

cant."[96] However, as seen previously in this study, the concept of 'symbol' alone is not adequate here, for "the mystery incorporated in the inscapes is not to be grasped as a straightforward picture."[97] For Hopkins, as von Balthasar writes, inscape as 'Christic' carries much more:

> The mystery of Christ is, on the one hand, of infinite depth, penetrating all the levels of being from flesh to spirit and beyond into the abyss of the Trinity; on the other, it is an infinitely dramatic event that in the kenotic descent into man and matter exalts and changes them, redeems and deifies them... The image that should interpret the mystery of Christ is, in itself as an image of nature, utterly overtaxed, but in so far as it is grounded in Christ as the presupposition of nature, it is allowed to say by grace of the archetype what it cannot say of itself.[98]

What this means is that when Hopkins takes some reality, a 'Windhover' for example,[99] he is not arbitrarily relating it through image to Christ; he does not intend allegory. Rather, as von Balthasar aptly states, the Hopkinsian background here is that there is a "fundamental, ontological presupposition of all natural processes that all, knowingly or not, intrinsically signify or intend by pointing beyond themselves...God is immediately visible in the beauty of the world."[100] Hopkins' poetic context is that "the world is charged with the grandeur of God. It will flame out, like shining

[96] Ibid., pp. 393-94.

[97] Ibid., p. 394.

[98] Ibid.

[99] See Norman H. MacKenzie, ed. *The Poetical Works of Gerard Manley Hopkins* (Oxford: Clarendon Press, 1990), p. 144.

[100] Hans Urs von Balthasar, *Studies in Theological Style: Lay Styles*, ed. John Riches, trans. Andrew Louth, John Saward, Martin Simon, and Rowan Williams, 7 vols., The Glory of the Lord: A Theological Aesthetics, vol. 3 (San Francisco: Ignatius Press, 1986), pp. 394-95.

from shook foil."[101] The reality of the divinely kenotic 'Great Sacrifice' is not projected onto reality for Hopkins; rather, in the encounter between nature and beholder, the 'Great Sacrifice' is immediately experienced in some sense and *may* be 'discovered' for what it is. For von Balthasar, this discovery requires a dogmatic, Christian perception, although he does admit that it is impossible to say where the natural perception of God in nature ceases and the supernatural, Christian perception begins.[102] He claims that the "transfer of interpretation from sacramental signs to the indwelling grace of faith proceeds imperceptibly."[103] In any case, given this context, one begins to see why it may be a valid interpretation of Hopkins to claim 'poetry as sacrament'.

Although von Balthasar's presentation of Hopkins within the context of a theological aesthetic is not systematic, he nonetheless offers a first effort at synthesis which highlights some key elements of import to this study. Ultimately, poetry, in Hopkins, is a conveyance and heightening of some experience of unique divine self-expressiveness, 'beauty', or inscape. As such, it allows an encounter which can 'lay claim to the whole person' and aid one in the assent of faith. This encounter takes place within the adapted Scotistic context of the 'Great Sacrifice', which is also the context allowing Hopkins to relate 'being' and 'wording'. The dynamic of 'the Glory of God', that is, God's self-expressive beauty or Christic inscape in matter, being gleaned through the senses, 'in-selved' via instress, and then 're-incarnated' as word, is

[101] Norman H. MacKenzie, ed. *The Poetical Works of Gerard Manley Hopkins* (Oxford: Clarendon Press, 1990), p. 139.

[102] Hans Urs von Balthasar, *Studies in Theological Style: Lay Styles*, ed. John Riches, trans. Andrew Louth, John Saward, Martin Simon, and Rowan Williams, 7 vols., The Glory of the Lord: A Theological Aesthetics, vol. 3 (San Francisco: Ignatius Press, 1986), p. 396.

[103] Ibid.

the content of Hopkins' theological aesthetic. In spite of some limitations placed on the 'usefulness' of this aesthetic by von Balthasar's somewhat 'Gnostic' approach, his foray into Hopkins' undeveloped concept of 'poetry as sacrament' leads us now to a study of the 'matter' of this sacrament — Hopkins' vision and use of language in poetry.

D. Language in the 'Poem as Sacrament'

D.1 Introduction

In our study of John Ruskin's influence upon Hopkins, we saw how Ruskin loosely associates art with religion in claiming that "all great art is praise." Furthermore, Ruskin only loosely associates the thing seen and the 'idea' of God in his writing. Hopkins transforms this 'loose' association, as we have seen above, through his assimilation of certain Ignatian emphases and Scotist theology. Hopkins brings 'the created' and 'the Creator' very close in his theory of poetry. This creative union is essentially based upon Hopkins' theology of the Incarnation which is ultimately a theology of creation. In this approach, creation is dependent upon the decree of the Incarnation and not vice-versa. For Hopkins, the world, as Christopher Devlin puts it, is created as a field for Christ in which to exercise his adoration of the Father.[104] This adoration of the Father, or 'Christic selving', can be found in the world in two ways by the poet. First, it may be found through the presence of God's design or inscape (that is, Christic form) in nature,[105] and second,

[104] Christopher Devlin, S.J., ed. *The Sermons and Devotional Writings of Gerard Manley Hopkins* (London: Oxford University Press, 1959), p. 109.

[105] Patricia M. Ball, *The Science of Aspects: The Changing Role of Fact in the Work of Coleridge, Ruskin and Hopkins* (London: The Athlone Press, 1971), p. 110.

it may be found in "the working out of that design in the minds and wills" of persons.[106] As Ball adeptly states, "in the 'selves' of the universe, and in man's striving to reach his true inscape...Hopkins finds the personality, the beauty of Christ."[107]

What Hopkins finds as Christic inscape is also what he intends to convey through the medium of words as poetry. Poetry is words 'framed' to carry inscape. When Hopkins studies the inscape of a bluebell, and states that he knows "the beauty of the Lord by it,"[108] he is speaking of a unique form of Christic selving which can not only be known, but can also be 'carried over' in poetry. The "stem of stress," as Hopkins states, between human beings and nature is the word itself:

> And what is Earth's eye, tongue, or heart else, where
> Else, but in dear and dogged man?[109]

As Miller writes, "words have for Hopkins a magic quality of attaining the object, wresting from it its meaning and making that meaning a permanent possession for man."[110] Words are meant, for Hopkins, to 're-create' the object, "to demonstrate that 'things are' and *how* they are."[111] For Hopkins, then, there is a close relationship between being, knowing, and wording:

[106] Christopher Devlin, S.J., ed. *The Sermons and Devotional Writings of Gerard Manley Hopkins* (London: Oxford University Press, 1959), p. 109.

[107] Patricia M. Ball, *The Science of Aspects: The Changing Role of Fact in the Work of Coleridge, Ruskin and Hopkins* (London: The Athlone Press, 1971), p. 111.

[108] Humphry House, ed. *The Journals and Papers of Gerard Manley Hopkins*, 2nd ed. (London: Oxford University Press, 1959), p. 199.

[109] Norman H. MacKenzie, ed. *The Poetical Works of Gerard Manley Hopkins* (Oxford: Clarendon Press, 1990), p. 171.

[110] Joseph Hillis Miller, *Victorian Subjects* (Durham: Duke University Press, 1991), p. 13.

[111] Patricia M. Ball, *The Science of Aspects: The Changing Role of Fact in the Work of Coleridge, Ruskin and Hopkins* (London: The Athlone Press, 1971), p. 112.

> To be and to know or Being and thought are the same. The truth in thought is Being, stress, and each word is one way of acknowledging Being and each sentence by its copula *is* (or its equivalent) the utterance and assertion of it.[112]

The context for this approach, as we have seen above, is Hopkins' adoption of a kind of intuitive cognition. This approach is in contrast to Thomism. The Thomistic tenet that a person knows only universals or essences pushes away the possibility of any kind of 'direct' or intuitive knowledge of a singular material thing. Thomism claims that the intellect cannot know material objects directly; rather, it knows them through phantasms in which the intellect apprehends the universals, because it has abstracted the intelligible species of a thing.[113] This theory of knowledge, which moves towards the essence of an object instead of the 'thisness' of an object, is in contrast to the epistemological context which makes Hopkins' poetry and his specific vision of words possible. This philosophical context he found in Duns Scotus.

The epistemology of Duns Scotus allowed Hopkins to justify his focus on the senses. In contrast to Thomistic epistemology which holds a material thing at a distance, so to speak, in order to discover its essence beyond its accidental properties, Scotism allows for a certain 'intimacy' with the thing known. The knower, in the Scotistic context, comprehends all that inheres in the subject "so

[112] Humphry House, ed. *The Journals and Papers of Gerard Manley Hopkins*, 2nd ed. (London: Oxford University Press, 1959), p. 129. This tenet of Hopkins probably started with "the pre-Socratic notion that the Being in things is the being in the self and the being in words — all the same universal being." We certainly know that Hopkins was drawn to Parmenides' philosophical fragments on the unity of being. Cf. David Anthony Downes, *Hopkins' Achieved Self* (Lanham: University Press of America, 1996), p. 54.

[113] Bernard J. Quint, "The Nature of Things: Hopkins and Scotus," in *Twilight of Dawn: Studies in English Literature in Transition*, ed. O. M. Brack, Jr. (Tucson: The University of Arizona Press, 1987), p. 85.

that the whole is the object of knowledge... What is real about the subject out there inheres in the subject as the object of knowledge."[114] What is known, in Scotesian philosophy, is the thing. In this context, the sensible 'accidents' of a thing are not qualities which disturb the search for essence in the abstractive process, rather they become necessary in the intuitive experience of a thing. The thing considered is something so individual in this context that it must stand radically so. This philosophical stress matched Hopkins' already existent aesthetic stress, adopted from Ruskin, perfectly. This movement toward accuracy and objectivity in the sensation of a thing became part of the inscaping process for Hopkins. It also influenced Hopkins' use of words in poetry.

For example, let us take some of the words from Hopkins' poem *That nature is a Heraclitean Fire and of the Comfort of the Resurrection*:

> Cloud-puffball, torn tufts, tossed pillows | flaunt forth, then
> chevy on an air-
> built thoroughfare; heaven-roysterers, in gay-gangs | they
> throng; they glitter in marches.
> Down roughcast, down dazzling whitewash, | wherever an
> elm arches,
> Shivelights and shadowtackle in long | lashes lace, lance, and
> pair...[115]

In a Thomistic context, these words represent a set of accidental conformations which cannot speak of the essence of the thing they describe nor be objective or realistic precisely because they do not deal with the essence. In Thomistic terms, Hopkins' understanding of a cloud is rather the use of the power of imagination. As Quint writes:

[114] Ibid., p. 86.

[115] Norman H. MacKenzie, ed. *The Poetical Works of Gerard Manley Hopkins* (Oxford: Clarendon Press, 1990), pp. 197-8.

> When the image is of a cloud or clouds, an individual thing or group of things is created. It is the image that is known, not the cloud or clouds. The knowledge of the cloud or clouds is only indirect because the conversion to a phantasm is what allows the mind to have any knowledge of the thing. When the imagination converts the thing to an image, the mind knows the thing indirectly only because the image alone is presented to the mind.[116]

The words of the poem take on a different nature in the context of Scotistic thought. For Scotus and Hopkins, the apprehension of an object is intense and real. The senses are central to the knowledge not only of the accidents of a thing but of the thing itself. As Quint summarizes:

> Although the senses apprehend the thing, the intellect knows intuitively what the senses apprehend. Even though the knowledge of a singular thing is imperfect, it is so because our intellectual operations are imperfect. The imperfection, however, does not prevent the intellectual intuition of a singular thing as existence.[117]

Therefore, Hopkins' words in this poem, such as "puffball," "tossed pillows," "chevy on an air-built thoroughfare," and "heaven-roysterers" are not simply descriptions of physical outlines or shapes that are accidental conformations of a thing. Rather, since the thing poeticized, in this case a cloud, has a unique specificity at a particular moment in time, the attempt to define the thing at a particular moment is an attempt to apprehend the 'thisness' which marks it essentially. Thus the words used in Hopkins' poem about clouds are more than a statement of an essence abstracted from material things. They are, rather, an example of "the way of realistically describing...the peculiar 'thisness' that belongs to

[116] Bernard J. Quint, "The Nature of Things: Hopkins and Scotus," in *Twilight of Dawn: Studies in English Literature in Transition*, ed. O. M. Brack, Jr. (Tucson: The University of Arizona Press, 1987), p. 88.

[117] Ibid.

clouds as those clouds without diminishing their nature as clouds."[118] The emphasis is on the object sensibly and intensely observed as present in its actual existence and none other. The context, then, is a kind of intuition that focuses sharply on actual existence and apprehends the present thing as existent. Hopkins spurns the distinction between essence and existence. He is not concerned with the universal, but with wording and thus poetically inscaping an individual existing thing in a particular moment of time — "I caught this morning morning's minion...dapple-dawn-drawn Falcon, in his riding..."[119]

Words, in this sense, bring about an incarnation, or better yet, a re-incarnation for Hopkins.[120] This certainly places unusual strain upon language, because for Hopkins, to word a thing in its inscape appropriately is potentially to 'selve' a thing in poetic language. Downes describes this dynamic well:

> The poet, as Hopkins put it, "takes in an inscape," which is to say that he apprehends ("instresses") a thing to the level of its actualized identification, so deep does its meaning penetrate, and then its singular form is poetically metamorphized ("stressed") in making the utterance, that is, its "inscaped" form is reexpressed in the verbal form of the poem, a new reincarnation of its "inscaped" ontic form. The poetic utterance in its verbal body is the very eidolon of the original "inscape" possessing in its mimetic symbolism the dynamic nature, ideality, and meaning of the thing.[121]

[118] Bernard J. Quint, "The Nature of Things: Hopkins and Scotus," in *Twilight of Dawn: Studies in English Literature in Transition*, ed. O. M. Brack, Jr. (Tucson: The University of Arizona Press, 1987), p. 88.

[119] Norman H. MacKenzie, ed. *The Poetical Works of Gerard Manley Hopkins* (Oxford: Clarendon Press, 1990), p. 144.

[120] Patricia M. Ball, *The Science of Aspects: The Changing Role of Fact in the Work of Coleridge, Ruskin and Hopkins* (London: The Athlone Press, 1971), p. 112.

[121] David Anthony Downes, *Hopkins' Achieved Self* (Lanham: University Press of America, 1996), p. 11.

However, whenever Hopkins speaks of poetic language, it is important to remember that he is not speaking only of 'written' language. In fact, Hopkins places great emphasis upon poetic speech, as opposed to the 'meaning' of the written word.[122] The 'wording' of inscape requires the actual 'making flesh' of the poetic language; i.e., it requires the vividness and heightened sense of poetic speech. As Ball writes, "he sees it epitomizing the verbal role and achieving the maximum emphasis, that extreme 'vividness of idea' which is the idea made flesh."[123] Declaiming a poem can make all the difference in its effectiveness as a mediator of inscape. Hopkins speaks of this declamation as follows:

> It is plain that metre, rhythm, rhyme, and all the structure which is called verse both necessitate and engender a difference in diction and in thought. The effect of verse is one on expression and thought, viz. *concentration* [my italics] and all which is implied by this. This does not mean terseness nor rejection of what is collateral nor emphasis nor even definiteness though these may be very well, or best, attained by verse, but mainly, though the words are not quite adequate, vividness of idea or... liveliness.[124]

The stimulation of the concentration, or 'contemplation', of the hearer is key (just as 'contemplation' is key for the seer of inscape). Thus, Hopkins defines poetry as 'speech in relation to inscape':

[122] Poetry carries 'meaning' that goes beyond its words for Hopkins. The sound and structure is also a kind of 'meaning' in poetry: "Poetry is speech framed for contemplation of the mind by way of hearing or speech framed to be heard for its own sake and interest even over and above its interest of meaning." Humphry House, ed. *The Journals and Papers of Gerard Manley Hopkins*, 2nd ed. (London: Oxford University Press, 1959), p. 289.

[123] Patricia M. Ball, *The Science of Aspects: The Changing Role of Fact in the Work of Coleridge, Ruskin and Hopkins* (London: The Athlone Press, 1971), p. 112.

[124] Humphry House, ed. *The Journals and Papers of Gerard Manley Hopkins*, 2nd ed. (London: Oxford University Press, 1959), p. 84.

> Poetry is in fact speech only employed to carry the inscape of speech for the inscape's sake — and therefore the inscape must be dwelt on. Now if this can be done without repeating it *once* of the inscape will be enough for art and beauty and poetry but then at least the inscape must be understood as so standing by itself that it could be copied and repeated. If not/ repetition, *oftening, over-and-overing, aftering* of inscape must take place in order to detach it to the mind and in this light poetry is speech which afters and oftens its inscape...[125]

Additionally, Hopkins emphasizes the verse, rhythm, metre, etc., all as ways of stressing the inscape. They are tools, or part of the 'matter', of the conveyance of inscape in poetic speech which allows the poem to impress on minds the inscapes of the world.[126] This 'impression of inscape' is, however, not the same as the impression of an intellectual 'meaning' for Hopkins. The poem in its entirety, in its words, pattern, sound, and technique, conveys inscape through a fusion of thought and feeling. This poetic fram-

[125] Humphry House, ed. *The Journals and Papers of Gerard Manley Hopkins*, 2nd ed. (London: Oxford University Press, 1959), p. 289.

[126] Patricia M. Ball, *The Science of Aspects: The Changing Role of Fact in the Work of Coleridge, Ruskin and Hopkins* (London: The Athlone Press, 1971), p. 113. Marucci makes a related point in speaking about the *figure* of the poem and its role in impressing meaning. He defines *figure* here as "not only those operations recognized and classified by classical rhetoric but also any *écart* manifest both in the 'inferior' levels — phonological, morphological, lexical, grammatical, syntactic, rhythmic, metrical, prosodic, etc. — as well as in the 'superior' levels of the poetic text — enunciative, semantic, symbolic, etc. The figure has...the task of semantically and emotionally intensifying a message that, without it, might become weaker or be condemned to 'dispersion' on its way from the sender to the receiver, as bearer of a new and unheard-of *sense* which might be easily taken for literal *counter-* or *non-sense*. The figure therefore rouses and keeps the reader's attention, saves him from intellectual routine, opens up deeper furrows for the penetration of the sense and helps the reader to appropriate it integrally." This is part of the 'bidding' function in Hopkins' poetry (see below). See Franco Marucci, *The Fine Delight That Fathers Thought: Rhetoric and Medievalism in Gerard Manley Hopkins* (Washington, D.C.: The Catholic University Press, 1994), pp. 20-21.

ing and conveyance of inscape in a poem should not be seen as some linear exercise in certainty. Key to Hopkins' use of poetic language and how he believed it could carry inscape is his concept of 'mystery'. To this subject we now turn.

D.2 'Mystery' and Poetic Language in Hopkins

As we have seen thus far in this study, Hopkins created a kind of theological aesthetic which bound together being, knowing, and wording. This synthesis was based upon the mystery of the Incarnation, in a Scotistic sense, which led to a radically sacramental vision of created reality.[127] This synthesis helped Hopkins 'define' and 'legitimize' the experience of beauty, as well as the poetic attempt to 'capture' and convey this beauty in the service of faith and praise. The 'bridges' within this synthesis, as von Balthasar holds, are Hopkins' concepts of inscape and instress.

In an attempt to delineate Hopkins' dense and varied explications of inscape and instress, there is a kind of mental momentum which desires to simplify in such a way that the movement from the intuitive cognition of the 'divine wording' of something to its ultimate 'capture' in language becomes almost mathematical in nature — "a trick found out" as Hopkins would say. Such an impression would neither honor Hopkins' point of view nor the way in which poetic language can be sacramental. A corrective to the possible oversimplification of Hopkins' meaning, as well as an interpretive key to his method as a poet, is his understanding of 'mystery'.[128]

[127] As Christopher Devlin made so clear in his writings, "behind and under the words [of Hopkins]," one finds "the philosophy of Duns Scotus." See Madeleine Devlin, *Christopher Devlin* (London: Macmillan Co., 1970), p. 30.

[128] This is emphasized by Hopkins' own words in *The Wreck of the Deutschland*:

Hopkins, unfortunately, did not say a great deal about mystery as such. However, in a debate with his longtime friend and agnostic, Robert Bridges, he paints the outline of his understanding and experience of mystery. I quote Hopkins at length here as he attempts to educate his poet friend:

> But by the way you say something I want to remark on: "Even such a doctrine as the Incarnation may be believed by people like yourself", as a mystery, till it is formulated, but as soon as it is it seems dragged down to the world of pros and cons, and "*as its mystery goes*, so does the hold on their minds." Italics the present writer's. You do not mean by mystery what a Catholic does. You mean an interesting uncertainty: the uncertainty ceasing interest ceases also. This happens in some things; to you in religion. But a Catholic by mystery means an incomprehensible certainty: without certainty, without formulation there is no interest (of course a doctrine is valuable for other things than its interest, its interestingness, but I am speaking now of that); the clearer the formulation, the greater the interest. At the bottom the source of interest is the same in both cases, in your mind and in ours; it is the unknown, the reserve of truth beyond what the mind reaches and still feels to be behind. But the interest a Catholic feels is, if I may say so, of a far finer kind than yours. Yours turns out to be a curiosity only; curiosity satisfied, the trick found out (to be a little profane), the answer heard, it vanishes at once. But you know, there are some solutions to, say, chess problems so beautifully ingenious, some resolutions of suspensions so lovely in music that even the feeling of interest is keenest when they are known and over, and for some time survives the discovery. How

Since, tho' he is under the world's splendour and wonder
His **mystery** must be instressed, stressed;
For I greet him the days I meet him, and bless when I understand.

See Norman H. MacKenzie, ed. *The Poetical Works of Gerard Manley Hopkins* (Oxford: Clarendon Press, 1990), p. 120. J.R. Watson claims that "mystery is the central instress of God" for Hopkins. "It must be the force which is felt by the believer" in poetry. See J.R. Watson, *The Poetry of Gerard Manley Hopkins* (London: Penguin Books, 1987), p. 34.

> must it then be when the very answer is the most tantalising statement of the problem and the truth you are to rest in the most pointed putting of the difficulty![129]

Here we begin to find hints of what Hopkins hopes for from his poetry and poetic language. Although he believes that this language is the best possibility of carrying and conveying inscape, or Christic beauty, and that what it conveys is in some sense very 'objective', nonetheless, he sees this in the context of mystery. Mystery, for Hopkins, is in part the knowing of something as an 'incomprehensible certainty' in such a way that one's interest, concentration, or even contemplation of the 'reserve of truth beyond' the mystery is held and intensified; i.e., "the clearer the formulation, the clearer the interest." The mystery of God takes form in the world through the Word, but to grasp the Word in the world as beauty and to convey this as word to others is not the same as wholly grasping God. As Ellis states, "a God who could be wholly grasped would not be God; a God who could be wholly named would not generate nor be worthy of either faith or poetry."[130] Hopkins continues in this vein with Bridges:

> For if the Trinity [for example]...could be explained by grammar and tropes, why then he could furnish explanations for himself; but then where wd. be the mystery? the true mystery, the incomprehensible one. At that pass one should point blank believe or disbelieve... There are three persons, each God and each the same, the one, the only God: to some people this is a 'dogma', dull algebra of schoolmen; to others it is news of their dearest friend or friends, leaving them all their lives balancing whether they have three heavenly friends or one — not that they have any doubt on the subject, but that their knowledge leaves their minds swinging; poised, but on a

[129] Claude Colleer Abbott, ed. *The Letters of Gerard Manley Hopkins to Robert Bridges*, 2nd ed. (London: Oxford University Press, 1955), pp. 186-87.

[130] Virginia Ridley Ellis, *Gerard Manley Hopkins and the Language of Mystery* (Columbia: University of Missouri Press, 1991), p. 20.

> quiver. And this might be the *ecstasy* [my italics] of interest, one would think.[131]

Extrapolating a bit, here it is implied that Hopkins means poetic language to convey an immediate experience of the beauty of God, as mystery, in the world. 'As mystery' refers not only to the inability of grasping God in whole, but also to the achievement of leaving the hearers' minds "swinging; poised, but on a quiver;" that is, allowing the hearers of the poetically worded inscape to encounter this sensible self-expressiveness, or beauty, of God in such a way that they can focus, concentrate, even contemplate the "reserve of truth beyond" a particular form of Christic selving.[132] In other words, as von Balthasar holds, the poetic stress of Hopkins here is to put forward the self-expressive Word, the 'ec-stasy of God', in such a way that the whole person confronts "the *visi-*

[131] Claude Colleer Abbott, ed. *The Letters of Gerard Manley Hopkins to Robert Bridges*, 2nd ed. (London: Oxford University Press, 1955), pp. 187-88.

[132] I believe this dynamic is otherwise expressed by Paul Tillich. He said: "The artist brings to our senses and through them to our whole being something of the depth of our world and of ourselves, something of the mystery of being. When we are grasped by a work of art things appear to us which were unknown before — possibilities of being, unthought-of powers, hidden in the depth of life which take hold of us." Paul Tillich, *On Art and Architecture* (New York: Crossroad Publishing Company, 1987), p. 247. From the hell that was World War I, Tillich gives a personal example. As a chaplain on the front, he used to thumb through magazines by candlelight during the lulls in fighting at Verdun. There were reproductions of famous paintings in them. One reproduction of *Madonna and Child with Singing Angels* by Botticelli especially comforted him. After the war, he went to the Kaiser Friedrich Museum in Berlin to see the original. He writes: "I felt a state approaching ecstasy... In the beauty of the painting there was Beauty itself. It shone through the colors of the paint as the light of day shines through the stained-glass windows of a medieval church. As I stood there, bathed in the beauty its painter had envisioned so long ago, something of the divine source of all things came through to me. I turned away shaken... That moment has affected my whole life, given me the keys for the interpretation of human existence, brought vital joy and spiritual truth." Ibid., p. 235.

bleness of the *Invisible* one, *the* definitive and determinant form of God in the world and the primal archetypal source of all worldly beauty,"[133] and in doing so, in coming to this "ecstasy of interest," may assent to the presence with a responding *ec-stasy* of faith, praise, and delight. Otherwise put, in poetic language, Hopkins hopes to convey mystery, which, as he says vis-à-vis the Trinity as an example, is not "dogma, the dull algebra of schoolmen," but rather "news of dearest friends." Poetic language, as mysterious in this sense, opens possibilities for a more humanly whole 'encounter' with God. This is brought out as Hopkins continues his explanation to Bridges:

> So too of the Incarnation, a mystery less incomprehensible, it is true: to you it comes to: Christ is in some sense God, in some sense he is not God — and your interest is in the uncertainty; to the Catholic it is: Christ is in every sense God and in every sense man, and the interest is in the locked and inseparable combination, or rather it is in the person in whom the combination has its place. Therefore we speak of the events of Christ's life as mysteries...the mystery being always the same, that the child in the manger is God, the culprit on the gallows God, and so on. Otherwise birth and death are not mysteries, nor is it any great mystery that a just man should be crucified, but that *God should* [my italics] fascinates — with the interest of awe, of pity, of shame, of every harrowing feeling. But I have said enough.[134]

Poetic language is meant to carry inscape. This presumes, in Hopkins' synthesis, that God is sensible in self-expressive, Christic forms in the world. Beauty, in this sense, is the visible form of

[133] John Coulson, "Hans Urs von Balthasar: Bringing Beauty Back to Faith," in *The Critical Spirit and the Will to Believe: Essays in Nineteenth-Century Literature and Religion*, ed. David Jasper and T.R. Wright (New York: St. Martin's Press, 1989), p. 230.

[134] Claude Colleer Abbott, ed. *The Letters of Gerard Manley Hopkins to Robert Bridges*, 2nd ed. (London: Oxford University Press, 1955), p. 188.

the invisible God, a 'comprehension' of the "incomprehensible certainty", in other words, 'mystery'. This "incomprehensible certainty" which takes form in the world as mystery requires and is receptive to formulation,[135] and as Hopkins states above, "the clearer the formulation the greater the interest." The context for Hopkins' vision of 'the poem as sacrament', and his claims for the possibilities of poetic language, ultimately acknowledge "the unknown, the reserve of truth beyond what the mind reaches and still feels to be behind." Thus, Hopkins' poetic language possesses a dynamic which is certainly 'sacramental' — it both reveals and obscures its 'subject'. As Ellis states, Hopkins' poetic language is "characterized on the one hand by concreteness, distinctness, precision, vividness of language, sharpness of form, and on the other by complexity, density, multiplicity of meaning, even obscurity of meaning."[136] She continues:

> Either deliberately, half-deliberately, or simply instinctively and inevitably, given [Hopkins'] poet-priest's vision of the world, [Hopkins' poetic method] is a sacramental method, a visible sign of the invisible God, in all its aspects working to incarnate the great mystery in words that give it precise "formulation," but a formulation that...never gives us the sense of "the trick found out," never fails to make us feel that "the truth you are to rest in [is] the most pointed putting of the difficulty." In method as well as content, the Real Presence is thus celebrated as real. Divinity in Hopkins' poems informs what represents it, never wholly or neatly contained by its "symbols" of language, rhythm, form, movement, music, but so permeating them that it is felt to be both tangibly immanent and ungraspably transcendent. As a result, the effect of this poetry...is "to leave our minds swinging; poised, but on the quiver."[137]

[135] Virginia Ridley Ellis, *Gerard Manley Hopkins and the Language of Mystery* (Columbia: University of Missouri Press, 1991), p. 19.

[136] Ibid., pp. 19-20.

[137] Ibid., p. 20.

The poem as sacrament then becomes, in all its aspects, a working to incarnate the great mystery in words that give it precise formulation. Clearly, as mentioned above, this places a heightened stress on words and what words in their fullness may convey. Let us now look at Hopkins' approach to the nature of words in the poem as sacrament.

D.3 Hopkins' Theory of Words in the Poem as Sacrament

Hopkins holds that the 'subject' of poetry as mystery and "incomprehensible certainty" can only be insulted and shrunk by mere clarity of words. Hopkins expresses this thought in a variety of ways. For example, Hopkins insists that there are "excellences higher than clearness at first reading" in poetic language.[138] In other words, 'obscurity' also has a purpose in the poetic conveyance of inscape.[139] As Hopkins states, the reader or listener of poetry should have one of two kinds of clearness — "either the meaning to be

[138] Claude Colleer Abbott, ed. *The Letters of Gerard Manley Hopkins to Robert Bridges*, 2nd ed. (London: Oxford University Press, 1955), p. 54.

[139] Marucci would not agree with my choice of words here. He argues that obscurity, in Hopkins, is a major poetic vice. 'Realism' is the key word for Marucci in describing the character of Hopkins' poetry. I do not think we are in total disagreement here. The 'obscurity' I speak of is methodological; i.e., it is for the purpose of a deeper and more impressive clarity and perception of the poetic inscape. Marucci is saying the same thing in a different manner. He argues that Hopkins' wordplay and rhetoric is after a clarity of perception, a realistic 'sensation' of the object poeticized. "In the domain of poetry, realism becomes a synonym for…palpability, vividness, graphicness, and highlighting of particulars, so as to bring them into relief… " See Franco Marucci, *The Fine Delight That Fathers Thought: Rhetoric and Medievalism in Gerard Manley Hopkins* (Washington, D.C.: The Catholic University Press, 1994), pp. 27-29. I agree that Hopkins wants to create vivid figures before the mind's eye, but he does not necessarily believe that this must be done 'on the run'. Putting off immediate clarity to achieve a more powerful overall grasping and 'perceptive explosion' is a higher poetic good.

felt without effort as fast as one reads or else, if dark at first reading," the meaning had when the whole is grasped, so that, "once made out," the meaning "explodes."[140] The latter is a preferred method of 'clearness' in poetry. Hopkins attempts to explain this to Bridges:

> Epic and drama and ballad and many, most, things should be at once intelligible; but everything need not and cannot be. Plainly [in poetry] if it is possible to express a subtle and recondite thought on a subtle and recondite subject in a subtle and recondite way and with great felicity and perfection, in the end, something must be sacrificed, with so trying a task, in the process, and this may be the being at once, nay perhaps even the being without explanation at all, intelligible.[141]

In otherwords, to achieve greatest 'clarity' in poetry which honors the mystery of its most recondite subject (God), what is at times sacrificed is the immediate comprehension of the language used to 'word' the subject. This approach touches upon Hopkins' thoughts on the nature of language in general and of words in particular.

Hopkins contends that English, as a poetic language, has a too-specific and too-limiting tendency: "This seems in English a point craved for and insisted on, that words shall be single and specific marks for things, whether self-significant or not."[142] This is not an inclination that Hopkins adopts in his use of poetic language. Rather, Hopkins postulates a theory of the "moments of words" which "reconciles richness and multiplicity of suggestion

[140] Claude Colleer Abbott, ed. *The Letters of Gerard Manley Hopkins to Robert Bridges*, 2nd ed. (London: Oxford University Press, 1955), p. 90. This corresponds, I believe, to Coleridge's point of view that words in a poem take their meaning as components in a whole field of force — a poem succeeds only when the appreciation of it as a whole is compatible with an appreciation of its component parts.

[141] Ibid., pp. 265-66.

[142] Ibid., p. 165.

and meaning with precision of meaning and wholeness of impact."[143]

This 'reconciliation' or expansion is already hinted at in passages from one of Hopkins' undergraduate essays. Hopkins writes that "all words mean either things or relation of things: you may also say then substances or attributes or again wholes or parts."[144] Here he implies that there is a distinction between words which is more than the part of speech that they play (e.g., verb or noun). Rather, as Ellis contends, distinction among words for Hopkins is also "based on the relative fullness, substance, and density of the ideas conveyed in individual words."[145] In Hopkins' poems, words are used to function as 'multi-dimensional wholes', and not only parts. In this 'holistic' approach to poetic words, words "that mean things" such as nouns and verbs ("acts that are things") are especially central. Hopkins develops this approach in the following passages where he describes the key 'moments' or 'terms' of a word that means a thing — "prepossession"and "definition"[146]:

> To every word meaning a thing and not a relation belongs a passion or prepossession or *enthusiasm* [my italics] which it has the power of suggesting or producing but not always or in everyone. This *not always* refers to its evolution in the man and secondly in man historically.[147]

[143] Virginia Ridley Ellis, *Gerard Manley Hopkins and the Language of Mystery* (Columbia: University of Missouri Press, 1991), p. 45.

[144] Humphry House, ed. *The Journals and Papers of Gerard Manley Hopkins*, 2nd ed. (London: Oxford University Press, 1959), p. 125.

[145] Virginia Ridley Ellis, *Gerard Manley Hopkins and the Language of Mystery* (Columbia: University of Missouri Press, 1991), p. 46.

[146] Hopkins refers also to a third 'moment', the "application" of a word. However, he does not expand on this moment. See footnote 156 below.

[147] Humphry House, ed. *The Journals and Papers of Gerard Manley Hopkins*, 2nd ed. (London: Oxford University Press, 1959), p. 125.

For Hopkins, words have a "passion, prepossession, or enthusiasm" which allow great freedom of association and use while having some kind of objective origin and root. In other words, the prepossession of a word "is everything a word originally meant and still may mean and convey emotionally, in itself or to an individual."[148] To describe this 'moment' of a word, Hopkins uses the analogy of the soul. This is a word's "living, almost spiritual, power and specialness."[149] In fact, Hopkins seems to hold that this aspect of a word is "some remnant of the original power that first matched [the word] with reality."[150] He also refers to this as the 'form' of a word.

A word also has a "definition."[151] A word's definition is first, but not only, its pure denotation, or, as Ellis states, its "abstraction uncluttered and unenriched by associations or extensions, something cerebrally grasped, equivalent to the most limited 'meaning'."[152] She goes onto write:

> [Definition] is distinguished both from 'prepossession', which is "not a word but something connotatively meant by it," and from the third

[148] Virginia Ridley Ellis, *Gerard Manley Hopkins and the Language of Mystery* (Columbia: University of Missouri Press, 1991), p. 46. F. D. Martin refers to the same thing when he writes the following: "Words are haunted with meanings from the past. The poet... has piety towards words. [The poet] arranges them in such a way that the words reveal, if we attend, something of their past through how and what they designate. What they designate in part is something of the depth dimension of the things they name..." See F. David Martin, *Art and the Religious Experience: The "Language" of the Sacred* (Lewisburg: Bucknell University Press, 1972), p. 184.

[149] Virginia Ridley Ellis, *Gerard Manley Hopkins and the Language of Mystery* (Columbia: University of Missouri Press, 1991), p. 46.

[150] Christopher Devlin, S.J., "The Image and the Word–I," *The Month (ns)* 3, no. 2 (1950), p. 115.

[151] Humphry House, ed. *The Journals and Papers of Gerard Manley Hopkins*, 2nd ed. (London: Oxford University Press, 1959), p. 125.

[152] Virginia Ridley Ellis, *Gerard Manley Hopkins and the Language of Mystery* (Columbia: University of Missouri Press, 1991), p. 46.

> moment, 'application or extension', which is "not a word but a thing meant by it," the outward impetus of the word, what it points to, its designation of "the concrete things coming under it."[153]

'Definition', then, is the 'moment' of a word that is, in Hopkins' theory, the word understood as a conception. However, 'definition' has another layer of meaning more vital to this study. It is also "expression," i.e., the inward or outward "uttering of the idea in the mind."[154] This idea expressed by the word is not only an abstract concept for Hopkins but also an image or involvement "of sight or sound or *scapes* of the other senses."[155] In other words, the 'definition' of a word involves the calling up of instinctive images and has what Ellis refers to as a "physical even visceral" quality.[156]

[153] Virginia Ridley Ellis, *Gerard Manley Hopkins and the Language of Mystery* (Columbia: University of Missouri Press, 1991), p. 46. F.W. Farrar offered a related parsing of the nature of words: "We can distinguish three factors: (I) the sound, which is the incarnation of the thought; (ii) the inner form of the word, or the special method of this incarnation; and (iii) the meaning, i.e. the intuitions and concepts which the word expresses. In this respect a word resembles a work of art, which also contains three elements: e.g., the material of this statue is marble; the form of it is a virgin figure with sword and scales; and it *represents* justice." See F.W. Farrar, *Chapters on Language*, (London, 1865), pp. 287-88 quoted in Giuseppe Castorina, "The Science of Language and the Distinctive Character of Hopkins's Poetry and Poetics," in *Gerard Manley Hopkins: Tradition and Innovation*, ed. G. Marra P. Bottalla, F. Marucci (Ravenna: Longo Editore, 1991), p. 88.

[154] Humphry House, ed. *The Journals and Papers of Gerard Manley Hopkins*, 2nd ed. (London: Oxford University Press, 1959), p. 125.

[155] Ibid.

[156] Virginia Ridley Ellis, *Gerard Manley Hopkins and the Language of Mystery* (Columbia: University of Missouri Press, 1991), p. 47. It is difficult to distinguish what Hopkins sees as the real distinctions between 'definition' and 'prepossession' at times. The terms seem to be used for emphasis of particular aspects of a word as opposed to true distinctions. Additionally, Hopkins refers to a third 'moment' — the "application" of a word. However, he does not expand much on the 'moment of application' as distinct from the 'moment of definition'. In one way or another, the moment of 'definition' certainly seems to overlap with the other 'moments' of a word.

In this light, a word for Hopkins is not a symbol as much as it is a visceral, organic 'thing' and living power.[157] For Hopkins, words are not simply conceptions to be manipulated. Rather, they are allowed a "vibrant life of their own" which is related to "the images they invoke, their sound, and their 'uttering' in the mind of the speaker."[158]

[157] Cf. Margaret R. Ellsberg, *Created to Praise: The Language of Gerard Manley Hopkins* (New York: Oxford University Press, 1987), pp. 79-80. Here Ellsberg echoes a line of thought developed by W. A. M. Peters. In his 1948 work, he wrote: "... to Hopkins a word was very much more than a sign for a thing. To this poet a word was as much an individual as any other thing; it had a self as every other object, and consequently just as he strove to catch the inscape of a flower or a tree or a cloud, he similarly did not rest until he knew the word as a self. He attended to the various meanings this word might have, he let its sounds grow upon him and take hold of his ear, he realized its likeness in sound to other words, he felt its instress, in brief, he caught its inscape. And once a word had been inscaped, it was no longer merely a name for a thing... it functioned [rather] with its most complete being; it 'dealt out its own being'..., that is, it was functional in the phrase with its various meanings, its suggestive power, its connotative value, &c. His words now have a body and bodily they are placed in the text." W. A. M. Peters, *Gerard Manley Hopkins: A Critical Essay towards the Understanding of his Poetry* (London: Oxford University Press, 1948), pp. 141-42.

[158] See Howard Fulweiler, *Letters from the Darkling Plain: Language and the Grounds of Knowledge in the Poetry of Arnold and Hopkins* (Columbia: University of Missouri Press, 1972), p. 89. This vision of language as a 'physical' thing with a kind of "refined energy accenting the nerves" [cf. Humphry House, ed. *The Journals and Papers of Gerard Manley Hopkins*, 2nd ed. (London: Oxford University Press, 1959), p. 125] is the reason Yvor Winters dismisses Hopkins as a poet of any value. For Winters, "words are primarily conceptual: the words *grief, tree, poetry, God*, represent concepts; they may communicate some feeling and remembered sensory impression as well, and they may be made to communicate a great deal of these, but they will do it by virtue of their conceptual identity." Yvor Winters, "Gerard Manley Hopkins," in *Hopkins: A Collection of Critical Essays*, ed. Geoffrey Hartman (Englewood Cliffs: Prentice-Hall, Inc., 1966), pp. 37-8. Hopkins rejects this vision of words which defines them as solely representing concepts.

The 'definition' of a word also involves the energy of the individual mind adding its own private shapes. In short, here Hopkins, in a nascent form, attempts to reconcile a word's "potentially dense, multiple, and even private suggestiveness with its essential fidelity to a precise and objective reality, precisely apprehended but not narrowly comprehended."[159]

Here is a connection between Hopkins' visions of word and world. In each case, a singular entity, in word or world, is uniquely rich and layered, "offering free play to the mind, yet precise, and precisely governed."[160] Words used to mean things must not be merely suggestive; their connotations must not be separated from

[159] Virginia Ridley Ellis, *Gerard Manley Hopkins and the Language of Mystery* (Columbia: University of Missouri Press, 1991), p. 47. In describing a word's 'definition' or 'abstraction', Hopkins explains that the 'definition' can be subdivided into two "terms": "the image (of sight or sound or *scapes* of the other senses), which is in fact physical and a refined energy accenting the nerves, a word to oneself, an inchoate word, and secondly the conception." Humphry House, ed. *The Journals and Papers of Gerard Manley Hopkins*, 2nd ed. (London: Oxford University Press, 1959), p. 125. As Marucci indicates, even though Hopkins does not use the words 'inscape' or 'instress' here, he is "clearly trying to describe the moment in which the purely sensible perception ("*scapes* of the...senses") produces in some special cases ("not always or in everyone") an instantaneous "quasi-" or "pre-knowledge," which is neither of a sensible nature nor already intellectualized, and which afterwards is dissolved and resolved into a pure "conception" through abstraction." Franco Marucci, *The Fine Delight That Fathers Thought: Rhetoric and Medievalism in Gerard Manley Hopkins* (Washington, D.C.: The Catholic University Press, 1994), p. 180. Hopkins continues: "Works of art of course like words utter the idea and in representing real things convey the prepossession with more or less success." *The Journals and Papers of Gerard Manley Hopkins*, 2nd ed. (London: Oxford University Press, 1959), p. 126. Here we find a connection between Hopkins' acceptance of the possibility of intuitive cognition and the 'formulation' of this 'prepossession' in words.

[160] Virginia Ridley Ellis, *Gerard Manley Hopkins and the Language of Mystery* (Columbia: University of Missouri Press, 1991), p. 47.

their precise meaning in poetry.[161] This certainly parallels Hopkins' approach to inscape which, while allowing for free play of the mind, insists on a vital connection to a thing through careful 'seeing' and 'oftening'. Poetic language, while remaining true to the 'roots' of the words being used, must in its richness and complexities of meaning also allow the apprehension of a 'whole in the parts', i.e., must allow for the clarity of meaning that comes through the 'explosion' of the grasping of the whole, the grasping, in fact, of the inscape of the thing poetically conveyed. Divine eloquence in the world must find human eloquence in poetic language. Hopkins writes:

> Even in the successive arts as music, for full enjoyment, the synthesis of the succession should give, unlock, the contemplative enjoyment of the unity of the whole... The more intellectual, less physical, the spell of contemplation the more complex must be the object, the more close and elaborate must be the comparison the mind has to keep making between the whole and the parts, the parts and the whole. For this reference or comparison is what the sense of unity means; mere sense that a thing is one and not two has no interest or value except accidentally... The further in anything, as a work of art, the organisation is carried out, the deeper the form penetrates, the prepossession flushes the matter, the more effort will be required in apprehension, the more power of comparison, the more capacity for receiving that synthesis of impressions which gives us the unity with the prepossession conveyed by it.[162]

In other words, the language in the poem as a sacramental vehicle must not only develop idea, feeling, and imagery in a "dramatic and linear" way, but must also allow the reader/hearer to "receive the synthesis of succession" both horizontally and vertically,

[161] Humphry House, ed. *The Journals and Papers of Gerard Manley Hopkins*, 2nd ed. (London: Oxford University Press, 1959), p. 126.

[162] Ibid.

"apprehending many meanings, dimensions, and levels of insight, in one moment of perception."[163] The poem will thus be like a simultaneous layering of multiple exposures so as to create a whole out of many different dimensions — a kind of 'hologram' in motion.

Yet in attempting this heightened use of language and words in particular, the poet must also have respect for clarity and 'formulation' (as we have seen). As Ellis states, "without initial precision of observation, without intensive and minute study of the world's variety, the observer cannot do justice to either its richness or its underlying laws, and the poet cannot create a style adequate to his simultaneous sense of mystery and formulated pattern."[164] Thus, we come full circle to Hopkins' early Ruskinian roots which encourage the study of the 'laws' and 'mechanics' of things in order to discover what is within and beyond them.[165] This capturing of inscape, or 'instressing of inscape', can then begin to find 'formulation' in the mind of the poet through the association of words, central imagery, sound patterns, meter, etc., always working toward "the complex unity, music, and multidimensional qualities of language that create the voice of 'incomprehensible certainty'."[166]

The words which capture inscape must also beg attention or compel the hearer's attention. Hopkins calls this '*bidding*' and states that "it is the art or virtue of saying everything right *to* or *at*

[163] Virginia Ridley Ellis, *Gerard Manley Hopkins and the Language of Mystery* (Columbia: University of Missouri Press, 1991), p. 48. Here Ellis is conflating Hopkins' ideas in Humphry House, ed. *The Journals and Papers of Gerard Manley Hopkins*, 2nd ed. (London: Oxford University Press, 1959), p. 126.

[164] Virginia Ridley Ellis, *Gerard Manley Hopkins and the Language of Mystery* (Columbia: University of Missouri Press, 1991), p. 49.

[165] Humphry House, ed. *The Journals and Papers of Gerard Manley Hopkins*, 2nd ed. (London: Oxford University Press, 1959), p. 252.

[166] Virginia Ridley Ellis, *Gerard Manley Hopkins and the Language of Mystery* (Columbia: University of Missouri Press, 1991), p. 51.

the hearer, interesting him, holding him in the attitude of correspondent...making it [poetry] everywhere an act of intercourse."[167] This 'bidding' is at the root of von Balthasar's interpretation of Hopkins' poetic accomplishment, which is to allow the 'hearer' to encounter, in intensified and heightened fashion, the *doxa* of God in the Christic forms of the world, and thus potentially to respond in faith, praise, and delight.

Words, in their fullness of 'moments', as well as rhetorical ingenuity,[168] attempt to facilitate this encounter in Hopkins' poetry. As Boyle claims, Hopkins is not after presenting something that can be grasped only by the hearer's abstractive intellect, rather, he attempts "to drag the intellect into the imagination...to force the intellect to peer into material being just as it is..." and see "that there is in that material being what is clearly not really there."[169] This is another way of stating Ellis' point; namely, that Hopkins' is after conveying in his poetic language the "comprehension" of "incomprehensible certainty."

D.3.1 The 'Rhyme-Principle' in Hopkins

Hopkins did not develop his thoughts on language and the poetic word fully. Nonetheless, from his early writings through his mature poetry, Hopkins demonstrated a running theme in his 'the-

[167] Claude Colleer Abbott, ed. *The Letters of Gerard Manley Hopkins to Robert Bridges*, 2nd ed. (London: Oxford University Press, 1955), p. 160.

[168] See Franco Marucci, *The Fine Delight That Fathers Thought: Rhetoric and Medievalism in Gerard Manley Hopkins* (Washington, D.C.: The Catholic University Press, 1994). Marucci does an excellent job showing how rhetorical technique in Hopkins is an important aspect of the 'matter' of the poem as sacrament in Hopkins. Also see Philip Ballinger, "Review Article," *Studies* 85, no. 338 (1996): 195-97.

[169] Robert Boyle, *Metaphor in Hopkins* (Chapel Hill: University of North Carolina Press, 1961), p. 178.

ory of words' which Joseph Hillis Miller refers to as the "rhyme-principle":

> This relation may be found everywhere in the universe: in words which resemble one another without being identical, in trees or clouds which have similar but not identical patterns, and so on. The universe, although no two things in it are exactly alike, is full of things which rhyme, and by extending the range of observed rhymings who knows how many things may ultimately be brought into harmony?[170]

This effort to "reorganize the universe" begins from the "realm of words" for Hopkins.[171] There are etymological speculations throughout Hopkins' writings which, as Miller contends, appear to be "reconstructions of the world through the discovery of rhymes."[172] For example, in an 1863 journal entry, Hopkins makes the following etymological speculation:

> *Grind, gride, gird, grit, groat, greet,* krouein, *crush, crash,* krotein, etc. Original meaning to *strike, rub,* particularly *together*. That which is produced by such means is the *grit*, the *groats* or crumbs, like *fragmentum* from *frangere, bit* from *bite*. *Crumb, crumble* perhaps akin. To *greet*, to strike the hands together(?). *Greet*, grief, wearing, *tribulation*. *Grief* possibly connected. *Gruff*, with a sound as of two things rubbing together. I believe these words to be onomatopoetic. *Gr* common to them all representing a particular sound. In fact I think the onomatopoetic theory has not had a fair chance. Cf. *Crack, creak, croak, crake, graculus, crackle*. These must be onomatopoetic.[173]

[170] Joseph Hillis Miller, *The Disappearance of God: Five Nineteenth-Century Writers* (Cambridge: Harvard University Press, 1963), p. 277.

[171] Franco Marucci, *The Fine Delight That Fathers Thought: Rhetoric and Medievalism in Gerard Manley Hopkins* (Washington, D.C.: The Catholic University Press, 1994), p. 128.

[172] Joseph Hillis Miller, *The Disappearance of God: Five Nineteenth-Century Writers* (Cambridge: Harvard University Press, 1963), p. 279.

[173] Humphry House, ed. *The Journals and Papers of Gerard Manley Hopkins*, 2nd ed. (London: Oxford University Press, 1959), p. 5. These kinds of speculations are found throughout the early journal entries of Hopkins. Hopkins also con-

In opposition to a vision of things in the world and of words for those things as being unrelated, Hopkins asserts a basic connection or 'organization' between things, their 'words', and their qualities and actions. For Hopkins, "words... similar in sound... will also be similar in meaning. Hopkins assumes that a group of words of similar sound are variations of some *ur*-word and root meaning."[174] This is what he refers to as a version of the onomatopoetic approach to words. Marucci describes this approach to words as a form of 'diatonism'; i.e., Hopkins sees each word "as having an autonomy of its own but also as being placed at a fixed interval from other similar words and being therefore able to chime with them, both in sound and in mean-

nects these etymological 'rhymings' with the 'rhymings' of physical things. For example, here are excerpts from his speculations on the 'word' (in its fullest sense) *Horn*: "The various lights under which a horn may be looked at have given rise to a vast number of words of language. It may be regarded as a projection, climax, a badge of strength, power or vigour, a tapering body, a spiral, a wavy object, a bow, a vessel to hold withal or to drink from, a smooth hard material not brittle, stony, metallic, or wooden, something sprouting up, something to thrust or push with, a sign of honour or pride, an instrument of music, etc. From the curve of a horn, *koronis, corona, crown*. From the spiral *crinis*, meaning ringlets, locks. From its being the highest point comes our *crown* perhaps, in the sense of the top of the head, and the Greek *keras*, horn, and *kara*, head, were evidently identical; then for its sprouting up and growing, compare *keren, cornu, keras*, horn with grow, *cresco, grandis,* grass, great, *groot*. For its curving, *curvus* is probably from the root *horn* in one of its forms. *Korone* in Greek and *corvus, cornix* in Latin and *crow* (perhaps also *raven*, which may have been *craven* originally) in English bear a striking resemblance to *cornu, curvus*. So also *geranos, crane, heron, herne*. Why these birds should derive their names from *horn* I cannot presume to say. The tree *cornel*, Latin *cornus* is said to derive its name from the hard horn-like nature of its wood, and the *corns* of the foot perhaps for the same reason. *Corner* is so called from its shape, indeed the Latin is *cornu*." Hopkins continues on in this vein. Humphry House, ed. *The Journals and Papers of Gerard Manley Hopkins*, 2nd ed. (London: Oxford University Press, 1959), p. 4.

[174] Joseph Hillis Miller, *The Disappearance of God: Five Nineteenth-Century Writers* (Cambridge: Harvard University Press, 1963), p. 280.

ing."[175] Thus each word has its own richness of 'definition' (see above), but also has the magnifying possibility of 'chiming' with the sound and 'definitions' of other words, thus allowing for a communication of inscape with great depth.

This 'diatonic' approach to language paralleled Hopkins' 'diatonic' approach to physical reality. In contemplating an apparently 'chromatic', flux-full nature that seemed to be, as Pater put it, a stream "of impressions, unstable, flickering, inconsistent, which burn and are extinguished with our consciousness of them,"[176] Hopkins' chose to stress the 'fixed points' or 'breaks' amidst the phenomenal flow. The chromatic continuity of world and words could be organized into diatonic intervals as along a musical scale.[177] Thus Hopkins from early on, as we have seen, gleaned nature for 'forms' and structure. Eventually he learned "to organize groups of phenomena into an ordered whole;" the chromatic could become diatonic "if a series of impressions were grouped, like tones, into a 'chord'."[178] Hopkins demonstrates this approach in an early journal poetic snippet:

> Or else their cooings came from bays of trees,
> Like a contented wind, or gentle shocks

[175] Franco Marucci, *The Fine Delight That Fathers Thought: Rhetoric and Medievalism in Gerard Manley Hopkins* (Washington, D.C.: The Catholic University Press, 1994), p. 128.

[176] Quoted by Anselm Hufstader, O.S.B., in "The Experience of Nature in Hopkins' Journals and Poems," *Downside Review* 84, no. 275 (1966), p.129.

[177] This 'diatonism' led Hopkins to a vision of words which not only claimed that they can imitate things and re-present them in different forms, but also that they can rescue things from the ceaselessly moving realm of nature. Words can, as Hopkins held, 'catch' things, 'stall' them and transform them into spiritual stuff. In this way, Hopkins held that words can carry an object alive into the heart. Cf. Joseph Hillis Miller, *Victorian Subjects* (Durham: Duke University Press, 1991), p. 13.

[178] Anselm Hufstader, O.S.B., in "The Experience of Nature in Hopkins' Journals and Poems," *Downside Review* 84, no. 275 (1966), p. 129.

Of falling water. This and all of these
We tuned to one key and made their harmonies.[179]

Effectively, Hopkins' onomatopoetic approach to language allows the word to imitate "in its substance and inscape the substance and inscape of the thing it names."[180] Ultimately, for Hopkins, there is a universal 'rhyme-element' which connects all things and even words for those things. This may be called the definitive *Ur-Word* and thus unifying *Ur-Inscape*. This is, for Hopkins, the Incarnate Word. It is from the Logos in the world that all words derive their origin as intelligible 'beings'.[181] The intelligibility of all being, of which Hopkins is convinced, including the intelligibility of the sound of language,[182] is rooted in the Incarnate Logos. Christ is the Logos of words as well as of nature.[183]

[179] Humphry House, ed. *The Journals and Papers of Gerard Manley Hopkins*, 2nd ed. (London: Oxford University Press, 1959), p. 34.

[180] Joseph Hillis Miller, *The Disappearance of God: Five Nineteenth-Century Writers* (Cambridge: Harvard University Press, 1963), p. 285.

[181] John T. Netland, "Linguistic Limitation and the Instress of Grace in "The Wreck of the Deutschland"," *Victorian Poetry* 27, no. 2 (1989), p. 191. What this notion infers is that, in a general sense, all being is a form of language. Cf. David Anthony Downes, *Hopkins' Achieved Self* (Lanham: University Press of America, 1996), p. 97.

[182] Cf. Bernadette Ward, "Philosophy and Inscape: Hopkins and the Formalitas of Duns Scotus," *Texas Studies in Literature and Language* 32, no. 2 (1990), pp. 227-8.

[183] Downes offers a variant interpretation of the same stress in Hopkins: "Hopkins understood that the making of meaning of the world (his hermeneutics of 'instressing inscapes') is to begin to grasp the ordered unity of Creation. The basic principle undergirding this unity (Nature, self, idea, word) is Hopkins' hermeneutic notion of God as the maker of the primal utterance known as Creation. It is God who uttered himself as Creation and thereby surcharged all existence as his 'texts'. His original utterances are foundational to making sense of everything, hence God is the source of all hermeneutical consciousness." David Anthony Downes, *Hopkins' Achieved Self* (Lanham: University Press of America, 1996), pp. 14-5.

One thing is clear for Hopkins — **all things 'rhyme in Christ'**.[184] There is a Christic forming of and form in all things. In a sense, the enfleshed Logos 'vibrates' or chimes simultaneously at all the possible 'pitches' of creation. In this sense, Hopkins "sees the universe as a vast interlocking harmony, full of fraternal echoes and resonances."[185] Miller writes:

> [Hopkins'] vision of Christ as the common nature is the culmination of [his] gradual integration of the world. Christ is the model for all inscapes, and can vibrate simultaneously at all frequencies. He is the ultimate guarantee for the validity of metaphor. It is proper to say that one thing is like another only because all things are like Christ.[186]

[184] Cf. Joseph Hillis Miller, *The Disappearance of God: Five Nineteenth-Century Writers* (Cambridge: Harvard University Press, 1963), p. 313. Interestingly, Hopkins already tied his concept of 'rhyming' with his idea of 'beauty' in an early undergraduate essay. He says that "all beauty may by a metaphor be called rhyme..." Cf. Humphry House, ed. *The Journals and Papers of Gerard Manley Hopkins*, 2nd ed. (London: Oxford University Press, 1959), p. 102. The 'rhyming' and 'beauty' Hopkins connects in his essay are not yet 'Christic'. Nonetheless, early on he saw some kind of metaphysical 'common denominator' in being as well as a linkage between being and wording.

[185] Joseph Hillis Miller, *Victorian Subjects* (Durham: Duke University Press, 1991), p. 65.

[186] Joseph Hillis Miller, *The Disappearance of God: Five Nineteenth-Century Writers* (Cambridge: Harvard University Press, 1963), p. 313. Miller's terms 'metaphor' and 'like' may not always be sufficient when interpreting Hopkins. In 'doing' their being, things and humans express their inscapes and reveal their Christic stem. In 'speaking' their own 'names' they also 'speak' Christ's 'name'. This is particularly true of human beings. In the poem *As Kingfishers catch fire* (Norman H. MacKenzie, ed. *The Poetical Works of Gerard Manley Hopkins* [Oxford: Clarendon Press, 1990], p. 141), Hopkins expresses these themes forcefully:

> As kingfishers catch fire, dragonflies draw flame;
>   As tumbled over rim in roundy wells
>   Stones ring; like each tucked string tells, each hung bell's
> Bow swung finds tongue to fling out broad its name;
> Each mortal thing does one thing and the same:

Created reality finds its intelligibility in its Christic root. World and words also 'rhyme' in Christ. As Ellis perceptively speculates, "perhaps all things also parse in Christ; if Christ is the Logos in whom all words find their origin, he is perhaps also the underlying structure of all *logoi*, of all ordered utterance."[187] For Hopkins, everything radiates back to a Christic center, including word and world. Although God is one, God is expressed in language and in creation, in word and world, as multiplicity; yet for Hopkins, all variety, 'piedness', seeming chaos, or variant motion find unity, metaphysically and verbally, in the Word made flesh. The inscapes of world and word, nature and language, flow from

Deals out that being indoors each one dwells;
Selves — goes itself; *myself* it speaks and spells,
Crying *What I do is me: for that I came.*

I say more: the just man justices;
Keeps grace: that keeps all his goings graces;
Acts in God's eye what in God's eye he is —
Christ. For Christ plays in ten thousand places,
Lovely in limbs, and lovely in eyes not his
To the Father through the features of men's faces.

Additionally, in some spiritual notes commenting on grace, Hopkins writes: "It is as if a man said: That is Christ playing at me and me playing at Christ, only that it is no play, but truth; That is Christ *being me* and me being Christ." Christopher Devlin, S.J., ed. *The Sermons and Devotional Writings of Gerard Manley Hopkins* (London: Oxford University Press, 1959), p. 154.

[187] Virginia Ridley Ellis, *Gerard Manley Hopkins and the Language of Mystery* (Columbia: University of Missouri Press, 1991), p. 62. Eleanor McNees refers to this line of thought as the "poetics of Real Presence." She writes: "If words are generated through the one Word of Christ, all language that man speaks links him literally to Christ. The equation of Christ with Word refutes metaphor and upholds Hopkins's theory that separate words and objects are internally linked to each other through their origin in Christ the Logos. Language is charged with a sacramental reality that finds its most significant symbol in the Eucharist..." See Eleanor McNees, "Beyond 'The Half-way House': Hopkins and Real Presence," *Texas Studies in Literature and Language* 31, no. 1 (1989), p. 100.

Christ.[188] He is what "Heaven and earth are word of, worded by."[189] He is the Ur-Word.

E. The Poem as Sacrament

We have seen how throughout his mature writings, Hopkins used his ciphers of inscape and instress as bridges between being, knowing, and wording. The inscape of a thing ultimately came to mean the utterly individual and distinctive form of Christic selving that marks a thing's existence in the world.[190] Instress came to mean, in part, "the fusion of the inscape of a given being with a given human consciousness in contact at a given moment with that being in all its uniqueness."[191] Hopkins further held that inscape could be instressed in others through poetic language. This means that the Incarnate Word, as self-expressive beauty in a particular sensible reality,[192] could be instressed and worded by the poet, and thus could also be communicated in a heightened, 'bidding' fashion to hearers of the poetic word. In eucharistic terms, a blessed instress occurs in this dynamic at the moment the worded, poetically

[188] Joseph Hillis Miller, *The Disappearance of God: Five Nineteenth-Century Writers* (Cambridge: Harvard University Press, 1963), p. 317.

[189] Norman H. MacKenzie, ed. *The Poetical Works of Gerard Manley Hopkins* (Oxford: Clarendon Press, 1990), p. 126.

[190] Cf. Walter J. Ong, S.J., *Hopkins, the Self, and God* (Toronto: University of Toronto Press, 1986), p. 156.

[191] Ibid.

[192] Two examples in Hopkins' journals: "I do not think I have ever seen anything more beautiful than the bluebell I have been looking at. I know the beauty of our Lord by it. Its inscape is a mix of..." Humphry House, ed. *The Journals and Papers of Gerard Manley Hopkins*, 2nd ed. (London: Oxford University Press, 1959), p. 199. "As we drove home the stars came out thick: I leant back to look at them and my heart opening more than usual praised out Lord to and in whom all that beauty comes home." Ibid., p. 254.

inscaped Word is offered under the elements[193] of the inscaping poem to the communicant.[194] This sacramental dynamic is an example of the theological dictum *sacramenta continent quae significant* in a unique way. To carry the eucharistic analogy further, the goal of the poem, then, "is to operate like the Prayer of Consecration,

[193] T.L. Thibodeaux speaks of the 'materiality' of Hopkins' vision of language. In this sense, language truly is an 'element' for Hopkins. Language has a "material existence beyond that which it represents." "The signifier has an existence that is independent of the signified..." See Troy L. Thibodeaux, "The Resistance of the Word: Hopkins's "(Carrion Comfort)"," *The Hopkins Quarterly* 22, no. 3-4 (1995), pp. 87 and 88. Bruns speaks of Hopkins' vision of words in a similar vein: "Words [for Hopkins] are like bodies...they are physical quantities, but not inert extensions of matter capable of change only in relation to space; rather, they possess both mass *and* energy, in the sense that they are capable of changes in form, state, equilibrium, and intensity... [This] makes more sense if we remember that Hopkins is thinking of spoken rather than written language." See Gerald L. Bruns, "The Idea of Energy in the Writings of Gerard Manley Hopkins," *Renascence* 29, no. 1 (1976), p. 34. Also see Clive Scott, *A Question of Syllables: Essays in Nineteenth-Century French Verse* (Cambridge: Cambridge University Press, 1986), pp. 138-40. Ellsberg echoes this view and writes that "by endowing words with substance and energy, Hopkins gave them...the enriched nature of a sacramental element." Margaret R. Ellsberg, *Created to Praise: The Language of Gerard Manley Hopkins* (New York: Oxford University Press, 1987), p. 81. Lichtmann attaches this 'material' aspect of language to its ability to carry inscape and, thus, Christ: "Characteristically, Hopkins submits spirit to matter, trusting that its stress might the more 'flame out' from matter's density. While the signifiers of his poems draw us in by their fleshiness and opacity, it is within their austere materiality that the signified emerges. The very palpability of his signs, their rich alliterative and assonantal quality, and the material structure of his poetry, its texture of repeated phrases, formed what he calls the 'inscape' of the poem." "Hopkins inscapes Christ in his poems in the way he found him in the quintessential Incarnation of Eucharist in matter. That is, Christ lies hidden, and inert in the matter of his poems just as he first lies within the matter of the Eucharist." See Maria R. Lichtman, "The Incarnational Aesthetic of Gerard Manley Hopkins," *Religion and Literature* 23, no. 1 (1991), pp. 47 and 43.

[194] Eleanor McNees, "Beyond 'The Half-way House': Hopkins and Real Presence," *Texas Studies in Literature and Language* 31, no. 1 (1989), p. 102, endnote #3.

to effect the Real Presence of Christ within the reader"[195] (or better — the hearer). Miller says something similar by stating that in the poem "the inscapes of words, the inscapes of nature, [and] the inscape of the self can be expressed at once as the presence of Christ."[196]

It is debatable whether Hopkins intended such an interpretation of the poem.[197] On one level, I do not believe he did, certainly not explicitly or consciously. I realize that my conclusion here appears to create a paradox for this study. On the one hand, I claim certain theological views and approaches for Hopkins, while on the other hand, I conclude that he did not consciously hold these views. Perhaps the paradox can be solved to some degree by stressing that there is a considerable 'écart' between Hopkins the academic and Hopkins the artist. An analysis of Hopkins' theology, philosophy,

[195] Ibid., p. 89.

[196] Joseph Hillis Miller, *The Disappearance of God: Five Nineteenth-Century Writers* (Cambridge: Harvard University Press, 1963), p. 323.

[197] For example, ultimately J. Hillis Miller would not see this as a possibility even if it was some kind of intention in Hopkins. He sees an irrevocable tension between the all-encompassing unity of the Logos and the hopeless multiplicity of language: "There is no masterword for the Word, only metaphors of it, for all words are metaphors..." See Joseph Hillis Miller, "The Linguistic Moment in 'The Wreck of the Deutschland'," in *The New Criticism and After*, ed. Thomas Daniel Young (Charlottesville: University Press of Virginia, 1976), p. 55. Paul Mariani also concludes that Hopkins general attempt to 'en-word the Word' was a failure. Hopkins' poems are a battle between "lexical plenitude" and "lexical spareness" which "reflects Hopkins's consciously failing attempts to utter what can, finally, only be imperfectly uttered no matter how rich the verbal lode one has to command." See Paul Mariani, "'O Christ, Christ, Come Quickly!': Lexical Plenitude and Primal Cry at the Heart of 'The Wreck'," in *Readings of "The Wreck": Essays in Commemmoration of the Centenary of G.M. Hopkins' "The Wreck of the Deutschland"*, ed. Peter Milward (Chicago: Loyola University Press, 1976), p. 40. Also see Rachel Salmon's reflection on this topic in her article "Hopkins and the Rabbis: Christian Religious Poetry and Midrashic Reading," in *Rereading Hopkins: Selected New Essays*, ed. Francis L. Fennell, *English Literary Studies* (Victoria: University of Victoria, 1996), pp. 97-8.

and linguistics has led me to believe that his vision as an artist, that is, the view implied by his poetic practice, seems far ahead of his consciousness and piecemeal 'academic' reflections. In light of this écart, I feel justified in deducing a certain 'systematic' theology from Hopkins' poetry while recognizing that Hopkins himself probably did not consciously adhere to such a theology. Therefore, with some extrapolation, I hold that Hopkins' theological aesthetic may be specifically developed into a view of the poem as sacrament.

A hint in this direction is given by Hopkins himself in a letter to Bridges. Bridges received Hopkins' first poetic effort after a seven year hiatus, *The Wreck of the Deutschland*, and apparently informed Hopkins that he did not like it. Hopkins, in response, writes:

> I think if you will study what I have here said you will be much more pleased with it and may I say? converted to it. You ask may you call it 'presumptuous jugglery'. No... but... I cannot think of altering anything. Why shd. I? I do not write for the public. You are my public and I hope to convert you.[198]

Hopkins is using double entendre here. The poem becomes, in part, a 'sacramental opportunity' for the reader/hearer. If the reader is open to it, he or she can be spiritually transformed by the poetically inscaped Christ. Referring to *The Wreck of the Deutschland*, McNees writes that Hopkins attempts to construe *The Wreck* as a "poetic Eucharist in which words actually embody the things they signify."[199] As we have seen, this sacramental vision of poetry is possible for Hopkins because Christ is the ground of sensible reality and of language. Using the language of Transubstantiation, as

[198] Claude Colleer Abbott, ed. *The Letters of Gerard Manley Hopkins to Robert Bridges*, 2nd ed. (London: Oxford University Press, 1955), p. 46.

[199] Eleanor McNees, "Beyond 'The Half-way House': Hopkins and Real Presence," *Texas Studies in Literature and Language* 31, no. 1 (1989), p. 90.

Hopkins would have, the 'accidents' of words, such as rhetoric, rhythm, rhyme, and style, can carry the 'substance' of the inscaped Christ. In this sense, language is incarnational for Hopkins. Physical realities and the words which inscape them can lead to a corresponding instress in the seer/hearer.

This incarnational, even eucharistic, approach to poetry brings us full circle back to Hopkins' interpretation of Duns Scotus. As we have seen, based on his reading of Scotus, Hopkins accepted a kind of aeonian incarnation of Christ into matter which he referred to as the *ensarkosis*. This belief in the emptying of the Word into matter, or "preexistent eucharistic Incarnation" (as Lichtmann would say),[200] forms Hopkins' vision of poetry. Poetry, via inscape, may incarnate Christ. Let us develop this dynamic further.

In a Scotistically influenced vision, Hopkins sees all created nature as consecrated in a kind of eucharistic incarnation. This is the 'Great Sacrifice' we have explained above. Through the 'Great Sacrifice', Christ is found incarnate in creation; Christ is 'enfleshed' in a eucharistic paradigm. This *ensarkosis* sanctifies matter. As Lichtmann puts it, "Christ's body inscapes the world as Eucharist."[201] This is an interpretation of what Hopkins means when he states that the world is "charged with the grandeur of God."[202] Thus, Hopkins is able "to entrust to matter his sense that the world carries the divine" in some way.[203]

The effect this foundational doctrine has on Hopkins' poetry is significant. In his contemplation of nature, Hopkins seeks to cap-

[200] Maria R. Lichtman, "The Incarnational Aesthetic of Gerard Manley Hopkins," *Religion and Literature* 23, no. 1 (1991), p. 40.

[201] Ibid., p. 41.

[202] Norman H. MacKenzie, ed. *The Poetical Works of Gerard Manley Hopkins* (Oxford: Clarendon Press, 1990), p. 139.

[203] Maria R. Lichtman, "The Incarnational Aesthetic of Gerard Manley Hopkins," *Religion and Literature* 23, no. 1 (1991), p. 41.

ture the sanctification of matter. Via inscape, he seeks to experience the unique incarnations of the Word in matter. As Lichtmann explains, "inscape expresses the eucharistic status of a nature somehow taken up in Christ's first Incarnation, his *ensarkosis*."[204] Hopkins may imply this in an explanation he gives to a fellow poet, Coventry Patmore:

> It is certain that in nature outward beauty is the proof of inward beauty, outward good of inward good. Fineness, proportion, of feature comes from a moulding force which succeeds in asserting itself over the resistance of cumbersome or restraining matter...[205]

Hopkins, as we have seen, believes that an experience of a thing's inscape is possible, if not common. The spiritual, forming core, or inscape, of a particular material reality may be sensed and thus instressed or enselved. Hopkins gives an example of this in a journal entry describing his experience of water flowing at St. Winifred's well:

> The sight of the water in the well as clear as glass, greenish like beryl or aquamarine, trembling at the surface with the force of the springs, and shaping out the five foils of the well quite drew and held my eyes to it...The strong unfailing flow of the water...took hold of my mind with wonder at the bounty of God in one of His saints, *the sensible thing so naturally and gracefully uttering the spiritual reason of its being* [my italics]...and the spring in place leading back the thoughts by its spring in time to its spring in eternity: even now the stress and buoyancy and abundance of the water is before my eyes.[206]

[204] Ibid.

[205] Claude Colleer Abbott, ed. *Further Letters of Gerard Manley Hopkins Including his Correspondence with Coventry Patmore*, 2nd ed. (London: Oxford University Press, 1956), p. 306.

[206] Humphry House, ed. *The Journals and Papers of Gerard Manley Hopkins*, 2nd ed. (London: Oxford University Press, 1959), p. 261. Hopkins' experience of the flowing water at the well led him to write, or 'inscape' his experience, in a poem.

Christ inscapes the particulars of being thus becoming available to one open to being instressed — "Since, though he is under the world's splendour and wonder, / His mystery must be instressed, stressed..."[207] As Lichtmann extrapolates, "instress...opens one to the mystery of inscape. The poem is the stress that follows instress."[208] Thus the poem becomes a kind of instressing of inscape, or incarnation of inscape. Hopkins writes that "I have often felt when I have...felt the depth of an instress or how fast the inscape holds a thing that nothing is so pregnant and straightforward to the truth as simple *yes* and *is*."[209] Hopkins' poetry becomes a responding '*yes* and *is*' to the mystery of the inscaped Christ he finds in nature. This assent to the mystery on his part finds expression in the poem, where "each word is one way of acknowledging Being and each sentence by its copula *is*...the utterance and assertion of it."[210]

Therefore, as Lichtmann writes, "Hopkins inscapes Christ in his poems in the way he found him in the quintessential Incarnation of Eucharist in matter."[211] The Christ of Hopkins' poetry is often the Christ of the *ensarkosis* — the Christ incarnated into matter and thus a Christ whose divine and human spirit are 'mute'. Hopkins' poems give voice to this 'eucharistically incarnate' and muted Christ. This poetically inscaped Christ is the Christ "enfleshed in

[207] Norman H. MacKenzie, ed. *The Poetical Works of Gerard Manley Hopkins* (Oxford: Clarendon Press, 1990), p. 120.

[208] Maria R. Lichtman, "The Incarnational Aesthetic of Gerard Manley Hopkins," *Religion and Literature* 23, no. 1 (1991), p. 42.

[209] Humphry House, ed. *The Journals and Papers of Gerard Manley Hopkins*, 2nd ed. (London: Oxford University Press, 1959), p. 127.

[210] Ibid. Cf. Maria R. Lichtman, "The Incarnational Aesthetic of Gerard Manley Hopkins," *Religion and Literature* 23, no. 1 (1991), p. 42.

[211] Maria R. Lichtman, "The Incarnational Aesthetic of Gerard Manley Hopkins," *Religion and Literature* 23, no. 1 (1991), p. 43.

bare matter" where his presence is almost absence.[212] Hopkins' poetic images for Christ in matter bear this out:

> Even the images that bear the charge of God in "God's Grandeur" are inanimate — shaken tinfoil and squeezed oil. In "The Windhover," dedicated...to "Christ our Lord," the poet leaves us with ploughed silt and burnt-out embers as the matter in which to find Christ's spirit; in another poem, insects, stones, and bells...carry Christ's spirit until...it breaks forth in "ten thousand places"... In "Ribblesdale," it is "Earth, sweet Earth, sweet landscape" which, like Christ in the Eucharist, "canst but only be." In Hopkins's idea of the Incarnation, there is, at least at first, very little triumph of spirit. If the Incarnation succeeds, it succeeds against all odds. Christ lies dormant in the most viscous, murky, opaque of substances, "the ooze of oil, / Crushed" until, because of his "billion times told" greater fire, he "gash[es] gold-vermillion."[213]

Hopkins' sacramental view of language enables him to trust language to carry the enfleshed Word he finds in nature. The poem becomes, in this sense, the Body of Christ. As Lichtmann concludes:

> [The poem] is Eucharist in the sense of bearing the motionless, lifeless Real Presence of Christ, of acting with sacramental, transforming instress on the reader as Hopkins has himself been instressed by nature, and of being, as in the meaning of the word "Eucharist" itself, a thanksgiving.[214]

The poet incarnates in poetry the Incarnate Christ, the enfleshed Logos who is the foundation of world and word. As Hopkins proclaims, "All things are charged with love, are charged with God and if we know how to touch them give off sparks, take fire, yield drops and flow, ring and tell of

[212] Ibid., p. 44.
[213] Ibid.
[214] Ibid.

him."[215] McNees states the following in her development of this sacramental theme in Hopkins:

> The poet's task is to inscape the Word in the world and to return the Word to God. The poem is the record of this inscaping process. The reader's corresponding task is to capture the inscaped Word in the act of reading and to realize Christ's presence. The poem is analagous to the Prayer of Consecration in which Christ's words are recalled and offered to the people. In the Eucharist the priest inscapes Christ in the elements through the Prayer of Consecration. In the poem the poet seeks to consecrate words, to reanimate them by means of rhetorical devices which effect a divine presence. The rhetorical devices operate like sudden charges of divine energy. They temporarily dislocate ordinary syntactical patterns and enable one to see beyond the temporal sequence in things.[216]

As we have seen above through the contexts of key influences upon Hopkins and significant interpretations of Hopkins, the ultimate cipher for this sacramental dynamic is the Incarnation of the Word, Hopkins' *ensarkosis*, the 'translation' of the Word into Christic selvings or inscapes. The shining out and perception of these inscapes is the dynamic of beauty. Beauty as divine self-expressiveness, in this sense, is what Hopkins the poet seeks to inscape and communicate through his poetry. By recreating, wrenching, and opening the 'matter' of poetic language to new forms and new 'chimings', Hopkins seeks to communicate an 'inscaped Christ' that may then be 'instressed' in the hearer.

In this light, Hopkins saw poetry as language meant to carry inscape, and many of his poems have an experience of inscape as their theme. Conveying inscape and the beauty of Christ is ulti-

[215] Christopher Devlin, S.J., ed. *The Sermons and Devotional Writings of Gerard Manley Hopkins* (London: Oxford University Press, 1959), p. 195.

[216] Eleanor McNees, "Beyond 'The Half-way House': Hopkins and Real Presence," *Texas Studies in Literature and Language* 31, no. 1 (1989), p. 92.

mately what Hopkins aimed for in poetry.[217] More than simply conveying inscape, a poem sustains its own inscape.[218] For Hopkins, poetry is a creation of a pattern of words so highly 'pointed' that it cannot resemble any other pattern of words.[219] It must have "its own distinct self-taste, must represent the one set of words that alone can utter a particular experience" and inscape.[220] The words of the poem bring about a kind of incarnation. The realization of inscape through words can engender a selving, or instress an inscape, in the hearer. Through extrapolation, it may be said then that the poem incarnates Christ sacramentally. As J. Sandman writes:

> The poet, functioning in a way that parallels the creative work of God, can shape a poem — another kind of inscape — that has the capability to interact with readers so that they too are instressed with a manifestation of the sacred... [The poet] mirrors the incarnation by 'enfleshing' in his poetry the perfect inscape of Christ himself, who sustains all other inscapes.[221]

Although Hopkins does not say the following directly, I believe that he saw the poet's task as crafting poetic language in such a way that it, like the sacraments, both signified the presence of

[217] Cf. Claude Colleer Abbott, ed. *The Letters of Gerard Manley Hopkins to Robert Bridges*, 2nd ed. (London: Oxford University Press, 1955), p. 66.

[218] David Downes affirms this in writing that "in this hermeneutic transfer of 'inscape' into metaphoric symbol, the poem itself, in its narrative pattern, image variety, thematic components, its elaborated rhythm and sound systems, becomes a new eidetic entity itself, an 'inscape' in its own right open to textual appropriation by readers." David Anthony Downes, *Hopkins' Achieved Self* (Lanham: University Press of America, 1996), p. 11.

[219] Cf. Joseph Hillis Miller, *The Disappearance of God: Five Nineteenth-Century Writers* (Cambridge: Harvard University Press, 1963), p. 281.

[220] Hilary Fraser, *Beauty and Belief: Aesthetics and Religion in Victorian Literature* (Cambridge: Cambridge University Press, 1986), p. 101.

[221] Joseph Sandman, "Sacramental Dimensions in the Poetry of Gerard Manley Hopkins" (Ph.D. Dissertation, Notre Dame, 1991), pp. 118 and 182.

Christ and encouraged the acknowledgment and reception of this presence by the hearer. This is done by the poet through poetically instressing Christic inscapes in such a way that they 'bid' the hearer to enter more deeply into the life of grace. This is an objective, so to speak, of the sacraments in Christian faith. It is in this sense I conclude that Hopkins saw the poem as sacrament.

The presence of the poetically inscaped Christ is analogous to that of the Eucharistically present Christ. Like the elements of the Eucharist, the words of the poem are meant to transform their own inscape as poem into a spiritual reality which establishes a deeper level of communion between God and the hearer.[222] For Hopkins, it may be said that a successful poem "enacts the eucharistic process."[223]

It is fitting to conclude this reflection on the 'poem as sacrament' with one of Hopkins' mature, sacramental works — *Hurrahing the Harvest*:

> Summer ends now; now, barbarous in beauty, the stooks rise
> Around; up above, what wind-walks! what lovely behaviour
> Of silk-sack clouds! has wilder, wilful-wavier
> Meal-drift moulded ever and melted across skies?
>
> I walk, I lift up, I lift up heart, eyes,
> Down all that glory in the heavens to glean our Saviour;
> And, eyes, heart, what looks, what lips yet gave you a
> Rapturous love's greeting of realer, of rounder replies?
>
> And the azurous hung hills are his world-wielding shoulder
> Majestic — as a stallion stalwart, very-violet-sweet! —
> These things, these things were here and but the beholder
> Wanting; which two when they once meet,

[222] Joseph Sandman, "Sacramental Dimensions in the Poetry of Gerard Manley Hopkins" (Ph.D. Dissertation, Notre Dame, 1991), p. 213.

[223] Ibid., p. 94.

The heart rears wings bold and bolder
And hurls for him, O half hurls earth for him off under his feet.[224]

In this poem, as in others, Hopkins strives to make Christ present in the minds and hearts of his hearers, in imitation of, as Ward says, "the supreme mode in which the Word of God is present through the language of men: the Eucharist."[225]

[224] Norman H. MacKenzie, ed. *The Poetical Works of Gerard Manley Hopkins* (Oxford: Clarendon Press, 1990), pp. 148-9.

[225] Bernadette Ward, "Means and Meaning: Gerard Manley Hopkins' Scotist Poetic of Revelation and Matthew Arnold's Poetic of Social Control" (Doctoral Dissertation, Stanford University, 1990), p. 235.

CONCLUSION

Gerard Manley Hopkins, priest and poet, attempted to join his aesthetics with his theological vision of reality. Unfortunately, nowhere in his writings did Hopkins attempt a synthesis of his theological aesthetic. In fact, there is not even "one central comprehensive philosophical formulation of his theory of inscape,"[1] the heart of his theological aesthetic. The scholar is therefore required to construct a synthetic whole from bits and pieces framed in the various contexts of Hopkins' poems, sermons, letters, and personal musings. In this process, I believe there emerges the beginning of a comprehensive theological aesthetic.

Hopkins desired to make sense of his keen aesthetic experience of the world in light of the essential truths of Christian revelation — particularly the Incarnation. Reflecting on the Incarnation as central to his life, he wrote:

> Our lives... are in their whole direction, not only inwardly but most visibly and outwardly, shaped by Christ's. Without that even outwardly the world could be so different that we cannot even guess it. And my life is determined by the Incarnation down to most of the details of the day. Now this being so that I cannot even stop it, why should I not make the cause that determines my life, both as a whole and in much detail, determine it in greater detail still and to the greater efficiency of what I in any case should do, and to my greater happiness in doing it?[2]

[1] Hilary Fraser, *Beauty and Belief: Aesthetics and Religion in Victorian Literature* (Cambridge: Cambridge University Press, 1986), p. 68.

[2] Christopher Devlin, S.J., ed. *The Sermons and Devotional Writings of Gerard Manley Hopkins* (London: Oxford University Press, 1959), p. 263.

In the light of the Incarnation, even aesthetic experience for Hopkins is somehow Christened. In this light too, within the context of Duns Scotus' influence, Hopkins perceived nature to be Christically expressive. The Incarnate Word is the ultimate pattern and intelligibility of created being in its various and singular manifestations. As Hilary Fraser says, Hopkins explained "the distinctiveness of certain forms, of certain experiences of beauty… by their resemblance to the uniqueness of Christ's inscape."[3]

Within the context of the primacy of the Incarnation of the Word, inscape became the common link between Hopkins' religious belief, aesthetic theory, and poetic expression. Christ as the arch- and ultimate inscape of created being is therefore the archetype of created beauty. Beauty is Christic self-expressiveness in the forms of the created world.

As we have seen in this study, in his theory of inscape, and its corollary instress, Hopkins linked perception, expression, and elements of Christian faith to create a highly sophisticated and unique, though nascent, theological aesthetic. For Hopkins, the Great Sacrifice, or *ensarkosis*, of the Word sealed his theory of inscape, spiritualized the aesthetic philosophy he inherited in part from John Ruskin, "intellectually authorised the sanctification of sensory experience,"[4] and justified his sacramental vision of poetic language. In this vision of an aeonian Incarnation, whereby the Word becomes 'eucharistically' present in matter, Hopkins saw that God, humanity, nature, and language participate in one other and are bound together. In this sense, as Miller indicates, the Eucharist became a kind of archetype for Hopkins whereby created things participate in the divine reality they sig-

[3] Hilary Fraser, *Beauty and Belief: Aesthetics and Religion in Victorian Literature* (Cambridge: Cambridge University Press, 1986), p. 70.

[4] Ibid., p. 94.

nify.[5] Hopkins could thereby take joy in the perception of natural forms or inscapes because they were a medium through which he could instress Christ:

> "God's utterance of himself in himself is God the Word, outside himself is this world. This world then is word, expression, news of God. Therefore its end, its purpose, its purport, its meaning, is God and its life or work to name and praise him.[6]

Through the senses, Hopkins believed that Christ could be instressed and that this process, this 'instressing of inscape', was in part intuitive. Christ as beauty, as self-expressiveness, as inscape, could be 'gleaned' or 'wafted' out of nature.[7] This beauty could then be returned to God, "beauty's self and beauty's giver."[8] In an Ignatian dynamic, Hopkins' believed that this process was a function of human duty. Human beings are the voice of the world and are created to praise, to return 'beauty', Christ, back to God:

> The sun and the stars shining glorify God. They stand where he placed them, they move where he bid them. 'The heavens declare the glory of God.' They glorify God, *but they do not know it*. The birds sing to him, the thunder speaks of his terror, the lion is like his strength, the sea is like his greatness, the honey like his sweetness; they are something like him, they make him known, they tell of him, they give him glory, but they do not know they do, they do not know him, they never can... But AMIDST THEM ALL IS MAN... Man was created. Like the rest then to praise, reverence, and serve God; to give him glory. He does so, even by his being, beyond all visible creatures: ...But man can know God, *can mean to give him glory*.

[5] Joseph Hillis Miller, *The Disappearance of God: Five Nineteenth-Century Writers* (Cambridge: Harvard University Press, 1963), p. 3.

[6] Christopher Devlin, S.J., ed. *The Sermons and Devotional Writings of Gerard Manley Hopkins* (London: Oxford University Press, 1959), p. 129.

[7] Norman H. MacKenzie, ed. *The Poetical Works of Gerard Manley Hopkins* (Oxford: Clarendon Press, 1990), pp. 149 and 120.

[8] Ibid., p. 170.

> This then was why he was made, to give God glory and to mean to give it.[9]

This 'returning of beauty back to God' is not only a matter of gleaning Christ's inscaped presence in nature; it is also a matter of completing Christ's 'design' in human beings. Through 'instressing' Christ via the mystery of grace and will, thus making self-sacrifice to God, men and women may also participate in the Great Sacrifice, in the incarnating of Christ in the World. Thus, we who are "Jack, joke, poor potsherd, patch, matchwood" may also be "immortal diamond," for we can become what Christ is:[10]

> I say more: the just man justices;
> Keeps grace: that keeps all his goings graces;
> Acts in God's eye what in God's eye he is —
> Christ. For Christ plays in ten thousand places,
> Lovely in limbs, and lovely in eyes not his
> To the father through the features of men's faces.[11]

As we have seen, Hopkins saw poetry as language meant to carry inscape, and many of his poems have an experience of inscape as their theme. Conveying inscape and instressing the beauty of Christ is ultimately what Hopkins aimed for in poetry.[12] Furthermore, although Hopkins does not indicate it directly in his writings, I believe that he saw the poet's task as crafting poetic language in such a way that it sacramentally 'carried' the presence of Christ and encouraged the reception of this presence by the hearer. In

[9] Christopher Devlin, S.J., ed. *The Sermons and Devotional Writings of Gerard Manley Hopkins* (London: Oxford University Press, 1959), p. 239.

[10] Norman H. MacKenzie, ed. *The Poetical Works of Gerard Manley Hopkins* (Oxford: Clarendon Press, 1990), p. 198.

[11] Ibid., p. 141.

[12] Cf. Claude Colleer Abbott, ed. *The Letters of Gerard Manley Hopkins to Robert Bridges*, 2nd ed. (London: Oxford University Press, 1955), p. 66.

Hopkins' mind, this is accomplished by the poet through poetically instressing Christic inscapes in such a way that they 'bid' the hearer to enter more deeply into the life of grace. This parallels the objective, so to speak, of the sacraments in Christian faith. It is in this sense I have concluded that Hopkins saw the poem as sacrament.

Furthermore, also in a kind of sacramental fashion, I do not believe Hopkins held that faith was necessary either to inscape Christic beauty poetically or to be instressed by poetically inscaped Christic beauty. The *ensarkosis* and its consequences are ontologically rooted for Hopkins and are therefore ultimately independent of whether they are perceived as Christic or not. This, as I have held, is a corrective to von Balthasar's 'fideistic' interpretation of Hopkins.

Hopkins' theological aesthetic, centered upon his theory of inscape, has been criticized and even rejected. J. Hillis Miller sees it ultimately as a failure in wording the unwordable: "In the end, after a lifetime spent in God's service, [Hopkins] has not gone beyond his beginning, which was to know God as the deity of Isaiah, the God who hides himself."[13] Hilary Fraser claims that Hopkins' theory of inscape failed to meet his deepest needs: "The theory of inscape, so brilliant and flawless, so complete, remained merely an intellectual solution."[14] Finally, Joseph Feeney concludes that Hopkins essential aesthetic 'optimism' about God and the world 'broke down' towards the end of his life with a return to his pre-Jesuit period moralism.[15] In short, key Hopkins scholars claim

[13] Joseph Hillis Miller, *The Disappearance of God: Five Nineteenth-Century Writers* (Cambridge: Harvard University Press, 1963), p. 359.

[14] Hilary Fraser, *Beauty and Belief: Aesthetics and Religion in Victorian Literature* (Cambridge: Cambridge University Press, 1986), p. 106.

[15] See Joseph Feeney, S.J., "The Collapse of Hopkins' Jesuit Worldview: A Conflict between Moralism and Incarnationalism," in *Gerard Manley Hopkins Annual 1992*, ed. Michael Sundermeier (Omaha: Creighton University Press, 1992), 105-126. Feeney offers Hopkins' 'terrible sonnets' (see the introduction of

that his attempt at a theological aesthetic is overly optimistic about the possibility of gleaning the divine from nature.

Certainly, Hopkins himself did not explicitly address the issue of how his aesthetic would remain meaningful in a natural or social context in which the 'beautiful' is absent and in which suffering or even evil seem dominant. Still, Hopkins did not limit the realm of his theological aesthetic to the world of 'nature'. Indeed, although I have not made Hopkins' aesthetic of humanity's 'inner landscape', of grace, a major theme of this dissertation, one could argue that ultimate 'beauty' and divine self-expressiveness for Hopkins find their most intense expression and realization within the hearts and wills of human beings responding to, or 'seconding', their Christic inscapes. For Hopkins, perhaps the aesthetic of the 'Great Sacrifice' finds its highest expression in natural and social contexts in which human hearts and wills are most pressed in 'seconding' the kenotic Incarnation; i.e., in becoming 'after-Christs'. Perhaps in this light we can understand Hopkins' words in his poem *To what serves Mortal Beauty?*:

> Our law says: Love what are | love's worthiest, were all known;
> World's loveliest — men's selves. Self | flashes off frame and face.
> What do then? how meet beauty? | Merely meet it; own,
> Home at heart, heaven's sweet gift; | then leave, let that alone.
> Yea, wish that though, wish all, | God's better beauty, grace.[16]

this study for examples of these sonnets) as proof that Hopkins' 'Jesuit worldview' deteriorated into a kind of forlorn spiritual isolation in which the world was 'dry' and God was absent. I, rather, see these sonnets as being on the continuum of Hopkins' consistent theological aesthetic. The 'terrible sonnets' may not reflect the shining optimism of Hopkins' earlier nature sonnets, but they are still faithful to the expression of inscape as Christic self-expressiveness. They word and convey the 'terrible beauty' of Christ's self-sacrifice on the Cross as it is seconded in Hopkins' own life. Thus, Hopkins could hold that "I am what Christ is since Christ was what I am." Afterall, the Great Sacrifice at the heart of Hopkins' aesthetic finds its most pointed expression in the Cross.

[16] Norman H. MacKenzie, ed. *The Poetical Works of Gerard Manley Hopkins* (Oxford: Clarendon Press, 1990), p. 183.

This aesthetic of the inner life is again highlighted in Hopkins' poem in honor of St. Alphonsus Rodriguez, laybrother of the Society of Jesus:

> Honour is flashed off exploit, so we say;
> And those strokes once that gashed flesh or galled shield
> Should tongue that time now, trumpet now that field,
> And, on the fighter, forge his glorious day.
> On Christ they do and on the martyr may;
> But be the war within, the brand we wield
> Unseen, the heroic breast not outward-steeled,
> Earth hears no hurtle then from fiercest fray.
> Yet God (that hews mountain and continent,
> Earth, all, out; who, with trickling increment,
> Veins violets and tall trees makes more and more)
> Could crowd career with conquest while there went
> Those years and years by of world without event
> That in Majorca Alfonso watched the door.[17]

Even if one does not accept the inscaping and instressing process by which Hopkins moves from being to knowing to wording, and even if one does not accept his sacramental vision of world and word, still it must be said that Hopkins highlights much of great value.

First of all, Hopkins insists that we take the experience of the senses seriously. Even before his centering upon the Incarnation via the influence of Ignatius and, even more importantly, Duns Scotus, Hopkins heard the admonition of John Ruskin — the artist must take physical reality and its observance at least as seriously as the scientist. This admonition may be heard with good effect by the theologian as well. If creation in all of its particular and unique manifestations is a 'holy book', in Ruskin's term, then we should read it. However, if furthermore, as Hopkins came to hold, creation

[17] Ibid., p. 199.

is Christic in its core, how much more does Ruskin's admonition take force? No wonder that Hopkins reflected in sadness upon how much the beauty of inscape was unknown and buried away from people even though it was near at hand for those that had eyes to see it. Acts of sensuous perception take on great value in light of Hopkins' theological aesthetic. As we have seen, for Hopkins all things are charged with love, are charged with God, and if we know how to touch them give off sparks and take fire, yield drops and flow, ring and tell of the divine.

Even more, the *ensarkosis* sanctifies created reality and places much more stress upon the role of the senses. Hopkins held that the intelligibility of world and word is rooted in the Word made flesh — the Logos emptied into matter. Christ is the very "ground of being, and granite of it."[18] This radical Incarnationalism, which Hopkins derived from Duns Scotus, is the root of any theological aesthetic. It is also the root of beauty and beauty's place in the life of faith and of coming to faith. The spiritual dynamic, which this approach to the 'self-expressiveness' of created reality, of nature, allows, is a gift. God is a God of beauty and energy suffusing all created forms.

A corollary to this Hopkinsian stress upon the value of the senses and of sensuous perception is the value of created reality itself. If creation, especially in its particularity, is Christic and thereby potentially revelatory, then the approach of humanity to nature must be affected. Hopkins has been called the 'environmental poet' for his insistence on a kind of 'devotional' respect for creation.[19] We find this sentiment expressed in Hopkins' poem, *God's Grandeur*:

[18] Norman H. MacKenzie, ed. *The Poetical Works of Gerard Manley Hopkins* (Oxford: Clarendon Press, 1990), p. 127.

[19] Cf. Michael Sundermeier, "Of Wet and Wildness: Hopkins and the Environment," in *Gerard Manley Hopkins Annual 1992*, ed. Michael Sundermeier (Omaha: Creighton University Press, 1992), pp. 59-81.

The world is charged with the grandeur of God.
  It will flame out, like shining from shook foil;
  It gathers to a greatness, like the ooze of oil
Crushed. Why do men then now not reck his rod?
Generations have trod, have trod, have trod;
  And all is seared with trade; bleared, smeared with toil;
  And wears man's smudge and shares man's smell: the soil
Is bare now, nor can foot feel, being shod.

And for all this, nature is never spent;
  There lives the dearest freshness deep down things;
And though the last lights off the black West went
  Oh, morning, at the brown brink eastward, springs —
Because the Holy Ghost over the bent
  World broods with warm breast and with ah! bright wings.[20]

Hopkins asks and proclaims:

What would the world be, once bereft
Of wet and of wildness? Let them be left,
O let them be left, wildness and wet;
Long live the weeds and the wilderness yet.[21]

Just as Hopkins' aesthetic asks the theologian to pay at least as much attention to nature as the scientist does, so it also asks us to remove our 'intellectual sandals' in our approach to nature, for we move on holy ground, on ground 'eucharistically' incarnated by the Word.

The nature of the potential role of the artist, much less the poet, is also a lesson Hopkins offers. The concept of the artist as one who 'captures' inscape ("I caught this morning morning's minion...")[22] and 'instresses' it is of great value in a utilitarian world.

[20] Norman H. MacKenzie, ed. *The Poetical Works of Gerard Manley Hopkins* (Oxford: Clarendon Press, 1990), p. 139.

[21] Ibid., p. 168.

[22] Ibid., p. 144.

The artist, for Hopkins, is one who "instresses" then "stresses" inscape:

> I kiss my hand
> To the stars, lovely-asunder
> Starlight, wafting him out of it; and
> Glow, glory in thunder;
> Kiss my hand to the dappled-with-damson west:
> Since, tho' he is under the world's splendour and wonder,
> His mystery must be instressed, stressed;
> For I greet him the days I meet him, and bless when I understand.[23]

The Christian artist is one who 'words' the intuitive experience of inscape by Christ:

> Ah! there was a heart right!
> There was single eye!
> Read the unshapeable shock night
> And knew the who and the why;
> Wording it how but by him that present and past,
> Heaven and earth are word of, worded by?[24]

This vision of the Hopkinsian artist approaches, perhaps, the vision of the iconographer in Christian Orthodoxy. In any case, the artist's role of 'capturing' and 'heightening' beauty cannot be diminished in light of Hopkins' aesthetic.

The Beautiful, just as the Good and the True, is an attribute of the Divine in Hopkins' eyes. Furthermore, beauty is given in creation. It is there to be known and to be worded. It is there 'to bid' us to faith and 'to bid' us to confirm our own Christic inscape, for we can become what Christ is. We, who are "Jack, joke, poor potsherd" can become "immortal diamond."[25] In this sense, Hopkins indeed brings beauty back to faith.

[23] Norman H. MacKenzie, ed. *The Poetical Works of Gerard Manley Hopkins* (Oxford: Clarendon Press, 1990), p. 120.

[24] Ibid., p. 126.

[25] Ibid., p. 198.

BIBLIOGRAPHY

I. *Editions of Works by Gerard Manley Hopkins*

Abbott, Claude Colleer, ed. *The Correspondence of Gerard Manley Hopkins and Richard Watson Dixon*. London: Oxford University Press, 1955.

Abbott, Claude Colleer, ed. *The Letters of Gerard Manley Hopkins to Robert Bridges*. London: Oxford University Press, 1955.

Abbott, Claude Colleer, ed. *Further Letters of Gerard Manley Hopkins Including his Correspondence with Coventry Patmore*. London: Oxford University Press, 1956.

Devlin, Christopher, S.J., ed. *The Sermons and Devotional Writings of Gerard Manley Hopkins*. London: Oxford University Press, 1959.

Gardner, W.H., and N.H. MacKenzie, ed. *The Poems of Gerard Manley Hopkins*. London: Oxford University Press, 1967.

House, Humphry, ed. *The Journals and Papers of Gerard Manley Hopkins*. London: Oxford University Press, 1959.

II. *Books about Gerard Manley Hopkins and Related Topics*

Allsopp, Michael E., and David Anthony Downes, ed. *Saving Beauty: Further Studies in Hopkins*. Edited by Todd K. Bender. Vol. 6, *Origins of Modernism: Garland Studies in British Literature*. New York: Garland Publishing, Inc., 1994.

Allsopp, Michael E., and Michael W. Sundermeier, ed. *Gerard Manley Hopkins (1844-1889): New Essays on His Life, Writing, and Place in English Literature*. Vol. 1, *Studies in British Literature*. Lewiston: The Edwin Mellen Press, 1989.

Ball, Patricia M. *The Science of Aspects: The Changing Role of Fact in the Work of Coleridge, Ruskin and Hopkins*. London: The Athlone Press, 1971.

Balthasar, Hans Urs von. *Seeing the Form*. Translated by Leiva-Merikakis, Erasmo. Vol. 1. 7 vols. *The Glory of the Lord: A Theological Aesthetics*, ed. Joseph Fessio and John Riches. Edinburgh: T. & T. Clark, 1982.

Balthasar, Hans Urs von. *Studies in Theological Style: Clerical Styles*. Translated by Andrew Louth, Francis McDonagh, and Brian McNeil. Vol. 2. 7 vols.

The Glory of the Lord: A Theological Aesthetics, ed. Joseph Fessio and John Riches. San Francisco: Ignatius Press, 1984.

Balthasar, Hans Urs von. *Studies in Theological Style: Lay Styles*. Translated by Andrew Louth, John Saward, Martin Simon, and Rowan Williams. Vol. 3. 7 vols. *The Glory of the Lord: A Theological Aesthetics*, ed. John Riches. San Francisco: Ignatius Press, 1986.

Barolsky, Paul. *Walter Pater's Renaissance*. University Park: The Pennsylvania State University Press, 1987.

Belitt, Ben. *The Forged Feature: Toward a Poetics of Uncertainty*. New York: Fordham University Press, 1995.

Bender, Todd K. *Gerard Manley Hopkins: The Classical Background and the Critical Reception of His Work*. Baltimore: The Johns Hopkins University Press, 1966.

Bergonzi, Bernard. *Gerard Manley Hopkins*. New York: Macmillan Publishing Co., Inc., 1977.

Bottalla, P., G. Marra, F. Marucci, ed. *Gerard Manley Hopkins: Tradition and Innovation*. Ravenna: Longo Editore, 1991.

Boyle, Robert. *Metaphor in Hopkins*. Chapel Hill: University of North Carolina Press, 1961.

Brown, Frank Burch. *Religious Aesthetics: A Theological Study of Making and Meaning*. Princeton: Princeton University Press, 1989.

Chadwick, Owen. *The Mind of the Oxford Movement*. Stanford: Stanford University Press, 1960.

Christ, Carol T. *Victorian and Modern Poetics*. Chicago: University of Chicago Press, 1984.

Cohen, Edward H. *Works and Criticism of Gerard Manley Hopkins*. Washington, D.C.: The Catholic University of America Press, 1969.

Copleston, Frederick, S.J. *Mediaeval Philosophy: Albert the Great to Duns Scotus*. Vol. 2, *A History of Philosophy*. Garden City: Image Books, 1962.

Cotter, James Finn. *Inscape: The Christology and Poetry of Gerard Manley Hopkins*. Pittsburgh: University of Pittsburgh Press, 1972.

Coulson, John. *Newman and the Common Tradition: A Study in the Language of Church and Society*. Oxford: Clarendon Press, 1970.

Coulson, John. *Religion and Imagination*. Oxford: Clarendon Press, 1981.

De Angelis, Palmira. *L'Imagine Epifanica*. Rome: Bulzone Editore, 1989.

Dessain, Charles Stephen, ed. *The Letters and Diaries of John Henry Newman*. Oxford: Clarendon Press, 1973.

Devlin, Madeleine. *Christopher Devlin*. London: Macmillan Co., 1970.

Downes, David Anthony. *Victorian Portraits: Hopkins and Pater*. New York: Bookman Associates, Inc., 1965.

Downes, David Anthony. *The Great Sacrifice*. Lanham: University Press of America, 1983.

Downes, David Anthony. *Ruskin's Landscape of Beatitude*. New York: Peter Lang Publishing, Inc., 1984.

Downes, David Anthony. *Hopkins' Sanctifying Imagination*. Lanham: University Press of America, 1985.

Downes, David Anthony. *The Ignatian Personality of Gerard Manley Hopkins*. 2nd ed. Lanham: University Press of America, 1990.

Downes, David Anthony. *Hopkins' Achieved Self*. Lanham: University Press of America, 1996.

Dunne, Tom. *Gerard Manley Hopkins: A Comprehensive Bibliography*. Oxford: Clarendon Press, 1976.

Ellis, Virginia Ridley. *Gerard Manley Hopkins and the Language of Mystery*. Columbia: University of Missouri Press, 1991.

Ellsberg, Margaret R. *Created to Praise: The Language of Gerard Manley Hopkins*. New York: Oxford University Press, 1987.

Folz, William, and Todd K. Bender. *A Concordance to the Sermons of Gerard Manley Hopkins*. Vol. 1235 *Garland Reference Library of the Humanities*. New York: Garland Publishing, Inc., 1989.

Fraser, Hilary. *Beauty and Belief: Aesthetics and Religion in Victorian Literature*. Cambridge: Cambridge University Press, 1986.

Fulweiler, Howard. *Letters from the Darkling Plain: Language and the Grounds of Knowledge in the Poetry of Arnold and Hopkins*. Columbia: University of Missouri Press, 1972.

Ganss, George E., S.J., ed. *Ignatius of Loyola: The Spiritual Exercises and Selected Works* in *The Classics of Western Spirituality*. New York: Paulist Press, 1991.

Gardner, W.H. *Gerard Manley Hopkins (1844-1889): A Study of Poetic Idiosyncrasy in Relation to Poetic Tradition*. 2 vols. London: Secker and Warburg, 1948.

Goeme, Christine. *Jeans Duns Scot ou la révolution subtile*. Paris: Fac Editions, 1982.

Graves, Robert. *The White Goddess: A Historical Grammar of Poetic Myth*. American ed. New York: Farrar, Straus and Giroux, 1966.

Heuser, Alan. *The Shaping Vision of Gerard Manley Hopkins*. London: Oxford University Press, 1958.

Iynegar, K. R. Srinivasa. *Gerard Manley Hopkins: The Man and the Past*. New York: Haskell House Ltd., 1971.

James, David Gwilyn. *The Romantic Comedy: An Essay on English Romanticism*. London: Oxford University Press, 1948.

Johnson, Wendell Stacy. *Gerard Manley Hopkins: The Poet as Victorian*. Ithaca: Cornell University Press, 1968.

Keble, John. *Lectures on Poetry: 1832-1841*. Translated by Edward Kershaw Francis. 2 vols. Oxford: Clarendon Press, 1912.

Knickerbocker, William S. *Creative Oxford: Its Influence in Victorian Literature*. Syracuse: Syracuse Bookstore, 1925.

Küng, Hans. *Art and the Question of Meaning*. Translated by Edward Quinn. New York: Crossroad Publishing Company, 1981.

Lahey, Gerard F., S.J. *Gerard Manley Hopkins*. London: Oxford University Press, 1930.

Landow, George P. *The Aesthetic and Critical Theories of John Ruskin*. Princeton: Princeton University Press, 1971.

Mackenzie, Norman H. *A Reader's Guide to Gerard Manley Hopkins*. Ithaca: Cornell University Press, 1981.

MacKenzie, Norman H., ed. *The Poetical Works of Gerard Manley Hopkins*. Oxford: Clarendon Press, 1990.

Martin, F. David. *Art and the Religious Experience: The "Language" of the Sacred*. Lewisburg: Bucknell University Press, 1972.

Martin, Robert Bernard. *Gerard Manley Hopkins: A Very Private Life*. New York: G.P. Putnam's Sons, 1991.

Martz, Louis L. *The Poetry of Meditation*. 2nd ed. New Haven: Yale University Press, 1962.

Marucci, Franco. *The Fine Delight That Fathers Thought: Rhetoric and Medievalism in Gerard Manley Hopkins*. Washington, D.C.: The Catholic University Press, 1994.

Maurois, André. *The Quest for Proust*. Translated by Gerard Hopkins. London: Jonathan Cape, 1950.

Miller, Authur. *Insights of Genius: Imagery and Creativity in Science and Art*. New York: Copernicus, 1996.

Miller, Joseph Hillis. *The Disappearance of God: Five Nineteenth-Century Writers*. Cambridge: Harvard University Press, 1963.

Miller, Joseph Hillis. *Victorian Subjects*. Durham: Duke University Press, 1991.

Newman, John Henry. *Certain Difficulties Felt by Anglicans in Catholic Teaching*. Vol. 2. 2 vols. London: Longmans and Green, 1918.

Newman, John Henry. *Apologia Pro Vita Sua*, ed. Martin J Svalgic. Oxford: Clarendon Press, 1967.

Newman, John Henry. *An Essay in Aid of a Grammar of Assent*, ed. I. T. Ker. London: Oxford University Press, 1985.

Nixon, Jude V. *Gerard Manley Hopkins and His Contemporaries: Liddon, Newman, Darwin and Pater*. Vol. 5. 6 vols. *Origins of Modernism: Garland*

Studies in British Literature, ed. Todd K. Bender. New York: Garland Publishing, Inc., 1994.

Noppen, Leo van. *The Critical Reception of Gerard Manley Hopkins in The Netherlands and Flanders: 1908-1979*. The International Hopkins Association Monograph Series, ed. Richard F. Giles. Waterloo: The International Hopkins Association, 1980.

Ong, Walter J., S.J. *Hopkins, the Self, and God*. Toronto: University of Toronto Press, 1986.

Owen, W.J.B., and Jane Worthington Smyser, ed. *The Prose Works of William Wordsworth*. Vol. 1. Oxford: Clarendon Press, 1974.

Perkins, Mary Anne. *Coleridge's Philosophy: The Logos as Unifying Principle*. Oxford: Oxford University Press, 1994.

Peters, W.A.M. *Gerard Manley Hopkins: A Critical Essay towards the Understanding of his Poetry*. London: Oxford University Press, 1948.

Phare, E.E. *The Poetry of Gerard Manley Hopkins*. Cambridge: Cambridge University Press, 1933.

Pick, John. *Gerard Manley Hopkins: Priest and Poet*. Oxford: Oxford University Press, 1942.

Pick, John. *Gerard Manley Hopkins: Priest and Poet*. 2nd ed. New York: Oxford University Press, 1966.

Prickett, Stephen. *Romanticism and Religion: The Tradition of Coleridge and Wordsworth in the Victorian Church*. Cambridge: Cambridge University Press, 1976.

Proust, Marcel. *Remembrance of Things Past*. Translated by C.K. Scott Moncrieff. Vol. 1. 3 vols. New York: Random House, 1934.

Roberts, Louis. *The Theological Aesthetics of Hans Urs von Balthasar*. Washington, D.C.: The Catholic University of America Press, 1987.

Roston, Murray. *Victorian Contexts: Literature and the Visual Arts*. New York: New York University Press, 1996.

Rowell, Geoffrey. *The Vision Glorious: Themes and Personalities of the Catholic Revival in Anglicanism*. Oxford: Oxford University Press, 1983.

Rowell, Geoffrey, ed. *Tradition Renewed: The Oxford Movement Conference Papers*. Allison Park: Pickwick Publications, 1986.

Ruskin, John. *The Complete Works of John Ruskin*. 15 vols. *The Complete Works of John Ruskin*. New York: T.Y. Crowell and Co., 19—.

Ruskin, John. *The Diaries of John Ruskin*. Vol. 1. 3 vols., ed. J. Evans and J.H. Waterhouse. Oxford: Clarendon Press, 1956.

Scott, Clive. *A Question of Syllables: Essays in Nineteenth-Century French Verse*. Cambridge: Cambridge University Press, 1986.

Seelhammer, Ruth. *Hopkins Collected at Gonzaga.* Chicago: Loyola University Press, 1970.

Shmeifsky, Mavel. *Sense at War with Soul: English Poetics (1865-1900).* The Hague: Mouton, 1972.

Storey, Graham. *A Preface to Hopkins.* Second ed. London: Longman Group UK Limited, 1992.

Sulloway, Alison G. *Gerard Manley Hopkins and the Victorian Temper.* New York: Columbia University Press, 1972.

Sulloway, Alison G., ed. *Critical Essays on Gerard Manley Hopkins.* Edited by Zack Bowen, *Critical Essays on British Literature.* Boston: G.K. Hall & Co., 1990.

Thomas, Alfred, S.J. *Hopkins the Jesuit: The Years of Training.* London: Oxford University Press, 1969.

Thornton, R.K.R. *Gerard Manley Hopkins: The Poems.* London: Edward Arnold, 1973.

Tillich, Paul. *On Art and Architecture.* New York: Crossroad Publishing Company, 1987.

Walhout, Donald. *Send My Roots Rain: A Study of Religious Experience in the Poetry of Gerard Manley Hopkins.* Athens: Ohio University Press, 1981.

Watson, J. R. *The Poetry of Gerard Manley Hopkins.* London: Penguin Books, 1987.

Westerlinck, Albert. *Mens en Grens: Over het Mensbeeld in de Moderne Europese Literatuur.* Bruges: Orion, 1972.

White, Norman. *Hopkins: A Literary Biography.* Oxford: Clarendon Press, 1992.

Zonneveld, Sjaak. *The Random Grim Forge: A Study of Social Ideas in the Work of Gerard Manley Hopkins.* Assen: Van Gorcum, 1992.

III. *Articles and Book Sections about Gerard Manley Hopkins*

Abad, Pilar. "Hopkins and the Modern Sonnet Tradition: Dylan Thomas, W.H. Auden and Seamus Heaney." In *Gerard Manley Hopkins: Tradition and Innovation*, ed. G. Marra P. Bottalla, F. Marucci, 223-34. Ravenna: Longo Editore, 1991.

Abraham, John. "The Hopkins Aesthetic." *Continuum* 1 (1963): 32-9.

Abraham, John. "The Hopkins Aesthetic II." *Continuum* 1 (1963): 355-60.

Allsopp, Michael E. "Gerard Manley Hopkins: The Oxford Years (1863-1867)." *Gregorianum* 70 (1989): 661-87.

Allsopp, Michael E. "G.M. Hopkins, Narrative, and the Heart of Morality: Exposition and Critique." *The Irish Theological Quarterly* 60 (1994): 287-307.

Arkins, Brian. "Style in the Poetry of Hopkins." *Studies* 86 (1997): 135-43.

Ballinger, Philip. "Book Review of Franco Marucci's *The Fine Delight that Fathers Thought: Rhetoric and Medievalism in Gerard Manley Hopkins.*" *Studies* 85 (1996): 195-97.

Ballinger, Philip. "Ruskin: Hopkins' "Silent Don"." *Studies* 85 (1996): 116-24.

Ballinger, Philip. "Book Review of David Downe's *Hopkins' Achieved Self.*" *Studies* 86 (1997): 183-85.

Ballinger, Philip. "Book Review of Francis Fennell's *Rereading Hopkins: Selected New Essays.*" *Studies* 86 (1997): 185-89.

Ballinger, Philip. "Created to Praise: Gerard Manley Hopkins and Ignatius of Loyola." *Louvain Studies* 22 (1997): 153-80.

Barth, Robert J. "Hopkins as a Romantic: A Coleridgean's View." *The Wordsworth Circle* 25 (1994): 107-13.

Bender, Todd K. "The Architecture of Hopkins's Poetic Vocabulary." In *Rereading Hopkins: Selected New Essays*, ed. Francis L. Fennell, 69, 157-64. Victoria: University of Victoria, 1996.

Bernad, Miguel, S.J. "Hopkins' Pied Beauty: A Note on its Ignatian Inspiration." *Essays in Criticism* 12 (1962): 217-20.

Bérubé, Camille. "Dialogue of Scotus with Modern Culture on 'Regnum Hominis et Regnum Dei'." In *Regnum Hominis et Regnum Dei: Acta Quarti Congressus Scotistici Internationalis*, ed. Camille Bérubé, 1-8. Rome: Societas Internationalis Scotistica, 1978.

Billi, Mirella. "Hopkins and the Figurative Arts." In *Gerard Manley Hopkins: Tradition and Innovation*, ed. G. Marra, P. Bottalla, F. Marucci, 69-81. Ravenna: Longo Editore, 1991.

Boggs, Rebecca. "Poetic Genesis, the Self, and Nature's Things in Hopkins." *Studies in English Literature 1500-1900* 37 (1997): 831-55.

Boggs, Rebecca. " 'There lives the dearest freshness deep down things': Articulating the Distinctions Between Man and the Things of Nature." *The Hopkins Quarterly* 22 (1995): 53-77.

Bonadei, Rossana. "From Diaries to Poetry — Focusing on the Tree." In *Gerard Manley Hopkins: Tradition and Innovation*, ed. G. Marra, P. Bottalla, F. Marucci, 101-19. Ravenna: Longo Editore, 1991.

Bouchard, Gary M. "What Gets Said in a Narrow (ten-by-fourteen) Room: A reconsideration of Hopkins's Later Sonnets." In *Rereading Hopkins: Selected New Essays*, ed. Francis L. Fennell, 180-92. Victoria: University of Victoria, 1996.

Bouyer, Louis. "Newman et le platonisme de l'âme anglaise." *Revue de Philosophie* 6 (1936): 285-305.

Bowman, Leonard. "Bonaventure and the Poetry of Gerard Manley Hopkins." In *Philosophica*, ed. Jacques Guy Bougerol, 3, 553-70. Rome: Collegio S. Bonaventura Grottoferrata, 1973.

Bowman, Leonard. "Bonaventure's Symbolic Theology and Gerard Manley Hopkins' 'Inscapes'." A paper delivered at the Congresso internazionale per il settimo centenario di san Bonaventura da Bagnoregio, Rome, 1974.

Bowman, Leonard. "Another Look at Hopkins and Scotus." *Renascence* 29 (1976): 50-6.

Boyd, John D., S.J. "'I Say More': Sacrament and Hopkins's Imaginative Realism." *Renascence* 42 (1989): 51-64.

Boyd, John D., S.J. "Hopkins and the Paschal Action." *Thought* 65 (1990): 481-85.

Brittain, Clark M. "God's Better Beauty: Hopkins, Pusey, and Tractarian Aesthetics." *Christianity and Literature* 40 (1990): 7-22.

Bruce, Donald William. "Hopkins the Observer, 1844-1889." *Contemporary Review* 123 (1989): 247-53.

Bruns, Gerald L. "The Idea of Energy in the Writings of Gerard Manley Hopkins." *Renascence* 29 (1976): 25-42.

Bump, Jerome. "Hopkins' Drawings." In *All My Eyes See: The Visual World of Gerard Manley Hopkins*, ed. R.K.R. Thornton, 69-87. Tyne and Wear: Ceolfrith Press, 1975.

Burns, Chester A., S.J. "Gerard Manley Hopkins, Poet of Ascetic and Aesthetic Conflict." In *Immortal Diamond*, ed. S.J. Norman Weyand, 175-91. New York: Octagon Books, 1969.

Cane, Aleta. "Double Discourses in 'The Wreck of the Deutschland'." In *Rereading Hopkins: Selected New Essays*, ed. Francis L. Fennell, 165-79. Victoria: University of Victoria, 1996.

Caro, Robert V., S.J. "An Ignatian Meditation: 'Carrion Comfort'." *Studies* 84 (1995): 152-59.

Carroll, Martin C., S.J. "Gerard Manley Hopkins and the Society of Jesus." In *Immortal Diamond: Studies in Gerard Manley Hopkins*, ed. Norman Weyand, 3-50. New York: Octagon Books, 1969.

Casey, Gerard. "Hopkins — Poetry and Philosophy." *Studies* 84 (1995): 160-67.

Castorina, Giuseppe. "The Science of Language and the Distinctive Character of Hopkins's Poetry and Poetics." In *Gerard Manley Hopkins: Tradition and Innovation*, ed. G. Marra P. Bottalla, F. Marucci, 83-100. Ravenna: Longo Editore, 1991.

Caws, Mary Ann. "Cognitive Poetics and Passionate Reading in Ruskin and Hopkins." *Rivista di letterature moderne e comparate* 44 (1991): 241-62.

Cervo, Nathan. "Scotistic Elements in the Poetry of Hopkins." *Hopkins Quarterly* 10 (1983): 55-68.

Chen, Tung-jung. "Preoccupations of the poet: a reading of Gerard Manley Hopkins and Seamus Heaney." *Journal of Humanities East/West* 8 (1990): 155-76.

Ch'iu-lang, Chi. "Esthetics of Mystical Understanding: Joyce, Hopkins, and Tsung-ping." *Tamkang Review* 20 (1989): 88-109.

Cochran, Leonard, O.P. "Instress and Its Place in the Poetics of Gerard Manley Hopkins." *The Hopkins Quarterly* 6 (1980): 143-82.

Colley, Ann C. "Mapping in and out of the Borders of Time: Ruskin and Hopkins." *Victorian Literature and Culture* 19 (1991): 107-21.

Collins, James. "Philosophical Themes in G.M. Hopkins." *Thought* 22 (1947): 67-106.

Corrington, Robert S. "The Christhood of Things." *Drew Gateway* 52 (1981): 41-7.

Cotter, James Finn. "Rhetoric and Poetic in Hopkins." In *Rereading Hopkins: Selected New Essays*, ed. Francis L. Fennell, 143-56. Victoria: University of Victoria, 1996.

Coulson, John. "Hans Urs von Balthasar: Bringing Beauty Back to Faith." In *The Critical Spirit and the Will to Believe: Essays in Nineteenth-Century Literature and Religion*, ed. David Jasper and T.R. Wright, 218-32. New York: St. Martin's Press, 1989.

Coulthard, A.R. "Gerard Manley Hopkins: Priest vs. Poet." *The Victorian Newsletter* 88 (1995): 35-40.

Cowles, James R. "Hopkins and the Numinous: A Consideration of the Poetry in the Light of Rudolf Otto's 'The Idea of the Holy'." *The Hopkins Quarterly* 21 (1994): 43-76.

Cranny, Titus. "Modern Man: Scotus Through Hopkins." In *Regnum Hominis et Regnum Dei: Acti Quarti Congressus Scotistici Internationalis*, ed. Camille Bérubé, 625-40. Rome: Societas Internationalis Scotistica, 1978.

Danielou, Jean. "The Ignatian Vision of the Universe and of Man." *Cross Currents* 4 (1953): 357-66.

Devlin, Christopher, S.J. "Hopkins and Duns Scotus." *New Verse* 14 (1935): 12-5.

Devlin, Christopher, S.J. "The Ignatian Inspiration of Gerard Manley Hopkins." *Blackfriars* 16 (1935): 887-900.

Devlin, Christopher, S.J. "An Essay on Scotus." *The Month* 182 (1946): 456-66.

Devlin, Christopher, S.J. "Time's Eunuch." *The Month* 1 (1949): 303-12.

Devlin, Christopher, S.J. "Correspondence." *The Month (ns)* 4 (1950): 213-15.

Devlin, Christopher, S.J. "The Image and the Word—I." *The Month (ns)* 3 (1950): 114-27.

Devlin, Christopher, S.J. "The Image and the Word—II." *The Month (ns)* 3 (1950): 191-202.

Devlin, Christopher, S.J. "The Psychology of Duns Scotus: A Paper Read to the London Aquinas Society on 15 March, 1950." *The Aquinas Papers* 15 (1950).

Downes, David Anthony. "Beatific Landscapes in Hopkins." *The Hopkins Quarterly* 1 (1975): 185-201.

Downes, David Anthony. "Gerard Manley Hopkins' Christed Vision of Ultimate Reality and Meaning." *Ultimate Reality and Meaning* 12 (1989): 61-80.

Downes, David Anthony. "'Self Flashes': Ricoeur's Achieved Self in Hopkins." In *Rereading Hopkins: Selected New Essays*, ed. Francis L. Fennell, 46-62. Victoria: University of Victoria, 1996.

Doyle, Thomas, S.J. "'What I do is me': Scotist Elements in the Poetry of Gerard Manley Hopkins." *The Hopkins Quarterly* 20 (1993): 3-21.

Dulles, Avery, S.J. "St. Ignatius and the Jesuit Theological Tradition." *Studies in the Spirituality of Jesuits* 14 (1982): 1-26.

Earl, James W. "The One Rapture of Inspiration." *Thought* 65 (1990): 550-62.

Edwards, Paul. "Hopkins and the Exercitant." *The Way* 66 (1989): 42-51.

Egan, Desmond. "Hopkins' Influence on Poetry." In *Saving Beauty: Further Studies in Hopkins*, ed. Michael E. Allsopp, 295-311. New York: Garland Publishing, 1994.

Endean, Philip, S.J. "A Note on How Hopkins's Contemporaries Understood Jesuit Spirituality." *The Hopkins Quarterly* 7 (1981): 167-71.

Endean, Philip, S.J. "The Spirituality of Gerard Manley Hopkins." *The Hopkins Quarterly* 8 (1981): 107-29.

Endean, Philip, S.J. "How Should Hopkins Critics Use Ignatian Texts?" In *Gerard Manley Hopkins: Tradition and Innovation*, ed. G. Marra P. Bottalla, F. Marucci, 161-76. Ravenna: Longo Editore, 1991.

Feeney, Joseph, S.J. "Hopkins' Failure in Theology: Some New Archival Data and a Reevaluation." *The Hopkins Quarterly* 13 (1987): 99-114.

Feeney, Joseph, S.J. "Gerard Manley Hopkins, Poeta." *La Civilta Cattolica* 141 (1990): 442-53.

Feeney, Joseph, S.J. "The Collapse of Hopkins' Jesuit Worldview: A Conflict between Moralism and Incarnationalism." In *Gerard Manley Hopkins Annual 1992*, ed. Michael Sundermeier, 105-26. Omaha: Creighton University Press, 1992.

Feeney, Joseph, S.J. "Hopkins: A Religious and a Secular Poet." *Studies* 84 (1995): 120-9.

Feeney, Joseph, S.J. "My Dearest Father: Some unpublished letters of Gerard Manley Hopkins." *The Literary Supplement*, December 22 1995, 13-4.

Feeney, Joseph, S.J. "The Bischoff Collection at Gonzaga University: A Preliminary Account." *The Hopkins Quarterly* 23 (1996): 71-91.

Feeney, Joseph, S.J. "Four Newfound Hopkins Letters: An Annotated Edition, with a Fragment of another Letter." *The Hopkins Quarterly* 23 (1996): 3-40.

Feeney, Joseph, S.J. "Hopkins in Community: How His Jesuit Contemporaries Viewed Him." In *Saving Beauty: Further Studies in Hopkins*, ed. Michael E. Allsopp, 253-94. New York: Garland Publishing, 1994.

Fennell, Francis L. "Hopkins's 'Spring and Fall': An Approach from/to Reception Theory." In *Rereading Hopkins: Selected New Essays*, ed. Francis L. Fennell, 84-95. Victoria: University of Victoria, 1996.

Fletcher, Ian. "Walter Horatio Pater." In *British Writers: Elizabeth Gaskell to Francis Thompson*, ed. Ian Scott-Kilvert, 337-60. New York: Charles Scribner's Sons, 1982.

Fletcher, John Gould. "Gerard Manley Hopkins: Priest or Poet?" *American Review* 6 (1936): 331-46.

Flinn, Sean. "Scotus and Hopkins: Christian Metaphysics and Poetic Creativity." In *Annual Report of the Duns Scotus Philosophical Association*, ed. Lynn Behl, 26, 51-94. Cleveland: Our Lady of Angels Franciscan Seminary, 1962.

Fraser, Hilary. "Truth to Nature: Science, Religion and the Pre-Raphaelites." In *The Critical Spirit and the Will to Believe: Essays in Nineteenth-Century Literature and Religion*, ed. David Jasper and T.R. Wright, 53-68. New York: St. Martin's Press, 1989.

Fulweiler, Howard. "Sexual Sentimentality and 'the Woman Question'." In *Rereading Hopkins: Selected New Essays*, ed. Francis L. Fennell, 36-45. Victoria: University of Victoria, 1996.

Gallet, Réné. " 'The Windhover' and 'God's First Intention Ad Extra'." In *Gerard Manley Hopkins: Tradition and Innovation*, ed. G. Marra P. Bottalla, F. Marucci, 55-68. Ravenna: Longo Editore, 1991.

Gallet, Réné. "Hopkins: Introductions, Traductions, Reflets(1889-1989)." In *Traductions, Passages: Le Domaine Anglais*, ed. Stephen Romer, 77-92. Tours: Université de Tours, 1993.

Gardner, W.H. "A Note on Hopkins and Duns Scotus." *Scrutiny* 5 (1936): 61-6.

Gardner, W.H. "The Religious Problem in Hopkins." *Scrutiny* 6 (1937): 32-42.

Gardner, W.H. "Correspondence." *The Month (ns)* 4 (1950): 210-13.

Gerard, Albert. "Duns Scot et G.M. Hopkins." *Revue des langues vivantes* 12 (1946): 35-38.

Giacon, Carlo. "Il Significato dell'Individuale in Duns Scoto." In *Regnum Hominis et Regnum Dei: Acta Quarti Congressus Scotistici Internationalis*, ed. Camille Bérubé, 348-54. Rome: Societas Internationalis Scotistica, 1978.

Graziosi, Marco. "G.M. Hopkins' Aesthetic Theory." *The Hopkins Quarterly* 16 (1989): 71-88.

Hammerton, H.J. "The Two Vocations of G.M. Hopkins." *Theology* 87 (1984): 186-89.

Hawley, John, S.J. "Hopkins and the Christian Imagination." In *Gerard Manley Hopkins Annual 1993*, ed. Michael Sundermeier and Desmond Egan, 45-55. Omaha: Creighton University Press, 1993.

Heaney, Seamus. "The Fire I' The Flint." In *Chatterton Lecture on an English Poet 1974*, ed. The British Academy. London: Oxford University Press, 1975.

Herzburg, Sophie. "The Winter World of Gerard Manley Hopkins." *Faith and Freedom* 44 (1991): 17-23.

Higgins, Lesley. "'Bone-house' and 'lovescape': Writing the Body in Hopkins' Canon." In *Rereading Hopkins: Selected New Essays*, ed. Francis L. Fennell, 11-35. Victoria: University of Victoria, 1996.

Higgins, Lesley J. "Hopkins and 'The Jowler'." *Texas Studies in Literature and Language* 31 (1989): 143-67.

Hoagwood, Terence A. "Hopkins's Intellectual Framework: Newman, Pater, and the Epistemological Circle." *Studies in the Literary Imagination* 21 (1988): 23-40.

House, Humphry. "A Note on Hopkins' Religious Life." *New Verse* 14 (1935).

Hufstader, Anselm, O.S.B. "The Experience of Nature in Hopkins' Journals and Poems." *Downside Review* 84 (1966): 127-49.

Jayantha, R.A. "Gerard Manley Hopkins: Devotional Poet of 'Unpropitious' Times." *The Literary Criterion* 24 (1989): 133-46.

Johnson, Wendell Stacy. "From Ruskin to Hopkins: Landscape and Inscape." *The Hopkins Quarterly* 8 (1981): 89-106.

Johnson, Wendell Stacy. "Halfway to a New Land: Herbert, Tennyson and the Early Hopkins." *The Hopkins Quarterly* 10 (1983): 115-23.

Keating, Joseph, S.J. "Fr. Gerard Manley Hopkins and the Spiritual Exercises." *The Month* 166 (1935): 268-70.

Kelly, Hugh. "Gerard Manley Hopkins: Jesuit — Poet." *Studies* 31 (1942): 438-44.

Kim, Dal-Yong. "Gerard Manley Hopkins' 'Inscape'." *The Journal of English Language and Literature* 36 (1990): 627-42.

King, Peter. "Duns Scotus on the Common Nature and the Individual Differentia." *Philosophical Topics* 20 (1990): 51-76.

King, Thomas M. "The consecration of our world in 'The Spiritual Exercises of St. Ignatius'"."*Journal of Spiritual Formation* 14 (1994): 273-85.

Lichtman, Maria R. "The Incarnational Aesthetic of Gerard Manley Hopkins." *Religion and Literature* 23 (1991): 37-50.

Little, Arthur, S.J. "Hopkins and Scotus." *Irish Monthly* 71 (1945): 47-59.

Lynch, Michael. "Recovering Hopkins, Recovering Ourselves." *The Hopkins Quarterly* VI (1979): 107-17.

MacKenzie, Norman H. "In Memoriam: The Rev. Anthony D. Bischoff, S.J. (1910-1993)." *The Hopkins Quarterly* 23 (1996): 93-112.

Mariani, Paul. " 'O Christ, Christ, Come Quickly!': Lexical Plenitude and Primal Cry at the Heart of 'The Wreck'." In *Readings of "The Wreck": Essays in Commemoration of the Centenary of G.M. Hopkins' "The Wreck of the Deutschland"*, ed. Peter Milward. Chicago: Loyola University Press, 1976.

Mathai, Varhese. "Breath, Utterance, and the Word: Three Elements of the 'Arch and Original' Sound in Hopkins." In *Rereading Hopkins: Selected New Essays*, ed. Francis L. Fennell, 127-42. Victoria: University of Victoria, 1996.

McGlade, James A. "Gerard Manley Hopkins: Priest-Poet — A Conflict of Vocations." *The Australasian Catholic Record* 67 (1990): 309-18.

McNamee, Maurice B. "Hopkins: Poet of Nature and of the Supernatural." In *Immortal Diamond: Studies in Gerard Manley Hopkins*, ed. Norman Weyand, 222-51. New York: Octagon Books, 1969.

McNamee, R.B., S.J. "The Ignatian Meditation Pattern in the Poetry of Gerard Manley Hopkins." *The Hopkins Quarterly* 2 (1975): 21-8.

McNees, Eleanor. "Beyond 'The Half-way House': Hopkins and Real Presence." *Texas Studies in Literature and Language* 31 (1989): 85-104.

Mertens, Herman-Emiel. "His Very Name Is Beauty: Aesthetic Experience and Christian Faith." *Louvain Studies* 20 (1995): 316-31.

Meyers, Joanna Shaw. "Hopkins and Mrs. Humphry Ward's Helbeck of Bannisdale." In *Rereading Hopkins: Selected New Essays*, ed. Francis L. Fennell, 63-83. Victoria: University of Victoria, 1996.

Miller, Joseph Hillis. "The Linguistic Moment in 'The Wreck of the Deutschland'." In *The New Criticism and After*, ed. Thomas Daniel Young. Charlottesville: University Press of Virginia, 1976.

Miller, Joseph Hillis. "Hopkins." In *The Linguistic Moment: From Wordsworth to Stevens*, 229-66. New Jersey: Princeton University Press, 1985.

Miller, Joseph Hillis. "Naming and Doing: Speech Acts in Hopkins' Poems." *Religion and Literature* 22 (1990): 173-91.

Miller, Joseph Hillis. "The Creation of the Self in Gerard Manley Hopkins." In *Victorian Subjects*, 1-23. Durham: Duke University Press, 1991.

Miller, Joseph Hillis. "Nature and the Linguistic Movement." In *Victorian Subjects*, 201-12. Durham: Duke University Press, 1991.

Miller, Joseph Hillis. "The Theme of the Disappearance of God in Victorian Literature." In *Victorian Subjects*, 49-68. Durham: Duke University Press, 1991.

Milward, Peter. "'The Wreck' and the Exercises." *English Literature and Language* (1975): 1-19.

Monsman, Gerald. "Pater, Hopkins, and Fichte's Ideal Student." *The South Atlantic Quarterly* 70 (1971): 365-76.

Moore, Michael D. "Newman and the 'Second Spring' of Hopkins' Poetry." *The Hopkins Quarterly* VI (1979): 119-37.

Moran, Maureen. "Manl(e)y Mortal Beauty: Hopkins as Tractarian Aesthete." *The Hopkins Quarterly* 22 (1995): 3-29.

Morse, Donald E. "Hopkins' Poetry and Joyces' Ulysses." *Studies* 87 (1998): 164-70.

Murphy, Russell E. "Enough I Say/I Say Enough: Hopkins, Arnold, and Teilhard's Mind-Mastered Globe." In *Gerard Manley Hopkins Annual*, ed. Michael Sundermeier, 127-44. Omaha: Creighton University Press, 1992.

Murphy, Russell E. "Hopkins and the Unrevealed Christ: Towards a Catholic Aesthetics." *Studies* 84 (1995): 173-80.

Murphy, Russell E. "Nature and Nature's God in Hopkins' St. Beuno Sonnets." *Studies* 87 (1998): 171-77

Murray, Gerry. "Gerard Manley Hopkins: His Influence on John Berryman." *Studies* 85 (1996): 125-35.

Netland, John T. "Linguistic Limitation and the Instress of Grace in 'The Wreck of the Deutschland'." *Victorian Poetry* 27 (1989): 187-99.

Newman, John Henry. "John Keble." In *Essays Critical and Historical*, 2. London: Longmans and Green, 1919.

Nixon, Jude V. "The Kindly Light: A Reappraisal of the Influence of Newman on Hopkins." *Texas Studies in Literature and Language* 31 (1989): 105-42.

Nowottny, Winifred. "Hopkins's Language of Prayer and Praise." In *The Hopkins Society Fourth Annual Lecture*, 1-20. Worcester: The Hopkins Society, 1973.

Ong, Walter J., S.J. "Hopkins' Sprung Rhythm and the Life of English Poetry." In *Immortal Diamond*, ed. S.J. Norman Weyand, 93-174. New York: Octagon Books, 1969.

Papetti, Viola. "The Figure of Mary in the Poetry of Hopkins." In *Gerard Manley Hopkins: Tradition and Innovation*, ed. G. Marra, P. Bottalla, F. Marucci, 177-90. Ravenna: Longo Editore, 1991.

Parekh, Pushpa Naidu. "Poetry as Performance: Hopkins and Reader-Response." *Studies* 87 (1998): 183-89.

Parini, Jay Lee. "Ignatius and 'The Wreck of the Deutschland'." *Forum for Modern Language Studies* 11 (1975): 97-105.

Pick, John. "The Growth of a Poet: Gerard Manley Hopkins, S.J." *The Month* 175 (1940): 39-46, 106-13.

Proust, Marcel. "John Ruskin." In *Marcel Proust: A Selection from His Miscellaneous Writings*, ed. Gerard Hopkins. London: Allan Wingate, 1948.

Quennell, Peter. "John Ruskin." In *British Writers: Elizabeth Gaskell to Francis Thompson*, ed. Ian Scott-Kilvert, 5, 173-85. New York: Charles Scribner's Sons, 1982.

Quint, Bernard J. "The Nature of Things: Hopkins and Scotus." In *Twilight of Dawn: Studies in English Literature in Transition*, ed. O M Brack, Jr., 84-90. Tucson: The University of Arizona Press, 1987.

Ritz, Jean Georges. "Hopkins, Défenseur de la Singularité des Êtres." In *Du Verbe au Geste*, 223-30. Nancy: Presses Universitaires de Nancy, 1986.

Rogers, Robert. "Hopkins' Carrion Comfort." *The Hopkins Quarterly* 7 (1981): 143-65.

Salmon, Rachel. "Frozen Fire: The Paradoxical Equation of 'That Nature is a Heraclitean Fire and of the Comfort of the Resurrection'." In *Critical Essays on Gerard Manley Hopkins*, ed. Alison Sulloway, 21-34. Boston: G.K. Hall & Co., 1990.

Salmon, Rachel. "Hopkins and the Rabbis: Christian Religious Poetry and Midrashic Reading." In *Rereading Hopkins: Selected New Essays*, ed. Francis L. Fennell, 96-126. Victoria: University of Victoria, 1996.

Seelhammer, Ruth. "A Hopkins Bibliography: 1978." *The Hopkins Quarterly* VI (1979): 95-106.

Servotte, Herman. "A Deconstructionistic Reading of 'The Windhover'." *English Studies* 70 (1989): 253-55.

Slakey, Roger L. "'God's Grandeur' and the Divine Impersoning." *Victorian Poetry* 34 (1996): 73-85.

Stanford, Donald. "The Harried Life of Gerard Manley Hopkins." *Review* 16 (1994): 209-19.

Storey, Graham. "Gerard Manley Hopkins." In *British Writers: Elizabeth Gaskell to Francis Thompson*, ed. Ian Scott-Kilvert, 5, 361-82. New York: Charles Scribner's Sons, 1982.

Sulloway, Alison G. "St. Ignatius Loyola and the Victorian Temper: Hopkins' Windhover as Symbol of 'Diabolic Gravity'." *The Hopkins Quarterly* 1 (1974): 43-51.

Sundermeier, Michael. "Of Wet and Wildness: Hopkins and the Environment." In *Gerard Manley Hopkins Annual 1992*, ed. Michael Sundermeier, 59-81. Omaha: Creighton University Press, 1992.

Thibodeaux, Troy L. "The Resistance of the Word: Hopkins's '(Carrion Comfort)'." *The Hopkins Quarterly* 22 (1995): 79-97.

Thomas, Alfred, S.J. "Was Hopkins A Scotist Before He Read Scotus?" In *Scotismus Decursu Saeculorum*, 4, 617-29. Rome: Cura Commissionis Scotisticae, 1968.

Thomas, Hywel. "Gerard Manley Hopkins and John Duns Scotus." *Religious Studies* 24 (1988): 337-64.

Thornton, R.K.R. "Gerard Manley Hopkins: Aesthete or Moralist?" In *Saving Beauty: Further Studies in Hopkins*, ed. Michael E. Allsopp, 39-58. New York: Garland Publishing, 1994.

van der Walt, B.J. ""Regnum Hominis et Regnum Dei": Historical-Critical Discussion of the Relationship between Nature and Supernature According to Duns Scotus." In *Regnum Hominis et Regnum Dei: Acta Quarti Congressus Scotistici Internationalis*, ed. Camille Bérubé, 219-29. Rome: Societas Internationalis Scotistica, 1978.

Walhout, Donald. "Scotism in the Poetry of Hopkins." In *Saving Beauty: Further Studies in Hopkins*, ed. Michael Allsopp, 113-32. New York: Garland Publishing, 1994.

Ward, Bernadette. "Newman's Grammar of Assent and the Poetry of Gerard Manley Hopkins." *Renascence* 43 (1990): 105-20.

Ward, Bernadette. "Philosophy and Inscape: Hopkins and the Formalitas of Duns Scotus." *Texas Studies in Literature and Language* 32 (1990): 214-39.

Warren, Austin. "Instress of Inscape." *The Kenyon Review* 2 (1989): 216-24.

White, Norman. "The Context of Hopkins' Drawings." In *All My Eyes See: The Visual World of Gerard Manley Hopkins*, ed. R.K.R. Thornton, 53-67. Tyne and Wear: Ceolfrith Press, 1975.

White, Norman. "Poet and Priest: Gerard Manley Hopkins, Myth and Reality." *Studies* 79 (1990): 140-49.

White, Norman. "'He Played the Droll Jester': Hopkins the Unorthodox." In *Gerard Manley Hopkins: Tradition and Innovation*, ed. G. Marra, P. Bottalla, F. Marucci, 149-59. Ravenna: Longo Editore, 1991.

Winters, Yvor. "Gerard Manley Hopkins." In *Hopkins: A Collection of Critical Essays*, ed. Geoffrey Hartman. Englewood Cliffs: Prentice-Hall, Inc., 1966.

Yoder, Emily K. "Evil and Idolatry in the 'Windhover'." *The Hopkins Quarterly* 1 (1975): 33-46.

Young, R.V. "Hopkins, Scotus, and the Predication of Being." *Renascence* 42 (1989): 35-50.

Zaniello, Thomas. "Hopkins' Scientific Interests: 'Face to Face with the Sphinx'." *Thought* 65 (1990): 510-21.

Zaniello, Thomas. "Another Link Between Hopkins and Newman." *The Hopkins Quarterly* 22 (1995): 43-9.

Zaniello, Thomas. "The Tonic of Platonism: The Origins and Use of Hopkins' "Scape"." *The Hopkins Quarterly* 5 (1978): 5-16.

Zoghby, Mary, O.S.M. "The Cosmic Christ in Hopkins, Teilhard, and Scotus." *Renascence* 24 (1971): 33-46.

IV. *Dissertations and Theses about Gerard Manley Hopkins*

Abraham, John August. "Hopkins and Scotus: An Analogy between Inscape and Individuation." Doctoral Dissertation, The University of Wisconsin, 1959.

Agnew, Francis Henry. "The Poetic Theory of Gerard Manley Hopkins and the Philosophy of Duns Scotus." Master of Arts Thesis, De Paul University, 1964.

Allsopp, Michael E. "The Meaning of Human Life in the Writings of Gerard Manley Hopkins, S.J." Doctoral Dissertation, The Gregorian Pontifical University, 1979.

Barnes, Roslyn. "Gerard Manley Hopkins and Pierre Teilhard de Chardin: A Formulation of Mysticism for a Scientific Age." Master of Arts, State University of Iowa, 1962.

Brennan, Joseph Xavier. "Gerard Manley Hopkins, S.J.: A Critical Interpretation of His Poetry." Master of Arts Thesis, Brown University, 1949.

Brittain, Clark M. "Logos, Creation, and Epiphany in the Poetics of Gerard Manley Hopkins." Doctoral Dissertation, University of Virginia, 1988.

DiCicco, Mario. "Hopkins and the Mystery of Christ." Doctoral Dissertation, Case Western Reserve University, 1970.

Downes, David Anthony. "The Ignatian Spirit in Gerard Manley Hopkins." Doctoral Dissertation, University of Washington, 1955.

Fike, Francis George. "The Influence of John Ruskin upon the Aesthetic Theory and Practice of Gerard Manley Hopkins." Doctoral Dissertation, Stanford University, 1964.

Fleming, Lenore Moe. "The Influence of Duns Scotus on Gerard Manley Hopkins." Master of Arts Thesis, Loyola University, 1954.

Giles, Richard. "Nature, Human Nature, and the Style of Gerard Manley Hopkins." Doctoral Dissertation, University of Toronto, 1987.

Gleeson, William F. "Gerard Manley Hopkins and the Society of Jesus." Master of Arts Essay, Columbia University, 1950.

Higgins, Lesley Jane. "Hidden Harmonies: Walter Pater and Gerard Manley Hopkins." Doctoral Dissertation, Queen's University, 1987.

King, Shelley. "'But Meaning Motion': Movement and Change in the Poetry of G.M. Hopkins." Doctoral Dissertation, University of Toronto, 1988.

Lirette, M. Leander. "Pitch: The Principle of Individuation in the Writing of Gerard Manley Hopkins." Master of Arts Thesis, College of the Holy Names, 1961.

Nielsen, Anne. "The Scotist Element in Hopkins." Master of Arts Thesis, St. John's University, 1964.

Sandman, Joseph. "Sacramental Dimensions in the Poetry of Gerard Manley Hopkins." Doctoral Dissertation, University of Notre Dame, 1991.

Shrake, Joyce Rogers. "The Poetic of Gerard Manley Hopkins: A Study of the Relationship between the Philosophy of Duns Scotus and the Poetic of Gerard Manley Hopkins." Master of Arts Thesis, Texas Christian University, 1957.

Tadych, Renita. "The Franciscan Perspective in the Nature Poetry of Gerard Manley Hopkins Augmented by the Writings of John Duns Scotus." Doctoral Dissertation, Indiana University of Pennsylvania, 1992.

Travers-Lake, Alison K. "'Wrestling with (My God!) My God': George Herbert, Gerard Manley Hopkins, and the Poetry of Spiritual Conflict." Master of Arts, University of North Carolina, 1993.

Vaughan, Patricia A. "Convergences of Ignatian and Scotist Elements in the Poetry of Hopkins." Doctoral Dissertation, University of Notre Dame, 1984.

Ward, Bernadette. "Means and Meaning: Gerard Manley Hopkins' Scotist Poetic of Revelation and Matthew Arnold's Poetic of Social Control." Doctoral Dissertation, Stanford University, 1990.

Wear, Stephen Edward. "John Duns Scotus and Gerard Manley Hopkins: The Doctrine and Experience of Intuitive Cognition." Doctoral Dissertation, The University of Texas at Austin, 1979.